Methods in Enzymology

Volume 335
FLAVONOIDS AND OTHER POLYPHENOLS

METHODS IN ENZYMOLOGY

EDITORS-IN-CHIEF

John N. Abelson   Melvin I. Simon

DIVISION OF BIOLOGY
CALIFORNIA INSTITUTE OF TECHNOLOGY
PASADENA, CALIFORNIA

FOUNDING EDITORS

Sidney P. Colowick and Nathan O. Kaplan

Methods in Enzymology
Volume 335

Flavonoids and Other Polyphenols

EDITED BY

Lester Packer

UNIVERSITY OF CALIFORNIA
BERKELEY, CALIFORNIA

Editorial Advisory Board

Gary Beecher
Enrique Cadenas
Josiane Cillard
Fulvio Ursini
Myron Gross
Barry Halliwell
William Pryor
Catherine Rice-Evans
Helmut Sies

ACADEMIC PRESS
San Diego London Boston New York Sydney Tokyo Toronto

This book is printed on acid-free paper.

Copyright © 2001 by ACADEMIC PRESS

All Rights Reserved.
No part of this publication may be reproduced or transmitted in any form or by any means, electronic or mechanical, including photocopy, recording, or any information storage and retrieval system, without permission in writing from the Publisher.

The appearance of the code at the bottom of the first page of a chapter in this book indicates the Publisher's consent that copies of the chapter may be made for personal or internal use of specific clients. This consent is given on the condition, however, that the copier pay the stated per copy fee through the Copyright Clearance Center, Inc. (222 Rosewood Drive, Danvers, Massachusetts 01923), for copying beyond that permitted by Sections 107 or 108 of the U.S. Copyright Law. This consent does not extend to other kinds of copying, such as copying for general distribution, for advertising or promotional purposes, for creating new collective works, or for resale. Copy fees for pre-2000 chapters are as shown on the title pages. If no fee code appears on the title page, the copy fee is the same as for current chapters. /00 $35.00

Explicit permission from Academic Press is not required to reproduce a maximum of two figures or tables from an Academic Press chapter in another scientific or research publication provided that the material has not been credited to another source and that full credit to the Academic Press chapter is given.

Academic Press
A Harcourt Science and Technology Company
525 B Street, Suite 1900, San Diego, California 92101-4495, USA
http://www.academicpress.com

Academic Press
Harcourt Place, 32 Jamestown Road, London NW1 7BY, UK
http://www.academicpress.com

International Standard Book Number: 0-12-182236-2

PRINTED IN THE UNITED STATES OF AMERICA
01 02 03 04 05 06 07 SB 9 8 7 6 5 4 3 2 1

Table of Contents

CONTRIBUTORS TO VOLUME 335 ix
PREFACE xiii
VOLUME IN SERIES xiv

Section I. Sources, Characterization, and Analytical Methods

1. Overview of Methods for Analysis and Identification of Flavonoids	STEPHEN J. BLOOR	3
2. Analysis of Complex Mixtures of Flavonoids and Polyphenols by High-Performance Liquid Chromatography Electrochemical Detection Methods	PAUL E. MILBURY	15
3. Analysis of Flavonoids in Medicinal Plants	PIERGIORGIO PIETTA AND PIERLUIGI MAURI	26
4. High-Performance Liquid Chromatography/Mass Spectrometry Analysis of Proanthocyanidins in Food and Beverages	SHERYL A. LAZARUS, JOHN F. HAMMERSTONE, GARY E. ADAMSON, AND HAROLD H. SCHMITZ	46
5. Direct Thiolysis on Crude Apple Materials for High-Performance Liquid Chromatography Characterization and Quantification of Polyphenols in Cider Apple Tissues and Juices	SYLVAIN GUYOT, NATHALIE MARNET, PHILIPPE SANONER, AND JEAN-FRANÇOIS DRILLEAU	57
6. Enzymes Involved in Hydroxycinnamate Metabolism	DIETER STRACK	70
7. Estimation of Procyanidin Chain Length	VERONIQUE CHEYNIER, BENOIT LABARBE, AND MICHEL MOUTOUNET	82

Section II. Bioavailability

8. Determination of Flavonols in Body Fluids	PETER C. H. HOLLMAN	97
9. Determination of Quantity and Quality of Polyphenol Antioxidants in Foods and Beverages	JOE A. VINSON, JOHN PROCH, AND PRATIMA BOSE	103

v

10. Preparation and Characterization of Flavonoid Metabolites Present in Biological Samples	CHRISTINE MORAND, CLAUDINE MANACH, JENNIFER DONOVAN, AND CHRISTIAN REMESY	115
11. Caffeic Acid as Biomarker of Red Wine Intake	PAOLO SIMONETTI, CLAUDIO GARDANA, AND PIERGIORGIO PIETTA	122
12. Measurement of *trans*-Resveratrol, (+)-Catechin, and Quercetin in Rat and Human Blood and Urine by Gas Chromatography with Mass Selective Detection	GEORGE J. SOLEAS, JOE YAN, AND DAVID M. GOLDBERG	130
13. Absorption of *trans*-Resveratrol in Rats	GEORGE J. SOLEAS, MARK ANGELINI, LINDA GRASS, ELEFTHERIOS P. DIAMANDIS, AND DAVID M. GOLDBERG	145

Section III. Antioxidant Action

14. Galvinoxyl Method for Standardizing Electron and Proton Donation Activity	HONGLIAN SHI, NORIKO NOGUCHI, AND ETSUO NIKI	157
15. Structure–Activity Relationships Governing Antioxidant Capacities of Plant Polyphenols	WOLF BORS, CHRISTA MICHEL, AND KURT STETTMAIER	166
16. Antioxidant and Prooxidant Abilities of Foods and Beverages	LEE HUA LONG AND BARRY HALLIWELL	181
17. Metal Chelation of Polyphenols	ROBERT C. HIDER, ZU D. LIU, AND HICHAM H. KHODR	190
18. Mechanism of Antioxidant Effect of Catechins	KAZUNARI KONDO, MASAAKI KURIHARA, AND KIYOSHI FUKUHARA	203
19. Free Radical Scavenging by Green Tea Polyphenols	BAOLU ZHAO, QIONG GUO, AND WENJUAN XIN	217
20. Polyphenol Protection of DNA against Damage	GUANGJUN NIE, TAOTAO WEI, SHENGRONG SHEN, AND BAOLU ZHAO	232
21. Markers for Low-Density Lipoprotein Oxidation	MICHAEL AVIRAM AND JACOB VAYA	244

22. Antioxidant Activity of Hydroxycinnamic Acids on Human Low-Density Lipoprotein Oxidation	ANNE S. MEYER AND EDWIN N. FRANKEL	256
23. Rapid Screening Method for Relative Antioxidant Activities of Flavonoids and Phenolics	ANANTH SEKHER PANNALA AND CATHERINE RICE-EVANS	266
24. Nitric Oxide Formation in Macrophages Detected by Spin Trapping with Iron–Dithiocarbamate Complex: Effect of Purified Flavonoids and Plant Extracts	QIONG GUO, GERALD RIMBACH, AND LESTER PACKER	273
25. Redox Cycles of Caffeic Acid with α-Tocopherol and Ascorbate	JOÃO LARANJINHA	282
26. DNA Damage by Nitrite and Peroxynitrite: Protection by Dietary Phenols	KAICUN ZHAO, MATTHEW WHITEMAN, JEREMY P. E. SPENCER, AND BARRY HALLIWELL	296
27. Repair of Oxidized DNA by the Flavonoid Myricetin	ISABELLE MOREL, VALÉRIE ABALEA, PIERRE CILLARD, AND JOSIANE CILLARD	308

Section IV. Biological Activity

28. Binding of Flavonoids to Plasma Proteins	OLIVIER DANGLES, CLAIRE DUFOUR, CLAUDINE MANACH, CHRISTINE MORAND, AND CHRISTIAN REMESY	319
29. Protein Binding of Procyanidins: Studies Using Polyacrylamide Gel Electrophoresis and French Maritime Pine Bark Extract	HADI MOINI, QIONG GUO, AND LESTER PACKER	333
30. Characterization of Antioxidant Effect of Procyanidins	FULVIO URSINI, IVAN RAPUZZI, ROSANNA TONIOLO, FRANCO TUBARO, AND GINO BONTEMPELLI	338
31. Inhibition of in Vitro Low-Density Lipoprotein Oxidation by Oligomeric Procyanidins Present in Chocolate and Cocoas	DEBRA A. PEARSON, HAROLD H. SCHMITZ, SHERYL A. LAZARUS, AND CARL L. KEEN	350
32. Biological Actions of Oligomeric Procyanidins: Proliferation of Epithelial Cells and Hair Follicle Growth	TOMOYA TAKAHASHI	361

33. Effect of Polyphenolic Flavonoid Compounds on Platelets — DHANANSAYAN SHANMUGANAYAGAM AND JOHN D. FOLTS — 369

34. Assessing Bioflavonoids as Regulators of NF-κB Activity and Inflammatory Gene Expression in Mammalian Cells — CLAUDE SALIOU, GIUSEPPE VALACCHI, AND GERALD RIMBACH — 380

35. Interaction between Cultured Endothelial Cells and Macrophages: *In Vitro* Model for Studying Flavonoids in Redox-Dependent Gene Expression — GERALD RIMBACH, CLAUDE SALIOU, RAFFAELLA CANALI, AND FABIO VIRGILI — 387

36. Determination of Cholesterol-Lowering Potential of Minor Dietary Components by Measuring Apolipoprotein B Responses in HepG2 Cells — ELZBIETA M. KUROWSKA — 398

AUTHOR INDEX 405

SUBJECT INDEX 433

Contributors to Volume 335

Article numbers are in parentheses following the names of contributors.
Affiliations listed are current.

VALÉRIE ABALEA (27), *Laboratoire de Biologie Cellulaire et Végétale, UFR des Sciences Pharmaceutiques et Biologiques, Universite de Rennes I, Rennes Cedex 35043, France*

GARY E. ADAMSON (4), *Mars, Inc., Hackettstown, New Jersey 07840*

MARK ANGELINI (13), *Department of Laboratory Medicine and Pathobiology, University of Toronto, Toronto, Ontario M5G 1L5, Canada*

MICHAEL AVIRAM (21), *Lipid Research Laboratory, Rambam Medical Center, Haifa 31096, Israel*

STEPHEN J. BLOOR (1), *Industrial Research Limited, Lower Hutt, New Zealand*

GINO BONTEMPELLI (30), *Department of Chemical Sciences and Technology, University of Udine School of Food Science, Udine I-33100, Italy*

WOLF BORS (15), *Institut für Strahlenbiologie, GSF Forschungszentrum für Umwelt und Gesundheit, Neuherberg D-85764, Germany*

PRATIMA BOSE (9), *Department of Chemistry, University of Scranton, Scranton, Pennsylvania 18510*

RAFFAELLA CANALI (35), *National Institute for Food and Nutrition Research, Rome I-00178, Italy*

VERONIQUE CHEYNIER (7), *INRA, UMR Sciences pour l'Oenologie, Montpellier 34060, France*

JOSIANE CILLARD (27), *Laboratoire de Biologie Cellulaire et Végétale, UFR des Sciences Pharmaceutiques et Biologiques, Universite de Rennes I, Rennes Cedex 35043, France*

PIERRE CILLARD (27), *Laboratoire de Biologie Cellulaire et Végétale, UFR des Sciences Pharmaceutiques et Biologiques, Universite de Rennes I, Rennes Cedex 35043, France*

OLIVIER DANGLES (28), *Université d'Avignon, UMR A408, Avignon 84000, France*

ELEFTHERIOS P. DIAMANDIS (13), *Department of Laboratory Medicine and Pathobiology, University of Toronto, Toronto, Ontario M5G 1L5, Canada*

JENNIFER DONOVAN (10), *Unité des Maladies Métaboliques et Micronutriments, INRA Theix, St. Genès-Champanelle 63122, France*

JEAN-FRANÇOIS DRILLEAU (5), *Laboratoire de Recherches Cidricoles, Biotransformation des Fruits et Légumes, INRA, Le Rheu F-35650, France*

CLAIRE DUFOUR (28), *Securité et Qualité des Produits d'origine Végétale, INRA-Site Agroparc, Avignon 84914, France*

JOHN D. FOLTS (33), *Department of Medicine, Cardiovascular Medicine Section, University of Wisconsin Medical School, Madison, Wisconsin 53792*

EDWIN N. FRANKEL (22), *Department of Food Science and Technology, University of California, Davis, California 95616-8598*

KIYOSHI FUKUHARA (18), *Division of Organic Chemistry, National Institute of Health Sciences, Tokyo 158-8501, Japan*

CLAUDIO GARDANA (11), *Department of Food Science and Microbiology, Division of Human Nutrition, University of Milan, Milan 20133, Italy*

DAVID M. GOLDBERG (12, 13), *Department of Laboratory Medicine and Pathobiology, University of Toronto, Toronto, Ontario M5G 1L5, Canada*

LINDA GRASS (13), *Department of Pathology and Laboratory Medicine, Mount Sinai Hospital, Toronto, Ontario M5G 1X5, Canada*

QIONG GUO (19, 24, 29), *Department of Molecular and Cell Biology, University of California, Berkeley, California 94720-3200*

SYLVAIN GUYOT (5), *Laboratoire de Recherches Cidricoles, Biotransformation des Fruits et Légumes, INRA, Le Rheu F-35650, France*

BARRY HALLIWELL (16, 26), *Department of Biochemistry, National University of Singapore, 119260, Singapore*

JOHN F. HAMMERSTONE (4), *Mars, Inc., Hackettstown, New Jersey 07840*

ROBERT C. HIDER (17), *Department of Pharmacy, King's College London, London SE1 8WA, United Kingdom*

PETER C. H. HOLLMAN (8), *State Institute for Quality Control of Agricultural Products, RIKILT, Wageningen 6708 PD, The Netherlands*

CARL L. KEEN (31), *Department of Nutrition, University of California, Davis, California 95616*

HICHAM H. KHODR (17), *Department of Pharmacy, King's College London, London SE1 8WA, United Kingdom*

KAZUNARI KONDO (18), *Division of Foods, National Institute of Health Sciences, Tokyo 158-8501, Japan*

MASAAKI KURIHARA (18), *Division of Organic Chemistry, National Institute of Health Sciences, Tokyo 158-8501, Japan*

ELZBIETA M. KUROWSKA (36), *KGK Synergize, Inc., The University of Western Ontario Research Park, London N6G 4X8, Ontario, Canada*

BENOIT LABARBE (7), *INRA, UMR Sciences pour l'Oenologie, Montpellier 34060, France*

JOÃO LARANJINHA (25), *Laboratory of Biochemistry, Faculty of Pharmacy, and Center for Neurosciences, University of Coimbra, Coimbra 3000, Portugal*

SHERYL A. LAZARUS (4, 31), *Mars, Inc., Hackettstown, New Jersey 07840*

ZU D. LIU (17), *Department of Pharmacy, King's College London, London SE1 8WA, United Kingdom*

LEE HUA LONG (16), *Department of Biochemistry, National University of Singapore, 119260, Singapore*

CLAUDINE MANACH (10, 28), *Unité des Maladies Métaboliques et Micronutriments, INRA de Clermont-Ferrand/Theix, St. Genès-Champanelle 63122, France*

NATHALIE MARNET (5), *Laboratoire de Recherches Cidricoles, Biotransformation des Fruits et Légumes, INRA, Le Rheu F-35650, France*

PIERLUIGI MAURI (3), *Institute of Advanced Biomedical Technologies, National Council of Research, Segrate, Milan 20090, Italy*

ANNE S. MEYER (22), *Department of Biotechnology, Technical University of Denmark, Lyngby DK-2800, Denmark*

CHRISTA MICHEL (15), *Institut für Strahlenbiologie, GSF Forschungszentrum für Umwelt und Gesundheit, Neuherberg D-85764, Germany*

PAUL E. MILBURY (2), *Antioxidants Research Laboratory, USDA Human Nutrition Research Center on Aging at Tufts University, Boston, Massachusetts 02111*

HADI MOINI (29), *Department of Molecular and Cell Biology, University of California, Berkeley, California 94720-3200*

CHRISTINE MORAND (10, 28), *Unité des Maladies Métaboliques et Micronutriments, INRA de Clermont-Ferrand/Theix, St. Genès-Champanelle 63122, France*

ISABELLE MOREL (27), *Laboratoire de Biologie Cellulaire et Végétale, UFR des Sciences Pharmaceutiques et Biologiques, Universite de Rennes I, Rennes Cedex 35043, France*

MICHEL MOUTOUNET (7), *INRA, UMR Sciences pour l'Oenologie, Montpellier 34060, France*

GUANGJUN NIE (20), *Laboratory of Visual Information Processing, Department of Molecular and Cell Biophysics, Institute of Biophysics, Academia Sinica, Beijing 100101, Peoples Republic of China*

ETSUO NIKI (14), *Research Center for Advanced Science and Technology, University of Tokyo, Tokyo 153-8904, Japan*

NORIKO NOGUCHI (14), *Research Center for Advanced Science and Technology, University of Tokyo, Tokyo 153-8904, Japan*

LESTER PACKER (24, 29), *Department of Molecular and Cell Biology, University of California, Berkeley, California 94720-3200*

ANANTH SEKHER PANNALA (23), *Wolfson Centre for Age Related Diseases, GKT School of Biomedical Sciences, King's College London, London SE1 9RT, United Kingdom*

DEBRA A. PEARSON (31), *Department of Human Biology, ES 301, University of Wisconsin, Green Bay, Wisconsin 54311*

PIERGIORGIO PIETTA (3, 11), *Institute of Advanced Biomedical Technologies, National Council of Research, Segrate, Milan 20090, Italy*

JOHN PROCH (9), *Department of Chemistry, University of Scranton, Scranton, Pennsylvania 18510*

IVAN RAPUZZI (30), *Department of Chemical Sciences and Technology, University of Udine School of Food Science, Udine I-33100, Italy*

CHRISTIAN REMESY (10, 28), *Unité des Maladies Métaboliques et Micronutriments, INRA Theix, St. Genès-Champanelle 63122, France*

CATHERINE RICE-EVANS (23), *Wolfson Centre for Age Related Diseases, GKT School of Biomedical Sciences, King's College London, London SE1 9RT, United Kingdom*

GERALD RIMBACH (24, 34, 35), *School of Food Biosciences, Hugh Sinclair Human Nutrition Unit, University of Reading, Reading RG6 6AP, United Kingdom*

CLAUDE SALIOU (34, 35), *Department of Molecular and Cell Biology, University of California, Berkeley, California 94720-3200*

PHILIPPE SANONER (5), *Laboratoire de Recherches Cidricoles, Biotransformation des Fruits et Légumes, INRA, Le Rheu F-35650, France*

HAROLD H. SCHMITZ (4, 31), *Mars, Inc., Hackettstown, New Jersey 07840*

DHANANSAYAN SHANMUGANAYAGAM (33), *Department of Medicine, Cardiovascular Medicine Section, University of Wisconsin Medical School, Madison, Wisconsin 53792*

SHENGRONG SHEN (20), *Department of Tea, Zhejiang University, Hangzhou, 310029, China*

HONGLIAN SHI (14), *Research Center for Advanced Science and Technology, University of Tokyo, Tokyo 153-8904, Japan*

PAOLO SIMONETTI (11), *Department of Food Science and Microbiology, Division of Human Nutrition, University of Milan, Milan 20133, Italy*

GEORGE J. SOLEAS (12, 13), *Quality Assurance, Liquor Control Board of Ontario, Toronto, Ontario M5E 1A4, Canada*

JEREMY P. E. SPENCER (26), *International Antioxidant Research Centre, King's College London, Guy's Campus, London SE1 8RT, United Kingdom*

KURT STETTMAIER (15), *Institut für Strahlenbiologie, GSF Forschungszentrum für Umwelt und Gesundheit, Neuherberg D-85764, Germany*

DIETER STRACK (6), *Abteilung Sekudärstoffwechsel, Leibniz-Institut für Pflanzenbiochemie, Halle (Saale) D-06120, Germany*

TOMOYA TAKAHASHI (32), *Tsukuba Research Laboratories, Kyowa Hakko Kogyo Co., Ibaraki 305-0841, Japan*

ROSANNA TONIOLO (30), *Department of Chemical Sciences and Technology, University of Udine School of Food Science, Udine I-33100, Italy*

FRANCO TUBARO (30), *Department of Chemical Sciences and Technology, University of Udine School of Food Science, Udine I-33100, Italy*

FULVIO URSINI (30), *Department of Biological Chemistry, University of Padova School of Medicine, Padova I-35121, Italy*

GIUSEPPE VALACCHI (34), *School of Medicine, University of California, Davis, California 95616*

JACOB VAYA (21), *Laboratory for Natural Medicinal Compounds, Migal-Galilee Technological Center, Kiriat-Shmona, Israel*

JOE A. VINSON (9), *Department of Chemistry, University of Scranton, Scranton, Pennsylvania 18510*

FABIO VIRGILI (35), *National Institute for Food and Nutrition Research, Rome 00178, Italy*

TAOTAO WEI (20), *Laboratory of Visual Information Processing, Department of Molecular and Cell Biophysics, Institute of Biophysics, Academia Sinica, Beijing 100101, Peoples Republic of China*

MATTHEW WHITEMAN (26), *International Antioxidant Research Centre, King's College London, Guy's Campus, London SE1 8RT, United Kingdom*

WENJUAN XIN (19), *Institute of Biophysics, Academia Sinica, Beijing 100101, Peoples Republic of China*

JOE YAN (12), *Quality Assurance, Liquor Control Board of Ontario, Toronto, Ontario, Canada M5E 1A4*

BAOLU ZHAO (19, 20), *Laboratory of Visual Information Processing, Department of Molecular and Cell Biophysics, Institute of Biophysics, Academia Sinica, Beijing 100101, Peoples Republic of China*

KAICUN ZHAO (26), *International Antioxidant Research Centre, King's College London, Guy's Campus, London SE1 8RT, United Kingdom*

Preface

Reactive oxygen and nitrogen species (ROS) and antioxidants are increasingly playing an important role in oxidative stress and disease. Certainly ROS are important in regulating oxidative processes in all biological systems. This volume of *Methods in Enzymology* on Flavonoids and Other Polyphenols was prepared in recognition of compelling evidence that these substances—important components of our food chain (such as fruits and vegetables and beverages such as tea or red wine)—have been reported to have health benefits. In pure form or as herbal extracts or plant products they have been reported to have antioxidant and cell regulation activity and to affect gene expression.

However, many unsolved problems exist with respect to the action of polyphenols and bioflavonoids in biological systems. There is a lack of information on bioavailability, metabolism, biochemical, and molecular biological effects on cell regulation and on effects on health. New methodologies described in this volume will aid progress in solving these unanswered questions.

In bringing this volume to fruition, credit must be given to the experts on various aspects of research in this field. Appreciation is extended to the contributors selected to contribute and to produce this state of the art volume. The topics included were chosen on the excellent advice of Drs. Gary Beecher, Enrique Cadenas, Josiane Cillard, Myron Gross, Barry Halliwell, William Pryor, Catherine Rice-Evans, Helmut Sies, and Fulvio Ursini. To these colleagues I extend my sincere thanks and appreciation.

LESTER PACKER

METHODS IN ENZYMOLOGY

VOLUME I. Preparation and Assay of Enzymes
Edited by SIDNEY P. COLOWICK AND NATHAN O. KAPLAN

VOLUME II. Preparation and Assay of Enzymes
Edited by SIDNEY P. COLOWICK AND NATHAN O. KAPLAN

VOLUME III. Preparation and Assay of Substrates
Edited by SIDNEY P. COLOWICK AND NATHAN O. KAPLAN

VOLUME IV. Special Techniques for the Enzymologist
Edited by SIDNEY P. COLOWICK AND NATHAN O. KAPLAN

VOLUME V. Preparation and Assay of Enzymes
Edited by SIDNEY P. COLOWICK AND NATHAN O. KAPLAN

VOLUME VI. Preparation and Assay of Enzymes (*Continued*)
Preparation and Assay of Substrates
Special Techniques
Edited by SIDNEY P. COLOWICK AND NATHAN O. KAPLAN

VOLUME VII. Cumulative Subject Index
Edited by SIDNEY P. COLOWICK AND NATHAN O. KAPLAN

VOLUME VIII. Complex Carbohydrates
Edited by ELIZABETH F. NEUFELD AND VICTOR GINSBURG

VOLUME IX. Carbohydrate Metabolism
Edited by WILLIS A. WOOD

VOLUME X. Oxidation and Phosphorylation
Edited by RONALD W. ESTABROOK AND MAYNARD E. PULLMAN

VOLUME XI. Enzyme Structure
Edited by C. H. W. HIRS

VOLUME XII. Nucleic Acids (Parts A and B)
Edited by LAWRENCE GROSSMAN AND KIVIE MOLDAVE

VOLUME XIII. Citric Acid Cycle
Edited by J. M. LOWENSTEIN

VOLUME XIV. Lipids
Edited by J. M. LOWENSTEIN

VOLUME XV. Steroids and Terpenoids
Edited by RAYMOND B. CLAYTON

VOLUME XVI. Fast Reactions
Edited by KENNETH KUSTIN

VOLUME XVII. Metabolism of Amino Acids and Amines (Parts A and B)
Edited by HERBERT TABOR AND CELIA WHITE TABOR

VOLUME XVIII. Vitamins and Coenzymes (Parts A, B, and C)
Edited by DONALD B. MCCORMICK AND LEMUEL D. WRIGHT

VOLUME XIX. Proteolytic Enzymes
Edited by GERTRUDE E. PERLMANN AND LASZLO LORAND

VOLUME XX. Nucleic Acids and Protein Synthesis (Part C)
Edited by KIVIE MOLDAVE AND LAWRENCE GROSSMAN

VOLUME XXI. Nucleic Acids (Part D)
Edited by LAWRENCE GROSSMAN AND KIVIE MOLDAVE

VOLUME XXII. Enzyme Purification and Related Techniques
Edited by WILLIAM B. JAKOBY

VOLUME XXIII. Photosynthesis (Part A)
Edited by ANTHONY SAN PIETRO

VOLUME XXIV. Photosynthesis and Nitrogen Fixation (Part B)
Edited by ANTHONY SAN PIETRO

VOLUME XXV. Enzyme Structure (Part B)
Edited by C. H. W. HIRS AND SERGE N. TIMASHEFF

VOLUME XXVI. Enzyme Structure (Part C)
Edited by C. H. W. HIRS AND SERGE N. TIMASHEFF

VOLUME XXVII. Enzyme Structure (Part D)
Edited by C. H. W. HIRS AND SERGE N. TIMASHEFF

VOLUME XXVIII. Complex Carbohydrates (Part B)
Edited by VICTOR GINSBURG

VOLUME XXIX. Nucleic Acids and Protein Synthesis (Part E)
Edited by LAWRENCE GROSSMAN AND KIVIE MOLDAVE

VOLUME XXX. Nucleic Acids and Protein Synthesis (Part F)
Edited by KIVIE MOLDAVE AND LAWRENCE GROSSMAN

VOLUME XXXI. Biomembranes (Part A)
Edited by SIDNEY FLEISCHER AND LESTER PACKER

VOLUME XXXII. Biomembranes (Part B)
Edited by SIDNEY FLEISCHER AND LESTER PACKER

VOLUME XXXIII. Cumulative Subject Index Volumes I-XXX
Edited by MARTHA G. DENNIS AND EDWARD A. DENNIS

VOLUME XXXIV. Affinity Techniques (Enzyme Purification: Part B)
Edited by WILLIAM B. JAKOBY AND MEIR WILCHEK

VOLUME XXXV. Lipids (Part B)
Edited by JOHN M. LOWENSTEIN

VOLUME XXXVI. Hormone Action (Part A: Steroid Hormones)
Edited by BERT W. O'MALLEY AND JOEL G. HARDMAN

VOLUME XXXVII. Hormone Action (Part B: Peptide Hormones)
Edited by BERT W. O'MALLEY AND JOEL G. HARDMAN

VOLUME XXXVIII. Hormone Action (Part C: Cyclic Nucleotides)
Edited by JOEL G. HARDMAN AND BERT W. O'MALLEY

VOLUME XXXIX. Hormone Action (Part D: Isolated Cells, Tissues, and Organ Systems)
Edited by JOEL G. HARDMAN AND BERT W. O'MALLEY

VOLUME XL. Hormone Action (Part E: Nuclear Structure and Function)
Edited by BERT W. O'MALLEY AND JOEL G. HARDMAN

VOLUME XLI. Carbohydrate Metabolism (Part B)
Edited by W. A. WOOD

VOLUME XLII. Carbohydrate Metabolism (Part C)
Edited by W. A. WOOD

VOLUME XLIII. Antibiotics
Edited by JOHN H. HASH

VOLUME XLIV. Immobilized Enzymes
Edited by KLAUS MOSBACH

VOLUME XLV. Proteolytic Enzymes (Part B)
Edited by LASZLO LORAND

VOLUME XLVI. Affinity Labeling
Edited by WILLIAM B. JAKOBY AND MEIR WILCHEK

VOLUME XLVII. Enzyme Structure (Part E)
Edited by C. H. W. HIRS AND SERGE N. TIMASHEFF

VOLUME XLVIII. Enzyme Structure (Part F)
Edited by C. H. W. HIRS AND SERGE N. TIMASHEFF

VOLUME XLIX. Enzyme Structure (Part G)
Edited by C. H. W. HIRS AND SERGE N. TIMASHEFF

VOLUME L. Complex Carbohydrates (Part C)
Edited by VICTOR GINSBURG

VOLUME LI. Purine and Pyrimidine Nucleotide Metabolism
Edited by PATRICIA A. HOFFEE AND MARY ELLEN JONES

VOLUME LII. Biomembranes (Part C: Biological Oxidations)
Edited by SIDNEY FLEISCHER AND LESTER PACKER

VOLUME LIII. Biomembranes (Part D: Biological Oxidations)
Edited by SIDNEY FLEISCHER AND LESTER PACKER

VOLUME LIV. Biomembranes (Part E: Biological Oxidations)
Edited by SIDNEY FLEISCHER AND LESTER PACKER

VOLUME LV. Biomembranes (Part F: Bioenergetics)
Edited by SIDNEY FLEISCHER AND LESTER PACKER

VOLUME LVI. Biomembranes (Part G: Bioenergetics)
Edited by SIDNEY FLEISCHER AND LESTER PACKER

VOLUME LVII. Bioluminescence and Chemiluminescence
Edited by MARLENE A. DELUCA

VOLUME LVIII. Cell Culture
Edited by WILLIAM B. JAKOBY AND IRA PASTAN

VOLUME LIX. Nucleic Acids and Protein Synthesis (Part G)
Edited by KIVIE MOLDAVE AND LAWRENCE GROSSMAN

VOLUME LX. Nucleic Acids and Protein Synthesis (Part H)
Edited by KIVIE MOLDAVE AND LAWRENCE GROSSMAN

VOLUME 61. Enzyme Structure (Part H)
Edited by C. H. W. HIRS AND SERGE N. TIMASHEFF

VOLUME 62. Vitamins and Coenzymes (Part D)
Edited by DONALD B. MCCORMICK AND LEMUEL D. WRIGHT

VOLUME 63. Enzyme Kinetics and Mechanism (Part A: Initial Rate and Inhibitor Methods)
Edited by DANIEL L. PURICH

VOLUME 64. Enzyme Kinetics and Mechanism (Part B: Isotopic Probes and Complex Enzyme Systems)
Edited by DANIEL L. PURICH

VOLUME 65. Nucleic Acids (Part I)
Edited by LAWRENCE GROSSMAN AND KIVIE MOLDAVE

VOLUME 66. Vitamins and Coenzymes (Part E)
Edited by DONALD B. MCCORMICK AND LEMUEL D. WRIGHT

VOLUME 67. Vitamins and Coenzymes (Part F)
Edited by DONALD B. MCCORMICK AND LEMUEL D. WRIGHT

VOLUME 68. Recombinant DNA
Edited by RAY WU

VOLUME 69. Photosynthesis and Nitrogen Fixation (Part C)
Edited by ANTHONY SAN PIETRO

VOLUME 70. Immunochemical Techniques (Part A)
Edited by HELEN VAN VUNAKIS AND JOHN J. LANGONE

VOLUME 71. Lipids (Part C)
Edited by JOHN M. LOWENSTEIN

VOLUME 72. Lipids (Part D)
Edited by JOHN M. LOWENSTEIN

VOLUME 73. Immunochemical Techniques (Part B)
Edited by JOHN J. LANGONE AND HELEN VAN VUNAKIS

VOLUME 74. Immunochemical Techniques (Part C)
Edited by JOHN J. LANGONE AND HELEN VAN VUNAKIS

VOLUME 75. Cumulative Subject Index Volumes XXXI, XXXII, XXXIV–LX
Edited by EDWARD A. DENNIS AND MARTHA G. DENNIS

VOLUME 76. Hemoglobins
Edited by ERALDO ANTONINI, LUIGI ROSSI-BERNARDI, AND EMILIA CHIANCONE

VOLUME 77. Detoxication and Drug Metabolism
Edited by WILLIAM B. JAKOBY

VOLUME 78. Interferons (Part A)
Edited by SIDNEY PESTKA

VOLUME 79. Interferons (Part B)
Edited by SIDNEY PESTKA

VOLUME 80. Proteolytic Enzymes (Part C)
Edited by LASZLO LORAND

VOLUME 81. Biomembranes (Part H: Visual Pigments and Purple Membranes, I)
Edited by LESTER PACKER

VOLUME 82. Structural and Contractile Proteins (Part A: Extracellular Matrix)
Edited by LEON W. CUNNINGHAM AND DIXIE W. FREDERIKSEN

VOLUME 83. Complex Carbohydrates (Part D)
Edited by VICTOR GINSBURG

VOLUME 84. Immunochemical Techniques (Part D: Selected Immunoassays)
Edited by JOHN J. LANGONE AND HELEN VAN VUNAKIS

VOLUME 85. Structural and Contractile Proteins (Part B: The Contractile Apparatus and the Cytoskeleton)
Edited by DIXIE W. FREDERIKSEN AND LEON W. CUNNINGHAM

VOLUME 86. Prostaglandins and Arachidonate Metabolites
Edited by WILLIAM E. M. LANDS AND WILLIAM L. SMITH

VOLUME 87. Enzyme Kinetics and Mechanism (Part C: Intermediates, Stereochemistry, and Rate Studies)
Edited by DANIEL L. PURICH

VOLUME 88. Biomembranes (Part I: Visual Pigments and Purple Membranes, II)
Edited by LESTER PACKER

VOLUME 89. Carbohydrate Metabolism (Part D)
Edited by WILLIS A. WOOD

VOLUME 90. Carbohydrate Metabolism (Part E)
Edited by WILLIS A. WOOD

VOLUME 91. Enzyme Structure (Part I)
Edited by C. H. W. HIRS AND SERGE N. TIMASHEFF

VOLUME 92. Immunochemical Techniques (Part E: Monoclonal Antibodies and General Immunoassay Methods)
Edited by JOHN J. LANGONE AND HELEN VAN VUNAKIS

VOLUME 93. Immunochemical Techniques (Part F: Conventional Antibodies, Fc Receptors, and Cytotoxicity)
Edited by JOHN J. LANGONE AND HELEN VAN VUNAKIS

VOLUME 94. Polyamines
Edited by HERBERT TABOR AND CELIA WHITE TABOR

VOLUME 95. Cumulative Subject Index Volumes 61–74, 76–80
Edited by EDWARD A. DENNIS AND MARTHA G. DENNIS

VOLUME 96. Biomembranes [Part J: Membrane Biogenesis: Assembly and Targeting (General Methods; Eukaryotes)]
Edited by SIDNEY FLEISCHER AND BECCA FLEISCHER

VOLUME 97. Biomembranes [Part K: Membrane Biogenesis: Assembly and Targeting (Prokaryotes, Mitochondria, and Chloroplasts)]
Edited by SIDNEY FLEISCHER AND BECCA FLEISCHER

VOLUME 98. Biomembranes (Part L: Membrane Biogenesis: Processing and Recycling)
Edited by SIDNEY FLEISCHER AND BECCA FLEISCHER

VOLUME 99. Hormone Action (Part F: Protein Kinases)
Edited by JACKIE D. CORBIN AND JOEL G. HARDMAN

VOLUME 100. Recombinant DNA (Part B)
Edited by RAY WU, LAWRENCE GROSSMAN, AND KIVIE MOLDAVE

VOLUME 101. Recombinant DNA (Part C)
Edited by RAY WU, LAWRENCE GROSSMAN, AND KIVIE MOLDAVE

VOLUME 102. Hormone Action (Part G: Calmodulin and Calcium-Binding Proteins)
Edited by ANTHONY R. MEANS AND BERT W. O'MALLEY

VOLUME 103. Hormone Action (Part H: Neuroendocrine Peptides)
Edited by P. MICHAEL CONN

VOLUME 104. Enzyme Purification and Related Techniques (Part C)
Edited by WILLIAM B. JAKOBY

VOLUME 105. Oxygen Radicals in Biological Systems
Edited by LESTER PACKER

VOLUME 106. Posttranslational Modifications (Part A)
Edited by FINN WOLD AND KIVIE MOLDAVE

VOLUME 107. Posttranslational Modifications (Part B)
Edited by FINN WOLD AND KIVIE MOLDAVE

VOLUME 108. Immunochemical Techniques (Part G: Separation and Characterization of Lymphoid Cells)
Edited by GIOVANNI DI SABATO, JOHN J. LANGONE, AND HELEN VAN VUNAKIS

VOLUME 109. Hormone Action (Part I: Peptide Hormones)
Edited by LUTZ BIRNBAUMER AND BERT W. O'MALLEY

VOLUME 110. Steroids and Isoprenoids (Part A)
Edited by JOHN H. LAW AND HANS C. RILLING

VOLUME 111. Steroids and Isoprenoids (Part B)
Edited by JOHN H. LAW AND HANS C. RILLING

VOLUME 112. Drug and Enzyme Targeting (Part A)
Edited by KENNETH J. WIDDER AND RALPH GREEN

VOLUME 113. Glutamate, Glutamine, Glutathione, and Related Compounds
Edited by ALTON MEISTER

VOLUME 114. Diffraction Methods for Biological Macromolecules (Part A)
Edited by HAROLD W. WYCKOFF, C. H. W. HIRS, AND SERGE N. TIMASHEFF

VOLUME 115. Diffraction Methods for Biological Macromolecules (Part B)
Edited by HAROLD W. WYCKOFF, C. H. W. HIRS, AND SERGE N. TIMASHEFF

VOLUME 116. Immunochemical Techniques (Part H: Effectors and Mediators of Lymphoid Cell Functions)
Edited by GIOVANNI DI SABATO, JOHN J. LANGONE, AND HELEN VAN VUNAKIS

VOLUME 117. Enzyme Structure (Part J)
Edited by C. H. W. HIRS AND SERGE N. TIMASHEFF

VOLUME 118. Plant Molecular Biology
Edited by ARTHUR WEISSBACH AND HERBERT WEISSBACH

VOLUME 119. Interferons (Part C)
Edited by SIDNEY PESTKA

VOLUME 120. Cumulative Subject Index Volumes 81–94, 96–101

VOLUME 121. Immunochemical Techniques (Part I: Hybridoma Technology and Monoclonal Antibodies)
Edited by JOHN J. LANGONE AND HELEN VAN VUNAKIS

VOLUME 122. Vitamins and Coenzymes (Part G)
Edited by FRANK CHYTIL AND DONALD B. MCCORMICK

VOLUME 123. Vitamins and Coenzymes (Part H)
Edited by FRANK CHYTIL AND DONALD B. MCCORMICK

VOLUME 124. Hormone Action (Part J: Neuroendocrine Peptides)
Edited by P. MICHAEL CONN

VOLUME 125. Biomembranes (Part M: Transport in Bacteria, Mitochondria, and Chloroplasts: General Approaches and Transport Systems)
Edited by SIDNEY FLEISCHER AND BECCA FLEISCHER

VOLUME 126. Biomembranes (Part N: Transport in Bacteria, Mitochondria, and Chloroplasts: Protonmotive Force)
Edited by SIDNEY FLEISCHER AND BECCA FLEISCHER

VOLUME 127. Biomembranes (Part O: Protons and Water: Structure and Translocation)
Edited by LESTER PACKER

VOLUME 128. Plasma Lipoproteins (Part A: Preparation, Structure, and Molecular Biology)
Edited by JERE P. SEGREST AND JOHN J. ALBERS

VOLUME 129. Plasma Lipoproteins (Part B: Characterization, Cell Biology, and Metabolism)
Edited by JOHN J. ALBERS AND JERE P. SEGREST

VOLUME 130. Enzyme Structure (Part K)
Edited by C. H. W. HIRS AND SERGE N. TIMASHEFF

VOLUME 131. Enzyme Structure (Part L)
Edited by C. H. W. HIRS AND SERGE N. TIMASHEFF

VOLUME 132. Immunochemical Techniques (Part J: Phagocytosis and Cell-Mediated Cytotoxicity)
Edited by GIOVANNI DI SABATO AND JOHANNES EVERSE

VOLUME 133. Bioluminescence and Chemiluminescence (Part B)
Edited by MARLENE DELUCA AND WILLIAM D. MCELROY

VOLUME 134. Structural and Contractile Proteins (Part C: The Contractile Apparatus and the Cytoskeleton)
Edited by RICHARD B. VALLEE

VOLUME 135. Immobilized Enzymes and Cells (Part B)
Edited by KLAUS MOSBACH

VOLUME 136. Immobilized Enzymes and Cells (Part C)
Edited by KLAUS MOSBACH

VOLUME 137. Immobilized Enzymes and Cells (Part D)
Edited by KLAUS MOSBACH

VOLUME 138. Complex Carbohydrates (Part E)
Edited by VICTOR GINSBURG

VOLUME 139. Cellular Regulators (Part A: Calcium- and Calmodulin-Binding Proteins)
Edited by ANTHONY R. MEANS AND P. MICHAEL CONN

VOLUME 140. Cumulative Subject Index Volumes 102–119, 121–134

VOLUME 141. Cellular Regulators (Part B: Calcium and Lipids)
Edited by P. MICHAEL CONN AND ANTHONY R. MEANS

VOLUME 142. Metabolism of Aromatic Amino Acids and Amines
Edited by SEYMOUR KAUFMAN

VOLUME 143. Sulfur and Sulfur Amino Acids
Edited by WILLIAM B. JAKOBY AND OWEN GRIFFITH

VOLUME 144. Structural and Contractile Proteins (Part D: Extracellular Matrix)
Edited by LEON W. CUNNINGHAM

VOLUME 145. Structural and Contractile Proteins (Part E: Extracellular Matrix)
Edited by LEON W. CUNNINGHAM

VOLUME 146. Peptide Growth Factors (Part A)
Edited by DAVID BARNES AND DAVID A. SIRBASKU

VOLUME 147. Peptide Growth Factors (Part B)
Edited by DAVID BARNES AND DAVID A. SIRBASKU

VOLUME 148. Plant Cell Membranes
Edited by LESTER PACKER AND ROLAND DOUCE

VOLUME 149. Drug and Enzyme Targeting (Part B)
Edited by RALPH GREEN AND KENNETH J. WIDDER

VOLUME 150. Immunochemical Techniques (Part K: *In Vitro* Models of B and T Cell Functions and Lymphoid Cell Receptors)
Edited by GIOVANNI DI SABATO

VOLUME 151. Molecular Genetics of Mammalian Cells
Edited by MICHAEL M. GOTTESMAN

VOLUME 152. Guide to Molecular Cloning Techniques
Edited by SHELBY L. BERGER AND ALAN R. KIMMEL

VOLUME 153. Recombinant DNA (Part D)
Edited by RAY WU AND LAWRENCE GROSSMAN

VOLUME 154. Recombinant DNA (Part E)
Edited by RAY WU AND LAWRENCE GROSSMAN

VOLUME 155. Recombinant DNA (Part F)
Edited by RAY WU

VOLUME 156. Biomembranes (Part P: ATP-Driven Pumps and Related Transport: The Na, K-Pump)
Edited by SIDNEY FLEISCHER AND BECCA FLEISCHER

VOLUME 157. Biomembranes (Part Q: ATP-Driven Pumps and Related Transport: Calcium, Proton, and Potassium Pumps)
Edited by SIDNEY FLEISCHER AND BECCA FLEISCHER

VOLUME 158. Metalloproteins (Part A)
Edited by JAMES F. RIORDAN AND BERT L. VALLEE

VOLUME 159. Initiation and Termination of Cyclic Nucleotide Action
Edited by JACKIE D. CORBIN AND ROGER A. JOHNSON

VOLUME 160. Biomass (Part A: Cellulose and Hemicellulose)
Edited by WILLIS A. WOOD AND SCOTT T. KELLOGG

VOLUME 161. Biomass (Part B: Lignin, Pectin, and Chitin)
Edited by WILLIS A. WOOD AND SCOTT T. KELLOGG

VOLUME 162. Immunochemical Techniques (Part L: Chemotaxis and Inflammation)
Edited by GIOVANNI DI SABATO

VOLUME 163. Immunochemical Techniques (Part M: Chemotaxis and Inflammation)
Edited by GIOVANNI DI SABATO

VOLUME 164. Ribosomes
Edited by HARRY F. NOLLER, JR., AND KIVIE MOLDAVE

VOLUME 165. Microbial Toxins: Tools for Enzymology
Edited by SIDNEY HARSHMAN

VOLUME 166. Branched-Chain Amino Acids
Edited by ROBERT HARRIS AND JOHN R. SOKATCH

VOLUME 167. Cyanobacteria
Edited by LESTER PACKER AND ALEXANDER N. GLAZER

VOLUME 168. Hormone Action (Part K: Neuroendocrine Peptides)
Edited by P. MICHAEL CONN

VOLUME 169. Platelets: Receptors, Adhesion, Secretion (Part A)
Edited by JACEK HAWIGER

VOLUME 170. Nucleosomes
Edited by PAUL M. WASSARMAN AND ROGER D. KORNBERG

VOLUME 171. Biomembranes (Part R: Transport Theory: Cells and Model Membranes)
Edited by SIDNEY FLEISCHER AND BECCA FLEISCHER

VOLUME 172. Biomembranes (Part S: Transport: Membrane Isolation and Characterization)
Edited by SIDNEY FLEISCHER AND BECCA FLEISCHER

VOLUME 173. Biomembranes [Part T: Cellular and Subcellular Transport: Eukaryotic (Nonepithelial) Cells]
Edited by SIDNEY FLEISCHER AND BECCA FLEISCHER

VOLUME 174. Biomembranes [Part U: Cellular and Subcellular Transport: Eukaryotic (Nonepithelial) Cells]
Edited by SIDNEY FLEISCHER AND BECCA FLEISCHER

VOLUME 175. Cumulative Subject Index Volumes 135–139, 141–167

VOLUME 176. Nuclear Magnetic Resonance (Part A: Spectral Techniques and Dynamics)
Edited by NORMAN J. OPPENHEIMER AND THOMAS L. JAMES

VOLUME 177. Nuclear Magnetic Resonance (Part B: Structure and Mechanism)
Edited by NORMAN J. OPPENHEIMER AND THOMAS L. JAMES

VOLUME 178. Antibodies, Antigens, and Molecular Mimicry
Edited by JOHN J. LANGONE

VOLUME 179. Complex Carbohydrates (Part F)
Edited by VICTOR GINSBURG

VOLUME 180. RNA Processing (Part A: General Methods)
Edited by JAMES E. DAHLBERG AND JOHN N. ABELSON

VOLUME 181. RNA Processing (Part B: Specific Methods)
Edited by JAMES E. DAHLBERG AND JOHN N. ABELSON

VOLUME 182. Guide to Protein Purification
Edited by MURRAY P. DEUTSCHER

VOLUME 183. Molecular Evolution: Computer Analysis of Protein and Nucleic Acid Sequences
Edited by RUSSELL F. DOOLITTLE

VOLUME 184. Avidin-Biotin Technology
Edited by MEIR WILCHEK AND EDWARD A. BAYER

VOLUME 185. Gene Expression Technology
Edited by DAVID V. GOEDDEL

VOLUME 186. Oxygen Radicals in Biological Systems (Part B: Oxygen Radicals and Antioxidants)
Edited by LESTER PACKER AND ALEXANDER N. GLAZER

VOLUME 187. Arachidonate Related Lipid Mediators
Edited by ROBERT C. MURPHY AND FRANK A. FITZPATRICK

VOLUME 188. Hydrocarbons and Methylotrophy
Edited by MARY E. LIDSTROM

VOLUME 189. Retinoids (Part A: Molecular and Metabolic Aspects)
Edited by LESTER PACKER

VOLUME 190. Retinoids (Part B: Cell Differentiation and Clinical Applications)
Edited by LESTER PACKER

VOLUME 191. Biomembranes (Part V: Cellular and Subcellular Transport: Epithelial Cells)
Edited by SIDNEY FLEISCHER AND BECCA FLEISCHER

VOLUME 192. Biomembranes (Part W: Cellular and Subcellular Transport: Epithelial Cells)
Edited by SIDNEY FLEISCHER AND BECCA FLEISCHER

VOLUME 193. Mass Spectrometry
Edited by JAMES A. MCCLOSKEY

VOLUME 194. Guide to Yeast Genetics and Molecular Biology
Edited by CHRISTINE GUTHRIE AND GERALD R. FINK

VOLUME 195. Adenylyl Cyclase, G Proteins, and Guanylyl Cyclase
Edited by ROGER A. JOHNSON AND JACKIE D. CORBIN

VOLUME 196. Molecular Motors and the Cytoskeleton
Edited by RICHARD B. VALLEE

VOLUME 197. Phospholipases
Edited by EDWARD A. DENNIS

VOLUME 198. Peptide Growth Factors (Part C)
Edited by DAVID BARNES, J. P. MATHER, AND GORDON H. SATO

VOLUME 199. Cumulative Subject Index Volumes 168–174, 176–194

VOLUME 200. Protein Phosphorylation (Part A: Protein Kinases: Assays, Purification, Antibodies, Functional Analysis, Cloning, and Expression)
Edited by TONY HUNTER AND BARTHOLOMEW M. SEFTON

VOLUME 201. Protein Phosphorylation (Part B: Analysis of Protein Phosphorylation, Protein Kinase Inhibitors, and Protein Phosphatases)
Edited by TONY HUNTER AND BARTHOLOMEW M. SEFTON

VOLUME 202. Molecular Design and Modeling: Concepts and Applications (Part A: Proteins, Peptides, and Enzymes)
Edited by JOHN J. LANGONE

VOLUME 203. Molecular Design and Modeling: Concepts and Applications (Part B: Antibodies and Antigens, Nucleic Acids, Polysaccharides, and Drugs)
Edited by JOHN J. LANGONE

VOLUME 204. Bacterial Genetic Systems
Edited by JEFFREY H. MILLER

VOLUME 205. Metallobiochemistry (Part B: Metallothionein and Related Molecules)
Edited by JAMES F. RIORDAN AND BERT L. VALLEE

VOLUME 206. Cytochrome P450
Edited by MICHAEL R. WATERMAN AND ERIC F. JOHNSON

VOLUME 207. Ion Channels
Edited by BERNARDO RUDY AND LINDA E. IVERSON

VOLUME 208. Protein–DNA Interactions
Edited by ROBERT T. SAUER

VOLUME 209. Phospholipid Biosynthesis
Edited by EDWARD A. DENNIS AND DENNIS E. VANCE

VOLUME 210. Numerical Computer Methods
Edited by LUDWIG BRAND AND MICHAEL L. JOHNSON

VOLUME 211. DNA Structures (Part A: Synthesis and Physical Analysis of DNA)
Edited by DAVID M. J. LILLEY AND JAMES E. DAHLBERG

VOLUME 212. DNA Structures (Part B: Chemical and Electrophoretic Analysis of DNA)
Edited by DAVID M. J. LILLEY AND JAMES E. DAHLBERG

VOLUME 213. Carotenoids (Part A: Chemistry, Separation, Quantitation, and Antioxidation)
Edited by LESTER PACKER

VOLUME 214. Carotenoids (Part B: Metabolism, Genetics, and Biosynthesis)
Edited by LESTER PACKER

VOLUME 215. Platelets: Receptors, Adhesion, Secretion (Part B)
Edited by JACEK J. HAWIGER

VOLUME 216. Recombinant DNA (Part G)
Edited by RAY WU

VOLUME 217. Recombinant DNA (Part H)
Edited by RAY WU

VOLUME 218. Recombinant DNA (Part I)
Edited by RAY WU

VOLUME 219. Reconstitution of Intracellular Transport
Edited by JAMES E. ROTHMAN

VOLUME 220. Membrane Fusion Techniques (Part A)
Edited by NEJAT DÜZGUÜNES

VOLUME 221. Membrane Fusion Techniques (Part B)
Edited by NEJAT DÜZGÜNES

VOLUME 222. Proteolytic Enzymes in Coagulation, Fibrinolysis, and Complement Activation (Part A: Mammalian Blood Coagulation Factors and Inhibitors)
Edited by LASZLO LORAND AND KENNETH G. MANN

VOLUME 223. Proteolytic Enzymes in Coagulation, Fibrinolysis, and Complement Activation (Part B: Complement Activation, Fibrinolysis, and Nonmammalian Blood Coagulation Factors)
Edited by LASZLO LORAND AND KENNETH G. MANN

VOLUME 224. Molecular Evolution: Producing the Biochemical Data
Edited by ELIZABETH ANNE ZIMMER, THOMAS J. WHITE, REBECCA L. CANN, AND ALLAN C. WILSON

VOLUME 225. Guide to Techniques in Mouse Development
Edited by PAUL M. WASSARMAN AND MELVIN L. DEPAMPHILIS

VOLUME 226. Metallobiochemistry (Part C: Spectroscopic and Physical Methods for Probing Metal Ion Environments in Metalloenzymes and Metalloproteins)
Edited by JAMES F. RIORDAN AND BERT L. VALLEE

VOLUME 227. Metallobiochemistry (Part D: Physical and Spectroscopic Methods for Probing Metal Ion Environments in Metalloproteins)
Edited by JAMES F. RIORDAN AND BERT L. VALLEE

VOLUME 228. Aqueous Two-Phase Systems
Edited by HARRY WALTER AND GÖTE JOHANSSON

VOLUME 229. Cumulative Subject Index Volumes 195–198, 200–227

VOLUME 230. Guide to Techniques in Glycobiology
Edited by WILLIAM J. LENNARZ AND GERALD W. HART

VOLUME 231. Hemoglobins (Part B: Biochemical and Analytical Methods)
Edited by JOHANNES EVERSE, KIM D. VANDEGRIFF, AND ROBERT M. WINSLOW

VOLUME 232. Hemoglobins (Part C: Biophysical Methods)
Edited by JOHANNES EVERSE, KIM D. VANDEGRIFF, AND ROBERT M. WINSLOW

VOLUME 233. Oxygen Radicals in Biological Systems (Part C)
Edited by LESTER PACKER

VOLUME 234. Oxygen Radicals in Biological Systems (Part D)
Edited by LESTER PACKER

VOLUME 235. Bacterial Pathogenesis (Part A: Identification and Regulation of Virulence Factors)
Edited by VIRGINIA L. CLARK AND PATRIK M. BAVOIL

VOLUME 236. Bacterial Pathogenesis (Part B: Integration of Pathogenic Bacteria with Host Cells)
Edited by VIRGINIA L. CLARK AND PATRIK M. BAVOIL

VOLUME 237. Heterotrimeric G Proteins
Edited by RAVI IYENGAR

VOLUME 238. Heterotrimeric G-Protein Effectors
Edited by RAVI IYENGAR

VOLUME 239. Nuclear Magnetic Resonance (Part C)
Edited by THOMAS L. JAMES AND NORMAN J. OPPENHEIMER

VOLUME 240. Numerical Computer Methods (Part B)
Edited by MICHAEL L. JOHNSON AND LUDWIG BRAND

VOLUME 241. Retroviral Proteases
Edited by LAWRENCE C. KUO AND JULES A. SHAFER

VOLUME 242. Neoglycoconjugates (Part A)
Edited by Y. C. LEE AND REIKO T. LEE

VOLUME 243. Inorganic Microbial Sulfur Metabolism
Edited by HARRY D. PECK, JR., AND JEAN LEGALL

VOLUME 244. Proteolytic Enzymes: Serine and Cysteine Peptidases
Edited by ALAN J. BARRETT

VOLUME 245. Extracellular Matrix Components
Edited by E. RUOSLAHTI AND E. ENGVALL

VOLUME 246. Biochemical Spectroscopy
Edited by KENNETH SAUER

VOLUME 247. Neoglycoconjugates (Part B: Biomedical Applications)
Edited by Y. C. LEE AND REIKO T. LEE

VOLUME 248. Proteolytic Enzymes: Aspartic and Metallo Peptidases
Edited by ALAN J. BARRETT

VOLUME 249. Enzyme Kinetics and Mechanism (Part D: Developments in Enzyme Dynamics)
Edited by DANIEL L. PURICH

VOLUME 250. Lipid Modifications of Proteins
Edited by PATRICK J. CASEY AND JANICE E. BUSS

VOLUME 251. Biothiols (Part A: Monothiols and Dithiols, Protein Thiols, and Thiyl Radicals)
Edited by LESTER PACKER

VOLUME 252. Biothiols (Part B: Glutathione and Thioredoxin; Thiols in Signal Transduction and Gene Regulation)
Edited by LESTER PACKER

VOLUME 253. Adhesion of Microbial Pathogens
Edited by RON J. DOYLE AND ITZHAK OFEK

VOLUME 254. Oncogene Techniques
Edited by PETER K. VOGT AND INDER M. VERMA

VOLUME 255. Small GTPases and Their Regulators (Part A: Ras Family)
Edited by W. E. BALCH, CHANNING J. DER, AND ALAN HALL

VOLUME 256. Small GTPases and Their Regulators (Part B: Rho Family)
Edited by W. E. BALCH, CHANNING J. DER, AND ALAN HALL

VOLUME 257. Small GTPases and Their Regulators (Part C: Proteins Involved in Transport)
Edited by W. E. BALCH, CHANNING J. DER, AND ALAN HALL

VOLUME 258. Redox-Active Amino Acids in Biology
Edited by JUDITH P. KLINMAN

VOLUME 259. Energetics of Biological Macromolecules
Edited by MICHAEL L. JOHNSON AND GARY K. ACKERS

VOLUME 260. Mitochondrial Biogenesis and Genetics (Part A)
Edited by GIUSEPPE M. ATTARDI AND ANNE CHOMYN

VOLUME 261. Nuclear Magnetic Resonance and Nucleic Acids
Edited by THOMAS L. JAMES

VOLUME 262. DNA Replication
Edited by JUDITH L. CAMPBELL

VOLUME 263. Plasma Lipoproteins (Part C: Quantitation)
Edited by WILLIAM A. BRADLEY, SANDRA H. GIANTURCO, AND JERE P. SEGREST

VOLUME 264. Mitochondrial Biogenesis and Genetics (Part B)
Edited by GIUSEPPE M. ATTARDI AND ANNE CHOMYN

VOLUME 265. Cumulative Subject Index Volumes 228, 230–262

VOLUME 266. Computer Methods for Macromolecular Sequence Analysis
Edited by RUSSELL F. DOOLITTLE

VOLUME 267. Combinatorial Chemistry
Edited by JOHN N. ABELSON

VOLUME 268. Nitric Oxide (Part A: Sources and Detection of NO; NO Synthase)
Edited by LESTER PACKER

VOLUME 269. Nitric Oxide (Part B: Physiological and Pathological Processes)
Edited by LESTER PACKER

VOLUME 270. High Resolution Separation and Analysis of Biological Macromolecules (Part A: Fundamentals)
Edited by BARRY L. KARGER AND WILLIAM S. HANCOCK

VOLUME 271. High Resolution Separation and Analysis of Biological Macromolecules (Part B: Applications)
Edited by BARRY L. KARGER AND WILLIAM S. HANCOCK

VOLUME 272. Cytochrome P450 (Part B)
Edited by ERIC F. JOHNSON AND MICHAEL R. WATERMAN

VOLUME 273. RNA Polymerase and Associated Factors (Part A)
Edited by SANKAR ADHYA

VOLUME 274. RNA Polymerase and Associated Factors (Part B)
Edited by SANKAR ADHYA

VOLUME 275. Viral Polymerases and Related Proteins
Edited by LAWRENCE C. KUO, DAVID B. OLSEN, AND STEVEN S. CARROLL

VOLUME 276. Macromolecular Crystallography (Part A)
Edited by CHARLES W. CARTER, JR., AND ROBERT M. SWEET

VOLUME 277. Macromolecular Crystallography (Part B)
Edited by CHARLES W. CARTER, JR., AND ROBERT M. SWEET

VOLUME 278. Fluorescence Spectroscopy
Edited by LUDWIG BRAND AND MICHAEL L. JOHNSON

VOLUME 279. Vitamins and Coenzymes (Part I)
Edited by DONALD B. MCCORMICK, JOHN W. SUTTIE, AND CONRAD WAGNER

VOLUME 280. Vitamins and Coenzymes (Part J)
Edited by DONALD B. MCCORMICK, JOHN W. SUTTIE, AND CONRAD WAGNER

VOLUME 281. Vitamins and Coenzymes (Part K)
Edited by DONALD B. MCCORMICK, JOHN W. SUTTIE, AND CONRAD WAGNER

VOLUME 282. Vitamins and Coenzymes (Part L)
Edited by DONALD B. MCCORMICK, JOHN W. SUTTIE, AND CONRAD WAGNER

VOLUME 283. Cell Cycle Control
Edited by WILLIAM G. DUNPHY

VOLUME 284. Lipases (Part A: Biotechnology)
Edited by BYRON RUBIN AND EDWARD A. DENNIS

VOLUME 285. Cumulative Subject Index Volumes 263, 264, 266–284, 286–289

VOLUME 286. Lipases (Part B: Enzyme Characterization and Utilization)
Edited by BYRON RUBIN AND EDWARD A. DENNIS

VOLUME 287. Chemokines
Edited by RICHARD HORUK

VOLUME 288. Chemokine Receptors
Edited by RICHARD HORUK

VOLUME 289. Solid Phase Peptide Synthesis
Edited by GREGG B. FIELDS

VOLUME 290. Molecular Chaperones
Edited by GEORGE H. LORIMER AND THOMAS BALDWIN

VOLUME 291. Caged Compounds
Edited by GERARD MARRIOTT

VOLUME 292. ABC Transporters: Biochemical, Cellular, and Molecular Aspects
Edited by SURESH V. AMBUDKAR AND MICHAEL M. GOTTESMAN

VOLUME 293. Ion Channels (Part B)
Edited by P. MICHAEL CONN

VOLUME 294. Ion Channels (Part C)
Edited by P. MICHAEL CONN

VOLUME 295. Energetics of Biological Macromolecules (Part B)
Edited by GARY K. ACKERS AND MICHAEL L. JOHNSON

VOLUME 296. Neurotransmitter Transporters
Edited by SUSAN G. AMARA

VOLUME 297. Photosynthesis: Molecular Biology of Energy Capture
Edited by LEE MCINTOSH

VOLUME 298. Molecular Motors and the Cytoskeleton (Part B)
Edited by RICHARD B. VALLEE

VOLUME 299. Oxidants and Antioxidants (Part A)
Edited by LESTER PACKER

VOLUME 300. Oxidants and Antioxidants (Part B)
Edited by LESTER PACKER

VOLUME 301. Nitric Oxide: Biological and Antioxidant Activities (Part C)
Edited by LESTER PACKER

VOLUME 302. Green Fluorescent Protein
Edited by P. MICHAEL CONN

VOLUME 303. cDNA Preparation and Display
Edited by SHERMAN M. WEISSMAN

VOLUME 304. Chromatin
Edited by PAUL M. WASSARMAN AND ALAN P. WOLFFE

VOLUME 305. Bioluminescence and Chemiluminescence (Part C)
Edited by THOMAS O. BALDWIN AND MIRIAM M. ZIEGLER

VOLUME 306. Expression of Recombinant Genes in Eukaryotic Systems
Edited by JOSEPH C. GLORIOSO AND MARTIN C. SCHMIDT

VOLUME 307. Confocal Microscopy
Edited by P. MICHAEL CONN

VOLUME 308. Enzyme Kinetics and Mechanism (Part E: Energetics of Enzyme Catalysis)
Edited by DANIEL L. PURICH AND VERN L. SCHRAMM

VOLUME 309. Amyloid, Prions, and Other Protein Aggregates
Edited by RONALD WETZEL

VOLUME 310. Biofilms
Edited by RON J. DOYLE

VOLUME 311. Sphingolipid Metabolism and Cell Signaling (Part A)
Edited by ALFRED H. MERRILL, JR., AND YUSUF A. HANNUN

VOLUME 312. Sphingolipid Metabolism and Cell Signaling (Part B)
Edited by ALFRED H. MERRILL, JR., AND YUSUF A. HANNUN

VOLUME 313. Antisense Technology (Part A: General Methods, Methods of Delivery, and RNA Studies)
Edited by M. IAN PHILLIPS

VOLUME 314. Antisense Technology (Part B: Applications)
Edited by M. IAN PHILLIPS

VOLUME 315. Vertebrate Phototransduction and the Visual Cycle (Part A)
Edited by KRZYSZTOF PALCZEWSKI

VOLUME 316. Vertebrate Phototransduction and the Visual Cycle (Part B)
Edited by KRZYSZTOF PALCZEWSKI

VOLUME 317. RNA–Ligand Interactions (Part A: Structural Biology Methods)
Edited by DANIEL W. CELANDER AND JOHN N. ABELSON

VOLUME 318. RNA–Ligand Interactions (Part B: Molecular Biology Methods)
Edited by DANIEL W. CELANDER AND JOHN N. ABELSON

VOLUME 319. Singlet Oxygen, UV-A, and Ozone
Edited by LESTER PACKER AND HELMUT SIES

VOLUME 320. Cumulative Subject Index Volumes 290–319

VOLUME 321. Numerical Computer Methods (Part C)
Edited by MICHAEL L. JOHNSON AND LUDWIG BRAND

VOLUME 322. Apoptosis
Edited by JOHN C. REED

VOLUME 323. Energetics of Biological Macromolecules (Part C)
Edited by MICHAEL L. JOHNSON AND GARY K. ACKERS

VOLUME 324. Branched-Chain Amino Acids (Part B)
Edited by ROBERT A. HARRIS AND JOHN R. SOKATCH

VOLUME 325. Regulators and Effectors of Small GTPases (Part D: Rho Family)
Edited by W. E. BALCH, CHANNING J. DER, AND ALAN HALL

VOLUME 326. Applications of Chimeric Genes and Hybrid Proteins (Part A: Gene Expression and Protein Purification)
Edited by JEREMY THORNER, SCOTT D. EMR, AND JOHN N. ABELSON

VOLUME 327. Applications of Chimeric Genes and Hybrid Proteins (Part B: Cell Biology and Physiology)
Edited by JEREMY THORNER, SCOTT D. EMR, AND JOHN N. ABELSON

VOLUME 328. Applications of Chimeric Genes and Hybrid Proteins (Part C: Protein-Protein Interactions and Genomics)
Edited by JEREMY THORNER, SCOTT D. EMR, AND JOHN N. ABELSON

VOLUME 329. Regulators and Effectors of Small GTPases (Part E: GTPases Involved in Vesicular Traffic)
Edited by W. E. BALCH, CHANNING J. DER, AND ALAN HALL

VOLUME 330. Hyperthermophilic Enzymes (Part A)
Edited by MICHAEL W. W. ADAMS AND ROBERT M. KELLY

VOLUME 331. Hyperthermophilic Enzymes (Part B)
Edited by MICHAEL W. W. ADAMS AND ROBERT M. KELLY

VOLUME 332. Regulators and Effectors of Small GTPases (Part F: Ras Family I)
Edited by W. E. BALCH, CHANNING J. DER, AND ALAN HALL

VOLUME 333. Regulators and Effectors of Small GTPases (Part G: Ras Family II)
Edited by W. E. BALCH, CHANNING J. DER, AND ALAN HALL

VOLUME 334. Hyperthermophilic Enzymes (Part C)
Edited by MICHAEL W. W. ADAMS AND ROBERT M. KELLY

VOLUME 335. Flavonoids and Other Polyphenols
Edited by LESTER PACKER

VOLUME 336. Microbial Growth in Biofilms (Part A: Developmental and Molecular Biological Aspects) (in preparation)
Edited by RON J. DOYLE

VOLUME 337. Microbial Growth in Biofilms (Part B: Special Environments and Physicochemical Aspects) (in preparation)
Edited by RON J. DOYLE

VOLUME 338. Nuclear Magnetic Resonance of Biological Macromolecules (Part A) (in preparation)
Edited by THOMAS L. JAMES, VOLKER DÖTSCH, AND ULI SCHMITZ

VOLUME 339. Nuclear Magnetic Resonance of Biological Macromolecules (Part B) (in preparation)
Edited by THOMAS L. JAMES, VOLKER DÖTSCH, AND ULI SCHMITZ

VOLUME 340. Drug-Nucleic Acid Interactions (in preparation)
Edited by JONATHAN B. CHAIRES AND MICHAEL J. WARING

VOLUME 341. Ribonucleases (Part A) (in preparation)
Edited by ALLEN W. NICHOLSON

VOLUME 342. Ribonucleases (Part B) (in preparation)
Edited by ALLEN W. NICHOLSON

VOLUME 343. G Protein Pathways (Part A: Receptors) (in preparation)
Edited by RAVI IENGAR AND JOHN D. HILDEBRANDT

VOLUME 344. G Protein Pathways (Part B: G Proteins and Their Regulators) (in preparation)
Edited by RAVI IYENGAR AND JOHN D. HILDEBRANDT

VOLUME 345. G Protein Pathways (Part C: Effector Mechanisms) (in preparation)
Edited by RAVI IYENGAR AND JOHN D. HILDEBRANDT

Section I

Sources, Characterization, and Analytical Methods

[1] Overview of Methods for Analysis and Identification of Flavonoids

By STEPHEN J. BLOOR

The flavonoids are a class of plant secondary metabolites derived from the condensation of a cinnamic acid with three malonyl-CoA groups. All flavonoids arise from this initial reaction, which is catalyzed by the chalcone synthase enzyme. The chalcone is usually converted rapidly into a phenylbenzopyran, and further modification leads to the flavones, isoflavones, flavonols, or anthocyanins (Fig. 1). Additional structural elaboration, mainly through glycosylation but also via acylation or alkylation, gives us the huge variety of flavonoid structures seen throughout the plant kingdom.

Many books and review articles have been written on the subject of flavonoids, their occurrence, and analysis. The series "The Flavonoids: Advances in Research"[1–3] provides the most comprehensive coverage of occurrence and structural variation, whereas more specific texts or articles relate to the analysis of flavonoids.[4–7] Although the general methodology used for analysis and identification of flavonoids has much in common with the techniques used for many other groups of natural products, a number of useful techniques have evolved that can provide shortcuts enabling the rapid identification of flavonoid type and substitution pattern. Despite the plethora of flavonoid structures presented in the scientific literature, the number of common, basic structural units remains limited; the flavone and flavonol compounds are by far the most common structural types and so are the main focus of this article. Another particular advantage the analyst has in flavonoid analysis is the distinctive UV (or UV–VIS) spectra of these

---

[1] J. B. Harborne and T. J. Mabry (eds.) "The Flavonoids-Advances in Research." Chapman and Hall, London, 1982.

[2] J. B. Harborne (ed.) "The Flavonoids-Advances in Research since 1980." Chapman and Hall, London, 1988.

[3] J. B. Harborne (ed.) "The Flavonoids-Advances in Research since 1986." Chapman and Hall, London, 1994.

[4] K. R. Markham and S. J. Bloor *in* "Flavonoids in Health and Disease" (C. A. Rice-Evans and L. Packer, eds.), pp. 1–33. Dekker, New York, 1998.

[5] T. J. Mabry, K. R. Markham, and M. B. Thomas "The Systematic Identification of Flavonoids." Springer-Verlag New York, 1970.

[6] K. R. Markham "Techniques of Flavonoid Identification." Academic Press, London, 1982.

[7] P. M. Dey and J. B. Harborne (eds.) "Methods in Plant Biochemistry," Vol. 1. Academic Press, London, 1989.

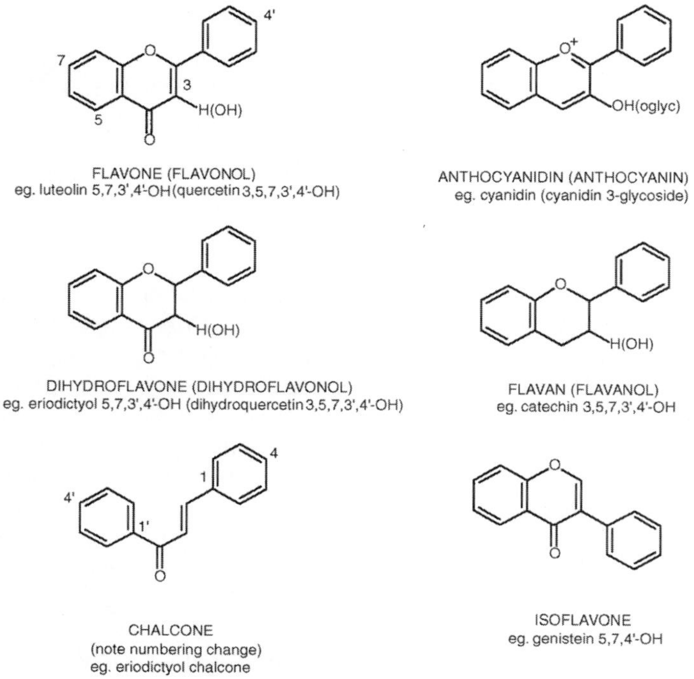

FIG. 1. Basic flavonoid structural types.

compounds where minor differences in structure are often seen as significant differences in their UV spectra, enabling rapid diagnosis of certain structural features. Modern instrumental techniques enable us to gain much information regarding the mass and UV–VIS spectra of individual components in a complex mixture. A combination of some more traditional analytical techniques combined with these modern techniques enables at least a partial identification of most flavonoid components without large-scale purification of the individual compounds. Although more specific examples of applications are described elsewhere in this volume, this article gives a general overview of the typical procedures used to determine type and quantities of flavonoid compounds.

For the purposes of analysis, the flavonoids can be basically classified into three types: flavonoid glycosides, nonpolar flavonoids (aglycones, methylated or alkylated flavonoids), and anthocyanins. Each type requires a different analytical technique. The proanthocyanidins,[8] especially the oligomeric forms, would constitute a fourth class but are not discussed in detail here.

---

[8] L. J. Porter *in* "The Flavonoids: Advances in Research since 1986" (J. B. Harborne, ed.), pp. 23–53. Chapman and Hall, London, 1994.

Chromatographic Analysis

Thin-Layer Chromatography (TLC) [or Paper Chromatography (PC)]

Historically, paper chromatography has been the preferred method for flavonoid analysis, and relative mobility data are available for a large variety of compounds.[9,10] However, not all laboratories are equipped to perform PC, but TLC on cellulose sheets (e.g., Schleicher and Schuell, Keene, NH, Avicel) is a useful alternative. Cellulose TLC is especially useful for quick analyses of materials containing flavonoids, especially flavonoid glycosides. The relative mobility and appearance of spots under UV, before and after spraying with various reagents, enable a good approximation of structural type. Two solvent systems are routinely used for such preliminary analyses: 15% acetic acid (acetic acid : H_2O, 15 : 85) and TBA (*t*-BuOH : acetic acid : H_2O, 3 : 1 : 1). Rough relative mobilities for various flavonoid types are shown in Table I. Typically, plastic-backed sheets are cut to a suitable size, e.g., 5 cm wide × 7 cm high, the samples are applied as small 1- to 2-mm spots, allowed to dry, and the sheets developed in a glass tank (a slide-staining jar is useful for small sheets). The relative mobilities in TBA and 15% acetic acid should give a rough guide as to flavonoid type. Sheets are dried and viewed under UV light (366 nm). The significance of the spot color and the behavior on exposure to ammonia vapor are detailed in Table II. The sheet is then sprayed with NA reagent (a 1 % solution of diphenylboric acid–ethanolamine complex in methanol), dried, and again viewed under UV light. Most flavonoids show some color, but most significantly 3', 4'-dihydroxyflavones or flavonols are orange and the 4'-hydroxy equivalents are yellow-green. If desired, a fingerprint type of two-dimensional chromatogram can be prepared by applying one sample only on a spot in one corner of the sheet, developing the sheet with TBA in one dimension, and then, after drying, developing in the second dimension with 15% acetic acid. Generally, most flavonoids will be separated on two-dimensional TLC and their spot characteristics can be noted. Silica TLC is also a useful screening system for flavonoid compounds.[11] Ethyl acetate–formic acid–acetic acid–H_2O (100 : 11 : 11 : 27) gives a good range of mobilities for flavone and flavonol glycosides (R_f : diglycosides < monoglycosides < aglycones) and, as discussed earlier, most spots can be visualized with the NA reagent. Anthocyanins are best analyzed by TLC using a quite acidic solvent system. A useful general eluting solvent for cellulose TLC of these compounds is HCl : formic acid : H_2O (1 : 1 : 2).[12] The intensely colored spots (violet for trihydroxylated B-ring, red for dihydroxy and orange for mono-) do not require spraying, and generally the R_f increases as the degree of glycosylation increases.

---

[9] K. R. Markham and R. D Wilson *Phytochem. Bull.* **20**, 8 (1988).
[10] K. R. Markham and R. D Wilson *Phytochem. Bull.* **21**, 2 (1989).
[11] H. Wagner, S. Bladt, and E. M. Zgainski "Plant Drug Analysis." Springer-Verlag, Berlin, 1983.
[12] O. M. Andersen and G. W. Francis *J. Chromatogr.* **318**, 450 (1985).

TABLE I
TWO-DIMENSIONAL PC MOBILITY^a OF VARIOUS FLAVONOID TYPES IN TBA AND 15% ACETIC ACID^b

15% Acetic acid mobility	TBA mobility	Flavonoid types commonly encountered
Low	Med-high	Flavone, flavonol, biflavone, chalcone, aurone aglycones, and anthocyanidins
Low+	Med-low	Flavone mono-C-glycosides
Low+	Med	Flavone mono-O-glycosides and flavonol 7-O-glycosides
Med-low	Low	Anthocyanidin 3,5-diglycosides (p-coumaroyl)
Med-low	Med-low	Anthocyanin 3-monoglycosides
Med-low	Med	Anthocyanin 5-monoglycosides
Med	Med-low	Flavone di-C-glycosides
Med	Med	Flavone C-glycosides (O-glycosides of)
Med	Med	Flavonol 3-diglycosides
Med	Med-high	Flavonol 3-monoglycosides
Med+	Low	Anthocyanidin 3,5-glycosides
Med+	High	Isoflavone, dihydroflavone, dihydroflavonal aglycones, and cinnamic acids
Med-high	Low	Flavonoid sulfates
Med-high	Med	Anthocyanidin 3,5-diglycosides (acetyl)
Med-high	Med-high	Flavonol 3,7-diglucosides
High	Med-low	Flavonol 3-triglycosides
High	Low	Flavonol 3-tetraglycosides
High	Med	Isoflavone mono- and diglycosides
High	Med-high	Dihydroflavonol 3-glycosides

^{a} Mobility code (typical R_f range): low, 0 to 33 med, 0.33 to 0.65; high, 0.66 to 1.0; "+" indicates high end of range. Hyphenated categories indicate a range covering both.

^{b} Reproduced from K. R. Markham and S. J. Bloor, in "Flavonoids in Health and Disease" (C. A. Rice-Evans and L. Packer, eds.), pp. 1–33. Dekker, New York, 1998, with permission.

HPLC

The distinctive UV–VIS spectra of most flavonoids and the widespread availability of high-performance liquid chromatography (HPLC) systems with multiwavelength capability, or the ability to record on-line spectra, has meant HPLC is now the method of choice for flavonoid analysis.[13–15] Most often separations are performed using a reversed-phase column (RP-18) and a gradient elution system starting with a predominantly aqueous phase and introducing an increasing proportion of an organic solvent such as methanol or acetonitrile. As most flavonoids are ionizable, some acid is usually added to the mobile phase to control the pH.

[13] K. Vande Casteele, C. Van Sumere, and H. Geiger *J. Chromatogr.* **240**, 81 (1982).
[14] C. Van Sumere, P. Fache, K. Vande Casteele, L. De Cooman, and E. Everaert *Phytochem. Anal.* **4**, 279 (1993).
[15] M. C. Pietrogrande and Y. D. Kahie *J. Liq. Chrom.* **17**, 3655 (1994).

TABLE II
INTERPRETATION OF TYPICAL FLAVONOID SPOT COLORS ON TWO-DIMENSIONAL PC[a]

Spot color (366 nm)	Spot color (NH$_3$)	Common structural types indicated[b]
Dark purple	Yellow or yellow-green shades	Flavones (5- and 4'-OH)
		Flavonols (3-OR and 4'-OH)
	Red or orange	Chalcones (2'-OH, with free 2- or 4-OH)
	Little change	Flavone C-glycosides (5-OH)
		Flavones (5-OH and 4'-OR)
		Flavones (5-OH and 3,4'-OR)
		Isoflavones (5-OH)
		Dihydroflavones (5-OH)
		Dihydroflavones (5-OH)
		Biflavonyls (5-OH)
Yellow fluorescent	Little change	Flavonols (3-OH)
	Red or orange	Aurones
Magenta, pink, yellow fluorescence	Blue with time	Anthocyanins/anthocyanidins (Pelargonidin-3, 5-OR, yellow fluorescence)
Blue fluorescence	Yellow-green or blue-green	Flavones (5-OR)
		Dihydroflavones
		Flavonols (3,5-OR)
	No change	Isoflavones (5-OR)
	Brighter blue	Cinnamic acids and derivatives

[a] Reproduced from K. R. Markham and S. J. Bloor, in "Flavonoids in Health and Disease" (C. A. Rice-Evans and L. Packer, eds.), pp. 1–33. Dekker, New York, 1998, with permission.

[b] -OR, O-glycoside or O-alkyl.

The order of elution from most polar through to least polar means triglycosides (and higher glycosides) are eluted early, along with most anthocyanin glycosides, followed by di- and monoglycosides and then acylated or alkylated glycosides and aglycones. The requirement for a low pH solvent system for anthocyanins, a high organic modifier content for aglycones, and an extended gradient system to separate the many possible mono- and diglycosidic combinations means no one solvent system will give optimal separation of all flavonoid types.

Three useful gradients solvent systems are outlined. All are designed for use with the same RP-18 column [Merck Lichrospher 100 RP-18 endcapped (5 μm, 11.9 × 4 cm) or Supersphere (4 μm)]. The first is a general-purpose solvent system with high acid content and relatively fast total analysis time for routine analysis of plant extract, including those containing anthocyanin pigments. Table III gives a list of retention times for a variety of flavonoid compounds using this system. Note the bunching together of the aglycones at the end of the chromatographic run. The second system is designed for maximum resolution of flavone and flavonol

TABLE III
HPLC Retention Times for Flavonoids[a,b]

Anthocyanidin glycosides	Anthocyanidins	Flavonoid aglycones	Flavone/flavonol glycosides	Retention time (mins.)
Delphinidin 3,5-di-O-glucoside				7.6
		Dihydromyricetin		8.9
Cyanidin 3,5-di-O-glucoside				9.7
Delphinidin 3-O-glucoside				11.2
Delphinidin 3-O-rutinoside				12.7
Cyanidin 3-O-glucoside				13.5
			Vicenin-2	14.2
Peonidin 3,5-di-O-glucoside				14.6
	Delphinidin			15.3
		Dihydroquercetin		15.5
Petunidin 3-O-glucoside				15.5
Malvidin 3,5-di-O-glucoside				15.6
			Quercetin 3-O-sophoroside	16.1
			Saponarin	17.0
Peonidin 3-O-glucoside				17.8
			Luteolin 3′,7-O-diglucoside	18.2
			Kaempferol 3-O-sophoroside	19.3
Malvidin 3-O-glucoside				19.3
	Cyanidin			19.5
			Luteolin 3′-O-glucoside	19.6
			Luteolin 5-O-glucoside	19.6
			Vitexin	19.8
			Luteolin 7-O-glucoside	20.4
		Dihydrokaempferol		20.7
			Quercetin 3-O-glucoside	21.0
			Kaempferol 3-O-glucoside 7-O-rhamnoside	21.0 21.4
	Petunidin			21.4
			Kaempferol 3-O-rutinoside	22.0
			Isovitexin	22.2
	Pelargonidin			23.5
Malvidin 3-O-glucoside, 5-O-(6-acetylglucoside)				23.8
		Myricetin		24.0
			Quercetin 3-O-rhamnoside	24.2
			Quercetin 3-O-rutinoside	24.9
			Kaempferol 3-O-glucoside	25.0
	Peonidin			25.7
	Malvidin			27.0
			Apigenin 7-O-neohesperidoside	27.7
			Kaempferol 3-O-rhamnoside	29.2
		Quercetin		31.7
		Luteolin		33.1
		Narigenin		33.3
		Kaempferol		38.4
		Apigenin		38.6
		Tricin		39.5
		Chrysoeriol		39.6
		Isorhamnetin		39.9

[a] Reproduced from K. R. Markham and S. J. Bloor, in "Flavonoids in Health and Disease" (C. A. Rice-Evans and L. Packer, eds.), pp. 1–33. Dekker, New York, 1998, with permission.
[b] HPLC column and conditions as outlined in text [solvent system (1)].

glycosides. A slow increase in acetonitrile content between 0 and 52 min is used, but most glycosides elute between 25 and 40 min. The third solvent system is useful for nonpolar flavonoids, such as those found on the leaf surface, and uses a high methanol content.

1. Elution (0.8 ml/min) is performed using a solvent system comprising solvents A (1.5% H_3PO_4) and B [acetic acid: $CH_3CN : H_3PO_4 : H_2O$ (20 : 24 : 1.5 : 54.5)] mixed using a gradient starting with 80% A, linearly decreasing to 33% A after 30 min, 10% A after 33 min, and 0% A after 39.3 min (column temperature 30°).

2. Solvents A (water adjusted to pH 2.5 with H_3PO_4) and B (CH_3CN) are mixed in a gradient (0.8 ml/min) starting with 100% A, linearly decreasing to 91% A after 12 min, 87% A after 20 min, 67% A after 40 min and then held at 67% A for 2 min, then a linear decrease to 57% A after 52 min, and then finally to 0% A at 55 min (column temperature 24°).

3. Elution (1.0 ml/min) is performed using a solvent system comprising 5% formic acid in water (A) and methanol mixed according to a gradient, starting with 65% A, linearly decreasing to 55% A after 10 min, held at 55% A until 20 min, then linearly decreased to 20% A at 55 min, and 5% A at 60 min (column temperature 30°).

The utility of HPLC is best illustrated by an example such as that shown in Fig. 2. A mixture of flower extracts has been used to present a broad spectrum of compound type. Detection in the range of 340–360 nm is suitable for flavones and flavonols (e.g., 352 nm for chromatograms in Fig. 2). In the crude mixture the series of peaks can be grouped based on analysis of on-line spectra (Fig. 3), as luteolin or apigenin flavone glycosides, kaempferol glycosides, an obvious chalcone glycoside, and possibly some aromatically acylated flavonoid glycosides. Alkaline hydrolysis (see later) removes any acyl groups, dramatically increasing the relative level of one of the kaempferol glycosides, confirming that the acylated late-eluting compounds are acyl derivatives of compound 3 (Fig. 3). Acid hydrolysis (see later) to cleave *O*-glycosides gives the expected peaks for luteolin and kaempferol, but the peaks assigned to apigenin glycosides are still present, indicating these are flavone *C*-glycosides. This example demonstrates the considerable amount of information that can be gleaned from a few small-scale experiments.

Quantification of flavonoids is another forte of HPLC in combination with UV detection. In Fig. 2a, a rough estimate of the flavonoid level can be arrived at by comparing integration data for that chromatogram with that from the injection of a known amount of a readily accessible standard such as rutin (quercetin 3-rutinoside) run under the same conditions. A more rigorous quantification will involve the use of several standard compounds; in this case, a flavone glycoside

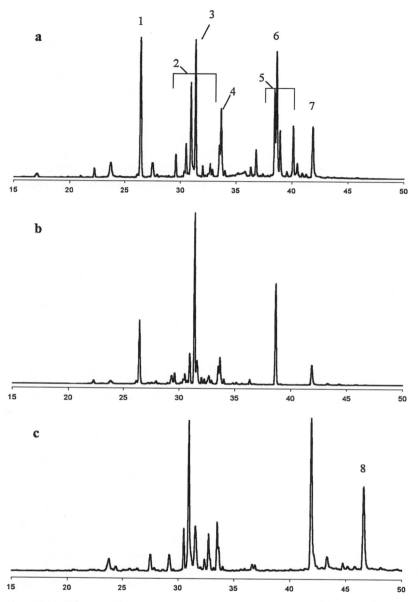

FIG. 2. HPLC chromatograms [absorbance at 352 nm vs. time (min)] of (a) a flavonoid mixture showing seven different groups of compounds: a kaempferol triglycoside [1], a set of apigenin glycosides [2], a kaempferol diglycoside [3], a luteolin glycoside [4], a set of acylated kaempferol glycosides [5], a chalcone [6], and luteolin [7]; (b) the alkaline hydrolysis product of the same mixture shows a large relative increase in peak 3 and loss of the acylated kaempferol glycoside peaks; and (c) the acid-hydrolyzed mixture showing luteolin and kaempferol [8]. Peaks due to apigenin glycosides are still present, showing these are apigenin C-glycosides.

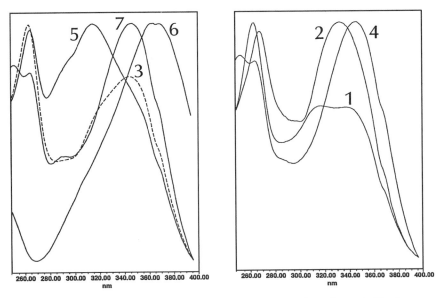

FIG. 3. On-line UV spectra of selected peaks from Fig. 2a. Peak numbers as in Fig. 2.

and chalcone glycoside should also be used. Suitable compounds are available commercially (e.g., Roth, Apin Chemicals).

Sample Preparation

Most flavonoids are extracted readily from source material by alcohol or alcohol/water mixtures. Prior knowledge of the nature of the flavonoids is of course helpful in determining the most efficient extraction technique. Aglycones and nonpolar flavonoids in a product such as bee propolis are best extracted by neat alcohol; plant surface flavonoids can be washed from leaves or stems with ethyl acetate. Glycosides, which are the major vacuolar flavonoids, require more water, and anthocyanins are usually extractable with water containing an organic acid (e.g., 5% formic acid). A useful general solvent for extraction of most flavonoid types from plant tissue is methanol : H_2O : acetic acid, 70 : 23 : 7. Solid source material should be well ground to facilitate maximum extraction. For most plant material, an extract of 50 mg of plant material with 2–5 ml of extraction solvent will yield an extract suitable for TLC or HPLC analysis. The addition of sufficient alcohol to denature the proteins (ca. 50%) to the extraction solvent will allay concerns regarding degradation due to enzyme action when extracting fresh tissue.

In many cases, some pretreatment of extracts is advisable prior to analysis. Chlorophyll can be removed by washing the aqueous alcoholic extract with hexane or diethyl ether, or the source material may be preextracted with such a solvent. Excessive amounts of sugars (e.g., in fruit extracts) may be removed by solid-phase extraction, e.g., on a RP SPE column. In this case, most of the alcohol in the extract should be evaporated, the sample made up to >90% water, and applied to the preconditioned cartridge. The cartridge is washed with water and the flavonoids eluted with alcohol. Some extracts have excessive quantities of nonflavonoid phenolic acids, which overwhelm the flavonoids, making analysis difficult. These can often be removed by SPE with cellulose. The sample is dried, reconstituted in 1% acetic acid in water, and applied to a small cellulose column; washing with 2% acetic acid removes most of the phenolic acids and the flavonoids are recovered by washing the column with alcohol.

Structural Analysis by Hydrolysis

The almost limitless range of flavonoid glycosides means that for most analyses the starting point for identification is the recognition of the flavonoid aglycone. Acid hydrolysis will cleave most sugars from flavonoid glycosides.[6,16] For this type of hydrolysis the flavonoid (0.25–0.5 mg) mixture is dissolved in 300 μl 2 N HCl : methanol (1 : 1 v/v) or in trifluoroacetic acid (TFA) : methanol (1 : 1 v/v), sealed in a screw-top polypropylene tube, and heated on a steam bath for 30 min. The sample can then be analyzed directly by TLC or HPLC, and the aglycone identified by comparison with common standard compounds. For TLC, better results are obtained if the posthydrolysis mixture is extracted with an equal volume of ethyl acetate or isoamyl alcohol. The upper layer is evaporated to dryness, redissolved in methanol, and analyzed. The lower aqueous layer can be used for analysis of sugars. Most common flavonoid O-glycosides undergo hydrolysis under these conditions; however, glycuronides generally require more extreme conditions, e.g., 2 hr at 100°. Flavonoid C-glycosides are not cleaved under these conditions, but the sugar can isomerize between the C-6 and the C-8 positions.[6]

Alkaline hydrolysis is generally only employed if the presence of acyl groups is suspected. To avoid oxidation of the flavonoid, these hydrolyses should be performed in the absence of air. A syringe or small sealed polypropylene vial may be used. An equal volume of 2 M NaOH is added to the flavonoid sample solution (generally the sample becomes intensely colored) and left at room temperature for 2 hr. The reaction is halted by neutralization with acid (returned to original color). The sample can then be analyzed directly by HPLC or TLC. Once again, better analytical results are achieved if the hydrolysis mixture is partially purified, in this case RP SPE is best. Comparison of chromatograms of crude and hydrolyzed

---

[16] J. B. Harborne, *Phytochemistry* **4**, 107 (1965).

material will show which flavonoids are acylated (see section on HPLC). Aromatic acyl groups, from acid or alkaline hydrolysis, such as *p*-coumaric or caffeic acid are detected easily by HPLC, whereas the less easily detected aliphatic acids are more suited to TLC or (after derivatization) gas chromatography.

Enzymatic hydrolysis is used occasionally with flavonoids and can provide selective cleavage of 3-*O*- or 7-*O*-glycosides.[6,13] These hydrolyses can be performed by dissolving the sample (0.3–0.5 mg) in water (500 μl), adding about 10 mg enzyme (e.g., β-glucosidase), and standing at room temperature. The sample is then analyzed as described earlier.

Analysis of Flavonoids by UV–VIS Spectroscopy

A special feature of most flavonoid compounds is the fact that their UV–VIS spectra can be very diagnostic. Often the basic flavonoid structural type is indicated by the spectra of individual compounds in a mixture obtained from on-line detection in HPLC or appearance under UV light on TLC or paper chromatograms. A considerable amount of additional information can be gained from the use of shift reagents; however, the use of these reagents is only really applicable for purified flavonoids.[5,6]

Individual flavonoids are dissolved in methanol (concentration such that the maximum absorbance is between 0.05 and 1.00 AU), and the basic spectrum is measured. Most flavonoids show a band in the 210- to 290-nm region (band II) and a second band at 320–380 nm (band I). Compilations of spectral data are available for comparison.[5,17] For anthocyanins, the latter band is in the visible region (490–540 nm).[18]

The procedure for shift reagent tests and the resultant shifts for flavones or flavonols[5,6] are as follows.

i. Two to three drops of 2.5% NaOMe in methanol are added directly to the cuvette containing the methanolic solution of the flavonoid. A 45- to 65-nm shift in band 1 of flavones and flavonols with no loss of intensity indicates a free 4′-OH. A decrease in intensity of this band indicates a substituted 4′-OH. If band I reduces or degrades after several minutes, then either a free 3,4′-OH or three adjacent OHs are likely, and if a new low intensity band appears at 320–335 nm, a free 7-OH is indicated.

ii. Several milligrams of solid sodium acetate are added to a fresh methanolic solution of the flavonoid. A shift of band II to longer wavelength indicates a free 7-OH.

---

[17] L. Jurd *in* "The Chemistry of Flavonoid Compounds" (T. A. Geissman, ed.), pp. 107–155. Pergamon Press, Oxford, 1962.
[18] F. J. Francis *in* "Anthocyanins as Food Colors" (P. Markakis, ed.), pp. 181–207. Academic Press, New York, 1982.

iii. Several milligrams of solid boric acid are added to the solution from step ii to diagnose for the presence of ortho di-OH groups. Without these groups the spectrum will revert to the original methanol spectrum. Movement of band I (12–36 nm) indicates an ortho di-OH group(s).

iv. Another methanolic flavonol solution is prepared and tested for response to $AlCl_3$. A few drops of 5% $AlCl_3$ in methanol solution are added, and a shift of band II by 20 to 40 nm again indicates ortho di-OH group. Two to three drops of 20% aqueous HCl are then added to the methanol/$AlCl_3$ flavonoid solution, and a shift of 35 to 70 nm indicates a free 5- and/or 3-OH.

These shift reagents have proven to be extremely useful guides to substitution patterns for a variety of flavonoids. Little material is required and the tests are performed directly in the sample cuvette, taking only a few minutes to complete.

Quantification of flavonoids is often performed by UV–VIS spectroscopy. However, the contribution of interfering compounds in a mixture should be accounted for. Anthocyanins are generally free of interference if the measurements are made at the visible maximum (500–540 nm, measure at pH 1.0). A useful ε value is 30,175 for the common glycoside, cyanidin 3,5-diglucoside.[13] Flavones and flavonol glycosides can be measured at their band I maxima of 340–360 nm and a general ε value of 14,500 used (i.e., in a 1-cm path length cell, a 1 M solution will give a absorption value of 14,500 AU; specific values for many flavonoids are available).

For many plant extracts the presence of cinnamic acid derivatives are a problem when quantifying flavonoids using this wavelength range of the UV–VIS spectrum. Most cinnamic acids can be removed by SPE with cellulose (see earlier).

The overview presented here has only skimmed the surface of this subject. There are many steps separating a quick analysis of a mixture and the accurate determination of the structural features of each constituent flavonoid. A number of techniques used in flavonoid analysis such as nuclear magnetic resonance, (NMR), electrophoresis, and liquid chromatography-mass spectrometry (LC-MS) have not been discussed and the reader is referred to more detailed reviews (^{13}C NMR,[19] ^{1}H NMR[20], electrophoresis[21–24]). Some of these methods are also discussed in other articles in this volume that deal with more specific examples of flavonoid analysis.

---

[19] K. R. Markham and V. M Chari *in* "The Flavonoids: Advances in Research" (J. B. Harborne and T. J. Mabry, eds.), pp. 19–134. Chapman and Hall, London, 1982.

[20] K. R. Markham and H. Geiger *in* "The Flavonoids: Advances in Research since 1986" (J. B. Harborne, ed.), pp. 441–497. Chapman and Hall, London, 1993.

[21] F. A. Tomas-Barberan *Phytochem. Anal.* **6,** 177 (1995).

[22] P. Pietta, P. Mauri, A. Bruno, and C. Gardana *Electrophoresis* **15,** 1326 (1994).

[23] T. K. McGhie and K. R. Markham *Phytochem. Anal.* **5,** 121 (1994).

[24] K. R. Markham and T. K. McGhie *Phytochem. Anal.* **7,** 300 (1996).

[2] Analysis of Complex Mixtures of Flavonoids and Polyphenols by High-Performance Liquid Chromatography Electrochemical Detection Methods

By PAUL E. MILBURY

Introduction

Flavonoids and plant polyphenols have received attention as dietary constituents of potential importance to health and antioxidant defense mechanisms. The emergence of renewed interest in "traditional herbal medicinal extracts" and increased interest in manufacturing "healthful" foods have driven a need for better methods of rapidly assessing plant phytochemicals and extracts for quality control as well as efficacy studies in human and animal trials.

Flavonoids and polyphenols, derived from plant or animal sources, comprise an enormous number, exceeding 4000 individual analytes. As a group, flavonoids are plant secondary metabolites based on the structure of a pyran ring flanked by at least two phenyl rings, designated as A and B rings (Fig. 1). They vary from one another subtly in the degree of unsaturation, the pattern of hydroxylation or methylation, and type of sugar attached or the degree of polymerization. Flavonoids in most abundance are classified as flavonols, anthocyanidins, and proanthocyanidins.

Considering such numbers and complexity, even the best high-performance liquid chromatography (HPLC) methods cannot resolve and separate all phenolic compounds from complex mixtures and often elaborate preparative procedures are necessary to investigate some selected and unique compounds. Further complicating analysis is the limited number of authentic or purified standards available for many of the flavonoids. Fortunately, the majority of the flavonoids constituent in significant amounts in the human diet and in "natural" medicinal preparations number far fewer than the total number known to exist.

Separation methods have been reviewed[1] and equally adequate HPLC methods have been described previously.[2] The majority of chromatographic methods utilized for the study of these colorful and UV absorptive plant compounds have naturally favored UV/visible detection. Photodiode array detectors facilitate collection of spectra of unknown flavonoids during chromatographic analysis. The absorption spectrum can help identify a compound if it is cleanly resolved and if

---

[1] D. J. Daigle and E. J. Conkerton, *J. Liquid Chrom.* **11**, 309 (1988).
[2] A. L. Waterhouse, S. F. Price, and J. D. McCord, *Methods Enzymol.* **299**, 113 (1999).

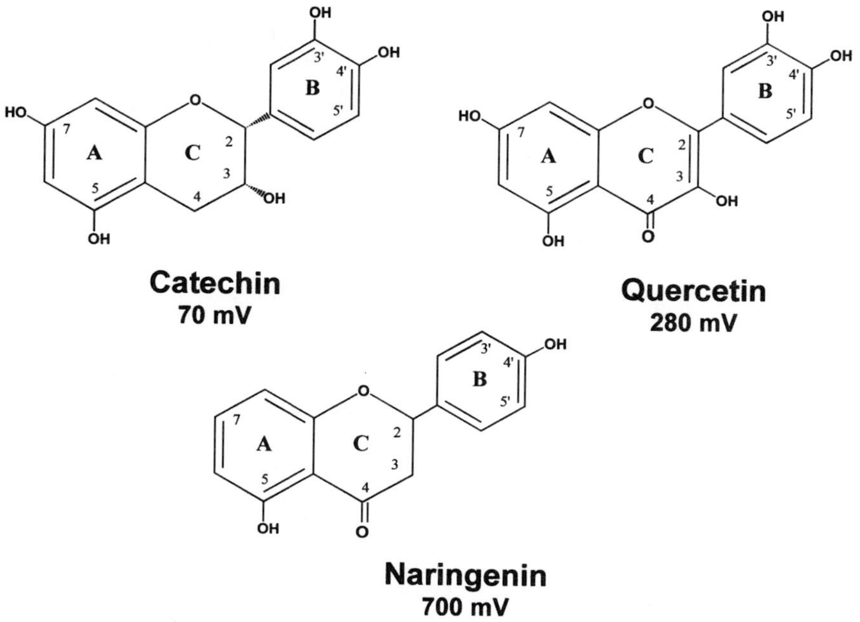

Catechin
70 mV

Quercetin
280 mV

Naringenin
700 mV

FIG. 1. Flavonoid structures of catechin, the flavonol quercetin, and the flavanone naringenin. The influence of the hydroxyl substitution and A, B, and C ring structures on the oxidation potential.

pure standards are available for comparison. Often the chromatography of complex mixtures is plagued by coelution of absorbing materials other than flavonoids and polyphenolics. The hydroxyl and methyl substitutions characteristic of individual flavonoids afford access to additional identification information through the use of electrochemical array detection.

A significant body of literature has accumulated, primarily from *in vitro* investigations, regarding the antioxidant properties of flavonoids and other plant polyphenolics.[3] As the essence of redox chemistry involves electron transfer, it would seem natural that electrochemical detection would rival UV, visible, and fluorescent detection for those compounds that are purported antioxidants. Until recently, technical difficulties inherent in the instrumental methods required to operate electrochemical detectors made their use, to say the least, inconvenient. With improvements over the last decade in both electrochemical detector geometries and electronics, coupled with a requirement for increased sensitivity, as the field moves from analysis of plant composition to bioavailablility and bioactivity studies, the use of electrochemical detectors may offer significant additional advantages when combined with traditional UV/VIS detection in the analysis of plant polyphenols.

[3] C. A. Rice-Evans, N. J. Miller, and George Paganga, *Free Radic. Biol. Med.* **20,** 933 (1996).

This article focuses on the use of the electrochemical array in HPLC analysis of flavonoids and other electron-donating polyphenols.

Methods

Chemicals and Standards

Chromatographic solvents were obtained as follows: 2-propanol is from C.M.S. (Houston, TX), whereas methanol and acetonitrile are from EM Science (Gibbstown, NJ). Lithium hydroxide, sodium phosphate, sodium acetate, and trichloroacetic acid are from Sigma (St. Louis, MO), whereas glacial acetic acid and pentanesulfonic acid sodium salt are from J. T. Baker (Phillipsburg, NJ). Flavonoids and other phenolic compounds are either from Indofine Chemical Company, Inc. (Somerville, NJ) or from Sigma; whenever possible, the same compounds are from both sources for comparison.

Flavonoid and Polyphenol Extraction and Sample Preparation

Williams and Harborne[4] and Hertog *et al.*[5] have previously reported established acid hydrolysis conditions for extraction of conjugated flavonoids from fruits and vegetables.[4] Fresh fruits or juices are from local markets. Concord grape juice and fresh cranberries are used in the demonstration of these methods. Wet weights of 5 g of fruit are extracted with 5 ml of 2 M HCl in a methanol:water (60 : 40, v/v) mixture. The extract is heated to 90° for 2 hr, cooled, and centrifuged in a Jouan table-top centrifuge at 12,000 rpm for 15 min. The supernatant is diluted 10-fold in mobile phase A, and the pH is checked and adjusted if needed to pH 3.50. This extraction results in a hydrolysis of proanthocyanidins and converts glycosylated compounds to their corresponding aglycone. Milder extraction in 1% acidic methanol : water (60 : 40, v : v) will not result in hydrolysis. The use of formic, trifluoroacetic, or acetic acid will minimize hydrolysis. Wines and juice mixtures are either directly injected or diluted 10-fold with mobile-phase A just prior to injection. Preliminary tests of the extraction efficiency and recoveries are always a prudent practice, as some plant materials are more fibrous than others. Recovery of catechin, spiked into four independent cranberry extract processings, is 97.5 ± 4.8% (mean ± SD).

Sample (especially fruit) can be prepared using a freeze-drying protocol. Whole fruit is frozen in liquid nitrogen and ground to a powder. Nitrogen is then allowed to evaporated at −70° prior to weighing and freeze dried for a 4-day period. One gram of freeze-dried powder is subjected to two 30-min extractions with 9 ml of acidic acetone. In each extraction step, the mixture is sonicated briefly. The combined

---

[4] C. A. Williams and J. B. Harborne, *in* "Methods in Plant Biochemistry" (J. B. Harborne, ed.), p. 357. Academic Press, London 1989.
[5] M. G. L. Hertog, P. C. H. Hollman, and D. P. Venema, *J. Agric. Food Chem.* **40**, 1591 (1992).

extractions are then centrifuged 15 min at 12,000 rpm at 4° in a table-top centrifuge, and the supernatant is diluted two- to fivefold for injection on the column.

If chosen chromatographic conditions cannot be modified to resolve a flavonoid of interest from other electrochemically active compounds, then a more elaborate sample preparation may be required. One milliliter of sample extract can be placed on an SPE Sep-Pak cartridge, and 3 ml water wash is used to remove sugars and organic acids. Application of 1 ml of 1 M ammonium hydroxide will remove phenolic acids. Flavonoids can then be eluted using 1 ml of methanol. A more detailed elution protocol for separation of the classes has been described previously[6] and outlined in another volume of this series.[7]

Instrumentation and Chromatography

A survey of the literature reveals that the majority of investigators separating flavonoids are using C_{18} columns, protected with guard columns. Occasionally, polystyrene divinylbenzene is used as the solid phase to extend pH capabilities. The literature mobile phases are generally acidic aqueous methanol mixtures or acidic aqueous acetonitrile mixtures. An ion-pairing agent is also used, especially when it becomes necessary to separate metabolic products generated by glycosylation and sulfation. The examples presented here are generated from one of three methods, which are by no means ideal for all separations but have proved useful for studies with both plant and animal samples. In all methods described here, gradient operation is provided by two Schimadzu LC-10AD HPLC pumps. Whenever possible the HPLC systems with electrochemical detectors are kept running for extended periods for more stable results. When this is not possible, at least two cycles through the gradient are performed prior to analytical runs. Both methods described here are gradient methods that reach significant organic concentrations prior to return to initial aqueous conditions. This not only cleans the column but also cleans the graphite detectors. Near the end of each run, during the reequilibration stage, the array cells are boosted to 1 V to "clean" off any biomaterial accumulating during the analysis. Detectors are reset automatically to initial array conditions and stabilized prior to the next injection. All injections, typically 20 μl, are made into the gradient for sharper resolution.

Method 1

Sample extracts are separated on a Meta-250, 4.6 mm × 25 cm, C_{18} 5-μm column obtained from ESA, Inc. (Chelmsford, MA). C_{18} guard columns are installed both upstream of a static gradient mixer and before the analytical column. Gradient mixer, pulse damper, columns, and detectors are contained within a temperature-controlled enclosure maintained at 35°. Analyte detection is accomplished using

---

[6] J. Oszmianski, T. Ramos, and M. Bourzeix, *Am J. Enol Vitic.* **39,** 259 (1988).
[7] D. M. Goldberg and G. J. Soleas, *Methods Enzymol.* **299,** 122 (1999).

14 channels of a coulometric electrode array system (CEAS) (ESA, Inc.) incremented from −100 to 990 mV versus paladium. Method 1 mobile phase A contains 11 g/liter pentanesulfonic acid sodium salt (PSA) in water at pH 3.0 adjusted with acetic acid. All solutions are made with 18 MΩ-cm deionized double glass distilled water from a Barnstead water system (Barnstead Inc., MA) and filtered trough 0.2-μm nylon hydrophilic membrane filters obtained from Schleicher & Schuell (Keene, NH). Mobile phase B is 0.1 M lithium acetate at pH 3.0 adjusted with acetic acid in methanol/acetonitrile/2-propanol (80 : 10 : 10, v : v : v). The PSA, delivered as an ion-pairing agent in mobile phase A, demonstrates an improved ability to solublize and remove protein and peptide fragments from both C_{18} columns and coulometric detectors, whereas the high organic modifier, delivered as mobile phase B, effectively removes residual lipids and polysaccharides. Sulfonic acids are, however, inherently contaminated, necessitating obtaining the highest grade PSA available. In practice, cleaning on a pyrolytic graphite electrode at a potential of 1000 mV versus paladium (H) is effective, but is not accomplished easily in most laboratories. Installation of a C_{18} guard column unit upstream of the gradient mixer and injector of the HPLC system serves as a trap for electroactive PSA-derived contaminants on the upstream C_{18} during the aqueous portion of the chromatography. Contaminants are released by the increasing organic gradient into the flow stream prior to the injector and enter the detector array to appear as a broad, slight rise in the background current on top of which analytes of interest can be measured. The chromatographic method involves a 120-min complex gradient from 0% B to 100% B, with the flow rate adjusted to compensate for aziotropic viscosity effects during the transition to the organic phase. The gradient method is depicted in Table I.

Method 2

An MCM, 4.6 mm × 15 cm, 5-μm, C_{18} column is obtained from ESA, Inc. Method 2 mobile phase A consists of 0.1 M monobasic sodium phosphate containing 10 mg/liter sodium dodecyl sulfate (SDS) and adjusted to pH 3.35 with phosphoric acid. Mobile phase B consists of 0.1 M monobasic sodium phosphate containing 50 mg/liter SDS in acetonitrile and methanol (10 : 60 : 30, v/v/v) adjusted to pH 3.45 with phosphoric acid. Fifteen channels of a CEAS instrument are set in array with redox potentials from 0 to 840 mV versus palladium at intervals of 60 mV. The gradients are depicted in Table I.

Discussion

In amperometric electrochemical detection the fluid steam from HPLC is passed over or between or over flat electrode surfaces. There are still many applications for this technology; however, the geometry of these detectors permits oxidation within the boundary layer of only a portion of potentially oxidizable

TABLE I
HPLC GRADIENTS AND FLOW RATES

	Method 1[a]			Method 2[b]	
Time (min)	% Mobile phase B	Flow rate (ml/min)	Time (min)	% Mobile phase B	Flow rate (ml/min)
0	1.5	1	0	6	1
7	1.5	1	10	6	1
30	10	1		Linear gradient to	
45	30	1	30	30	1
	Linear gradient to			Linear gradient to	
55		0.8	40	100	1
57		0.8	45	100	1
75		0.9	46	6	1
90	100	1	55	Next analysis	1
95	100	1.2			
100	100	1.2			
101.1	0	1.2			
104		1.2			
107		1			
108		1			
114	0	1			
125	1.5	1			
Next analysis					

[a] Method 1: Column, Meta-250, 4.6 mm × 25 cm C_{18}; mobile phase A, 11 g/liter pentane sulfonic acid sodium salt in water at pH 3.00; mobile phase B, 0.1 M lithium acetate at pH 3.00 adjusted with acetic acid in methanol/acetonitrile/2-propanol (80 : 10 : 10, v : v : v).

[b] Method 2: Column, MCM 4.6 mm × 15 cm C_{18}; mobile phase A, 0.1 M sodium phosphate and 10 mg/liter SDS at pH 3.35; mobile phase B, 0.1 M sodium phosphate and 10 mg/liter SDS at pH 3.35 in methanol; acetonitrile (30 : 10 : 60, v/v/v).

analytes that flow by the detectors. The coulometric electrodes utilized in the CEAS and Coularray instruments are constructed from porous graphite, and the instrument flow stream is directed through the electrode rather than over it. Vastly increased electrode areas and a flow turbulence that disturbs and mixes the boundary layer permit near coulometric conditions. With efficiencies of nearly 100% analyte oxidation, lower detection limits are possible than those achieved in prior electrochemical detectors.

Data generated from an electrochemical array can be thought of in much the same was as that obtained from a photodiode array. Analytes are resolved by chromatographic retention time while the current is monitored from each detector in the series and each detector can be set to distinct redox potentials. A compound entering the array is oxidized upon reaching the first electrode in the array that is set to a redox potential capable of oxidizing it. When the array is configured with

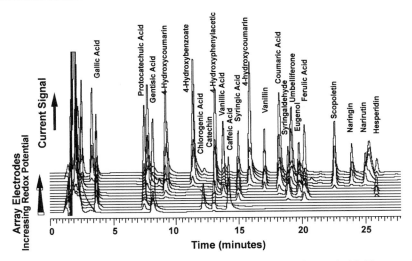

FIG. 2. A 15-channel chromatogram from the CEAS instrument running method 2. The array increases from 0 to 840 mV and is displayed from the lower to the upper (back) trace. The vertical response represents current signal in nanoampere (nA).

increasing potentials, a three-dimensional retention time/redox potential/current response display can be generated (Fig. 2). If resolved cleanly, each analyte will exhibit a characteristic retention time and a response across the array as depicted for catechin (Fig. 3). When a pure standard analyte is available, setting the array at increments of less than 10 mV permits very accurate determination of the redox potential versus paladium at the pH of the mobile phase.

To achieve the widest possible range of potentials between the operating limits of the detectors (less than 1000 mV) and to operate within the expected range of the analytes being measured, increments of 60 to 70 mV are typically used for a 16-channel array. In practice the majority of the signal response will be generated by the electrode set closest to the analyte redox potential while adjacent detectors will exhibit responses with ratios to the dominant detector that are characteristic for a given analyte. The electrochemical properties responsible for these distinct ratios can be understood from study of the current voltage curves (Fig. 4). Channel ratios can be used in much the same manner as spectra to verify peak purity. To quantify an analyte for which there are authentic standards, responses from the array are summed and compared with the responses for the authentic standard analyzed under the same conditions.

Table II shows dominant channel oxidation potentials for authentic flavonoid and phenolic standards. In the case of phenolic and polyphenolic compounds, the electrochemical behavior is directly related to the structure of the compound.

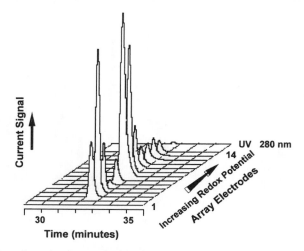

FIG. 3. A three-dimensional representation of retention time, array channel potentials, and channel current response for catechin. The first and second oxidation potentials are evident. The last (back) channel display is the output of a UV detector monitoring 280 nm.

Resolution across the electrochemical array is therefore also dependent on differences in the chemical structures of the oxidizable moieties of compounds that may coelute on HPLC. A compound-like catechin (Fig. 1, Fig. 3), which exhibits two oxidation waves across the array, may preclude accurate quantitative measurement of coeluting compounds, which may oxidize at a higher potential. Compounds like

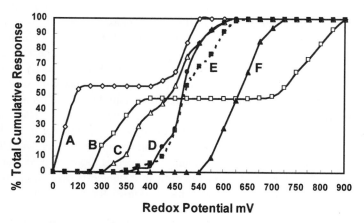

FIG. 4. Hydrodynamic voltammograms of representative flavonoids and phenolics. A, catechin; B, hespiridin; C, genistein; D, equol; E, daidzein; F, naringin. Hydrodynamic voltammograms from coulometric array data by plotting the cumulative current to a given potential as a function of the total signal across the array.

TABLE II
OXIDATION POTENTIALS OF FLAVONOID AND PHENOLIC COMPOUNDS VERSUS PALADIUM

Method 1		Method 2	
Compound	Oxidation potential (mV)	Compound	Oxidation potential (mV)
Caffeic acid	70	Caffeic acid	60
Catechin	70	Gallic acid	60
Cyanidin	70	Gentisic acid	60
Epicatechin	70	Catechin	120
Ideain chloride	70	Chlorogenic acid	120
Caffeic acid	140	Protocatechuic acid	120
Callistephin cloride	140	Ferulic acid	300
Protocatechuic add	210	Syringic acid	300
Sinapinic acid	210	Hesperidin	360
Quercetin	280	Syringaldehyde	360
Ferulic acid	350	Eugenol	420
Malvidin	350	Vanillic acid	420
Peonidin	350	4-Hydroxyphenylacetate	540
Syringic acid	350	Coumaric acid	540
Malvidin 3-galactoside	420	Tyrosine	540
Pelargonidin	420	4-Aminobenzoic acid	600
Rutin	420	Scopoletin	600
Vanillic acid	420	4-Hydroycoumarin	660
Daidzein	490	Naringenin	660
Epicatechin	490	Narirutin	660
Gallocatechin gallate	490	Umbelliferone	660
Hesperidin	490	4-Hydroxybenzoic acid	720
p-Coumaric acid	490	7-Methoxycoumarin	780
Peonidin chloride	490		
Genistein	500		
Resveratrol	560		
Equol	560		
Tyrosine	560		
4-Hydroxybenzoic acid	630		
Callistephin	700		
Naringenin	700		
4-Hydroxycoumarin	770		
Eriodictyol	770		
Fustin	770		
Kaempferol	700–770		

caffeic acid (Fig. 1), which are completely oxidized at a low potential to a stable radical or rearranges to a nonelectroactive compound, will afford the opportunity to measure other coeluting compounds oxidizing at a higher potential.

Examination of Table II reveals a relationship between aromatic substitutions and voltammetric properties. Catechin, caffeic acid, and chlorogenic acid (compounds with a catechol groups) are easily oxidized at lower potentials. The redox potentials of the flavonoids follow a pattern based on the nature of the hydroxyl substitutions on the A and B rings where the B ring substitutions have more influence on the redox potential than those of the A ring. Flavonols, anthocyanidin structures, and simple phenolics that possess hydroxyl substitutions, especially in the 3' and 4' positions, have lower redox potentials. Monophenolic compounds, such as tyrosine and *p*-coumaric acid that do not readily yield up their electrons and therefore have higher redox potentials, have been shown to be less effective antioxidants than those having *o*-diphenolic structures. The flavanones, such as hespiredin, with only one hydroxyl group in the B ring, oxidize at an even higher potential high. Although possessing a monohydroxy substitution, ferulic acid also has a methoxy substitution in the ortho position that results in resonance delocalization, permitting a lower oxidation potential than *p*-coumaric acid. A methoxy substitution on the B ring of hesperidin may, in a similar manner to ferulic acid, result in resonance delocalization and permit a lower redox potential.

The structural–antioxidant relationship of flavonoids and phenolics has been reviewed,[3] which shows a growing consensus that the *o*-dihydroxy structure to the B ring confers electron delocalization and results in a more stable radical on oxidation. Furthermore, those compounds with 3- and 5-hydroxyl groups with 4-oxo function in the A and C rings have higher radical-scavenging capability. Additionally, a 2,3 double bond with a 4-oxo function in the C ring contributes to B ring destabilization that again results in more stable phenoxyl radicals. Under these assumptions, quercetin should be a more effective antioxidant than catechin, as quercetin has a 2,3 double bond in the C ring, which catechin does not. By measure of redox potential alone, catechin is oxidized more easily than quercetin. Using the electrode array, it is possible to set initial electrodes to oxidize compounds such as catechin and then, on subsequent electrodes, reduce the quinone formed by the original oxidation. Compounds such as myricetin, which undergo irreversible oxidation, will not be detectable under reduction. The oxidization of quercetin forms quinones that can be reduced. The quinones are thought to arise through a second electron transfer via the phenoxyl radical. In those cases where the redox potential of a flavonoid does not directly correlate with its ranking by a radical absorbency assay, the redox potential and stability of the resulting phenoxyl radical must be considered. The results shown here confirm that the redox potential of flavonoids is generally dependent on the presence of a catechol structure in the B ring of flavonoids. Quercetin, and possibly myricetin, may represent an exception where the C ring has more influence on the antioxidant capacity of the flavonol than is the case for other flavonoids. The influence of the 2,3 double bond in the C ring on the redox potential of quercetin requires more study. Overall, the raking of flavonoids and other phenolics by redox potentials, as measured by the coulometric electrode array, correlates well with their antioxidant capacity.

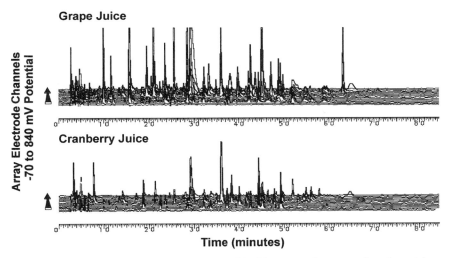

FIG. 5. Two electrochemical array tracings from the CEAS instrument for grape and cranberry juice.

A traditional, and indeed perhaps original, use for HPLC (using UV/VIS detector methods) was in the aid of identification and authentication efforts. Using patterns of anthocyanidins and other polyphenols, it is possible to create databases capable of identifying grape varietals. When prior database libraries are created, it is also possible to identify vineyard sources of unmixed wines.

This technique of pattern generation, sometimes referred to as "fingerprinting," has been used to verify plant powder materials in the herbal extraction industry. Comparisons of array chromatographic patterns have also been used to identified place of origin and species for various ginseng materials.

Electrochemical array chromatograms of 20-μl direct injections of grape and cranberry juice are shown in Fig. 5. Grape juice, by cursory observation, exhibits more electroactive analytes than cranberry juice.

Because electrochemical detection depends on redox chemistry, the theory evolved that there should be a relationship between sample total redox signal and total oxidative radical absorbency capacity as measured by a TEAC or ORAC assay. This relationship has been verified for many fruits and vegetables,[8] and the relationship holds for the comparison of grape juice with cranberry juice where the ORAC values are 14.70 (Concord grape) and 1.59 (cranberry).[9] The summation of all current responses across the array for the entire analysis is made possible due to

---

[8] C. Cuo, G. Cao, E. Sofic, and R. L. Prior, *J. Agric. Food Chem.* **45,** 1787 (1997).

[9] R. Prior USDA, Human Nutrition Research Center on Aging at Tufts University, *personnal communication* (2000).

the wide dynamic range inherent automatic gain ranging capability of the CEAS electronics. While the detection limit for most flavonoids is approximately 10–20 pg on column, the CEAS and Coularray instruments can reasonably quantify, depending on the analyte, as much as 500 μg of a compound injected on column. This capability permits simultaneous analysis of mixture components that may be present in vastly different concentrations.

When combined with photodiode array detection, electrochemical array detection provides additional parameters necessary for single analysis verification of identity and resolution purity from HPLC while also providing information regarding structure and antioxidant properties.

A great deal of interest has been exhibited lately in flavonoid and phenolic food composition studies, especially from the perspective of dietary influences on oxidative stress in human health and disease. Due to detection limits, which in many cases are 100-fold lower than UV/VIS detection, electrochemical detection may be the best detector choice in bioavailablility studies as sample sizes and flavonoid concentrations will be low.

[3] Analysis of Flavonoids in Medicinal Plants

By PIERGIORGIO PIETTA and PIERLUIGI MAURI

Introduction

Flavonoids are a group of plant polyphenols that have the common skeleton of the flavan nucleus (Fig. 1). Differences between classes consist of changes in ring C (presence of a double bond, a 3-hydroxy group, and/or a 4-oxo group) and in the number and position of hydroxyl and methoxyl groups in rings A and B. Besides being components of the human diet,[1] flavonoids have gained increasing interest because of their pharmacological activity.[2] Approximately 50 species of plants, from *Achillea millefolium* to *Viola tricolor* have been used as herbal remedies based on their flavonoid content. These preparations have been reported to be effective for the treatment of disorders of peripheral circulation and for the improvement of aquaresis. In addition, flavonoid-based herbal medicines are available in different countries as anti-inflammatory, antispasmodic, antiallergic, and antiviral remedies. The pharmacological effects of flavonoids are thought to be due to their functions

---

[1] J. Kuhnau, *J. World Rev. Nutr. Diet.* **24,** 117 (1976).
[2] E. Middleton, Jr., and C. Kandaswami, *in* "The Flavonoids: Advances in Research since 1986" (J. B. Harborne, ed.), p. 619. Chapman and Hall, London, 1994.

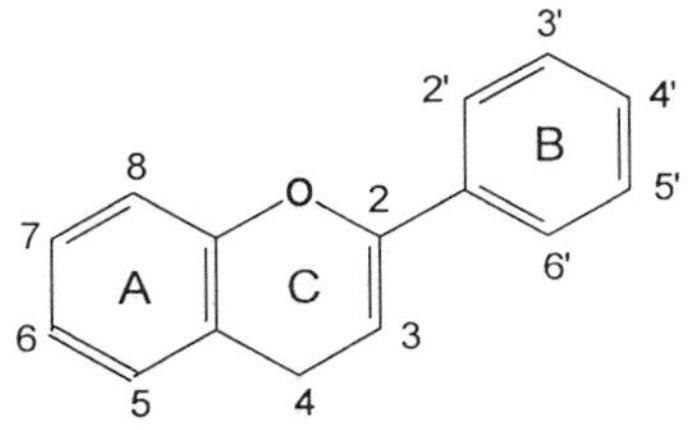

FIG. 1. Flavan nucleus.

as radical scavengers,[3] reductants,[4] metal chelators,[5] and due to their alternative nonantioxidant functions, including the interaction with different enzymes,[6] the inhibition of calcium ion influx into the cells,[7] and the regulation of cell signaling and gene expression.[8] However, the health-benefit properties of most medicinal plants high in flavonoids cannot be ascribed exclusively to these compounds as other components present in the phytocomplex may either directly contribute or display a "permissive" role that enhances the effects of flavonoids. A typical example is *Ginkgo biloba,* whose therapeutic effects have been associated with the presence of both flavonoids (flavonols of the rutin type and proanthocyanidins) and terpene lactones (ginkgolides and bilobalide).[9] The flavonoidic constituents reduce capillary fragility and protect epithelial cells against the attack of reactive oxygen species, thereby preserving the structure of the blood vessels. Terpene lactones inhibit platelet aggregation and are neuroprotective. The combined actions account for the improved peripheral and cerebral cerebral circulation and menthal performance associated with *G. biloba* therapy.[10] Other examples can also be considered, including different aquaretic, anti-inflammatory, sedative, and antispasmodic herbs, and it can be shown that the observed pharmacological effects are due to flavonoidic and nonflavonoidic constituents.

Many factors affect the quality of medicinal plants, including specie variation, environmental conditions, time of harvesting, storage, and processing. Therefore, the quality control of herbs is an important issue to guarantee the efficacy and

---

[3] Y. Hanasaki, S. Ogawa, and S. Fukui, *Free Radic. Biol. Med.* **16,** 845 (1994).
[4] P. G. Pietta, *J. Nat. Prod.,* **63,** 1035 (2000).
[5] J. E. Brown, H. Khodr, R. C. Rider, and C. A. Rice-Evans, *Biochem. J.* **330,** 1173 (1998).
[6] W. Bors, W. Heller, C. Michel, and K. Stettmaier, *in* "Handbook of Antioxidants" (E. Cadenas and L. Packer, eds.), p. 409. Dekker, New York, 1996.
[7] H. Voruela, P. Voruela, K. Tornquist, and S. Alaranta, *Phytomedicine* **2,** 167 (1997).
[8] S. Roy, H. Kobuchi, C. K. Sen, M. T. Droy-Lefaix, and L. Packer, *in* "Antioxidant Food Supplements in Human Health" (L. Packer, M. Hiramatsu, and T. Yoshikawa, eds.), p. 359. Academic Press, San Diego, 1999.
[9] T. A. van Beek, E. Bombardelli, P. Morazzoni, and F. Peterlongo, *Fitoterapia* **69,** 195 (1998).
[10] F. V. De Feudis, "Ginkgo biloba extract (Egb 761): From chemistry to clinic." Ullstein Medical, Wiesbaden, 1998.

the safety of herbal medicines. This is not an easy task because herbs and their preparations are (unlike the synthetic drugs) complex mixtures of components with different physicochemical (i.e., analytical) characteristics. High-performance liquid chromatography (HPLC) and capillary electrophoresis (CE), coupled with on-line UV detection and/or mass spectrometry, have proved highly valuable in the analysis of herbal ingredients. These techniques are particularly appropriate for flavonoid-containing medicinal plants.[11] Flavonoid aglycones and glycosides can be separated by HPLC or CE under relatively simple conditions. These compounds can be monitored and identified easily using UV detection, thanks to their high and specific absorption in the UV. Flavonoids can be detected by mass spectrometry either post-HPLC separation or by direct infusion.

This article oulines briefly the chromatographic, electrophoretic, ultraviolet, and mass spectrometric behavior of flavonoids commonly found in medicinal plants. Then, some examples of the analytical approaches presently applied in our laboratory are considered, with emphasis on electrospray mass spectrometry by the direct infusion mode.

Chromatographic Behavior

Flavonoids present in medicinal plants with proven therapeutic activity include flavonols, flavones, isoflavones, catechins, flavanolignans, and anthocyanins (Fig. 2). These flavonoid can be separated by HPLC in the reversed-phase mode. Under these conditions (i.e., using C_8 or C_{18} stationary phases with different organic modifiers, including methanol, acetonitrile, 2-propanol, propanol, and tetrahydrofuran), the elution order of flavonoid classes is dictated by the polarity. Within the same class, compounds with different number and position of hydroxyl groups can be separated, as both hydroxylation degree and site influence the mobility. As a general rule, retention times decrease with the increase of the hydroxyl number. When C_8 columns with neutral eluents are used, myricetin (3′,4′,5′-trihydroxyflavonol) elutes earlier than quercetin (3′,4′-dihydroxyflavonol), which in turn precedes isorhamnetin (3′-methoxy-4′-hydroxyflavonol); the slowest is kaempferol (4′-hydroxyflavonol) (Fig. 3). However, this tendency may be modified when using C_{18} columns and slightly acidic eluents (kaempferol elutes in front of isorhamnetin). The occurrence of hydrogen bonds may also influence retention times. This is the case with flavonols, which form an internal hydrogen bond between the carbonyl group at C-4 and the hydroxyl group at C-3. Consequently, flavonols elute slightly later than the related flavones, which lack the 3-OH group. Thus, the flavone chrysin (5,7-dihydroxyflavone) elutes earlier than the related flavonol galangin (5,7-dihydroxyflavonol). Methylation of the hydroxyl groups lowers the polarity, resulting in increased elution times as a function of the

---

[11] P. G. Pietta, in "Flavonoids in Health and Disease" (C. A. Rice-Evans and L. Packer, eds.), p. 61. Dekker, New York, 1998.

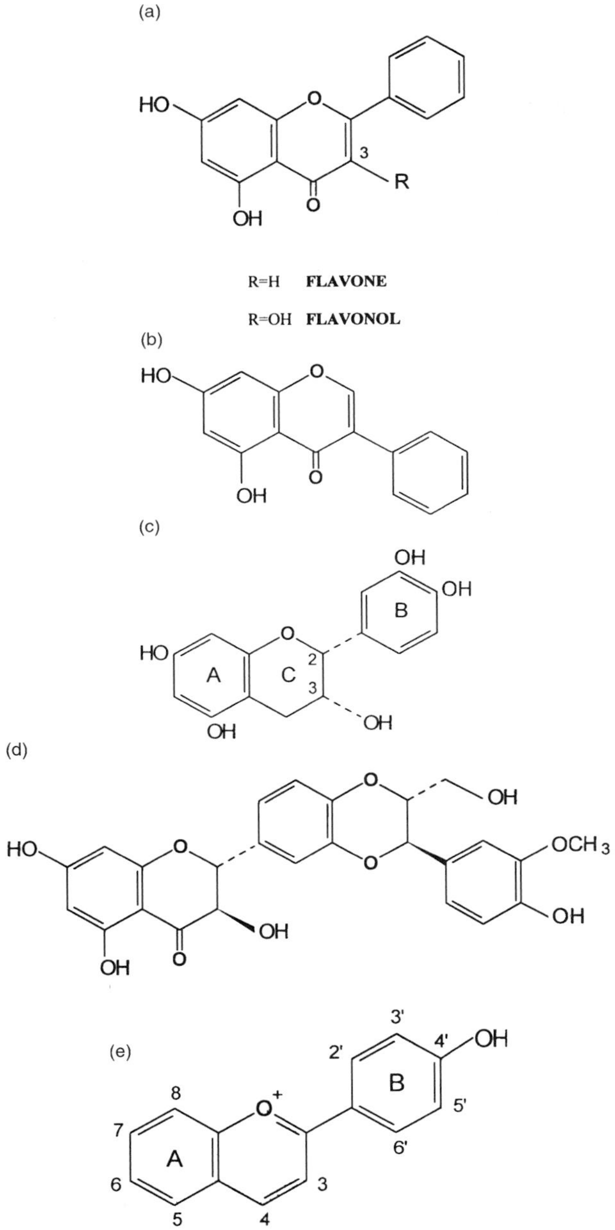

FIG. 2. Structures of main flavonoids: (a) flavones and flavonols, (b) isoflavones, (c) catechins, (d) flavanolignans, and (e) anythocyanins.

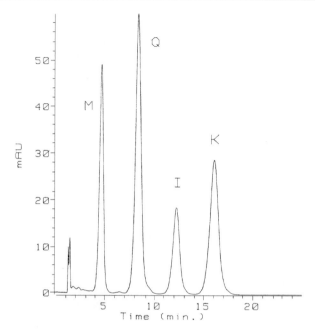

FIG. 3. HPLC of myricetin (M), quercetin (Q), isorhamnetin (I), and kaempferol (K).

methoxyl/hydroxyl ratio. However, glycosides elute earlier than the corresponding aglycones, and increasing glycosylation lowers retention times. Acylation and conversion into glucuronide and sulfate conjugates produce more polar derivatives that elute earlier than the parent compounds.

Electrophoretic Behavior

Capillary electrophoresis represents a valuable alternative for the analysis of flavonoids. The driving force in CE is the electroosmotic flow (EOF), which is the bulk flow of the background electrolyte from anode to cathode. Due to its strength, the EOF forces all the analytes (independently from their charge) to move toward the cathode. Cationic species are the fastest, as their tendency to reach the cathode is enhanced by the EOF. In contrast, anionic species are the slowest, as they are prone to move in the opposite direction of EOF. Neutral species move to the cathode at the same rate of the EOF. Depending on their molecular weight, the number and pK of hydroxyl groups, and the type and degree of glycosylation, flavonoids have been separated by CE.[12] However, the best results are achieved

[12] P. G. Pietta, in "Handbook of Capillary Electrophoresis Application" (H. Shintani and J. Polonsky, eds.), p. 324. Chapman and Hall, London, 1997.

by micellar electrokinetic chromatography (MEKC). In this CE mode, a surfactant, that forms negatively charged micelles is added to the alkaline background electrolyte. Flavonoids interact with these anionic and hydrophobic micelles to a different extent as a function of their hydrophobicity. Consequently, their migration to the cathode is influenced, and a resolution higher than that achievable by reversed-phase HPLC is obtained.[13] However, this technique is limited because the reproducibility of migration times is not as good as that of retention times in HPLC, making the identification of the analytes difficult when comparing their migration times with those of standards. Although this inconvenience may be limited by coupling CE with on-line UV detection, HPLC remains the method of choice for routine analysis of plant polyphenols. In addition, HPLC coupling with mass spectrometry (MS) is a validated and common technique, whereas interfacing of CE with MS is presently problematic.

Ultraviolet Behavior

The class of the flavonoid and the specific compound within the class may be indicated by on-line UV spectra, and this information, combined with chromatographic behavior, is of prime importance for a fast and reliable identification of flavonoids.

The UV spectrum in the 300- to 400-nm region reflects the substitution pattern of ring C and is indicative of the class, as shown in Fig. 4, for flavones (FN), flavonols (FL), and isoflavones (IF). Similarly, flavanones (FLN) may be easily discriminated from their precursors chalcones (CHL), as the latter show a strong absorption at 350–380 nm (Fig. 5). Catechins (CAT) have characteristic UV profiles with absorption at 270 nm, whereas anthocyanins (ANT) present an additional maximum at 520 nm (Fig. 6)

Within the same class, different members can be differentiated by comparing their spectra in the 240- to 300-nm region, as the shape of the maximum is strictly related to the substitution pattern of ring B. For example, within the class of flavonols, the profiles of kaempferol, quercetin, and isorhamnetin glycosides are distinctly different (Fig. 7). Flavones behave similarly. Thus, apigenin (4'-hydroxyflavone) and luteolin (3',4'-dihydroxyflavone) have different maximum profiles in the 240- to 300-nm region. However, these flavones show maximum shapes similar to those of their flavonol counterparts, i.e., kaempferol (4'-hydroxyflavonol) and quercetin (3',4'-dihydroxyflavonol), as the B ring has the same substitution pattern. Catechins and anthocyanins have characteristic ultraviolet spectra, as illustrated in Fig. 7. Concerning the influence of the sugar moiety, the intensity of the absorption at 350 nm is affected by the linked carbohydrate. Finally, if the sugar is acylated by cinnamoyl derivatives, the maximum at 350 nm is shifted to a shorter wavelength (about 320 nm).

[13] P. G. Pietta, P. L. Mauri, A. Rava, and G. Sabbatini, *J. Chromatogr.* **549,** 367 (1991).

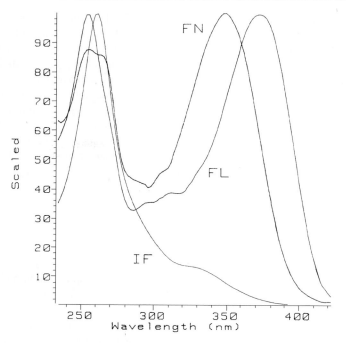

FIG. 4. Typical UV spectra of flavones (FN), flavonols (FL), and isoflavones (IF).

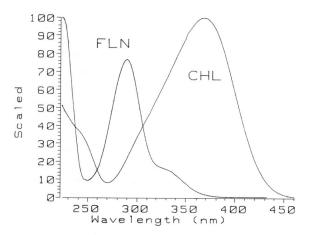

FIG. 5. Typical UV spectra of flavanones (FLN) and chalcones (CHL).

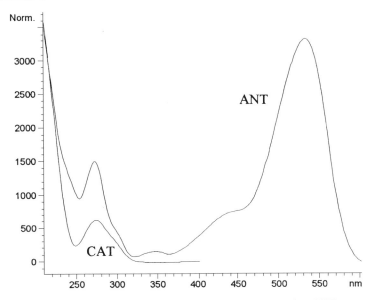

FIG. 6. Typical UV spectra of catechins (CAT) and anthocyanins (ANT).

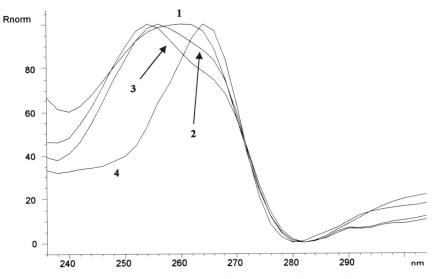

FIG. 7. Profiles of maxima at 240–300 nm for different flavonol 3-O-glycosides. (1) Myricetin, (2) quercetin, (3) isorhamnetin, and (4) kaempferol glycosides.

All these data are provided by both HPLC and MEKC when coupled with a diode-array ultraviolet detector. However, spectra acquired in an electrophoretic run show maxima shifted to longer (about 10 nm) wavelengths as compared to those obtained by HPLC. This difference is understandable, as MEKC utilizes an alkaline buffer instead of the slightly acidic eluents used in HPLC.

Mass Spectrometric Characteristics

Flavonoids have been studied by (MS) since 1970, and many papers have been published on off-line mass spectrometry, including fast atom bombardment (FAB), plasma desorption (PD), chemical ionization (CI), and electron impact (EI). The introduction of thermospray (TSP) and electrospray (ESI) interfaces allowed direct coupling of mass spectrometry with liquid chromatography, and this permitted studying different flavonoid-containing plants.[14] Both TSP and ESI are "soft" ionization techniques and allow the analysis of flavonoids in their native form without derivatization.

Flavonol monoglycosides analyzed by TSP-MS yield the ions corresponding to the aglycone $[A + H]^+$, and a small molecular ion $[M + H]^+$. Diglycosides may lose a sugar unit and yield the resulting fragments. This may help in evaluating whether the sugars are linked as disaccharides or as individual monosaccharides bound to different positions. ESI differs from TSP in that primarily molecular ions are produced. These ions are present as sodium or potassium adducts in the case of flavonol 3-O-glycosides (Fig. 8). In contrast, protonated molecular ions are obtained from flavonol 4'-O-glycosides, flavone glycosides, and flavonol/flavone aglycones (Fig. 9). In addition, the loss of the sugar unit from flavonol 3-O-glycosides to form the aglycone ion is reduced, suggesting that these compounds may form a crown-embedded cationic ion, which stabilizes the adduct and limits its fragmentation. For the same reason, isoflavones form only cationic molecular ions, and this may be ascribed to the presence in position 3 of the ring B, which may play the role of the sugar in flavonol 3-O-glycosides. Due to low fragmentation, ESI-MS is particularly suitable for direct analysis of samples without any preliminary chromatographic separation, and fingerprints specific for the examined plant extracts are obtained.

Analytical Approaches to Detect Flavonoids in Medicinal Plants

The analysis of flavonoids in medicinal plants has been the subject of a recent review[11] and some representative examples are described here.

---

[14] P. L. Mauri, L. Iemoli, C. Gardana, P. Riso, M. Porrini, and P. G. Pietta, *Rapid Commun. Mass Spectrom.* **13,** 924 (1999).

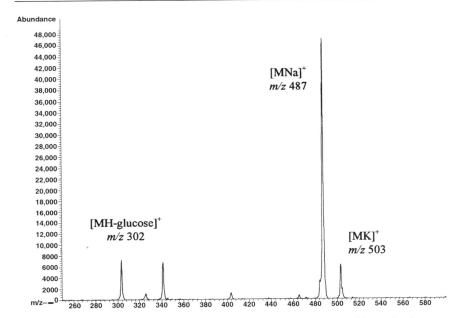

FIG. 8. Typical positive ESI-MS mass spectrum of quercetin 3-*O*-glucoside (molecular mass 464 Da).

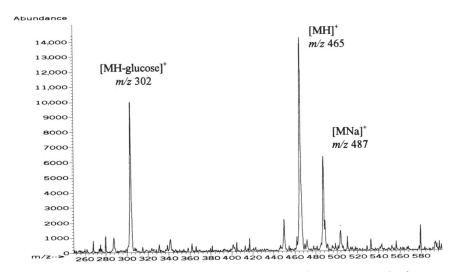

FIG. 9. Typical ESI-MS positive mass spectrum of quercetin 4'-*O*-glucoside (molecular mass 464 Da).

Standards

The following are obtained from Extrasynthese (Genay, France): quercetin, kaempferol, isorhamnetin, rutin, isoquercitrin, kaempferol 3-O-rutinoside, isorhamnetin 3-O-rutinoside, chrysin, galangin, daidzein, genistein, cyanidin, delphinidin, petunidin, malvidin, cyanidin 3-O-arabinoside, cyanidin 3-O-glucoside, petunidin 3-O-arabinoside, delphinidin 3-O-glucoside, mavidin 3-O-arabinoside, petunidin 3-O-glucoside, and malvidin 3-O-glucoside. (+)-Epicatechin (EC), (−)-epicatechin gallate (ECg), (−)-epigallocatechin (EGC), and (−)-epigallocatechin gallate (EGCg) are from Sigma-Aldrich (Milan, Italy).

Samples

Ginkgo biloba (Ginkgoselect), Green tea (Greenselect), *Vaccinium myrtillus* (Mirtoselect), and *Glycine max* (Soyselect) extracts are from Indena S.p.A (Milan, Italy). *Hypericum perforatum* extract and dewaxed propolis extract (EPID) are from Specchiasol (Verona, Italy). Green tea (leaves) is from a local supermarket.

Sample Preparation

For HPLC analyses, the sample (0.5–1.5 mg) is dissolved in 1 ml of 30% aqueous methanol (v/v) and filtered through 0.45-μm polytetrafluoroethylene (PTFE) syringe tip filters into glass HPLC vials. In the case of green tea leaves, an infusion is prepared (2 g in 250 ml of water at 80° for 10 min). An aliquot (1 ml) of the filtrate is diluted 10-fold with water and filtered through syringe tip filters into glass HPLC vials.

For MS analyses, the sample (0.3–1.2 mg) is dissolved in 1 ml of 50% aqueous methanol for negative mode or 0.5% acetic acid in 50% aqueous methanol (v/v) for positive mode.

HPLC Coupled with "On-Line" UV Detection

Ginkgoselect. A Hewlett-Packard (Waldbronn, Germany) Model 1090 high-performance liquid chromatograph with a photodiode array ultraviolet detector coupled to HPChemistation is used. The column is a C_8 Aquapore RP-300 (4.6 × 220, 7-μm particle size) with an Aquapore RP-300 guard column (4.6 × 30, 7-μm particle size) from Alltech Italia (Milan, Italy). The flow rate of the mobile phase is 1.0 ml/min and the volume injected is 25 μl. The mobile phase consists of two solvents: solvent A, 2-propanol and water (5 : 95, v/v) and solvent B, 2-propanol–tetrahydrofuran–water (40 : 10 : 50, v/v/v). The gradient profile is linear from 20 to 60% B in 40 min. Detection is at 360 nm.

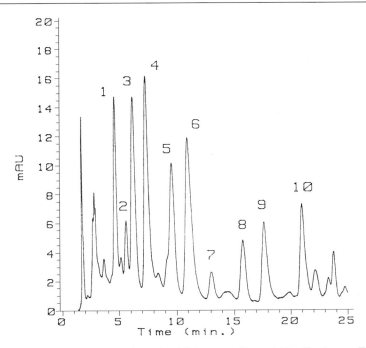

FIG. 10. Typical HPLC at 360 nm of a *Ginkgo biloba* extract. For peak identification, see Table I.

Chromatographic Resolution. Figure 10 demonstrates the satisfactory resolution achieved for the major flavonol glycosides of *G. biloba*. Rutin (4), isorhamnetin 3-*O*-rutinoside (5), and kaempferol 3-*O*-rutinoside (6) are identified by comparing their retention times and on-line ultraviolet spectra with those of standards. The identity of the compounds 1–3 and 8–10 (Table I) has been established by UV and MS characterization after their semipreparative HPLC isolation.[15]

Quantitation. The quantitation of flavonoids in their native form (i.e., as glycosides) is often critical because of the lack of and the high cost of reference compounds. In addition, the flavonol-glycosidic pattern is frequently complex, and an accurate evaluation of the components may be difficult. To overcome these difficulties, the samples are hydrolyzed under controlled acid conditions to convert a large array of glycosides into few aglycones (e.g., quercetin, kaempferol, and isorhamnetin for *G. biloba*). These are quantified easily by means of calibration plots obtained with jstandard aglycones, which are available and inexpensive.

[15] P. G. Pietta, P. L. Mauri, A. Bruno, and A. Rava, *J. Chromatogr.* **553**, 223 (1991).

TABLE I
FLAVONOL GLYCOSIDES OF *Ginkgo biloba*

Peak[a]	Compound
1	Quercetin 3-*O*-rhamnosyl-(1 → 2)-rhamnosyl-(1 → 6)-glucoside
2	Isorhamnetin 3-*O*-rhamnosyl-(1 → 2)-rhamnosyl-(1 → 6)-glucoside
3	Kaempferol 3-*O*-rhamnosyl-(1 → 2)-rhamnosyl-(1 → 6)-glucoside
4	Rutin
5	Isorhamnetin 3-*O*-rutinoside
6	Kaempferol 3-*O*-rutinoside
7	Astragalin
8	Kaempferol glycoside
9	Quercetin 3-*O*-[6'''-*p*-coumaroylglucosyl-(1 → 2)-rhamnoside]
10	Kaempferol 3-*O*-[6'''-*p*-coumaroylglucosyl-(1 → 2)-rhamnoside]

[a] See Fig. 10.

The hydrolysis is performed by adding 2 M HCl (0.1 ml) and methanol (0.15 ml) to the sample solution (0.25 ml). The vial is sealed under nitrogen and is maintained at 100° for 20 min. The hydrolyzed sample is diluted with water (1.5 ml) and applied to a previously activated (3 ml of methanol followed by 6 ml of water) Sep-Pak C_{18} cartridge (Waters, Milford, MA). After washing with water (3 ml), the aglycones are eluted with methanol (3 ml). The eluate is evaporated to dryness, and the residue is redissolved in methanol (1 ml). The recovery is evaluated for each aglycone by adding two concentrations to the sample and performing triplicate assays before and after each addition. The recovery (mean ± SD) is 92 ± 4.4%. Calibration curves for quercetin, kaempferol, and isorhamnetin are obtained from replicate injections ($n = 4$) of known amounts of the corresponding standards. Linearity is in the range of 5–150 µg/ml ($r^2 = 0.997 - 0.999$) at 360 nm.

Green Tea. The HPLC system consists of two Model 510 pumps (Waters) connected with a Model 996 photodiode array ultraviolet detector and a Millenium workstation (Waters). The injection is by means of a Model 7125 Rheodyne injector (Waters), and the volume injected is 20 µl. A symmetry C_{18} column (220 × 4.6 mm, 5-µm particle size) with a C_{18} guard column (Waters) is used. Elution (2 ml/min) is performed with a mobile phase composed of two solvents: solvent A, 0.03% trifluoroacetic acid (TFA) and solvent B, acetonitrile and 0.025% TFA. The gradient is linear from 100% A to 25% B in 30 min. Detection is at 270 nm.

Chromatographic Resolution. Figure 11 demonstrates the satisfactory resolution achieved for the major catechins of green tea. (−)-Epigallocatechin, (+)-epicatechin, (−)-epigallocatechin gallate, and (−)-epicatechin gallate are identified

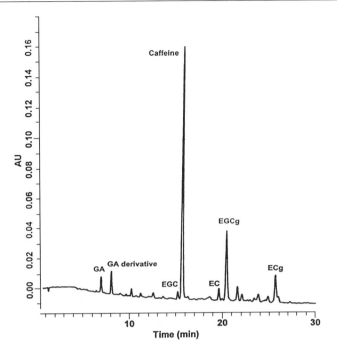

FIG. 11. Typical HPLC at 270 nm of a green tea infusion. EGC, epigallocatechin; EC, epicatechin; EGCg, epigallocatechin gallate; ECg, epicatechin gallate; GA, gallic acid.

by comparing their retention times and on-line ultraviolet spectra with those of standards. The chromatogram shows other peaks, including gallic acid (GA), a gallic acid derivative (GA derivative), and caffeine.

Quantitation. Calibration curves for EC, ECg, EGC, and EGCg are obtained from replicate injections ($n = 4$) of known amounts in the concentration range of 5–100 μg/ml ($r^2 = 0.996$–0.999).

Propolis EPID. The HPLC system and the column are the same used for green tea. The volume injected is 50 μl. Elution (1.5 ml/min) is performed with a mobile phase composed of two solvents: solvent A, 0.03% TFA and solvent B, acetonitrile and 0.025% TFA. The gradient is linear from 10 to 40% B in 40 min and from 40 to 55% B in 10 min Detection is at 360 nm.

Chromatographic Resolution. Figure 12 demonstrates the satisfactory resolution achieved for the major components of propolis EPID. Quercetin (peak 1, 26.1 min), kaempferol (peak 2, 32 min), isorhamnetin (peak 3, 32.9 min), chrysin (peak 4, 43.2 min), and galangin (peak 5, 45.2 min) are identified by comparing their retention times and on-line ultraviolet spectra with those of standards.

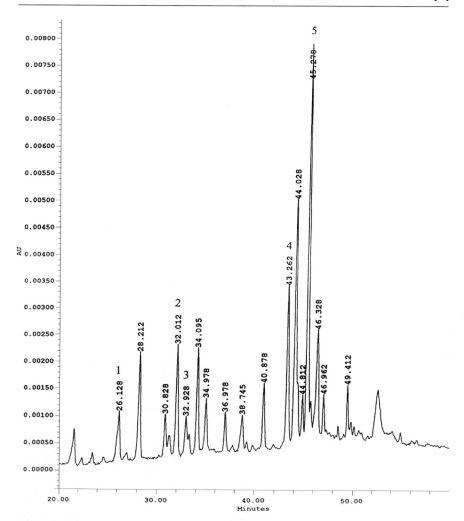

FIG. 12. Typical HPLC at 360 nm of propolis EPID. Peak 1, quercetin; peak 2, kaempferol; peak 3, isorhamnetin; peak 4, chrysin; peak 5, galangin.

Judging from their ultraviolet spectra, peaks at 28.2, 34, 36.1, 38.7, and 40.8 min are not yet identified flavonoids. Peaks eluting at 14.8, 21.6, 44, and 49.4 min are all hydroxycinnamates, as suggested by their ultraviolet spectra.

Quantitation. Calibration curves for quercetin, chrysin, and galangin are obtained from replicate injections ($n = 4$) of known amounts of the corresponding standards (3–80 μg/ml, $r^2 = 0.997$–0.999).

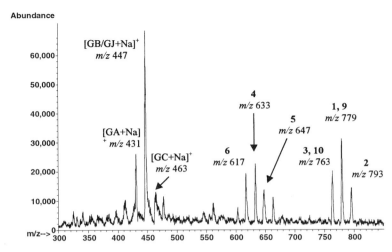

FIG. 13. Typical positive ESI-MS mass spectrum of *Ginkgo biloba* extract (Ginkgoselect) GA, ginkgolide A; GB, ginkgolide B; GC ginkgolide C; GJ, ginkgolide J. For flavonoid peaks, see Table I.

Electrospray Ionization Mass Spectrometry

Electrospray mass spectrometric analyses are performed on a Hewlett-Packard 5989A equipped with an electrospray interface 59987A. Nitrogen is used as nebulizing gas at a pressure of 50 psi and a temperature of 300°.

The samples are analyzed by direct infusion in ESI-MS by means of a syringe pump (Harvard Apparatus Inc., Natick, MA) at a flow rate of 10 μl/min in scan mode.

Ginkgo biloba (Ginkgoselect). *Ginkgo biloba* extracts contain ginkgoflavonol glycosides and terpenoids (ginkgolides and bilobalide). The fingerprint of *G. biloba* extracts obtained by direct infusion in ESI-MS is shown in Fig. 13. The ions (m/z 617-779) refer to different flavonol glycosides. In particular, the ions m/z 617, 633, and 647 are due to the sodium adducts of kaempferol 3-*O*-rutinoside, quercetin 3-*O*-rutinoside, and isorhamnetin 3-*O*-rutinoside, respectively. The ions m/z 763, 779, and 793 correspond to the sodium adducts of the 3-*O*-[rhamnosyl-(1 → 2)-rhamnosyl-(1 → 6)-glucoside] derivatives of kaempferol, quercetin, and isorhamnetin, respectively. Ions m/z 763 and 779 also account for the 3-*O*-[6'''-*p*-coumaroylglucosyl-(1 → 2)rhamnoside] derivatives of kaempferol and quercetin. Finally, the ions m/z 431, 447, and 463 correspond to sodium adducts of ginkgolide A, ginkgolide B/ginkgolide J, and ginkgolide C, respectively.

Hypericum perforatum. In addition to flavonoids, *H. perforatum* extracts contain naphthodianthrones and phloroglucinols. Figure 14a shows a typical negative ion mass spectrum of a sample of *H. perforatum* extract. The major ions at

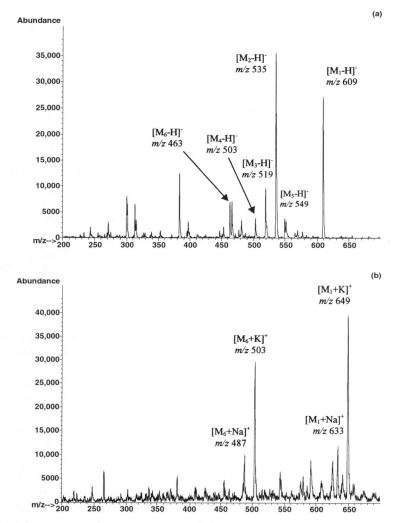

FIG. 14. Typical negative (a) and positive (b) ESI-MS mass spectra of *Hypericum perforatum* extract. M_1, rutin; M_2, hyperforin; M_3, pseudohypericin; M_4, hypericin; M_5, adhyperforin; M_6, isoquercitrin/hyperoside.

m/z 535 and 609 are due to $[M - H]^-$ ions of hyperforin and rutin, respectively. Hypericin (m/z 503), pseudohypericin (m/z 519), and adhyperforin (m/z 549) are present at lower abundance. $[M - H]^-$ ions of other known flavonoids can be detected, such as those of isoquercitrin and hyperoside (m/z 463). As illustrated

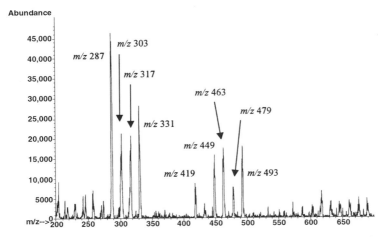

FIG. 15. Typical positive ESI-MS mass spectrum of *Vaccinium mirtyllus* extract (Mirtoselect).

in Fig. 14b, ESI-MS in positive mode shows the presence of cation adducts of isoquercitrin ($[M + Na]^+ = 487$; $[M + K]^+ = 503$) and rutin ($[M + Na]^+ = 639$; $[M + K]^+ = 649$). Thus, ESI-MS by direct infusion mode permits simultaneous detection of compounds representative of three different classes (flavonoids, naphthodianthrones, and phloroglucinols), providing a fingerprint of the *H. perforatum* extract.

Vaccinium myrtillus (Mirtoselect). Vaccinium myrtillus is characterized by the presence of anthocyanins and their aglycones (anthocyanidins). A typical ESI-MS spectrum of Mirtoselect obtained in positive detection is shown in Fig. 15. The most abundant ions are due to the molecular ions of anthocyanidins. In particular, cyanidin (m/z 287), delphinidin (m/z 303), petunidin (m/z 317), and malvidin (m/z 331) are detected. In addition, the following anthocyanins are evidenced: cyanidin 3-*O*-arabinoside m/z 419), cyanidin 3-*O*-glucoside/petunidin 3-*O*-arabinoside (m/z 449), delphinidin 3-*O*-glucoside/malvidin 3-*O*-arabinoside (m/z 463), petunidin 3-*O*-glucoside (m/z 479), and malvidin 3-*O*-glucoside (m/z 493).

Camellia sinensis (Greenselect). Figure 16 shows the ESI-MS positive mass spectrum of a decaffeinated green tea extract (Greenselect). The most abundant ion m/z 481 is due to the sodium adduct of EGCg. The ions m/z 465, 329, and 291 are present at lower abundance and correspond to the sodium adducts of ECg, EGC, and epicatechin/catechin (EC/C), respectively. Adding to the extract cesium chloride (2 mM), the resulting mass spectrum (Fig. 17) shows the cesium adducts of EGCg (m/z 591), ECg (m/z 575), and EGC (m/z 439), thereby confirming the identity of these catechins.

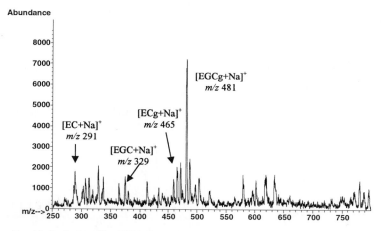

FIG. 16. Typical positive ESI-MS mass spectrum of green tea extract (Greenselect).

Glycine max (Soyselect). Soybean contains high concentrations of isoflavones either as highly polar glycoside conjugates (daidzin and genistin) or in the free form, e.g., daidzein and genistein. Figure 18 shows the positive mass spectrum of Soyselect. The main ion (m/z 255) is due to the aglycone daidzein, whereas glycoside daidzin (m/z 417) and genistein (m/z 271) are present at lower abundances.

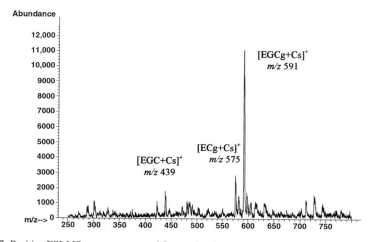

FIG. 17. Positive ESI-MS mass spectrum of Greenselect in the presence of cesium chloride (2 mM).

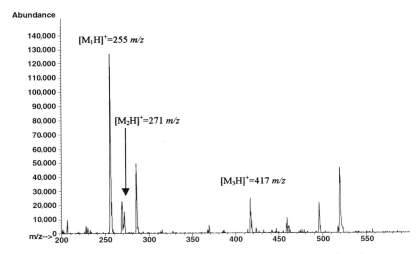

FIG. 18. Typical positive ESI-MS mass spectrum of soybean extract (Soyselect).

Based on these data, it may be concluded that the ESI-MS approach by direct infusion allows one to detect flavonoids in complex matrices. This method does not require prepurification steps and provides fingerprints of herbal extracts, as different constituents present in the sample are detected simultaneously. In addition, the HPLC conditions described in this article permit one to perform LC-MS analyses, thereby making it possible to discriminate compounds with the same m/z values and to obtain quantitative data.[14]

Acknowledgments

We acknowledge Indena S.p.A. (Milano) and Specchiasol S.r.l. (Verona) for funding this research.

[4] High-Performance Liquid Chromatography/Mass Spectrometry Analysis of Proanthocyanidins in Food and Beverages

By SHERYL A. LAZARUS, JOHN F. HAMMERSTONE, GARY E. ADAMSON, and HAROLD H. SCHMITZ

Introduction

A compound consisting of one aromatic ring that contains at least one hydroxyl group is classified as a simple phenol. In comparison, a polyphenol consists of more than one aromatic ring with each containing at least one hydroxyl group. Flavonoids are polyphenols that have a diphenylpropane (C_6-C_3-C_6) skeleton structure[1] and are found ubiquitously in the plant kingdom. The class of flavonoids called the proanthocyanidins are oligomers of flavan-3-ol monomer units most frequently linked either $4 \rightarrow 6$ or $4 \rightarrow 8$. The most common classes of proanthocyanidins are the procyanidins, which are chains of catechin, epicatechin, and their gallic acid esters, and the prodelphinidins, which consist of gallocatechin, epigallocatechin, and their galloylated derivatives as the monomeric units[2] (Fig. 1). Structural variations to proanthocyanidin oligomers may also occur with the formation of a second interflavanoid bond by C–O oxidative coupling to form A-type oligomers (Fig. 2). Due to the complexity of this conversion, A-type proanthocyanidins are not encountered as frequently in nature in comparison to B-type (i.e., singly linked) oligomers.[3]

Proanthocyanidins can be found in several foods commonly consumed in the diet, including apples, almonds, barley, grapes, tea, maize, cinnamon, cocoa, peanuts, wine, and strawberries.[4–7] Increasingly, these compounds are attracting attention as a result of research suggesting they may act as potent antioxidants and/or may modulate key biological pathways.[8,9] However, qualitative and

---

[1] E. Haslam, *in* "Practical Polyphenolics from Structure to Molecular Recognition and Physiological Action" (E. Haslam, ed.), p. 10. Cambridge Univ. Press, Cambridge, 1998.

[2] L. J. Porter, *in* "The Flavonoids: Advances in Research since 1986" (J. B. Harborne, ed.), p. 23. Chapman and Hall, London, 1994.

[3] S. Morimoto, G.-I. Nonaka, and I. Nishioka, *Chem. Pharm. Bull.* **33,** 4338 (1985).

[4] R. S. Thompson, D. Jacques, E. Haslam, and R. J. N. Tanner, *J. C. S. Perkin I.* **11,** 1387 (1972).

[5] G. Chiavari, P. Vitali, and G. C. J. Galletti, *J. Chromatogr.* **392,** 426 (1987).

[6] I. McMurrough and T. Baert, *J. Inst. Brew.* **100,** 409 (1994).

[7] S. A. Lazarus, G. E. Adamson, J. F. Hammerstone, and H. H. Schmitz, *J. Agric. Food Chem.* **47,** 3693 (1999).

[8] C. Rice-Evans and L. Packer (eds.) "Flavonoids in Health and Disease." Dekker: New York, 1998.

[9] L. Packer, G. Rimbach, and F. Virgili, *Free Radic. Biol. Med.* **27,** 704 (1999).

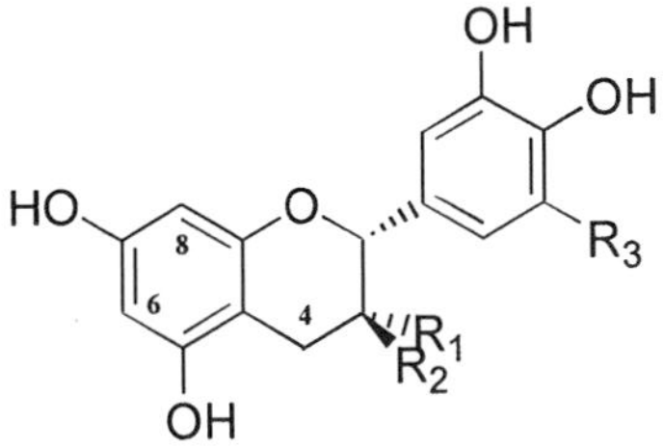

a) Procyanidins:
R_1=OH, R_2=H, R_3=H; Epicatechin
R_1=H, R_2=OH, R_3=H; Catechin
R_1=gallic acid ester, R_2=H, R_3=H; Epicatechin gallate
R_1=H, R_2=gallic acid ester, R_3=H; Catechin gallate

b) Prodelphinidins:
R_1=OH, R_2=H, R_3=OH; Epigallocatechin
R_1=H, R_2=OH, R_3=OH; Gallocatechin
R_1=gallic acid ester, R_2=H, R_3=OH; Epigallocatechin gallate
R_1=H, R_2=gallic acid ester, R_3=OH; Gallocatechin gallate

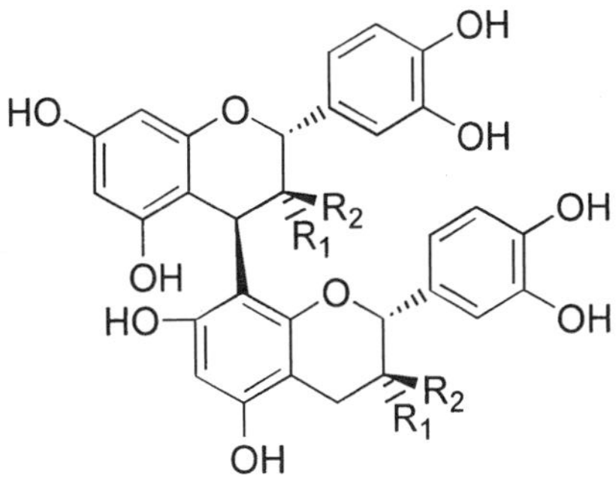

Procyanidin (4β→8)-Dimers

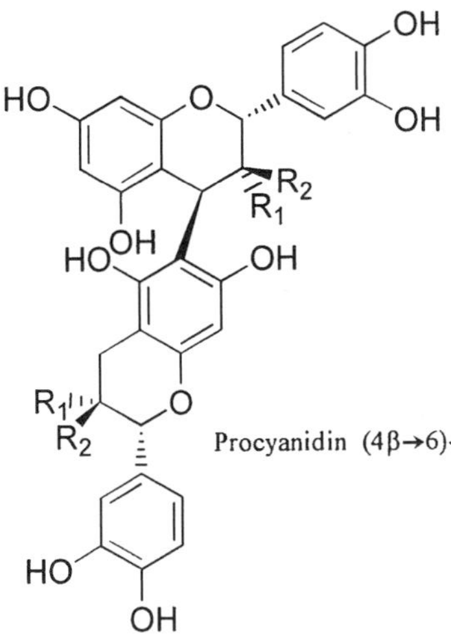

Procyanidin (4β→6)-Dimers

FIG. 1. Representative structures for proanthocyanidin monomers and dimers.

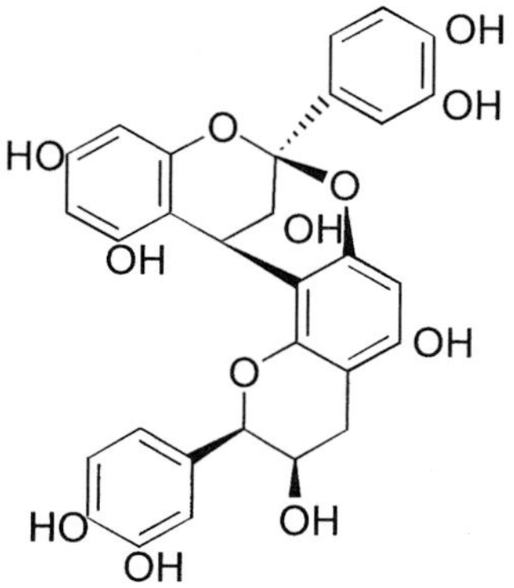

FIG. 2. Representative structure of doubly linked dimer.

quantitative information on the proanthocyanidin profiles in food products is especially lacking,[10] due in large part to the lack of appropriate analytical methodology and commercially available standards for the proanthocyanidin oligomers.

Regarding analytical methods, reversed-phase high-performance liquid chromatography (HPLC) has been the primary method of analysis for the quantification of proanthocyanidins in food samples. Although reversed-phase columns have the ability to separate oligomers of equivalent molecular mass into their isomers, analysis of higher oligomeric proanthocyanidins (i.e., ≥tetramers) is not feasible, as the number of isomers increases with increasing degrees of polymerization.[11,12] This effect results in a retention time overlap of isomers containing differing degrees of polymerization, causing the higher oligomers (>trimer) to coelute as a large unresolved peak.[13] Thus, quantification of oligomeric and polymeric proanthocyanidins is underestimated because only the resolved peaks corresponding to dimers and trimers are considered.

For quantitative standards, researchers have used (+)-catechin or (−)-epicatechin as an external standard and estimated oligomers as catechin equivalents.[14,15]

---

[10] J. Peterson and J. Dwyer, *J. Am. Diet. Assoc.* **98**, 677 (1998).
[11] E. L. Wilson, *J. Sci. Food Agric.* **32**, 257 (1981).
[12] J. F. Hammerstone and S. A. Lazarus, in "Caffeinated Beverages: Health Benefits, Physiological Effects and Chemistry" (T. H. Parliament, C. T. Ho, and P. Schieberle, eds.). American Chemical Society, Washington, D.C., 2000.
[13] S. Guyot, T. Doco, J.-M. Souquet, M. Moutounet, and J.-F. Drilleau, *Phytochemistry* **44**, 351 (1997).
[14] J. Clapperton, J. F. Hammerstone, L. J. Romanczyk, S. Yow, D. Lim, and R. Lockwood, "Proceedings of the 16th International Conference of Groupe Polyphenols, Lisbon, Portugal," Groupe Polyphenols, Norbonne, France, **Tome II**, 112 (1992).
[15] G. A. Spanos and R. E. Wrolstad, *J. Agric. Food Chem.* **38**, 1565 (1990).

However, use of catechins as external standards only allows for the estimation of oligomeric forms, as monomers have different detection response factors than their oligomeric counterparts. Therefore, synthesis[16,17] and rigorous purification techniques[6,18,19] have been developed for obtaining oligomers for use as quantitative standards.

Qualitative Identification and Characterization of Proanthocyanidins in Foods

For the analysis of proanthocyanidin oligomers in food products, the normal-phase HPLC method reported by Rigaud *et al.*[20] has been modified for improved separation based on the degree of polymerization. Proanthocyanidin oligomers are identified and characterized using diode array detection and by coupling the chromatographic method with API-ES mass spectrometry (MS). Normal-phase separations are best performed on a Phenomenex Luna or Lichrosphere silica column (250 × 4.6 mm; 5-μm particle size) at 37°. In general, traditional type A silica columns (e.g., Lichrosphere) produce better results than the type B silica packings (e.g., Luna). This may in part be due to the higher acidity of the type A packings. Type B columns often create additional separations of the isomers within each oligomeric class, resulting in decreased sensitivity, especially for the higher oligomeric forms (i.e., $\geq$ hexamer). The ternary mobile phase consists of (*A*) dichloromethane, (*B*) methanol, and (*C*) acetic acid and water (1 : 1, v/v). Separations are affected by a series of linear gradients of *B* into *A* with a constant 4% *C* at a flow rate of 1 ml/min as follows: elution starting with 14% *B* in *A*; 14–28.4% *B* in *A*, 0–30 min; 28.4–50% *B* in *A*, 30–60 min; 50–86% *B* in *A*, 60–65 min; 65–70 min isocratic.

Separations are monitored using UV at 280 nm, and spectrum is stored for all wavelengths between 220 and 600 nm. Spectra can be used to provide structural information regarding the proanthocyanidin class present in the food products. As can be seen in Fig. 3, UV spectra for (−)-epicatechin and its gallic acid ester have the same maximum absorption at 278 nm; however, the galloylated derivative is broader, with a tail at 320 nm, which is indicative of gallic acid. Furthermore, the prodelphinidin monomer, (−)-epigallocatechin, has a distinctively

---

[16] M.-H. Salagoïty-Auguste and A. Bertrand, *J. Sci. Food Agric.* **35**, 1241 (1984).

[17] J. Oszmianski and C. Y. Lee, *Am. J. Enol. Vitic.* **41**, 204 (1990).

[18] S. Carando, P.-L. Teissedre, L. Pascual-Martinez, and J.-C. Cabanis, *J. Agric. Food Chem.* **47**, 4161 (1999).

[19] G. E. Adamson, S. A. Lazarus, A. E. Mitchell, R. L. Prior, G. Cao, P. H. Jacobs, B. G. Kremers, J. F. Hammerstone, R. B. Rucker, K. A. Ritter, and H. H. Schmitz, *J. Agric. Food Chem.* **47**, 4184 (1999).

[20] J. Rigaud, M. T. Escribano-Bailon, C. Prieur, J. M. Souquet, and V. Cheynier, *J. Chromatogr. A* **654**, 255 (1993).

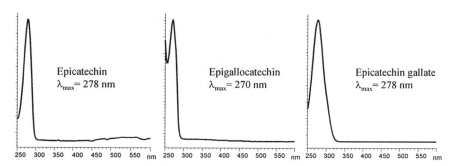

FIG. 3. UV spectra for proanthocyanidin monomers.

lower maximum absorption (270 nm) compared to the procyanidin monomers (278 nm).

For further structural information, the HPLC is interfaced to a mass spectrometer equipped with an atmospheric pressure ionization electrospray chamber in which the spray vector is orthogonal to the entrance of the mass analyzer. MS data are collected using both the scan mode (m/z 100–3000) and selected ion monitoring (SIM). SIM offers the advantage of looking for only those ions of interest and eliminates background from other components present in the food matrix. It is preferable to use the negative ion mode because ionization efficiency decreases rapidly with increasing degree of polymerization when using the positive ion mode. Conditions for analysis in the negative ion mode include a capillary voltage of 3.0–3.5 kV, a fragmentor voltage of 75–85 V, a nebulizing pressure of 25 psig, and the drying gas temperature at 350°. A buffering reagent is necessary to assist ionization and is added via a tee in the eluant stream of the HPLC just prior to the mass spectrometer. Ammonium hydroxide ($\sim$0.75 M at 0.04 ml/min)[7] and ammonium acetate (0.01 M at 0.1 ml/min) in methanol[12] have been used as buffering reagents. Ammonium hydroxide is not recommended because it may be highly corrosive to some HPLC pumps.

Mass spectral data are used to assign the degree of polymerization to each of the peaks in the HPLC chromatograms. For the procyanidin oligomers in cocoa and chocolate, molecular ions $(M-H)^-$ are obtained for the monomers through hexamers with the ions for each oligomer at m/z 289, 577, 865, 1153, 1441, and 1729, respectively. In addition to the molecular ions, multiply charged ions are also observed for the higher oligomers ($\geq$tetramers). Interestingly, no molecular ions are observed for the heptamers through the decamers. In fact, the doubly charged species is the most abundant ion for the hexamers through octamers (m/z 864, 1008, and 1152, respectively). For nonamers and decamers, the triply charged species (m/z 864 and 960, respectively) are the major ions, whereas the doubly charged species (m/z 1296 and 1440, respectively) are relatively minor in abundance. The presence and relative abundance of the molecular ion and the multiple charges give a "fingerprint," which can be used to identify which oligomeric class a peak belongs.

Mass spectral data can also be used to characterize structural features of the proanthocyanidins, such as distinguishing between singly and doubly linked proanthocyanidins. In the presence of one A-type linkage, the molecular ion is 2 mass units less than for its singly linked counterpart. Additionally, mass spectral data can also distinguish between the proanthocyanidin monomers and give preliminary evidence for the monomeric units present in oligomers. However, in certain cases, oligomers of different classes may have the same molecular mass and require further structural identification with nuclear magnetic resonance or tandem MS. Indeed, a dimer consisting of an epigallocatechin and an epicatechin gallate has the same molecular mass (746) as a dimer composed of epicatechin and epigallocatechin gallate.

Figure 4 shows representative HPLC chromatograms for the separation of the procyanidins in cocoa, apple, and red wine. As can be seen, the normal-phase method is successful at separating the proanthocyanidin oligomers based on the degree of polymerization. Oligomers through the decamers are observed in the cocoa and apple samples, whereas wine contains oligomers through the hexamers. It is interesting to note that in normal-phase separations of proanthocyanidins, galloylated derivatives elute significantly earlier than their nongalloylated counterparts,[7,20] and similar observations have been made for A-type oligomers as compared to their singly linked counterparts.[7]

Isolation and Purification of Procyanidins for Use as Quantitative Standards

Extraction of Cocoa Procyanidins

Fresh or freeze-dried cocoa beans should be ground in a high-speed laboratory mill with liquid nitrogen until the particle size is reduced to approximately 90 μm. Lipids need to be reduced to less than 2% by weight using hexane or petroleum ether in order to maximize extraction efficiency of the procyanidins. Typically, three or four extractions are required using a 5 : 1 (v/w) ratio of solvent to solids. To obtain a procyanidin fraction, the lipid-free solids can be extracted with pure or aqueous methanol, ethanol, or acetone at either neutral or acidic pH using a 10 : 1 (v/w) ratio of solvent to solids. For the procyanidin oligomers, methanol has a higher extraction efficiency than ethanol, whereas acetone is superior to either alcohol. In general, aqueous solvent (70%, v/v) is superior to pure solvent and acidified (1%, v/v) is better than neutral. Additionally, extraction efficiency can be

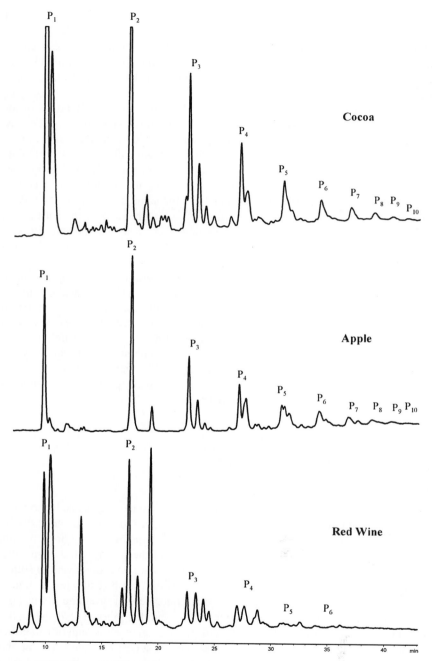

FIG. 4. HPLC chromatograms for proanthocyanidin separation in cocoa, apple, and red wine. P_1 through P_{10} correspond to monomers through decamers, respectively.

increased using warm solvent (50°) as compared to solvent at room temperature, and sonicating is superior to vortexing or shaking. Typically, the aqueous acetone layer is then reextracted with hexane (10 : 1, v/v) to remove residual lipids. The hexane layer is discarded, and the acetone extract is rotary evaporated, freeze dried, and stored in the dark at ambient temperature.

Gel-Permeation Chromatography of Cocoa Acetone Extract

The acetone extract prepared earlier is approximately 50% procyanidins, 11% methylxanthines, and 30% sugar by weight. In order to remove the major contaminants, the acetone extract is semipurified using gel-permeation chromatography, most commonly Sephadex LH-20, but other stationary phases have been used, such as Fractogel. The cocoa acetone extract is suspended in 70% aqueous methanol or ethanol, may be refrigerated overnight to partially precipitate theobromine, and centrifuged at 1500g at room temperature for 5 minutes before semipurification on a Sephadex LH-20 column (70 × 3 cm). Prior to sample loading, the GPC column is equilibrated previously with methanol at a flow rate of 2.5 ml/min. The acetone extract could be loaded in a single injection (2 g in 10 ml) or in multiple injections (10 g in 50 ml) spaced 15 min apart. The eluate is monitored for theobromine and caffeine by HPLC.[14] Once the theobromine and caffeine are eluted off the column, the remaining eluate is collected for at least an additional 6 hr or, more preferably, overnight at a flow rate of 0.5 ml/min, and rotary evaporated under partial vacuum at 40° to remove the organic solvent. Finally, the extract is suspended in water, freeze dried, and stored in the dark at ambient temperature.

Purification of Procyanidin Oligomers by Preparative Normal Phase HPLC

Procyanidin oligomer fractions are obtained using preparative normal-phase HPLC, which separates the procyanidins based on degree of polymerization similar to the analytical method described earlier. The separations are performed at ambient temperature using a 5-μm Supelcosil LC-Si 100 Å (50 × 2 cm) with a column load of 0.7 g of semipurified acetone extract, which has been dissolved in 7 ml of acetone : water : acetic acid (70 : 29.5 : 0.5, v/v). The procyanidin oligomers are eluted by a linear gradient of methanol into dichloromethane with a constant 4% aqueous acetic acid (1 : 1, v/v) under the conditions shown in Table I and monitored by UV at 280 nm. Fractions are collected at the valleys between the peaks corresponding to each oligomeric class. Fractions with equal retention times from several preparative separations are combined and rotary evaporated under partial vacuum. Given the large amount of acid present in the mobile phase (2% acetic acid, v/v), the temperature (40–45°) used during rotary evaporation is critical to the stability of the oligomeric fractions (unpublished data). Once the organic solvent has been removed, the fractions are freeze-dried and stored at ambient temperature over phosphorus pentoxide to ensure minimal moisture uptake. It has

TABLE I
GRADIENT PROFILE FOR PREPARATIVE NORMAL-PHASE HPLC

Time (min)	Dichloromethane (%)	Methanol (%)	Flow rate (ml/min)
0	88.8	7.2	10
10	88.8	7.2	40
30	87.8	8.2	40
145	74.9	21.1	40
150	13.4	82.6	40
155	13.4	82.6	50
180	0	96	50

been observed that storage of the fractions in a sealed jar at atmospheric humidity will result in approximately 7% moisture over a period of 6 months (unpublished data). When stored over phosphorus pentoxide, the fractions are stable for at least a year.

HPLC/MS Analysis of Oligomeric Fractions to Determine Purity

Oligomeric fractions are analyzed by HPLC/MS using the parameters described earlier for qualitative characterization to determine purity (Table II). Peaks are assigned to an oligomeric class based on mass spectral data. The peak area using UV at 280 nm is determined, and relative purities are calculated under the assumption that the molar absorptivity for each oligomeric class is the same. In the cases where a peak is a mixture of two oligomers, the ratio between ion abundances in mass spectra for that peak is used to determine how much of the UV peak area

TABLE II
RELATIVE PERCENTAGE PURITIES FOR OLIGOMERIC FRACTIONS FROM PREPARATIVE NORMAL-PHASE HPLC[a]

Fraction	Monomer	Dimer	Trimer	Tetramer	Pentamer	Hexamer	Heptamer	Octamer	Nonamer	Decamer
Dimer	1.0	**99.0**								
Trimer		5.2	**94.7**	0.1						
Tetramer		0.5	4.1	**95.4**						
Pentamer		0.3	0.6	5.8	**92.0**	1.3				
Hexamer		0.2	0.4	2.9	7.2	**86.2**	1.3	0.8	1.0	
Heptamer		0.4	0.4	1.5	4.9	25.1	**61.0**	2.8	3.9	
Octamer		0.3	0.3	0.8	1.4	9.1	12.9	**64.6**	10.6	
Nonamer		0.3	0.3	0.9	1.4	5.5	4.1	10.8	**76.7**	
Decamer		0.4	0.3	0.9	1.2	3.4	2.9	8.0	27.0	**55.9**

[a] Bold-face numbers designate the major oligomer of the fraction. Purity of a fraction is indicated from left to right.

should be attributed to each oligomeric class. For example, mass spectral analysis indicates that a peak at 26.2 min in the trimer fraction is a mixture of trimer and tetramer with a ratio between ion abundances of 4 : 1. As the area of this peak by UV is 13,390, when calculating purities, 10,715 is assigned to the area sum for trimer and 2675 to tetramer.

Quantitative HPLC Analysis of Procyanidins in Food Products

Standard Stock Solutions

A composite standard is made using commercially available (−)-epicatechin for the monomer and the purified procyanidin fractions described earlier for dimers through decamers. The oligomeric profile for the composite standard is shown in Table III. Stock solutions of the composite standard are made in acidified aqueous acetone (70% acetone and 0.5% acetic acid, v/v) at five different concentrations ranging from 0.4 to 20 mg/ml. The standard stock solutions are stable for at least 9 months when stored in the freezer (−15°) and approximately 3 months when stored in the refrigerator (5°). At ambient temperature, the solutions are only stable for 1–2 days (unpublished data).

Sample Preparation for Quantitative Analysis of Procyanidins

Sample preparation will vary depending on the food matrix. For some foods, grinding prior to procyanidin extraction may be required to ensure sample homogenization and/or reduce particle size. Experience in this laboratory suggests that it is necessary to reduce the particle size to less than 90 μm in order to

TABLE III
OLIGOMERIC PROFILE OF COMPOSITE STANDARD

Oligomer	Profile (% by weight)
Monomer	9.82
Dimer	13.25
Trimer	9.85
Tetramer	10.49
Pentamer	10.51
Hexamer	12.68
Heptamer	7.98
Octamer	8.44
Nonamer	11.56
Decamer	5.42

ensure the greatest extraction efficiency of the procyanidins. In some cases, such as fruits, sample homogenization may be facilitated by removing the moisture prior to grinding, which is best achieved by freeze drying.

When the lipid content is greater than 2% by weight, removal of the fat is necessary before analysis. In cases where the sample contains >10% fat by weight, it may be necessary to freeze the sample in liquid nitrogen prior to grinding to prevent a paste from forming. In these cases, it is preferable to do a crude grinding with liquid nitrogen, and then partially defat the sample with two extractions with hexane before grinding the sample to a particle size less than 90 μm. Once the particle size is reduced, the remaining lipids can be removed. Certain foods, such as chocolate, already have a particle size less than 90 μm and grinding is not required. However, lipids still need to be removed to ensure proper extraction efficiency of the procyanidins, which can be done by directly extracting the sample (8 g) with hexane (45 ml) or petroleum ether. When the fat content is less than 35% by weight, three extractions may be required to reduce the lipid content to less than 2%. Typically, when the fat content is greater than 35% by weight (e.g., peanuts), a fourth repetition is necessary.

The procyanidins can then be extracted from the defatted or dry materials using an extraction solvent of acetone, water, and acetic acid (70 : 29.5 : 0.5, v/v, respectively). Typically, 1–2 g of sample is extracted with 5 ml of solvent, depending on the procyanidin level in the sample. Extraction is best when sonicated for 5 min at 50° prior to centrifuging. Samples should be filtered through a 0.45-μm nylon filter membrane prior to HPLC analysis. Independent analysis may be required to determine the fat or moisture content so that the procyanidin content can be calculated back to the original starting material.

For beverages, samples typically require a concentration that can be achieved via rotary evaporation or solid-phase extraction. However, in the case of juices, removal of oligosaccharides (e.g., corn syrup) is necessary because they have been shown to reduce column life and this should be done using solid-phase extraction. Rotary evaporation should be done at 40–45° to ensure maximum recovery of the procyanidins. The samples should be brought to a standard volume using acetone to minimize solvent effects in the chromatography. Alternatively, samples could be concentrated using solid phase extraction. Sephadex LH-20 (5 g; 2 × 8 cm) or prepacked C_{18} cartridges could be used and the procyanidins eluted with the acidified aqueous acetone extraction buffer. Pure methanol could also be used to elute the procyanidins from the Sephadex LH-20; however, more solvent is required to completely recover the procyanidins as compared to the acidified aqueous acetone.

Quantification of Procyanidins

For the quantification of procyanidins, the normal-phase HPLC method described earlier is used, incorporating a column reequilibration between injections by rinsing the column with the equivalent of 25 ml (10 column volumes) of the initial mobile phase. Additionally, fluorescence detection (excitation 276 nm and emission 316 nm) is used because it is more sensitive than UV at 280 nm.[7] However, it should be noted that fluorescence detection is insensitive to proanthocyanidins containing a gallic acid ester and/or the gallocatechins as a monomeric unit and hence should only be used to quantify procyanidins consisting of (+)-catechin or (−)-epicatechin as the monomeric units.

Calibration curves are made from the stock solutions using a quadratic fit for the relationship of area sum versus concentration for the peaks corresponding to each oligomeric class as determined earlier using HPLC/MS. The areas for the peaks corresponding to each oligomeric class are summed in order to include the contributions from all of the isomers within an oligomeric class. In the cases where A-type procyanidins are present, area sum is more difficult, as the doubly linked oligomers elute significantly earlier than their singly linked counterparts, which can result in overlap between oligomeric classes.

Limit of detection and quantification has been determined to be less than 250 and 500 ng on column, respectively, for each oligomeric class. Additionally, this quantitative method has been found to be reliable and reproducible with the variability between duplicate samples and between laboratories analyzing the same samples typically less than 10%, except in the cases where the procyanidin concentration approaches the limit of quantification.[19]

[5] Direct Thiolysis on Crude Apple Materials for High-Performance Liquid Chromatography Characterization and Quantification of Polyphenols in Cider Apple Tissues and Juices

By SYLVAIN GUYOT, NATHALIE MARNET, PHILIPPE SANONER, and JEAN-FRANCOIS DRILLEAU

Introduction

Polyphenols are major constituents of apples that affect storage, processing, and sensory qualities of the fruits and derived products. In the cider industry, the global tannin content, combined with must acidity, is used to classify cider apple varieties into bitter, bittersweet, sweet, and sharp categories.[1] Several classes of

---

[1] T. P. Barker and L. F. Burroughs, *in* "Science and Fruit" (T. Wallace R. W. Marsh, eds.), p. 45. J. W. Arrowsmith Ltd., Bristol, 1953.

polyphenols are present in apples. Hydroxycinnamic acids, flavan-3-ols (including monomeric catechins and procyanidins), and dihydrochalcones are present in all parts of the fruit, whereas flavonols are essentially located in the peel.[2] Reversed-phase high-performance liquid chromatography (RP-HPLC) is probably the most widely used method to characterize the polyphenol profiles of apple varieties and apple products. Depending on the raw materials (juices, ciders, slices, etc.), HPLC analyses are performed by direct injection[3] or after extraction with organic solvents.[4,5] In most cases, direct RP-HPLC analysis of polyphenol extracts does not give complete information on the phenolic composition because compounds of the procyanidin class are underestimated: polymeric apple procyanidins may be associated with insoluble cell wall polysaccharides[6] and they may not be wholly recovered when solvent extractions are performed. In addition, only dimers, at best tetramers, of procyanidins give sufficient well-resolved peaks on chromatograms to be assayed precisely. In most cases, only dimer B2 and trimer C1 are considered, although other oligomeric and polymeric procyanidin forms may be preponderant. In a paper[7] dealing with the characterization of the polyphenol profile of 16 cider apple varieties and 1 dessert apple variety (Golden delicious), procyanidins were shown to be the predominant class in apples even for the dessert apple Golden delicious. A precise estimation of procyanidins in apple and apple products is all the more important as these compounds are largely involved in the quality of apples and apple products. In cider, procyanidins are partly responsible for bitter and astringent tastes that contribute to the "overall mouth feel".[8] In addition, they are receiving increasing attention as natural antioxidants with potential health protecting action.

The purpose of this article is to show the benefits as well as the limits of the application of the thiolysis reaction directly on crude apple materials as a precolumn reaction to RP-HPLC analysis for the characterization of polyphenols in crude apple tissues and apple juices without prior purification stages.

Procedures

Sampling of Raw Materials and Preparation of Samples

Attention must be paid in the preparation of fruit samples for comparative studies because of the variability encountered in the orchard (fruit size and maturity,

---

[2] S. Guyot, N. Marnet, D. Laraba, P. Sanoner, and J. F. Drilleau, *J. Agric. Food Chem.* **46**, 1698 (1998).
[3] B. Suarez, A. Pinicelli, J. Moreno, and J. J. Mangas, *J. Sci. Food Agric.* **78**, 461 (1998).
[4] M. J. Amiot, M. Tacchini, S. Aubert, and J. Nicolas, *J. Food Sci.* **57**, 958 (1992).
[5] B. Fernandez de Simon, J. Perez-Ilzarbe, T. Hernandez, C. Gomez-Cordoves, and I. Estrella, *J. Agric. Food Chem.* **40**, 1531 (1992).
[6] S. Guyot, L. Delalande, H. Matinier, P. Massiot, and J.-F. Drilleau, *in* "Polyphenols Communications 98" (F. Charbonnier, J.-M. Delacotte, and C. Rolando, eds.), Vol. 2, p. 399. Groupe Polyphenols, Bordeaux, France, 1998.

position of fruits on the trees, etc.). After harvest, fruits are freeze-dried rapidly, which is a good way to protect plant material from biochemical modifications before analysis.

Crude Apple Tissues. A batch of 30 kg of apple fruits from five trees is harvested in an experimental orchard. Fruits of the predominant size are selected and three batches having 10 fruits are used. Fruits of each batch are peeled and cored. The tissues are frozen in liquid nitrogen. Samples are weighed and freeze-dried. Homogeneous powders (three batches) are then obtained by crushing the dried tissues in closed vials to avoid hydration. Powders are kept in a desiccator until analysis.

Crude Apple Musts. Three batches of 1 kg of apple fruits of the predominant size are used. Each batch is milled and pressed to obtain the crude juice. A solution of diluted sodium fluoride (50 ml, 1 g/liter in water)[9] is added to the apple pulp before pressing to avoid oxidation. When oxidation is needed for comparative studies, only water is added (50 ml), and factors such as length of processing and temperature of fresh raw materials must be controlled carefully because phenolic compounds are largely transformed by the enzymatic oxidation phenomena. Crude apple musts may be centrifuged (6000g for 15 min at 4°) to obtain clarified apple juices. Then, juice yields are measured and aliquots (3 × 500 μl) are freeze-dried in 5-ml polyethylene vials. Freeze-dried samples are kept in a desiccator until analysis.

Thiolysis of Freeze-Dried Samples

The thiolysis reaction leads to the depolymerization of proanthocyanidin structures. The acid-catalyzed cleavage of the interflavanyl linkages of procyanidins converts the flavanol extender units into their carbocations, whereas the terminal units are liberated as monomeric flavanols.[10] The carbocations combine with toluene α-thiol, leading to the formation of thioether adducts. The reaction must be performed on dried materials and in a dry medium because the presence of water lowers thiolysis yields significantly.[11]

Freeze-dried apple powders are precisely weighed (50–100 mg) in 1.5-ml Eppendorf vials. Acidic methanol [3.3% (v/v) HCl, 400 μl] and toluene α-thiol (5% in methanol, 800 μl) are added. Vials are closed and incubated at 40° for

---

[7] P. Sanoner, S. Guyot, N. Marnet, D. Molle, and J.-F. Drilleau, *J. Agric. Food Chem.* **47**, 4847 (1999).
[8] A. G. H. Lea, *in* "Bitterness in Food and Beverages" (R. L. Roussef, ed.), p. 137. Elsevier, Oxford, 1990.
[9] This added volume is subtracted for the yield calculations and a correction factor is applied for the calculations of polyphenol concentrations.
[10] R. S. Thompson, D. Jacques, H. Haslam, and R. N. J. Tanner, *J. Chem. Soc. Perkin Trans. I* 1387 (1972).
[11] S. Matthews, I. Mila, A. Scalbert, B. Pollet, B. C. Lapierre, C. L. M. Hervé du Penhoat, C. Rolando, and D. M. X. Donnelly, *J. Agric. Food Chem.* **45**, 1195 (1997).

30 min with agitation on a vortex every 10 min. Then, the vials are cooled in an ice bath for at least 5 min, and 200 µl of the mixture is immediately filtered through a polytetrafluoroethylene (PTFE) membrane (0.45 µm) into an insert vial, which is closed with a butyl-Teflon cap and stored at 4° until RP-HPLC analysis. Samples are stable for 12 hr when kept at 4°. For analysis of, freeze-dried juices, the same procedure is applied to freeze-dried aliquots corresponding to 500 µl of fresh liquid. Because the extraction stages can be eliminated for the preparation of the samples, this procedure is more rapid and economical in comparison to our previous method,[2] which included successive solvent extraction with methanol and aqueous acetone before thiolysis and RP-HPLC analysis.

Direct Solvent Extraction of Polyphenols from Freeze-Dried Materials

When thiolysis is performed, no distinction can be made between the monomeric flavanols (i.e., catechins) from the terminal units of procyanidins and the native catechins already present before thiolysis. Therefore, a solvent extraction without prior thiolysis must be performed to allow the HPLC assay of the native catechins. Acidic methanol (1% acetic acid, v/v) is preferred to pure methanol because phenolics are more stable in slightly acidic media. Aliquots of freeze-dried powders or freeze-dried liquids are prepared as described earlier and extracted by sonication for 15 min in acidic methanol (1% acetic acid, 1.2 ml). After filtration (PTFE, 0.45 µm), the liquid phase is ready to be analyzed by RP-HPLC.

Reversed-Phase HPLC Analysis of Thiolysis Media and Crude Solvent Extracts

Chromatograms are obtained by injecting 10 µl of the filtered sample (thiolysis reaction media or methanol extract) onto a Purospher RP18 end-capped column, 250 × 4 mm, 5 µm (Merck, Darmstadt, Germany). The HPLC apparatus is a Waters (Milford, MA) system (717 plus autosampler, 600E multisolvent system, 996 photodiode array detector, and the Millenium 2010 manager system). The 717 autosampler is equipped with a cooling system (set at 4°) to increase the stability of the thiolysis derivatives, thus allowing a larger series of injections. The solvent system is a gradient of solvent A (aqueous acetic acid, 2.5%, v/v) and solvent B (acetonitrile) and the following gradient is applied: initial, 3% B, 0–5 min, 9% B linear; 5–15 min, 16% B linear; and 15–45 min, 50% B linear, followed by washing and reconditioning the column. HPLC peaks are identified on chromatograms according to their retention times and their UV–visible spectra by comparison with available standard compounds, as described by Guyot *et al.*[2] and Sanoner *et al.*[7] Quantification is performed by reporting the measured integration area in the calibration equation of the corresponding standard. Integration is performed at 280 nm for flavanol monomers, thioether adducts, and dihydrochalcones at 320 nm for hydoxycinnamic compounds or at 350 nm for flavonols.

Discussion

Chromatograms

Chromatographic profiles are presented in Fig. 1. Chromatogram A corresponds to the methanol extract of the apple powder (cortex of the Kermerrien variety), whereas chromatogram B corresponds to the same apple powder submitted to thiolysis.

The chromatograms clearly show evidence of the improvement of the chromatographic resolution when thiolysis is performed. Without thiolysis (chromatogram A), oligomeric and polymeric procyanidins (peaks x), which are eluted on a wide retention time window, cause the baseline deviation, which renders a precise integration of the main chromatographic peaks more difficult. With thiolysis (chromatogram B), procyanidins are converted into monomeric units, giving well-resolved peaks on the chromatograms. Peak 9 corresponds to (−)-epicatechin benzylthioether, which results from the extension units of procyanidin structures. No other benzylthioether derivative was observed on chromatograms in the study dealing with the characterization of the polyphenol profile of 16 different apple varieties.[7] Terminal units of apple procyanidins are (−)-epicatechin and (+)-catechin corresponding to peaks 4 and 1 on chromatogram B, respectively. These two peaks also include genuine monomeric catechins, which are assayed on chromatograms of crude methanol extracts (chromatogram A). One part of the peak 1 area on chromatogram B corresponds to (−)-catechin, resulting from epimerization of (−)-epicatechin.[12] (−)-Catechin is not separated from (+)-catechin in the HPLC system. To estimate the proportion of the epimerized form, incubations of pure (−)-epicatechin were performed under thiolysis conditions showing a maximum of 3.5% of conversion into the epimerized form.

Under thiolysis conditions, phenolic compounds of the hydroxycinnamic acid class are strongly modified (see Fig. 1). The presence of methanol and hydrochloric acid in thiolysis media is favorable to the methylesterification of compounds, which present carboxylic group in their structure. Chlorogenic acid (peak 2′, Fig. 1) and p-coumaroylquinic acid (peak 5′, Fig. 1) are thus largely converted into their corresponding methyl ester (peaks 2′ and 5′, respectively). These structures have been characterized by mass spectrometry.[7] However, this side reaction does not affect the quantification of these compounds (see later) because the conversion into their respective ester derivatives is quantitative.[2] Moreover, UV–visible spectra of the ester derivatives are similar to those of the genuine hydroxycinnamic acid. Consequently, the response factor of the native hydroxycinnamic acids can be used for their quantification even when they are present in their methyl ester forms.

[12] P. Kiatgrajai, J. D. Wellons, L. Gollob, and J. D. White, *J. Org. Chem.* **47**, 2910 (1982).

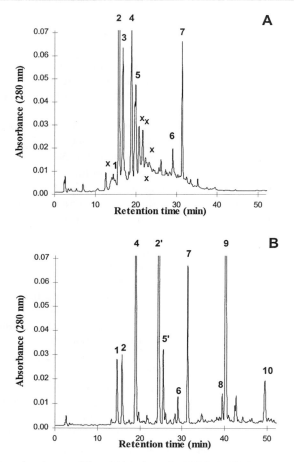

FIG. 1. Direct methanol extract (A) and thiolysis media (B) RP18-HPLC chromatograms (280 nm) of a cider apple powder (cortex of the Kermerrien variety). Peaks are numbered as follow: 1, (+)-catechin; 2, chlorogenic acid; 2', methyl ester of chlorogenic acid; 3, procyanidin B2; 4, (−)-epicatechin; 5, p-coumaroylquinic acid; 5', methyl ester of p-coumaroylquinic acid; 6, phloretin xyloglucoside; 7, phloridzin; 8, phloretin; 9, (−)-epicatechin benzylthioether, 10, toluene α-thiol; and x, procyanidins.

Characterization and Quantification of Procyanidins

The use of thiolysis is of main interest because it allows the simultaneous characterization and quantification of procyanidins, which are major phenolic compounds in cider apples.

Structural Aspects. By making the distinction between terminal (Fig. 1B; peaks 1 and 4) and extension units (Fig. 1B; peak 9), HPLC analysis of thiolysis

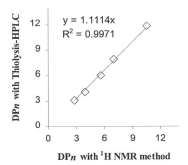

FIG. 2. DP_n values of pure apple procyanidin fractions: comparison of thiolysis-HPLC and ^{1}H NMR methods (according to Guyot et al.,[14])

media allows the determination of the nature and the proportion of the constitutive units of procyanidins. In addition, it gives access to the calculation of their average degree of polymerization (DP_n).[13] This criteria is of great importance for procyanidin characterization because it is strongly associated with their properties, such as their ability to interact with proteins. DP_n are thus obtained by calculating the molar ratio of all the flavan-3-ol units (benzylthioether adducts + terminal units) to the sum of (−)-epicatechin and (+)-catechin (terminal units). When genuine monomeric catechins are present in the sample, they must be assayed separately by HPLC without thiolysis and their proportions must be subtracted in the calculation of procyanidins DP_n. In another study,[14] similar DP_n were obtained when the ^{1}H nuclear magnetic resonance (NMR) and the thiolysis–HPLC methods were compared to the DP_n calculation of pure procyanidin fractions isolated from cider apple tissues (Fig. 2). Therefore, thiolysis–HPLC may be considered an appropriate method for DP_n estimation.

Results[7] showed that French cider apple procyanidins were essentially of the (−)-epicatechin series, with (+)-catechin being present in low proportion as a terminal unit. As a whole, DP_n values ranged between 4.2 and 7.5, but exceptionally high DP_n values (around 50) were observed for some varieties. In other work,[15] the relative standard deviation for the DP_n estimation was 6.0%, showing that

---

[13] J. Rigaud, J. Perez-Ilzarbe, J. M. Ricardo da Silva, and V. Cheynier, *J. Chromatogr.* **40,** 401 (1991).

[14] S. Guyot, C. Le Guernevé, N. Marnet, and J.-F. Drilleau, in "Plant Polyphenols 2: Chemistry, Biology, Pharmacology, Ecology" (G. G. Gross, R. W. Hemingway, and T. Yoshida, eds.), p. 211. Kluwer Academic/Plenum Publishers, New York, 1999.

[15] Unpublished results correspond to the direct HPLC and thiolysis–HPLC assays of 45 samples of apple powders. The cortex of five cider apple varieties in three states of maturity were analyzed according to the method that is described in the procedure section of this article. RSD were calculated by using ANOVA as a method of estimation [P. Dagnelie, *in* "Théorie et Méthodes Statistiques" (J. Duculot, ed.), Vol. II, p. 121, Gembloux, Belgium (1970)] and data were processed with SIGMASTAT software (Jandel Scientific, Germany).

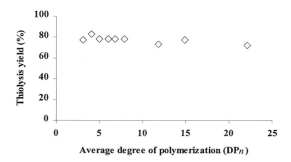

FIG. 3. Thiolysis yields for pure procyanidin fractions with different DP_n (according to Guyot et al.,[14])

thiolysis–HPLC is an accurate method for DP_n calculation in apple materials for comparative studies.

Quantitative Aspects. (+)-Catechin, (−)-epicatechin, and (−)-epicatechin benzylthioether are quantified on the thiolysis chromatogram according to their respective response factors. The amount of (−)-epicatechin benzylthioether is then converted into catechin equivalents. Thus, the total amount of procyanidins is obtained by summing all catechin equivalents assayed on the thiolysis chromatogram after subtracting the amounts of native catechins assayed on the chromatogram without thiolysis.

Thiolysis Yield. The question of the yield is of importance when precolumn reactions are used in HPLC quantification methods. Thiolysis yields depend on the conditions that are used for the reaction. Factors such as acidity (strong or weak acid), temperature, and the presence of water[11] may alter the yields of thiolysis because of the occurrence of side reactions[16,17] or the instability of reaction products.[9] The yields have been studied for thiolysis conditions by applying the reaction to pure procyanidin fractions with different average degrees of polymerization (DP_n). These fractions have been purified from the Kermerrien cider apple variety by semipreparative HPLC as described elsewhere.[14] For each fraction, the yield was calculated as the ratio of the molar quantity of reaction products (i.e., terminal flavanol units and extension flavanol units) assayed by HPLC after thiolysis to the weighted molar quantity (expressed in catechin equivalent) of the pure procyanidin fraction that was submitted to thiolysis.

Figure 3 shows that yields range around a value of 75–80% and do not depend on the polymerization state. In addition, yield values are probably underestimated because hydration water may be associated with pure oligomeric and

---

[16] B. R. Brown and M. R. Shaw, *J. Chem. Soc. Perkin Trans I* 2036 (1974).
[17] G. W. McGraw, J. P. Steynberg, and R. W. Hemingway, *Tetrahedron Lett.* **34**, 987 (1993).

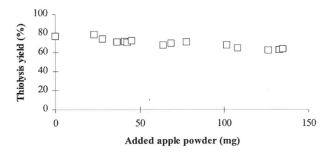

FIG. 4. Effect of the addition of freeze-dried apple powder (var. Golden delicious) in thiolysis media on thiolysis yields of a pure procyanidin fraction.

polymeric procyanidins (2.5–3 molecules of water per monomeric unit according to Czochanska *et al.*[18]). This may contribute to a 13–15% underestimation of yields.

Another question still remains regarding the real yield values when thiolysis is performed on complex apple samples (apple powders or freeze-dried juices) that contain many other materials, such as sugars, organic acids, and cell wall polysaccharides. To answer this question, we used a standard addition method.

A series of test samples were prepared by adding increasing amounts of freeze-dried apple powder (from 20 to 140 mg of parenchyma of the Golden delicious variety) to freeze-dried aliquots (1 mg) of a pure procyanidin fraction ($DP_n = 9.7$) that corresponded to a purified aqueous acetone extract of the Kermerrien variety as already described.[14] Three samples containing only pure procyanidins (1 mg) and three samples containing only apple powder (50 mg) were also prepared. All samples were submitted to thiolysis and HPLC analysis using standard conditions (as described earlier). Flavan-3-ols were quantified and, for each test sample, the corrected quantity of pure procyanidins was recalculated by subtracting the average quantity of flavanols obtained from the added powder. Yields were then, calculated as described earlier.

Yield values range between 80 and 65% (Fig. 4) and a yield reduction is observed when increasing quantities of apple powder are incorporated in the thiolysis media. Globally, this yield reduction is low but correlates with the quantity of apple powder that was added in the thiolysis media, thus showing that some apple powder constituents may alter the thiolysis reaction. Yields may be reduced because of the presence of water in apple powder. Water is a factor that contributes to yield reduction of the thiolysis reaction,[11] and freeze-dried apple powder is highly hygroscopic because of its high sugar concentration. The small amount of water of hydration that is added in the thiolysis media when apple powder is used may be sufficient to cause a slight reduction in the yield. Samples of apple powders or

---

[18] Z. Czochanska, L. Y. Foo, R. H. Newman and L. J. Porter, *J. Chem. Soc. Perkin Trans. I* 2278 (1980).

freeze-dried juices must be kept in a desiccator until they are used in the thiolysis reaction.

Although 100% yields are not obtained, thiolysis–HPLC constitutes a suitable method for the quantification of procyanidins in apples and apple products. In work[15] dealing with HPLC assays of polyphenols in five cider apple varieties and three states of maturity, we found a relative standard deviation (RSD) of 6.2% for procyanidin quantification. In most cases, when direct HPLC analysis is used, only dimeric procyanidin B2 and sometimes trimers and tetramers are considered for quantification. More polymerized procyanidins that do not give well-resolved peaks on chromatograms are not taken into account. Therefore, the quantification of total procyanidins may be largely underestimated. In a previous study,[7] the use of thiolysis–HPLC showed evidence that procyanidins corresponded to the predominant polyphenol class in the cortex of 16 apple varieties, even for the Golden delicious dessert apple. Procyanidin concentrations varied from 0.5 to 4.7 g per kilogram of fresh cortex tissue depending on the variety. Depending on the variety, procyanidin B2 that was assayed separately on chromatograms without thiolysis accounted for 1 to 29% of total procyanidins. In the Golden delicious variety, procyanidin B2 accounted for only 15% of total procyanidins.

Quantification of Native Monomeric Catechins by Direct HPLC

Native catechins in apple materials are not quantified in thiolysis media because they cannot be distinguished from catechins that are formed from the terminal units of procyanidins (peaks 1 and 4 in Fig. 1B). Nevertheless, they are highly soluble in acidified methanol. Therefore, they are assayed separately on chromatograms corresponding to direct HPLC of methanol extracts of the apple powders without thiolysis (peaks 1 and 4 in Fig. 1A). Our experiments showed that one stage of methanol extraction was sufficient to recover 95% of the monomeric catechins under experimental conditions (see the procedure section). (−)-Epicatechin is always the major monomeric catechin in French cider apple varieties.[7] However, (+)-catechin may account for 20% of the total catechins in some varieties. In another work,[14] the calculated RSD were 16.3 and 3.9% for (+)-catechin and (−)-epicatechin, respectively. RSD was clearly higher for (+)-catechin, probably because of the low concentrations of this compound in the samples. Moreover, (+)-catechin often coelutes with other minor unidentified compounds as observed in Fig. 1A (peak 1).

Comparison of Methods for HPLC Characterization of Phenolic Compounds in Freeze-Dried Apple Powders

Direct thiolysis on freeze-dried apple material combined with HPLC analysis may also be a suitable method for the quantification of non-flavan-3-ol phenolic compounds. Three methods are compared: direct HPLC analysis of acidic methanol extracts without thiolysis, thiolysis–HPLC of dry methanol and aqueous acetone

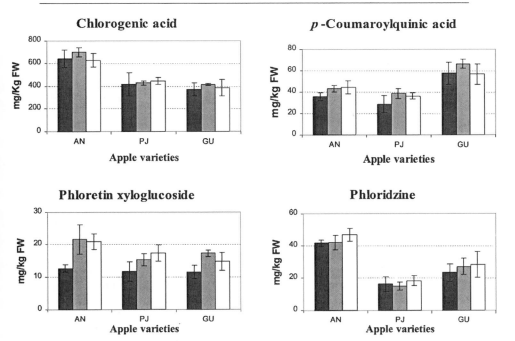

FIG. 5. Comparison of methods for the quantification of non-flavan-3-ol phenolic compounds in the cortex of three cider apple varieties (AN, Antoinette; PJ, Petit Jaune; GU, Guillevic): ■, thiolysis–HPLC with successive solvent extractions (according to Guyot *et al.*, 1998); ■, HPLC following direct thiolysis on freeze-dried tissues; □, direct HPLC of methanol extracts of freeze-dried tissues without thiolysis.

extracts obtained by successive solvent extractions from freeze-dried cider apple tissues,[2] and direct thiolysis and further HPLC analysis of freeze-dried tissues without any prior solvent extraction procedure.

The comparison of three methods for the quantification of dihydrochalcones (i.e., phloridzin and phloretin xyloglucoside) and hydroxycinnamic acids (i.e., chlorogenic acid and *p*-coumaroylquinic acid) is presented in Fig. 5 for three cider apple varieties. The three methods give values that are quite similar. Therefore, it can be deduced that the conversion of some hydroxycinnamic compounds into their methyl ester derivatives during thiolysis does not influence their overall quantification. Nevertheless, slightly lower concentrations are measured when the successive solvent extraction method is used because part of the compounds may be lost during the extraction steps of the procedure. In addition, dihydrochalcones are slightly underestimated when thiolysis–HPLC is used, probably because these compounds are partly converted into the corresponding aglycone (i.e., phloretin) as mentioned previously.[2]

TABLE I
RELATIVE STANDARD DEVIATIONS FOR POLYPHENOL ASSAYS IN APPLE VARIETIES[a]

Methods	Relative standard deviation (%)			
	CQA	p-CoQA	XPL	PL
Direct HPLC without thiolysis	4.5	10.7	22.7	13.1
Direct Thiolysis-HPLC	6.9	6.7	7.1	8.5

[a] For each method, RSD were calculated from HPLC analyses of 45 freeze-dried apple powder samples prepared as described and corresponding to 15 apple batches (different varieties and different maturity states). CQA, chlorogenic acid; p-CoQA, p-coumaroylquinic acid; XPL, phloretin xyloglucoside; PL, phloridzin.

HPLC methods with or without thiolysis are then compared on the basis of the relative standard deviations corresponding to the variations of the global methods of assay from sampling of the raw materials to HPLC analysis (Table I).

The relative standard deviations (RSD) vary from 4.5 to 22.7%, depending on the method and the compound. Lower RSD, which are obtained by thiolysis–HPLC, may be explained by the improvement in the resolution for chromatograms of thiolysis media (see Fig. 1). However, low RSD is obtained for the quantification of chlorogenic acid by direct HPLC because this compound corresponds to one of the highest peaks on the chromatograms, which may reduce integration error for these compounds. On the whole, RSD values show that thiolysis–HPLC is an accurate method for the quantification of hydroxycinnamic and dihydrochalcone compounds.

Thiolysis–Reversed-Phase HPLC of Cider Apple Musts

No particular difference is observed when thiolysis is carried out on freeze-dried apple juices vs. freeze-dried powders. Figure 6 shows chromatograms corresponding to oxidized and nonoxidized cider apple juices coming from the same apple batch (Kermerrien variety). Phenolic compounds that were measured in apple cortex tissue are also present in apple juices, but their proportions differ greatly.

The effect of oxidation is shown clearly on the chromatograms. All peaks are reduced but in different proportions, depending on the compound. A large peak reduction is observed for chlorogenic acid (peaks 2 and 4, Fig. 6, because of its partial conversion to methyl ether), which is a good substrate for apple polyphenol oxidase. In addition, peaks (1, 3, and 8) (Fig. 6) corresponding to flavan-3-ol units are strongly reduced on the chromatographic profile of the oxidized apple juice, showing that procyanidins are largely involved in the oxidation reactions. The zoom windows in Fig. 6 show evidence of the presence of oxidation products, which contribute to the baseline deviation on the chromatogram of oxidized apple

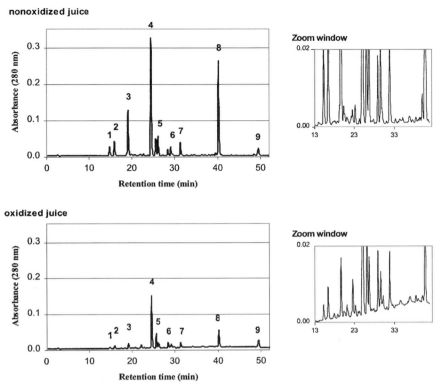

FIG. 6. RP18-HPLC chromatograms of thiolysis media of an oxidized and a nonoxidized cider apple juice (var. Kermerrien). 1, (+)-catechin; 2, chlorogenic acid; 3, (−)-epicatechin; 4, methyl ester of chlorogenic acid; 5, methyl ester of p-coumaroylquinic acid; 6, phloretin xyloglucoside; 7, phloridzin; 8, (−)-epicatechin benzylthioether; 9, toluene α-thiol.

juice. Calculated DP_n values for procyanidins are significantly different with values of 6.0 (±0.6) and 4.2 (±0.1) for nonoxidized and oxidized juices, respectively, showing that thiolysis–HPLC may be an accurate tool for showing qualitative differences in the phenolic composition of apple juices.

Conclusion

The benefits of the use of direct thiolysis on dried apple materials in combination with reversed-phase HPLC analysis can be listed as follow: (1) thiolysis reaction can be applied directly on solid crude samples without purification and extraction stages. Therefore, polyphenols such as polymerized procyanidins can be assayed even when they are associated with insoluble apple materials. (2) The method is not time-consuming and a large number of samples can be analyzed to obtain variability and comparison studies in apple materials. (3) The method may be applied to both freeze-dried apple tissues and juices.[19] (4) Chromatograms are well resolved, allowing accurate peak integration. Procyanidins may be quantified and characterized globally by their constitutive units and their average degree of polymerization. (5) The method is also convenient for the quantification of non-flavanol phenolic compounds. However, some limits of the method must also be noted. (1) HPLC analysis without thiolysis needs to be performed to assay native catechins and differentiate them from procyanidins. (2) Epimerization of catechins may occur in relatively low proportion (not more than 4% in our thiolysis conditions). (3) Procyanidins are not assayed individually and the method does not give information about their distribution and the type and position of their interflavanyl linkages.

[19] Some of our experiments (unpublished results, 2000) indicated that thiolysis could also be performed on freeze-dried apple pomaces and ciders without additional difficulties.

[6] Enzymes Involved in Hydroxycinnamate Metabolism

By DIETER STRACK

Introduction

Hydroxycinnamates (HCAs), i.e., 3-phenyl-2-propenoic acids, occur ubiquitously in higher plants. E-Caffeate is the most common HCA, followed by E-4-coumarate. E-ferulate, and E-sinapate,[1,2] derived from E-cinnamate, the product of L-phenylalanine ammonia-lyase (PAL) activity. HCAs are usually activated as coenzyme A (CoA) thioesters or 1-O-acylglucosides, being precursors of several thousands of phenolic compounds[3] (derivatives), e.g., the most widespread flavonoids[4,5] and lignins,[6,7] based on alterations of the propenoic acid side chain, such as elongation, degradation, or reduction.[3] However, HCAs accumulate as intact moieties in several hundred combined forms (conjugates) as esters or amides, rarely as glycosides. Conjugating moieties are carbohydrates, peptides, lipids, amino acids, amines, hydroxycarboxylic acids, terpenoids, alkaloids (including

[1] E. C. Bate-Smith, *J. Linn. Soc. (Bot.)* **58**, 95 (1962).
[2] J. B. Harborne, *Z. Naturforsch.* **21b**, 604 (1966).
[3] W. Barz, J. Köster, K.-M. Weltring, and D. Strack, *in* "Annual Proceedings of the Phytochemical Society of Europe" (C. F. Van Sumere and P. J. Lea, eds.), Vol. 25, p. 307. Clarendon Press, 1985.

betacyanins), or flavonoids. In addition, insoluble HCAs occur bound to polymers, such as cutins and suberins, lignins, and polysaccharides in cell wall fractions.

These HCA conjugates and derivatives are of considerable biological importance, e.g., as constituents of supporting tissues or as ecological agents. The conjugates are of prime ecological importance for plant survival,[8] e.g., in protecting plants against DNA-damaging ultraviolet (UV-B) light. Acylation of anthocyanins with HCAs results in an (intramolecular) copigmentation effect, protecting these pigments against degradation by water. Soluble and cell wall-bound HCAs participate in plant defense against microbial attack.

This article reviews some recent and earlier enzymatic work on the general phenylpropanoid pathway and the formation of HCA conjugates (esters and amides). Some of the techniques of enzyme preparation and activity determination are described in more detail than others. Standard column materials for enzyme purification can be purchased, e.g., from Sigma (St. Louis, MO), Pharmacia (Piscataway, NJ), Bio-Rad (Hercules, CA), or Merck (Darmstadt). The described procedures should be taken as the basis for the enzymologist to work out suitable methods applicable to the system in question.

Biosynthetic Enzymes of Hydroxycinnamates

HCAs are formed via the general phenylpropanoid pathway starting with the phenylalanine ammonia-lyase-catalyzed nonoxidative elimination of ammonia from L-phenylalanine, including the *pro-3S* hydrogen[9] with the formation of a double bond to give *E*-cinnamic acid. Subsequent hydroxylation and *O*-methylation of *E*-cinnamate lead to the different HCAs. Finally, HCA-specific CoA ligases and glucosyltransferases activate HCAs to form the acyl donors, which enter various HCA transferase reactions.[3]

Phenylalanine Ammonia-Lyase (PAL; EC 4.3.1.5)

Extractions and assays of PAL found in the literature are essentially according to the original description given by Koukol and Conn in 1961.[10]

---

[4] K. Hahlbrock, in "The Biochemistry of Plants" (P. K. Stumpf and E. E. Conn, eds.), Vol. 7, p. 425. Academic Press, London, 1981.
[5] J. B. Harborne and C. A. Williams, *Nat. Prod. Rep.* **15**, 631 (1998).
[6] H. Grisebach, in "The Biochemistry of Plants" (P. K. Stumpf and E. E. Conn, eds.), Vol. 7, p. 457. Academic Press, London, 1981.
[7] N. G. Lewis, L. B. Davin, and S. Sarkanen, in "Comprehensive Natural Products Chemistry" (Sir D. H. R. Barton, K. Nakanishi, and O. Meth-Cohn, eds.), Vol. 3, p. 617. Elsevier, Amsterdam, 1999.
[8] M. Petersen, D. Strack, and U. Matern, in "Biochemistry of Plant Secondary Metabolism" (M. Wink, ed.), Annual Plant Reviews, Vol. 2, p. 151. Sheffield Academic Press, Sheffield, 1999.
[9] K. R. Hanson and I. A. Rose, *Acc. Chem. Res.* **8**, 1 (1975).
[10] J. Koukol and E. E. Conn, *J. Biol. Chem.* **236**, 2692 (1961).

Extraction

Homogenize plant material in sodium borate buffer (100 mM, pH 8.5–8.8). Alternatively, extract the enzyme with borate buffer from acetone powder, which is prepared by the homogenization of plant material in cold acetone ($-10°$ to $-20°$). Dry the powder and store it in the cold. PAL activity is usually quite stable in this powder.[11] However, up to 40% loss of activity has been reported when stored for a week at $-15°$.[12]

Assay

Mix 100–200 mM sodium borate/NaOH buffer (pH 8.8) with 1–10 mM L-phenylalanine (cofactors are not required). Determine enzyme activity by measuring product (cinnamate) formation spectrophotometrically at 280–290 nm.

Problems in the spectrophotometric PAL assay can be caused by high protein concentrations or possible side reactions[13] due to the presence of low molecular weight contaminants. The latter can be avoided by filtration through Sephadex G-25 or ion exchangers.

Alternative and more reliable methods for measuring PAL activities are assays using radiolabeled L-phenylalanine followed by extraction of labeled cinnamate with diethyl ether[14] or diethyl ether/hexane (1 : 1; v/v)[15] and liquid scintillation counting. Other suitable methods, combined with product identification, are chromatographic techniques such as gas chromatography (GC)[16] or high-performance liquid chromatography (HPLC).[17,18] These methods are important to determine the PAL reaction with L-tyrosine leading to 4-coumarate. [The earlier notion that this reaction might be due to another enzyme, L-tyrosine ammonia-lyase (TAL), had to be abandoned.] Both products, cinnamate and 4-coumarate, can be separated by HPLC[17] and thin-layer chromatography (TLC).[18] The HPLC system is equipped with a C_{18} reversed-phase column using a linear gradient from 5% acetic acid to 20% acetic acid/40% acetonitrile. TLC is performed on aluminum sheets covered with silica gel 60 F_{254} (solvent: 1-butanol/acetic acid/water, 5 : 2 : 3, v/v/v).

Hydroxylases

In contrast to the widely accepted membrane-bound (microsomal) cytochrome P450 monooxygenase, cinnamate 4-hydroxylase, that catalyzes the formation of

---

[11] E. A. Havir and K. R. Hanson, *Biochemistry* **7**, 1896 (1968).
[12] V. H. Marsh, Jr., E. A. Havir, and K. R. Hanson, *Biochemistry* **7**, 1915 (1968).
[13] A. Erez, *Plant Physiol.* **51**, 409 (1973).
[14] R. Brödenfeldt and H. Mohr, *Z. Naturforsch.* **41c**, 61 (1986).
[15] N. Amrhein and M. H. Zenk, *Z. Pflanzenphysiol.* **64**, 145 (1971).
[16] U. Czichi and H. Kindl, *Hoppe Seyler's Z. Physiol. Chem.* **356**, 457 (1975).
[17] B. J. Murphy and C. A. Stutte, *Anal. Biochem.* **86**, 220 (1978).
[18] C. Appert, E. Logemann, K. Hahlbrock, J. Schmid, and N. Amrhein, *Eur. J. Biochem.* **225**, 491 (1994).

4-coumarate, the nature of the enzymes catalyzing the subsequent hydroxylation reactions is still a matter of dispute. In some earlier studies,[19,20] soluble phenolase activities, i.e., copper-containing polyphenol oxidases, have been described to catalyze the formation of caffeate. Further studies provided evidence that two alternative hydroxylation pathways might exist: one accepts 4-coumaroyl-CoA[21,22] and the other various 4-coumaroyl conjugates, such as 4-coumaroylquinate and -shikimate,[23,24] 4-coumaroylglucose,[25] or 4-coumaroyl mono- and dihydroxyphenyllactate.[26,27] The latter two are P450-dependent hydroxylases.

In addition to the two hydroxylase reactions leading to 4-coumarate and caffeate, there is a third one that hydroxylates ferulate to form 5-hydroxyferulate, reported to be another P450-dependent monooxygenase using a biochemical[28] and a molecular approach.[29] The latter has been corrected after biochemical characterization of the recombinant "ferulate 5-hydroxylase," as coniferaldehyde and coniferyl alcohol were found to be most likely the natural substrates, pointing to a new route of lignin biosynthesis.[30] This is supported by another paper that reports 5-hydroxylation and subsequent methylation of coniferaldehyde as the pivotal reactions in syringyl lignin biosynthesis in angiosperms.[31] This pathway is likely to be realized in lignin biosynthesis; however, analogous reactions at the level of free HCAs, HCA CoA thioesters, or HCA O-esters must exist in pathways leading to the various HCA conjugates. The biochemical approach to ferulate hydroxylation is not described in this article because it could not be reproduced following its discovery from *Populus* × *euramericana* stems.[28]

Cinnamate 4-Hydroxylase (EC 1.14.13.11)

As with other P450 enzymes, solubilization and reconstitution of cinnamate 4-hydroxylase for activity measurements are sometimes difficult to achieve. Gabriac et al.[32] were able to purify the enzyme from microsomes of manganese-treated

---

[19] P. F. T. Vaughan and V. S. Butt, *Biochem. J.* **113,** 109 (1969).
[20] G. P. Bolwell and V. S. Butt, *Phytochemistry* **22,** 37 (1983).
[21] J. Kamsteeg, J. Van Brederode, P. M. Verschuren, and G. Van Nigtevecht, *Z. Pflanzenphysiol.* **102,** 435 (1981).
[22] R. E. Kneusel, U. Matern, and K. Nicolay, *Arch. Biochem. Biophys.* **269,** 455 (1989).
[23] W. Heller and T. Kühnl, *Arch. Biochem. Biophys.* **241,** 453 (1985).
[24] T. Kühnl, U. Koch, W. Heller, and E. Wellmann, *Arch. Biochem. Biophys.* **258,** 226 (1987).
[25] M. Tanaka and M. Kojima, *Arch. Biochem. Biophys.* **284,** 151 (1991).
[26] M. Petersen, E. Häusler, B. Karwatzki, and J. Meinhard, *Planta* **189,** 10 (1993).
[27] M. Petersen, *Phytochemistry* **45,** 1165 (1997).
[28] C. Grand, *FEBS Lett.* **169,** 7 (1984).
[29] K. Meyer, J. C. Cusumano, C. Somerville, and C. C. S. Chapple, *Proc. Natl. Acad. Sci U.S.A.* **93,** 6869 (1996).
[30] J. M. Humphreys, M. R. Hemm, and C. Chapple, *Proc. Natl. Acad. Sci. U.S.A.* **96,** 10045 (1999).
[31] K. Osakabe, C. C. Tsao, L. Li, J. L. Popko, T. Umezawa, D. T. Carraway, R. H. Smeltzer, C. P. Joshi, and V. L. Chiang, *Proc. Natl. Acad. Sci. U.S.A.* **96,** 8955 (1999).
[32] B. Gabriac, D. Werck-Reichhart, H. Teutsch, and F. Durst, *Arch. Biochem. Biophys.* **288,** 302 (1991).

tuber tissues of *Helianthus tuberosus* in a three-step purification procedure with a yield of 2%.

Extraction and Purification. Step 1: Solubilize microsomes (2 mg protein ml^{-1}) with 2% (w/v) Triton X-114 for 10 min in 100 mM phosphate buffer (pH 7.4) containing 30% glycerol and 1.5 mM 2-mercaptoethanol and centrifuge at 100,000g for 60 min at 4°. Separate the spontaneously formed detergent-rich upper phase from the lower one. Step 2: Dilute the preparation from step 1 five times in 20 mM Tris–HCl buffer (pH 7.8) containing 20% glycerol, 1.5 mM 2-mercaptoethanol, and 0.1% (w/v) Emulgen 911 before loading onto a DEAE (diethylaminoethyl)-Trisacryl column (260 × 28 mm i.d.) equilibrated with Tris buffer containing 0.5% (w/v) Emulgen 911 and 0.5% (w/v) Triton X-114. (Emulgen avoids aggregation of Triton X-114 micelles.) A fraction that does not bind to the DEAE column should contain most of the enzyme activity. Step 3: Dialyze the enzyme fraction from Step 2 and apply to a hydroxyapatite column (50 × 16 mm, HTP-Bio-Rad). Elute the column first with a linear gradient (0 to 500 mM) of KCl and then with a 300 mM sodium phosphate step. Concentrate cytochrome P450-containing fractions by overnight dialysis against carboxymethyl(CM)-cellulose.

Quantitative measurement of cytochrome P450 can be achieved spectrophotometrically according to Omura and Sato[33] using an extinction coefficient of 91 mM^{-1} cm^{-1} ($A_{450-490\,nm}$). Purification of the enzyme can be monitored by spectral changes using type I substrate-binding properties of the enzyme.[32] Record absorption spectra between 370 and 500 nm 1 min after mixing a saturated dose of cinnamate to the sample fraction and express the results as ΔA between maximum near 390 nm and minimum at 420 nm.

4-Coumarate 3-Hydroxylase (Monophenol Monooxygenase; EC 1.14.18.1)

Despite the fact that it is a matter of dispute whether nonspecific copper-containing polyphenol oxidases (phenolase activities) are involved in the second hydroxylation step leading from 4-coumarate to caffeate, extraction and assay for phenolase activity are described.[34]

Extraction. Homogenize tissue sample in a chilled mortar in the presence of 10 mM citrate–phosphate buffer (pH 5.3) containing 700 mM sorbitol and 1 mM EDTA (ethylenediaminetetraacetic acid). Centrifuge the homogenate at 40g for 3 min to remove debris. Centrifuge the supernatant at 13,000g for 15 min and resuspend the precipitate in the extraction buffer. Comparison between tentoxin-treated plants with control ones should exclude most of the nonspecific polyphenol oxidases. (Tentoxin completely eliminates polyphenol oxidase-mediated 4-coumarate 3-hydroxylation.[35])

---

[33] T. Omura and R. Sato, *J. Biochem.* **239**, 2370 (1964).
[34] M. Kojima and W. Takeuchi, *J. Biochem.* **105**, 265 (1989).
[35] S. O. Duke and K. C. Vaughn, *Physiol. Plant.* **54**, 381 (1982).

Assay. Mix protein sample with 4-coumarate and ascorbate (3 mM each) and 500 mM $(NH_4)_2SO_4$ as the activator in 40 mM citrate–phosphate buffer (pH 5.3). Catalase may be added to prevent peroxidase-related hydroxylation.[34,35] Enzyme activity is determined by monitoring caffeate formation, either spectrophotometrically after chromatographic isolation from an ethyl acetate fraction[34] or by HPLC equipped with a photodiode array detector.[36]

4-Coumaroyl-CoA 3-Hydroxylase

Hydroxylase activity catalyzing the formation of caffeoyl-CoA from 4-coumaroyl-CoA was identified in crude extracts from cultured parsley cells.[22] To measure this enzyme activity it is essential to add optimal concentrations of ascorbate (50 mM) and Zn^{2+} (0.5 mM) to the assays.

Extraction and Assay. Homogenize frozen cells in a mortar in the presence of Dowex 1X2 and 50 mM MES buffer (pH 6.5) containing 50 mM ascorbate. Centrifuge the filtered homogenate and pass the supernatant through a Sephadex G-25 column. For measuring enzyme activity, mix 150 μM [β-^{14}C]4-coumaroyl-CoA, 0.5 mM $ZnSO_4$, and protein sample in 50 mM MES buffer (pH 6.5) and terminate the reaction at appropriate time intervals by the addition of 5 M NaOH (1 : 10). Heat this mixture for 15 min at 40°, acidify with HCl, extract the free HCAs with ethyl acetate (2 : 1), and separate caffeate from the remaining coumarate by TLC, e.g., on cellulose plates (Merck) using 1-butanol/acetic acid/water (5 : 2 : 3, v/v/v). Localize radioactivity with a TLC analyzer. It is advisable to carry out most operations in the dark to avoid *E/Z* isomerization.

Caffeate/5-Hydroxyferulate *O*-Methyltransferase (EC 2.1.1.68)

Work with *Populus nigra* shoots[37] is cited as a representative method for enzyme extraction.

Extraction

Homogenize young shoots (100 g) in a Waring blender in an equal weight of 100 mM Tris–HCl buffer (pH 7.3) containing 10 mg dithiothreitol (DTT) and 40 mg $NaBH_4$. Centrifuge the filtrate and bring the supernatant to 80% $(NH_4)_2SO_4$ saturation. Pass the redissolved protein through Sephadex G-25.

A convenient method for measuring enzyme activity includes *S*-adenosyl-L-[*methyl*-^{14}C]methionine (SAM) in the assay. This is exemplified with a methyltransferase from *Brassica oleracea* leaves.[38]

[36] P. B. Andrare, R. Leitao, R. M. Seabra, M. B. Oliveira, and M. A. Ferreira, *J. Liquid Chromatogr. Related Technol.* **20**, 2023 (1997).
[37] M. Shimada, H. Ohashi, and T. Higuchi, *Phytochemistry* **9**, 2463 (1970).
[38] E. De Carolis and R. K. Ibrahim, *Biochem. Cell Biol.* **67**, 763 (1989).

Assay

Mix 100 μM caffeate or 5-hydroxyferulate, 100 μM [*methyl*-^{14}C]SAM, 50 mM Tris–HCl buffer (pH 7.6) and add 50 μg protein sample into a final volume of 100 μl. Terminate the reaction by adding 10 μl 6 N HCl and extract the product in 250 μl ethyl acetate and measure the radioactivity by liquid scintillation counting. Proof of product identity can be achieved by TLC[38] or HPLC.[39]

Caffeoyl-CoA/5-Hydroxyferuloyl-CoA *O*-Methyltransferase

This *O*-methyltransferase has been described from elicitor-treated parsley cell suspension cultures.[40] The enzyme has been partially purified (82-fold) by standard column chromatographic techniques on Phenyl-Sepharose and Q-Sepharose, including a chromatofocusing step.

Assay

Mix 5 nmol caffeoyl-CoA, 7.5 nmol *S*-adenosyl-L-[*methyl*-^{14}C] SAM, and 5–20 μl protein sample with 50 μl 50 mM Tris–HCl buffer (pH 7.5) containing 200 mM MgCl$_2$, 2 mM DTT, and 10% (v/v) glycerol. Terminate the reaction by adding 5.5 μl 5 M NaOH. Heat this mixture for 15 min at 40°, acidify with HCl, extract into ethyl acetate, and measure the radioactivity of the product ferulate. This enzyme shows a narrow substrate specificity corresponding to the respective enzyme from carrot cells.[41]

4-Coumarate–CoA Ligase (4CL; EC 6.2.1.12)

Enzymes catalyzing the formation of HCA-CoA thioesters are collectively called 4-coumarate : CoA ligases. Enzymes from hybrid poplar and recombinant proteins converted HCAs in the order 4-coumarate > ferulate > caffeate, but did not accept sinapate.[42] Sinapoyl-CoA is probably formed via the alternative pathway involving hydroxylation and *O*-methylation of the HCA-CoAs[43] (see later).

Enzyme activity, isolated from soybean[44] or parsley,[45] can be measured using a spectrophotometric assay. The enzymes were isolated from the respective frozen cells by homogenization in phosphate (pH 7.8) or Tris–HCl buffer (pH 8.0)

---

[39] Y. Elkin, R. Edwards, M. Mavandad, S. A. Hedrick, O. Ribak, R. A. Dixon, and C. J. Lamb, *Proc. Natl. Acad. Sci. U.S.A.* **87**, 9057 (1990).

[40] A.-E. Pakusch, R. E. Kneusel, and U. Matern, *Arch. Biochem. Biophys.* **271**, 488 (1989).

[41] T. Kühnl, U. Koch, W. Heller, and E. Wellmann, *Plant Sci.* **60**, 21 (1989).

[42] S. M. Allina, A. Pri-Hadash, D. A. Theilmann, B. E. Ellis, and C. J. Douglas, *Plant Physiol.* **116**, 743 (1998).

[43] Z. H. Ye, R. E. Kneusel, U. Matern, and J. E. Varner, *Plant Cell* **6**, 1427 (1994).

[44] K.-H. Knobloch and K. Hahlbrock, *Eur. J. Biochem.* **52**, 311 (1975).

[45] K.-H. Knobloch and K. Hahlbrock, *Arch. Biochem. Biophys.* **184**, 237 (1977).

containing 28 or 14 mM 2-mercaptoethanol, precipitation with $MnCl_2$ and $(NH_4)_2SO_4$, followed by standard column chromatography using DEAE-cellulose, Sephadex G-100, Aminohexyl-Sepharose, and hydroxyapatite.

Assay

For measuring enzyme activity, mix 0.5 μmol of the HCA potassium salt, 0.3 μmol of CoA, 5 μmol of ATP (adjusted to pH 7–8 with KOH), 5 μmol of $MgCl_2$, 400 μmol Tris–HCl (pH 7.8), and protein samples in a total volume of 1 ml. Start the reaction by the addition of CoA and measure the change of long-wave absorbance at the appropriate wavelength,[46] i.e., 333, 363, and 345 nm for 4-coumaroyl-, caffeoyl-, and feruloyl-CoA, respectively.

To avoid possible interference with the optical test, analyze the reactions products by HPLC.[47] Stop the enzyme reaction by adding acetic acid, remove precipitated protein by centrifugation, and analyze the clear supernatant on a reversed-phase HPLC column (C_{18}). Apply at 1 ml min^{-1} a 20- to 30-min linear gradient from solvent A (1% phosphoric acid in water) to 50% solvent B (1% phosphoric acid in acetonitrile) in A+B. All the assay components, the substrates ATP, CoA, 4-coumarate, and the products AMP and HCA-CoA, are detectable spectrophotometrically at 270 nm.

Glucosyltransferases

The following extraction procedure is described for sinapate 1-glucosyltransferase (EC 2.4.1.120) from *Raphanus sativus* seedlings.[48]

Extraction and Assay

Treat 2-day-old seedlings with an Ultra Turrax homogenizer at 0° in the presence of PVP(2g/100 seedlings) and 100 mM phosphate buffer (pH 6.0) containing 10 mM DTT. Centrifuge the homogenate at 3000g for 15 min, fractionate the supernatant protein solution by $(NH_4)_2SO_4$ (30–60% saturation), and pass the redissolved protein through Sephadex G-25. For activity determination, mix protein sample with 1 mM HCA, 1 mM UDP-glucose, and 10 mM DTT in 100 mM phosphate buffer (pH 6.0).

HPLC analysis of the glucosyltransferase assays is a recommendable technique, not only to determine activity of ester formation, but also to distinguish formation of the *O*-esters from that of glucosides (glucose attachment at phenolic hydroxyl groups by a different enzyme).[49]

[46] G. G. Gross and M. H. Zenk, *Z. Naturforsch.* **21b**, 683 (1966).
[47] W. Knogge, G. Weissenböck, and D. Strack, *Z. Nataurforsch.* **36c**, 197 (1981).
[48] G. Nurmann and D. Strack, *Z. Pflanzenphysiol.* **102**, 11 (1981).
[49] P. A. Bäumker, M. Jütte, and R. Wiermann, *Z. Naturforsch.* **42c**, 1223 (1987).

Another example of a glucosyltransferase is cinnamate glucosyltransferase (EC 2.4.1.177) from *Ipomoea batatas*,[50] given as an example of crucial precaution in purification procedures. The enzyme was purified 540-fold using standard column chromatographic techniques. It was necessary to include phenylmethylsulfonyl fluoride (PMSF) and 2-mercaptoethanol during the purification procedures to avoid loss of activity. Storage of the partially purified enzyme at $-20°$ resulted in total loss of activity in 1 day. The enzyme could be stabilized, however, in the presence of 10% sorbitol.

Hydroxycinnamoyltransferases

Three types of HCA transferases have been described to date that catalyze the formation of HCA *O*-esters: (i) HCA-CoA thioester-, (ii) HCA 1-*O*-acylglucoside-, and (iii) HCA *O*-ester-dependent transferases. The pathway used is dependent on the plant species investigated. An example for converging lines for the formation of *O*-esters is *O*-caffeoylglucaric acid. In *Secale cereale* the biosynthesis proceeds via the caffeoyl-CoA thioester,[51] in *Cestrum elegans* via 1-*O*-caffeoylglucose,[52] and in *Lycopersicon esculentum* via 5-*O*-caffeoylquinic acid,[53] i.e., chlorogenic acid. With regard to the formation of HCA amides, there are only transferases described so far accepting HCA-CoA thioesters.

Coenzyme A Thioester Acyltransferase

A highly efficient method to purify CoA-ester-dependent HCA transferases is exemplified by hydroxycinnamoyl-CoA : tyramine hydroxycinnamoyltransferase (THT; EC 2.3.1.110) from cell suspension cultures of potato, including affinity chromatography on Reactive Yellow 3-Agarose using the acyl donor (feruloyl-CoA) as eluent.[54]

Extraction and Purification. Disrupt frozen potato cells in a chilled mortar or with an Ultra Turrax homogenizer in the presence of extraction buffer (buffer A): 50 mM Tris–HCl (pH 7.5), 1 mM EDTA, 5 mM DTT, 10 μg ml^{-1} α_2-macroglobulin (Sigma), and 20% (w/w) Polyclar AT (Serva, Heidelberg, Germany). After stirring for 90 min at 4°, centrifuge the homogenate for 20 min at 51,000g. Mix the supernatant with an equal volume of a potassium acetate solution (4 M, pH 7.5). After stirring for 30 min, centrifuge the mixture again for 20 min at 51,000g to remove precipitated proteins.

---

[50] T. Shimizu and M. Kojima, *J. Biochem.* **95**, 205 (1984).
[51] D. Strack, H. Keller, and G. Weissenböck, *J. Plant Physiol.* **131**, 61 (1987).
[52] D. Strack, W. Gross, J. Heilemann, H. Keller, and S. Ohm, *Z. Naturforsch.* **43c**, 32 (1988).
[53] D. Strack and W. Gross, *Plant Physiol.* **97**, 41 (1990).
[54] H. Hohlfeld, D. Scheel, and D. Strack, *Planta* **199**, 166 (1996).

Column Chromatography. Step 1: Apply the supernatant to a TSK-butyl-650 column (22.5 × 2.6 cm i.d.; Merck) equilibrated with 2 M potassium acetate in the extraction buffer (buffer B) and apply at a flow rate of 1.25 ml min^{-1} the following elution system: (i) 300 min buffer B; (ii) within 900 min from buffer B to buffer A (including 10% ethylene glycol); and (iii) 600 min buffer A. Step 2: Concentrate active protein obtained from step 1 by ultrafiltration (Diaflo PM30; Amicon, Beverly, MA) and transfer it to a Reactive Yellow 3-Agarose column (5 × 1.6 cm i.d.; Sigma) equilibrated with buffer A and elute the enzyme at 1 ml min^{-1} with the following elution system: (i) 3 min buffer A; (ii) elution stop for 30 to 60 min; (iii) 40 min buffer A; (iv) 40 min 0.1 M KCl in buffer A; (v) 40 min buffer A; 1 min buffer A containing 1 mM feruloyl-CoA; and (vi) 60 min buffer A. Final purification is achieved by Step 3: Concentrate active protein obtained from the last elution step (vi) of the previous chromatography by ultrafiltration and apply it to an FPLC-Superdex-74 column (60 × 1.6 cm i.d.; Pharmacia). Elute protein at 1 ml min^{-1} with 0.1 M NaCl in buffer A.

The affinity chromatography on Reactive Yellow 3-Agarose is the most effective step in purification protocols for these HCA transferases. This technique was also applied successfully for glucosyltransferases, i.e., betanidin 5-O-glucosyltransferase[55] and p-hydroxymandelonitrile O-glucosyltransferase,[56] and will probably be of great help in purification procedures of other glucosyl transferases and HCA transferases.

Assay.[57] Mix protein sample in a total volume of 30 μl with 250 mM phosphate buffer (pH 6.8), 0.5 mM feruloyl-CoA, and 20 mM tyramine. Product formation can be determined by HPLC as follows: on a 5-μm Nucleosil C$_{18}$ column (250 × 4 mm i.d.; Macherey-Nagel) with a 1-ml min^{-1} linear gradient elution from 60% solvent B (1.5% phosphoric acid, 20% acetic acid, 25% acetonitrile in water) in solvent A (1.5% phosphoric acid in water) within 6 min to 100% solvent B, followed by a 5-min isocratic elution with solvent B. The product, N-feruloyltyramine, can be detected spectrophotometrically at 320 nm.

1-O-Acylglucose Acyltransferase

The 1-O-sinapoylglucose:choline sinapoyltransferase (SCT; EC 2.3.1.91), widespread in members of the Brassicaceae,[58] will be taken as an example of 1-O-acylglucose-dependent transferases. The enzyme catalyzes the formation of O-sinapoylcholine (sinapine).

Extraction. Homogenize seeds (preferable immature seeds isolated from the pods) in the presence of liquid nitrogen, quartz sand, PVP, and 100 mM phosphate

---

[55] T. Vogt, E. Zimmermann, R. Grimm, M. Meyer, and D. Strack, *Planta* **203**, 349 (1997).
[56] P. R. Jones, B. L. Møller, and P. B. Høj, *J. Biol. Chem.* **274**, 35483 (1999).
[57] H. Hohlfeld, W. Schürmann, D. Scheel, and D. Strack, *Plant Physiol.* **107**, 545 (1995).
[58] J. Regenbrecht and D. Strack, *Phytochemistry* **24**, 407 (1985).

buffer (pH 7.0). Allow to stand with continuous stirring for 1 hr at 4°. Pass the homogenate through Miracloth, centrifuge at 48,000g at 4° and prepare enzyme activity from the supernatant by $(NH_4)_2SO_4$ precipitation (30–70% saturation) followed by filtration through Sephadex G-25.

Partial purification of the enzyme can be achieved on CM-Sepharose and Ultrogel AcA 44. The enzyme is fairly stable and allows heat treatment (65°, 10 min) in purification protocols.[59]

Assay. Mix 50 μl 2 mM sinapoylglucose, 10 μl 300 mM choline chloride with 30 μl 200 mM phosphate buffer (pH 7.0) and start the reaction by adding a 10 μl protein sample. Terminate the reaction after 2 hr by transferring the mixture to a freezer (−20°). Determine enzyme activity by HPLC analysis of sinapine formation. Use a C_{18} reversed-phase column and elute sinapine isocratically with a mixture of 30% acetonitrile, 15% acetic acid, and 1% phosphoric acid, and detect sinapine spectrophotometrically at 330 nm.

Preparation of Acyl Donors

Preparation of HCA-CoAs

It is easy to synthesize CoA esters by an ester exchange reaction via acyl N-hydroxysuccinimide esters.[60,61]

Substrate Preparation. To synthesize the HCA-CoA thioester in question, mix 15 mmol HCA, 15 mmol N-hydroxysuccinimide, and 17 mmol dicyclohexyl carbodiimide in ethyl acetate. After 24 hr, filter the reaction mixture and reduce to dryness under reduced pressure. Redissolve the ester in acetone and pour this solution dropwise into an aqueous solution of CoA (including $NaHCO_3$). After a reaction time of 12 hr under a continuous nitrogen stream, concentrate the reaction mixture under reduced pressure, centrifuge, and apply the clear supernatant onto a water-equilibrated polyamide column (24 × 2 cm i.d.; CC-Perlon) and purify the HCA-CoA thioester by a stepwise gradient[61]: 300–350 ml each of water, methanol, and increasing concentrations of aqueous ammonia in methanol (0.15, 0.44, 0.73, 1.5, 4.4, and 7.3%). The HCA-CoA thioesters elute in the range of 4.4 to 7.3%. This protocol gives recoveries of 30 to 40% as estimated by UV spectroscopy[60] and HPLC.[62] Prior to chromatography, the polyamide has to be washed carefully with 1 M aqueous ammonia in methanol, followed by water, 1 M aqueous HCl, and finally water.

---

[59] W. Gräwe and D. Strack, *Z. Naturforsch.* **41c,** 28 (1986).
[60] J. Stöckigt and M. H. Zenk, *Z. Naturforsch.* **30c,** 352 (1975).
[61] D. Strack, H. Keller, and G. Weissenböck, *J. Plant Physiol.* **131,** 61 (1987).
[62] D. Strack, A. Becher, S. Brall, and L. Witte, *Phytochemistry* **30,** 1493 (1991).

Preparation of HCA 1-O-glucosides

Although the HCA acylglucosides can be synthesized chemically,[63,50] it is much easier to isolate them from natural sources.[64,65] For example, 1-O-sinapoylglucose can be isolated easily from members of the Brassicaceae.

Substrate Preparation. Homogenize 3- to 4-day-old seedlings of *Raphanus sativus* or *Brassica napus* in 50% aqueous methanol. Centrifuge the homogenate at 10,000g and concentrate the supernatant at 30° under reduced pressure. Centrifuge again and apply 1- to 2-ml aliquots onto a Nucleosil 100-C_{18} column (Varioprep; 10 μm, 260 × 40 mm i.d.; Macherey-Nagel). Apply the following gradient: at a flow rate of 10 ml min^{-1} within 90 min a linear gradient from 10 to 70% methanol in 0.2% aqueous formic or acetic acid. Follow compound elution by UV detection at 330 nm and pool fractions showing high absorbance. Take pooled fractions to dryness at 30° under reduced pressure. Redissolve the residue in 50% aqueous methanol. Concentrate again to a small volume until precipitation of the compound is observed. Keep the sample overnight at 4°, centrifuge, and dry the pellet in a desiccator.

Another method is the production of HCA 1-O-acylglucosides using a HCA glucosyltransferase[48] (see earlier discussion). The most efficient procedure, however, is to produce the acylglucosides as detoxification products from cell cultures (glucose ester formation from the free HCAs). This is exemplified with a cell culture of *Chenopodium rubrum* for 1-O-caffeoylglucose production.[66] Feed 2-ml aliquots of an aqueous caffeate solution (2 mg ml^{-1}) to each of 70 ml of a 3-day-old cell culture in Erlenmeyer flasks. Harvest the cells after 2 days by filtration, extract 1-O-caffeoylglucose by treatment of the cells with methanol, and purify it on a polyamide column. Follow the same procedure as described for HCA-CoAs, however, omitting elution with methanol and ammonia/methanol mixtures. The glucose ester is obtained in the water eluate. Its elution can be monitored by an UV flow detector. Final purification is achieved either by chromatography on a Sephadex LH-20 column (90 × 2 cm i.d.) using methanol as solvent[67] or preparative HPLC as described earlier.

Acknowledgments

Research in the author's laboratory has been supported by the Deutsche Forschungsgemeinschaft (Bonn) and the Fonds der Chemischen Industrie (Frankfurt).

---

[63] L. Birkhofer, C. Kaiser, H. Kosmol, M. Donike, and G. Michaelis, *Liebigs Ann. Chem.* **699,** 223 (1966).
[64] D. Strack, *Z. Pflanzenphysiol.* **84,** 139 (1977).
[65] D. Strack, *Planta* **155,** 31 (1982).
[66] M. Bokern, V. Wray, and D. Strack, *Planta* **184,** 261 (1991).
[67] M. Linscheid, D. Wendisch, and D. Strack, *Z. Naturforsch.* **35c,** 907 (1980).

[7] Estimation of Procyanidin Chain Length

By VERONIQUE CHEYNIER, BENOIT LABARBE, and MICHEL MOUTOUNET

Introduction

Polyphenols show a great diversity of structure, from rather simple monomeric molecules to polymers. The latter are usually designated by the term tannins, referring to their ability to complex with proteins, and originally used in the formation of leather from hide. Among them, condensed tannins are of particular interest with respect to their large distribution in plants and numerous important properties. They are responsible for major organoleptic features of plant-derived foods, including astringency and haze development. In addition they show antinutritional properties and play a part in plant defense mechanisms. They are also receiving increasing attention as natural antioxidants and potential health-promoting agents.

Condensed tannins, also called proanthocyanidins because they release anthocyanidins when heated in acidic conditions, consist of chains of flavanol units linked by C-4–C-6 and/or C-4–C-8 bonds (B type) or doubly linked, with an additional C-2–O–C-7 linkage (A type) and eventually substituted (e.g., glycosylated, galloylated). Several classes can be distinguished on the basis of the hydroxylation pattern of the constitutive units, with the most common being procyanidins, based on (epi)catechin units. The structures of numerous oligomers have been elucidated[1] but they are usually present in relatively low concentrations compared to oligomers and polymers.[2] In addition, the degree of polymerization (*DP*) may vary greatly, as proanthocyanidins have been described having molecular weights up to 20,000.[3]

Tannin properties, including radical scavenging effects and protein-binding ability, largely depend on their structure, particularly on the nature and number of constitutive units. Formal identification of proanthocyanidins can be achieved by two-dimensional nuclear magnetic resonance (NMR) techniques.[4] However, this is restricted to pure compounds, which become increasingly difficult to isolate as their *DP* increases due to the larger number of possible isomers, smaller amounts of each individual compound, and poorer resolution of the chromatographic profiles. Therefore, several methods have been developed to determine the molecular weight and *DP* of proanthocyanidins in rather crude plant extracts and fractions.

---

[1] L. J. Porter, *in* "The Flavonoids: Advances in Research since 1984" (J. B. Harborne, ed.), p. 23. Chapman and Hall, London, 1994.
[2] Z. Czochanska, L. Y. Foo, R. H. Newman, and J. L. Porter, *J. Chem. Soc. Perkin Trans I*, 2278 (1980).
[3] E. Haslam and T. H. Lilley, *Crit. Rev. Food Sci. Nutr.* **27**, 1 (1988).
[4] L. Balas and J. Vercauteren, *Magn. Res. Chem.* **32**, 386 (1994).

Methods to Determine Average DP of Proanthocyanidin Extracts

A first group of methods aims at evaluating spectrophotometrically the mean DP (*mDP*) of proanthocyanidin crude extracts or fractions.

Most of them rely on the coupling of chemical reactions enabling them to specifically measure end units of flavanol chains, such as vanillin or dimethylaminocinnamaldehyde (DMCA) reactions, with methods allowing determination of all flavanol units. The rate and extent of color development after vanillin reaction are greatly solvent dependent.[5,6] In particular, the reaction is more sensitive in the absence of water. It is also less complex and specific of end units when performed in glacial acetic acid.[6] Reaction with DMAC is similar to that with vanillin, but appears more specific.[7,8] However, to our knowledge, it has not been tested with dihydrochalcones, flavanones, and flavanonols, which are the major compounds, other than flavanols, reported to react positively with vanillin.[9] Methods based on weight, Folin–Denis, or Folin–Ciocalteu and absorbance at 280 nm, using catechin as a standard, have been proposed to estimate flavanol content, but these methods lack specificity and can only be applied on isolated tannins.[10] Therefore, another more specific reaction, initially referred to as the leucocyanidin or Bate–Smith reaction, is usually preferred. It consists in depolymerization of condensed tannins by heating in butanol–HCl mixtures (Porter's reagent) releasing colored anthocyanidins, which can be assayed spectrophotometrically,[11] from polymer extension units.[12]

Each of these methods has been calibrated to determine proanthocyanidin concentrations.[13] However, the resulting conversion factors are seldom used in the calculation of *mDP*, which are usually given as arbitrary indexes, such as ratios of absorbance after vanillin[10,14] or DMCA[8] reaction to absorbance after the Bate-Smith reaction.

Concentrations of both extension and terminal units, and consequently *mDP*, can also be determined in a single reaction based on acid-catalyzed degradation in the presence of a nucleophilic agent,[15] followed by high-performance liquid

---

[5] R. B. Broadhurst and W. T. Jones, *J. Sci. Food Agric.* **29**, 788 (1978).
[6] L. G. Butler, M. L. Price, and J. E. Brotherton, *J. Agric. Food Chem.* **30**, 1087 (1982).
[7] C. W. Nagel and Y. Glories, *Am. J. Enol. Vitic.* **42**, 364 (1991).
[8] N. Vivas, Y. Glories, L. Lagune, C. Saucier, and M. Augustin, *J. Int. Sci. Vigne Vin* **28**, 319 (1994).
[9] S. K. Sarkar and R. E. Norwarth, *J. Agric. Food Chem.* **24**, 317 (1976).
[10] J. L. Goldstein and T. Swain, *Phytochemistry* **2**, 371 (1963).
[11] E. C. Bate-Smith, *Food* **23**, 124 (1954).
[12] L. J. Porter, L. N. Hrstich, and B. G. Chan, *Phytochemistry* **25**, 223 (1986).
[13] L. J. Porter, *in* "Methods in Plant Biochemistry (J. B. Harborne, ed.), Vol. 1, p. 389. Academic Press, London, 1989.
[14] P. Ribereau-Gayon and E. Stonestreet, *Chim. Anal.* **48**, 186 (1966).
[15] J. Thompson, D. Jacques, E. Haslam, and R. J. N. Tanner, *J. Chem. Soc., Perkin Trans. I,* 1387 (1972).

chromatography (HPLC)[16,17] or NMR[18] analysis of the resulting solution. Breakage of the interflavanic C–C bond under mild acidic conditions releases the terminal units as the corresponding flavanols and the upper and intermediate units as carbocations, which react with the nucleophilic reagent (usually phenylmethanethiol or phloroglucinol) to form stable adducts (e.g., benzylthioethers in the presence of phenylmethanethiol). This also gives access to the nature and proportion of the various constitutive units and can be used for quantitation purposes.

The average DP of a tannin extract can also be determined by NMR. The first method proposed[19,20] was based on ratios of ^{13}C signal areas corresponding to the C-3 of the lower units (located between 65 and 68 ppm) and those of the extension units (at 72–73 ppm). However, the accuracy of the method becomes lower as the DP increases. More recently, a ^1H NMR procedure based on the integration of the signals corresponding on one hand to all A-ring protons and, on the other hand, to H-4 protons of the terminal units has been developed and applied to epicatechin-based polymers extracted from cider apple.[21]

Another possibility is to use gel-permeation chromatography (GPC), which can supply the number average molecular weight. The major disadvantage of this technique is that it is usually performed on derivatized (acetylated) molecules from which the native compounds cannot be recovered.[22,23] The use of a dimethylformamide–3 M aqueous ammonium formate (95.5 : 0.5, v/v) solvent and polystyrene–divinylbenzene column has been proposed for rapid GPC analysis of native procyanidin oligomers,[24] but the resolution remains rather poor. Size-exclusion fractionation of native apple procyanidins has also been achieved using chromatography on Toyopearl with a mobile phase of acetone containing 8 M urea.[25]

Finally, analysis of tannins by electrospray ionization-mass spectrometry (ESI-MS) gives access to the molecular weights of the proanthocyanidin species

---

[16] Z. Shen, E. Haslam, C. P Falshaw, and M. J. Begley, *Phytochemistry* **25**, 2629 (1986).

[17] J. Rigaud, X. Perez-Ilzarbe, J. M. Ricardo da Silva, and V. Cheynier, *J. Chromatogr.* **540**, 501 (1991).

[18] Y. Cai, F. J. Evans, M. F. Roberts, J. D. Phillipson, M. H. Zenk, and Y. Y. Glebas, *Phytochemistry* **30**, 2033 (1991).

[19] Z. Czochanska, L. Y. Foo, R. H. Newman, L. Porter, W. A. Thomas, and W. T. Jones, *J. Chem. Soc. Chem. Commun.* 375 (1979).

[20] Z. Czochanska, L. Y. Foo, R. H. Newman, and L. Porter, *J. Chem. Soc., Perkin Trans. 1*, 2278 (1980).

[21] S. Guyot, C. Le Guerneve, N. Marnet, and J. F. Drilleau, in "Plant Polyphenols 2: Biogenesis, Chemical Properties, and Significance" (R. Hemingway, ed.). Plenum Press, New York, 2000.

[22] V. M. Williams, L. J. Porter, and R. W. Hemingway, *Phytochemistry* **22**, 569 (1982).

[23] C. Prieur, J. Rigaud, V. Cheynier, and M. Moutounet, *Phytochemistry* **36**, 781 (1994).

[24] Y. S. Bae, L. Y. Foo, and J. J. Karchesy, *Holzforschung* **48**, 4 (1994).

[25] A. Yanagida, T. Kanda, T. Shoji, M. Ohnishi Kameyama, and T. Nagata, *J. Chromatogr. A* **855**, 181 (1999).

present in a fraction or extract.[26] However, although, for a given compound, the signal intensity is proportional to the concentration, it depends on the capacity of the molecule to become ionized, which is influenced greatly by the structure. In particular, proanthocyanidins should be charged more easily as the degree of polymerization and/or number of galloyl substituents increase. In addition, as the molecular weight increases, several ions are formed from a single compound due to the occurrence of multiply charged ions. Therefore, the use of ESI-MS for quantitative purposes has not been attempted and would require a calibration curve to be established for each compound.

This article compares estimation of *mDPs* obtained using various methods on a series of fractions isolated from grape seeds and consisting of partially galloylated procyanidins.

Methods

Isolation of Procyanidin Fractions from Grape Seeds

Grapes seeds are obtained from berries of *Vitis vinifera,* var. Alicante Bouschet, harvested at commercial maturity. Ten grams of grape seeds is ground under liquid nitrogen, extracted with 60% acetone in water (100 ml), and centrifuged to remove plant debris. The acetone supernatant is taken to dryness under vacuum, dissolved in methanol (100 ml), and filtered to get a crude extract of procyanidins. Fractionation of this extract is achieved by normal-phase HPLC at the preparative scale using an ISA Jobin-Yvon system equipped with an axial compression column (240 × 40 mm) filled with Lichrospher Si60 phase (25–40 μm particle size, Merck, Darmstadt, Germany), a hydraulic compression module, and a manual injector and connected to an UV detector. The elution conditions are as follows: injected volume, 12 ml of the crude extract concentrated to 1.5 ml; flow rate, 24 ml $\cdot$ min^{-1}; column temperature 24°; solvent A: CH_2Cl_2–methanol–H_2O-trifluoroacetic acid (TFA) (10:86:2:0.005, v/v/v/v); solvent B: CH_2Cl_2–methanol–H_2O–TFA (82:18:2: 0.005, v/v/v/v); linear gradient from 0 to 40% A in 150 min, from 40 to 55% A in 15 min, and from 55 to 100% A in 15 min, followed by washing and reconditioning the column; detection, UV at 280 nm. Five fractions of 480 ml each are collected arbitrarily from 30 to 50 min (FI), from 60 to 80 min (FII), from 90 to 110 min (FIII), from 120 to 140 min (FIV) and from 150 to 170 min (FV). They are taken to dryness, dissolved in methanol, and repurified by chromatography on a Sephadex LH-20 column (Pharmacia, Sweden, 100 × 30 mm). After washing with 250 ml water, procyanidins are recovered from the column by elution with 60% acetone in water (200 ml), concentrated by rotary evaporation, and freeze-dried.

[26] V. Cheynier, T. Doco, H. Fulcrand, S. Guyot, E. Le Roux, J. M. Souquet, J. Rigaud, and M. Moutounet, *Analysis* **25** M32 (1997).

Standards

Catechin and epicatechin are from Merck (Darmstadt, Germany). Procyanidin dimers and trimers, epicatechin 3-gallate, phenylmethanethiol, and phloroglucinol derivatives of flavanols are purified in our laboratory as described earlier.[27]

Vanillin Assay

Fifteen microliters of a proanthocyanidin solution (1 g·liter^{-1} in methanol) is taken to dryness by rotary evaporation and dissolved in 2.5 ml of a solution containing 4% HCl and 0.5% vanillin in glacial acetic acid, prepared immediately before use. The absorbance is measured at 510 nm after 5 min of incubation. Under these conditions, the reaction can be used for specific determination of end groups.[6] Procyanidin B2 is used to establish standard curves.

DMCA Assay

Ten microliters of a proanthocyanidin solution (1 g·liter^{-1} in methanol) is taken to dryness by rotary evaporation and dissolved in 2.5 ml of a 0.1% solution of DMCA in methanol–12 N HCl (3 : 1, v/v). The absorbance is measured at 640 nm after 1 min of incubation. Catechin is used to establish the standard curves.

Depolymerization in Butanol-HCl (Porter's Reagent)

Twenty microliters of a proanthocyanidin solution (1 g·liter^{-1} in methanol) is taken to dryness by rotary evaporation, dissolved in 2.5 ml of a solution of butanol–12 N HCl (95 : 5, v/v), and added with 100 μl of $(NH_4)Fe(SO_4)_2$, 12 H_2O (2%, w/v in 2 N HCl). The test tubes are sealed with a Teflon-lined screw cap, mixed thoroughly, and heated for 30 min in a water bath at 95°. After cooling the solutions, the absorbance is measured at 550 nm. Procyanidin trimer C1 is used to establish the standard curve.

Depolymerization in Presence of Phenylmethanethiol (Thiolysis)

Fifty microliters of a proanthocyanidin solution (1 g·liter^{-1} in methanol) is put together with an equal volume of a 5% solution of phenylmethanethiol in methanol containing 0.2 N HCl into a glass ampoule. After sealing, the mixture is shaken and heated at 90° for 2 min. The solution is then analyzed immediately by reversed-phase HPLC using the following conditions: flow rate, 1 ml·min^{-1}; oven temperature, 30°; column, Spherisorb ODS-2 (250 × 4 mm); solvent A, 2.5% acetic acid in water; solvent B, acetonitrile–solvent A (80 : 20, v/v); linear gradient

---

[27] J. M. Ricardo da Silva, J. Rigaud, V. Cheynier, A. Cheminat, and M. Moutounet, *Phytochemistry* **30**, 1259 (1991).

from 5 to 50% B in 35 min and from 50 to 60% B in 5 min, followed by washing and reequilibrating the column; detection, UV at 280 nm. Calibration curves (based on peak areas) are established using flavanol and flavanol benzylthioether standards.

Depolymerization in Presence of Phloroglucinol

Fifty microliters of a proanthocyanidin solution (1 g·liter^{-1} in methanol) is put together with 25 μl of a pholoroglucinol solution (80 mg in 0.5 ml methanol) and 25 μl of 2% SO_2 in water into a glass ampoule. After sealing, the mixture is shaken and heated at 95° for 10 min. The solution is then analyzed immediately by reversed-phase HPLC using the following conditions: flow rate, 1 ml·min^{-1}; oven temperature, 30°; column, Spherisorb ODS-2 (250 × 4 mm); solvent A, 1.5% acetic acid in water; solvent B, acetonitrile–water (80 : 20, v/v); elution, isocratically with 90% A; detection, UV at 280 nm. Calibration curves (based on peak areas) are established using flavanol and phloroglucinol derivative standards.

Gel-Permeation Chromatography of Peracetate Derivatives

Acetylation is performed as described by Williams and co-workers[22] by incubating the tannin fractions in pyridine–acetic anhydride (1 : 1, v/v) at ambient temperature during 1 hr. The reagent in excess is destroyed by the addition of water, and the precipitate containing the acetylated molecules is recovered by centrifugation. After drying, the acetylated procyanidins are dissolved in tetrahydrofuran (THF) and analyzed directly by gel-permeation chromatography. The chromatographic system consists of two columns [TSK G2500 Hxl and G 3000 Hxl (particle size 6 μm, 7.8 × 300 mm each, TosoHaas, Philadelphia, PA)] connected in series. The elution is isocratic with THF at 1 ml·min^{-1} flow rate. Detection is at 254 nm. The GPC system is calibrated with 200 mg·liter^{-1} solutions of 1-liter polystyrene standards with molecular weights ranging from 162 to 50,000 in THF. The injection volume is 15 μl.

Results and Discussion

Calculation of Mean Degree of Polymerization (mDP) of Proanthocyanidin Extracts

The *mDP* of a given proanthocyanidin extract can be calculated from Eqs. (1)–(3).

$$mDP = [\text{tannin units}]_M/[\text{end units}]_M \tag{1}$$

which can also be written as

$$mDP = ([\text{upper and extension units}]_M + [\text{end units}]_M)/[\text{end units}]_M \tag{2}$$

or

$$mDP = [\text{tannin units}]_M / ([\text{tannin units}]_M - [\text{upper and extension units}]_M) \quad (3)$$

with all concentrations expressed on a molar basis.

The various methods described in this article give access to some of the terms of these three equations, as follows. Reactions with vanillin or DMCA allow determining the molar concentration of end units ($[\text{end units}]_M$); reaction with Porter's reagent allows determining $[\text{upper and extension units}]_M$, which theoretically equals the molar concentration of released cyanidin ($[\text{cya}]_M$); $[\text{tannin units}]_M$ can be calculated by gravimetry using Eq. (4).

$$[\text{tannin units}]_M = [\text{tannin units}]_W / mUMW \quad (4)$$

in which $mUMW$ is the mean unit molecular weight.

If the polymer is based on (epi)catechin units only, $mUMW$ can be calculated as $288 + 2/mDP$ or simply approximated to 288, as the term $2/mDP$ can be neglected. If the polymer contains gall% of galloylated units, mUMW = $288 + (\text{gall}\% \times 152) + 2/mDP$, or can be approximated to $288 + (\text{gall}\% \times 152)$.

Depolymerization in the presence of phloroglucinol or phenylmethanethiol gives access to $[\text{upper and extension units}]_M$, $[\text{end units}]_M$, and $[\text{tannin units}]_M$, as well as to $mUMW$ in a single analysis. Gel-permeation chromatography gives access to the mean molecular weight (mMW) of the acetylated derivatives, which is equal to $mDP \times 500$, if the polymer is based on (epi)catechin units, or to $mDP \times [500 + (\text{gall}\% \times 236)]$, if it contains gall% of galloylated moieties.

Results obtained on isolated procyanidins and procyanidin fractions from grape seeds by using these different methods are presented.

Depolymerization in Presence of Nucleophilic Agents

Depolymerization in the presence of nucleophilic agents such as phenylmethanethiol or phloroglucinol followed by HPLC analysis enables one to assay each constitutive unit individually and to also distinguish between end units, which are released in their native state, and all other units, which are found as derivatives. It thus gives access, after calibration of all released units, to the nature and average proportions of monomeric constituents of a given tannin or fraction and allows one to calculate its mDP using Eq. (2).

The values of mDP calculated from the results of thiolysis were 2.04, 2.08, and 3.12 respectively, for dimer B2, dimer B1 3-gallate, and trimer C1. However, in the case of B2 and C1, catechin was also detected among the thiolysis products due to epimerization of epicatechin in the thiolysis process. Although this does not interfere with the calculation of mDP, the occurrence of epimerization was checked by incubating each thiolysis product separately in the thiolysis reagent. After 2 min at 90°, 29% of epicatechin was converted to catechin, whereas only 4% of catechin was epimerized and epicatechin 3-gallate as well as all benzylthioethers

TABLE I
CHARCTERISTICS OF GRAPE SEED PROCYANIDIN FRACTIONS I TO V[a]

Characteristic	Fraction				
	I	II	III	IV	V
mDP (thiolysis)	2.3	3.6	5.4	7.8	15.1
% galloylation (thiolysis)	13	19	28	28	30
Reaction yield (thiolysis)	70	73	71	76	65
mDP (phloroglucinol)	2.5	4.8	7.7	12.2	19.1
% galloylation (phloroglucinol)	15	15	23	20	21
Reaction yield (phloroglucinol)	—	—	—	—	—

[a] Determined by depolymerization in the presence of phenylmethanethiol or phloroglucinol.

were stable. No epimerization occurred when thiolysis was carried out at room temperature for 24 hr.

The reaction yield, calculated as the ratio between the summed concentrations of all released units determined by HPLC and the initial procyanidin concentration, was 77% ($\pm 2\%$, $n = 6$) for procyanidin C1. The value of mDP, the proportion of galloylated units, and reaction yields obtained by thiolysis for fractions I to V are given in Table I. The yields calculated for fractions I to V were comprised between 65 and 76% and thus are similar to those measured for trimer C1 and to values reported in the literature,[17,19,28] meaning that all fractions were reasonably pure.

The values of mDP calculated after depolymerization in the presence of phloroglucinol were 2 and 2.81, respectively, for procyanidin dimer B2 and trimer C1. However, the phloroglucinol derivatives are highly unstable. Probably due to this high unstability, reaction yields comprised between 3 and 10% have been reported for this reaction,[29] which thus appears rather difficult to use.

Despite this difficulty, the values of mDP obtained by depolymerization in the presence of phloroglucinol were similar to those determined after thiolysis for all fractions (Table I). The proportions of galloylated units were also close to those estimated by thiolysis, except for the last two fractions.

Characterization of Acetylated Fractions Using Gel-Permeation Chromatography

The gel-permeation chromatography method was calibrated first with a range of 11 polystyrene standards (mMW from 162 to 50,000). The molecular weight

---

[28] A. Scalbert, in "Plant Polyphenols: Biogenesis, Chemical Properties, and Significance" (R. Hemingway, ed.), p. 259. Plenum Press, New York, 1992.
[29] S. Matthews, I. Mila, A. Scalbert, B. Pollet, C. Lapierre, C. L. M. Hervé du Penhoat, C. Rolando, and D. M. X. Donnelly, *J. Agric. Food Chem.* **45**, 1195 (1997).

TABLE II
CALIBRATION OF GPC METHOD WITH ACETYLATED
FLAVANOL MONOMERS AND PROCYANIDINS

mMW	Acetylated compound					
	Cat	Ec	B2	B2 3'G	C1	(Ec)$_4$
By GPC	503	510	974	1249	1472	1968
Exact	500	500	998	1234	1496	1994

can thus be related to the retention time following a negative regression of the form

$$\text{Retention time} = -0.24 \log mMW + 7.25 \quad (r^2 = 0.98)$$

Catechin, epicatechin, procyanidin dimers B2 [epicatechin-(4-8)-epicatechin] and B2 3'-gallate [epicatechin-(4-8)-epicatechin 3-gallate], procyanidin trimer C1 ([epicatechin-(4-8)-epicatechin-(4-8)-epicatechin], and procyanidin tetramer [epicatechin-(4-8)-epicatechin-(4-8)-epicatechin-(4-8)-epicatechin] were acetylated and injected onto the GPC system. The molecular weights determined by GPC analysis were very close to the actual molecular weights (500 for a fully acetylated monomer, 236 for an acetylated galloyl group), meaning that the regression equation established with the polystyrene standard was valid for acetylated procyanidins (Table II).

The molecular weights determined for each fraction were converted to mDP values, first assuming that all tannins in these fractions consisted of (epi)catechin units (Table III). Replacement of the arbitrary value of $mUMW$ (500) used in this calculation by the mean unit molecular weights determined from the percentage of galloylated units measured by depolymerization in the presence of either phenylmethanethiol, or phloroglucinol (cf. Table I), resulted only in a small decrease of the mDP values (Table III).

TABLE III
MOLECULAR WEIGHTS OF ACETYLATED FRACTIONS DETERMINED BY GPC[a]

Parameter	Fraction				
	I	II	III	IV	V
mMW (acetylated fractions)	1264	2312	3622	6458	9474
mDP ($mUMW = 500$)	2.5	4.6	7.2	12.9	18.9
mDP ($mUMW$ after thiolysis)	2.4	4.3	6.4	11.4	16.6
mDP ($mUMW$ after phloroglucinol)	2.4	4.3	6.5	11.8	17.3

[a] Values of mDP calculated by using 500 or 500 + (gall% × 236) as the mean acetylated unit molecular weight, with gall% as given in Table I.

TABLE IV
mDP Calculated for Fractions I to V from Results of Bate-Smith Reaction

Parameter	Fraction				
	I	II	III	IV	V
$E^{1\%}$	261	341	367	376	411
[cya] (mole/g)/yield (C1)	1.55×10^{-3}	2.08×10^{-3}	2.18×10^{-3}	2.23×10^{-3}	2.24×10^{-3}
mDP ($mUMW = 288$)	1.8	2.4	2.7	2.8	3.4
mDP ($mUMW$ by thiolysis)	1.9	2.8	3.6	3.8	5.4
mDP ($mUMW$ by phloroglucinol)	1.9	2.7	3.4	3.4	4.5

Depolymerization Using Porter's Reagent

Depolymerization using Porter's reagent releases cyanidin from upper and extension units of procyanidin oligomers and polymers. The molar concentration of upper and extension units thus theoretically equals that of released cyanidin (noted $[cya]_M$), which can be measured spectrophotometrically using the molar extinction coefficient of cyanidin (35,000). Values of mDP can thus be calculated from tannin weight and cyanidin molar concentration following Eq. (3).

Calibration with known solutions of procyanidin trimer C1 showed that the Beer–Lambert law was obeyed up to absorbance values of 0.6–0.7 ($Abs_{540} = 16,825[cya]_M + 0.0095$, $r^2 = 0.99$), as described in the literature.[28] The reaction yield was only 48%. However, this is slightly higher than the yields calculated from Porter's data on a series of oligomers and polymers (40 to 44%).[12]

The $E^{1\%}$ values increased from 260 to 413 from fraction I to fraction V (Table IV) and were thus of the same magnitude as those reported earlier.[12]

The mDP calculated by using Eq. (3) increased from fraction I to fraction V, but were lower than those determined using depolymerization in the presence of nucleophilic agents or GPC. Taking into account the percentage of gallates in the calculation of the mean unit molecular weight somewhat increased the values but they remained lower than those of calculated from the other methods. This suggests that the reaction yield is lower for all fractions than for procyanidin C1. In addition the yield seems constant (estimated at 35–40% in our series) as the molecular weight increases, but we noticed that a minor error on it has large consequences on the calculated mDP (especially for larger oligomers and polymers).

Vanillin Assay

Reaction with vanillin depends greatly on the solvent used, but has been reported to react specifically with proanthocyanidin end groups when performed in glacial acetic acid.[6] However, the molar extinction coefficients obtained under these conditions for flavanol monomers (9800 and 8500, respectively, for

TABLE V
CALCULATION OF mDP VALUES FROM RESULTS OF VANILLIN REACTION

Parameter	Fraction				
	I	II	III	IV	V
$E^{1\%}$	735	605	335	250	127
[end units](mole/g) ($\varepsilon_{B2} = 25200$)	2.9×10^{-3}	2.4×10^{-3}	1.0×10^{-3}	1.07×10^{-3}	0.5×10^{-3}
mDP (mUMW = 288)	1.2	1.4	2.6	3.5	6.9
mDP (mUMW by thiolysis)	1.1	1.3	2.3	3.0	6.0
mDP (mUMW by phloroglucinol)	1.1	1.3	2.3	3.2	6.2
mDP [Eq. (2)]	1.5	1.8	2.6	3.3	5.9

epicatechin and catechin) were lower than those obtained for procyanidin oligomers (11,200 and 12,300 for B2 and B3, 12,200 for C1), meaning that the use of catechin for standard curve results in the overestimation of tannin end units. The molar extinction coefficients that we determined for catechin and procyanidin B2 were 18,600 and 25,200, respectively, and thus approximately twice those published earlier.[6] The concentrations of end units were calculated for fractions I to V using the extinction coefficient determined for B2 (Table V). The values of *mDP* calculated using Eq. (1), with concentration of tannin units determined from the fraction weight, as described earlier, and Eq. (2), with extension and upper units determined with Porter's reagent, as given in Table IV, are reported in Table V.

The values obtained by both calculation methods were similar, but again lower than those obtained by depolymerization in the presence of nucleophilic agents and GPC, suggesting that the reaction is not specific of end units, at least for the larger molecular weight tannins. Replacement of 288 by the *mUMW* value estimated from the results of thiolysis and phloroglucinol analysis further decreased the *mDP* values calculated from vanillin and gravimetry data. The *mDP* calculations through Eq. (2) are based on both vanillin and Bate–Smith reactions: they do not require the knowledge of *mUMW* (which is dependent from the tannin composition such as galloylation extent).

DMCA Assay

The molar extinction coefficient at 640 nm determined for catechin after the DMCA reaction was 74,141 and thus was much higher than that of the vanillin reaction product, confirming the higher sensitivity of this reaction.[8]

The *mDP* values of fractions I to V were calculated as described earlier in the case of the vanillin reaction (Table VI). The absorbance values measured after the DMCA reaction decreased much less than those obtained by the vanillin

TABLE VI
CALCULATION OF mDP OF FRACTIONS I TO V USING DMCA REACTION AND WEIGHT[a]

	Fraction				
	I	II	III	IV	V
$E^{1\%}$	873	668	630	505	497
[end units](mole/g) ($\varepsilon_{cat} = 74141$)	1.18×10^{-3}	0.90×10^{-3}	0.85×10^{-3}	0.68×10^{-3}	0.67×10^{-3}
mDP ($mUMW = 288$)	3.0	3.9	4.1	5.1	5.2
mDP ($mUMW$ by thiolysis)	2.8	3.5	3.6	4.4	4.5
mDP ($mUMW$ by phloroglucinol)	2.7	3.6	3.6	4.6	4.7
mDP ([Eq. (2)])	2.3	3.3	3.6	4.3	4.7

[a] Or DMCA reaction and Porter's reagent.

reaction from fraction I to fraction V, indicating that it is less specific of end units. Consequently, the values of mDP thus determined showed little variation from fraction I to fraction V.

Conclusions

The values of mDP measured by all methods increased from fraction I to fraction V, but the values obtained by spectrophotometric methods were lower than those given by depolymerization in the presence of nucleophilic agents and GPC, especially for the highly polymerized fractions. In contrast, the values were quite consistent within each group of methods. In particular, results yielded by GPC, thiolysis, and depolymerization in the presence of phloroglucinol were very close for all fractions, although those determined by thiolysis were slightly lower. In addition, mDP values obtained by thiolysis have been shown earlier to be in good agreement with those determined by mass spectrometry[30] and by NMR.[21]

Although the procedures based on spectrophotometric methods are much faster than those involving an HPLC separation, they suffer from the following biases. Methods based on gravimetric determination do not allow accurate estimation of tannin molar concentration unless the nature of tannins is perfectly well known (e.g., nonsubstituted procyanidins). The yield of depolymerization with Porter's reagent is rather low (around 40%) and seems to be influenced by the presence of substituents (galloyl substituents in grape seed procyanidins). Because all anthocyanidins have different extinction coefficients at a given wavelength, the proportions of the various tannins units within a given polymer should be taken into account when converting the absorbance values to molar concentrations of upper

[30] E. Le Roux, T. Doco, P. Sarni Manchado, Y. Lozano, and V. Cheynier, *Phytochemistry* **48**, 1251 (1998).

and extension units, in the case of proanthocyanidins other than procyanidins. Finally, reactions with DMCA and, to a lesser extent, with vanillin do not seem to be specific of end units, especially in the case of higher polymer weight tannins.

In contrast, depolymerization in the presence of nucleophilic agents appears well suited for the determination of *mDP*, provided that all released units are properly identified and calibrated. It is worth noting that thiolysis sometimes failed to provide quantitative yields and that a number of by-products were identified in the reaction mixture.[31] However, this was probably due to the rather drastic conditions used (24 hr at 105°), as, under our milder conditions, 70% reaction yields were obtained and all peaks detected on the chromatogram could be attributed to the expected reaction products. Moreover, these techniques also give access to the nature and proportion of tannin constitutive units, which represent important information when studying structure–activity relationships. Finally, thiolysis can also be used for quantitation purposes, assuming a constant reaction yield.

Gel-permeation chromatography provides satisfactory estimations of molecular weight and should give access to polymer size distribution when applied to a proanthocyanidin mixture. However, conversion of these values to *mDP* values requires prior determination of the tannin composition (B-ring hydroxylation pattern, substituents). On addition, the procedure described herein requires a derivatization process, which is time-consuming and does not allow to recover the molecules for further use. Other GPC methods allowing one to separate procyanidins in their native state have been proposed.[24,25] However, a serious shortcoming common to all these procedures is the lack of high molecular weight proanthocyanidin standards to calibrate the separation and of easy mass detection methods. Coupling of normal-phase HPLC with ESI-MS has permitted the detection of epicatechin-based procyanidins from the dimer to the decamer in a cocoa extract.[32] However, as discussed in the introduction, ESI-MS does not provide quantitative data, unless a calibration curve can be established for each compound. Application of this technique should also prove very difficult in more complex proanthocyanidin mixtures, such as grape seed or wine extracts.

---

[31] G. W. Mc Graw, J. P. Steynberg, and R. W. Hemingway, *Tetrahedron Lett.* **34,** 987 (1993).
[32] J. F. Hammerstone, S. A. Lazarus, A. E. Mitchell, R. Rucker, and H. H. Schmitz, *J. Agric. Food Chem.* **47,** 490 (1999).

Section II

Bioavailability

[8] Determination of Flavonols in Body Fluids

By PETER C. H. HOLLMAN

Introduction

Flavonoids are polyphenolic compounds that occur ubiquitously in foods of plant origin. Flavonoids are categorized into flavonols, flavones, catechins, flavanones, anthocyanidins, and isoflavonoids. Potent biological effects have been described in many *in vivo* and *in vitro* studies.[1] Quercetin, the major representative of the flavonol subclass of flavonoids (Fig. 1), was inversely associated with subsequent coronary heart disease in most, but not all, observational studies.[2,3] The antioxidant properties of flavonoids[4] offer a plausible explanation for the effect found.[5] However, pharmacokinetic and pharmacodynamic studies are needed to investigate the fate of flavonoids in the body after their ingestion with the diet. This will allow unraveling of the potential role of flavonoids in the prevention of coronary heart disease.

For these studies, flavonols have to be determined in body fluids. High-performance liquid chromatography (HPLC) methods with UV detection[6] do not meet the requirements of sensitivity (limit of detection about 1 ng/ml in plasma) and specificity in these biological fluids. Thus, extensive sample enrichment and cleanup procedures would have to be considered. However, the enrichment attainable would probably not be sufficient for analyses in plasma. Consequently, detection techniques with enhanced sensitivity are needed. Electrochemical detection (coulometry) showed to offer sufficient sensitivity (limit of detection 0.5 ng/ml) for the detection of quercetin in plasma.[7] Fluorometry is another detection technique that could fulfill the requirements of sensitivity and specificity needed for body fluids.

---

[1] E. Middleton and C. Kandaswami, *in* "The Flavonoids: Advances in Research since 1986" (J. B. Harborne, ed.), p. 619. Chapman & Hall, London, 1994.
[2] P. C. H. Hollman and M. B. Katan, *Free Radic. Res.* **31,** S75 (1999).
[3] L. Yochum, L. H. Kushi, K. Meyer, and A. R. Folsom, *Am. J. Epidemiol.* **149,** 943 (1999).
[4] W. Bors, W. Heller, C. Michel, and M. Saran, *Methods. Enzymol.* **186,** 43 (1990).
[5] D. Steinberg, *N. Engl. J. Med.* **328,** 1487 (1993).
[6] M. G. L. Hertog, P. C. H. Hollman, and D. P. Venema, *J. Agric. Food Chem.* **40,** 1591 (1992).
[7] I. Erlund, G. Alfthan, H. Siren, K. Ariniemi, and A. Aro, *J. Chromatogr. B* **727,** 179 (1999).

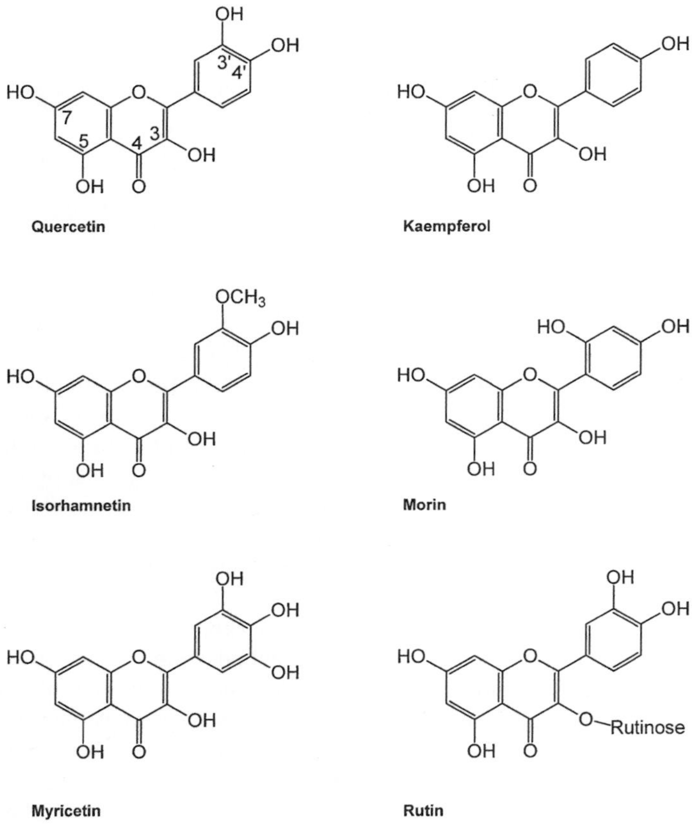

FIG. 1. Structures of flavonols.

We have developed a postcolumn derivatization technique for fluorescence detection of flavonols in HPLC.[8] This very sensitive method has been applied in a number of pharmacokinetic and absorption studies of flavonols in humans.[9–12]

---

[8] P. C. H. Hollman, J. M. P. van Trijp, and M. N. C. P. Buysman, *Anal. Chem.* **68,** 3511 (1996).
[9] P. C. H. Hollman, M. N. C. P. Buysman, Y. van Gameren, P. J. Cnossen, J. H. M. de Vries, and M. B. Katan, *Free Radic. Res.* **31,** 569 (1999).
[10] J. H. M. de Vries, P. C. H. Hollman, S. Meyboom, M. N. C. P. Buysman, P. L. Zock, W. A. van Staveren, and M. B. Katan, *Am. J. Clin. Nutr.* **68,** 60 (1998).
[11] P. C. H. Hollman, J. M. P. van Trijp, M. N. C. P. Buysman, M. S. van der Gaag, M. J. B. Mengelers, J. H. M. de Vries, and M. B. Katan, *FEBS Lett.* **418,** 152 (1997).
[12] P. C. H. Hollman, J. H. M. de Vries, S. D. van Leeuwen, M. J. B. Mengelers, and M. B. Katan, *Am. J. Clin. Nutr.* **62,** 1276 (1995).

Methods and Materials

Chemicals

All chemicals are reagent grade or HPLC grade.

Extraction of Plasma

Flavonols and their conjugates (glucuronides and sulfates) are extracted from plasma and simultaneously hydrolyzed to the aglycone form using 2 M HCl in aqueous methanol. For plasma, 1.00 ml of methanol containing 2 g/liter *tert*-butylhydroxyquinone (TBHQ) and 0.40 ml of 10 M HCl are added to 0.60 ml of plasma in a 4-ml vial (Waters, Milford, MA) and mixed. The vial is sealed tightly with a cap (Waters, Milford, MA) and septum (Waters, Milford, MA), inserted into a preheated aluminum block (Reacti-Block C-1, Pierce Europe, Oud-Beijerland, The Netherlands), heated in an oven at 90° for 5 hr, allowed to cool, and centrifuged at 1000g for 15 min after the addition of 2.00 ml of methanol containing 2 g/liter *tert*-butylhydroxyquinone (TBHQ). A volume of 1.5 ml of the upper phase is transferred into an HPLC vial, and 15 μl 100 g/liter ascorbic acid is added.

Extraction of Urine

Flavonol conjugates (glucuronides and sulfates) are extracted from urine and simultaneously hydrolyzed to the aglycone form using 2 M HCl in aqueous methanol. For urine, 15.0 ml methanol containing 2 g/liter TBHQ and 5 ml 10 M HCl are added to 5 g urine followed by mixing, the extract is refluxed at 90° for 8 hr with regular swirling, allowed to cool, and subsequently brought to a final volume of 50 ml with methanol containing 2 g/liter ascorbic acid. Urine extracts are sonicated for 5 min and filtered through a 0.45-μm filter for organic solvents (Acrodisc CR PTFE; Gelman Sciences, Ann Arbor, MI) before HPLC analysis.

Chromatography

For HPLC analysis, 20 μl of the sample is injected onto an Inertsil ODS-2 column (4.6 × 150 mm, 5 μm; GL Sciences Inc., Tokyo, Japan) protected by an MPLC Newguard RP-18 column (3.2 × 15 mm, 7 μm; Brownlee, Applied Biosystems Inc., Foster City, CA) using acetonitrile/0.025 M phosphate buffer, pH 2.4, 31 : 69 (v/v) as the mobile phase, at a flow rate of 1 ml/min using a Merck Hitachi L-6200 A pump (Hitachi Ltd., Tokyo, Japan). The columns are placed in a column oven set at 30°. The column effluent is mixed with 0.4 ml/min 1.5 M Al(NO$_3$)$_3$ in methanol containing 7.2% (v/v) acetic acid in a postcolumn reactor placed in a column oven. The postcolumn reactor consists of a 15-m (0.25 mm internal diameter) Teflon PTFE tubing (Upchurch Scientific Inc., Oak Harbor, WA) coiled to a diameter of 3 cm and connected to the HPLC column with a low dead volume tee (Upchurch Scientific Inc., Oak Harbor, WA). The column effluent and aluminum reagent enter the tee countercurrently at an angle of 180°.

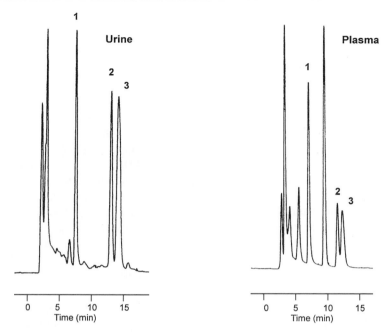

FIG. 2. Chromatograms of urine and plasma extracts: quercetin (1), kaempferol (2), and isorhamnetin (3).

A Merck Hitachi L-6000 A pump (Hitachi Ltd.) is used to generate the aluminum reagent. The fluorescence of the ensuing quercetin–metal complex is measured at 485 nm using a Jasco FP 920 (Jasco Corporation, Japan) fluorescence detector with an excitation wavelength set at 422 nm. The detector output is sampled using a Nelson (PE Nelson, Cupertino, CA) series 900 interface and Nelson integrator software (Model 2600, rev. 5), and the peak area of the quercetin–metal complex is determined.

Results

The high specificity of fluorescence detection produced "clean" chromatograms of plasma and urine extracts without interfering peaks (Fig. 2). Consequently, no sample cleanup was necessary.

Precision, Recovery, and Detection Limits

The relative standard deviation of duplicate analyses of quercetin was 4% for plasma and 6% for urine. The reproducibility of the quercetin assay was determined by duplicate analyses of 15 identical samples of plasma and of urine on 15 different days. The between series relative standard deviation of these quercetin determinations was 10% for plasma and 9% for urine.

Recovery of 100 ng quercetin aglycone added to 1 ml plasma was 88 ± 3% (six additions). Addition of 250 ng quercetin aglycone per gram of urine yielded a recovery of 99 ± 7% (three additions).

The limit of detection, i.e., the concentration producing a peak height three times the standard deviation of the baseline noise, was 2 ng/ml for quercetin in plasma and 3 ng/ml for quercetin in urine. The limits of detection of standard solutions were 0.15 ng/ml for quercetin, 0.05 ng/ml for kaempferol, 0.45 ng/ml for myricetin, and 0.05 ng/ml for isorhamnetin. The calibration graph for quercetin was linear up to 5000 ng/ml, showing an extended dynamic range.

Postcolumn Reaction Conditions

The fluorescence intensity of the various Al^{3+} complexes varied considerably (Table I). Although $Al(NO_3)_3$ concentrations in excess of 1.5 M were expected to increase the detector response, the high viscosity of the reagent would require unrealistically high pressures.[8] We observed that the addition of acetic acid to the aluminum reagent decreased the risk of clogging of the coil. The highest response was obtained at 34°, although this maximum was rather broad. At temperatures higher than 30°, the detector signal was less stable as spikes were observed frequently. An increase of the water fractions of methanol, as well as acetonitrile-based eluents, caused a dramatic drop in detector response.[8] It was shown previously[6] that a phosphate buffer (pH 2.4) improved the peak shape of flavonols in the reversed-phase chromatographic system used. This phosphate buffer also enhanced the detector response.

TABLE I
WAVELENGTHS OF EXCITATION AND EMISSION MAXIMA AND CORRESPONDING RELATIVE FLUORESCENCE INTENSITY OF VARIOUS FLAVONOLS

Flavonol	Maxima		Relative fluorescence intensity (%)
	λExcitation (nm)	λEmission (nm)	
Morin	418	490	100
Kaempferol	420	470	56
Isorhamnetin	430	480	54
Quercetin	430	480	21
Myricetin	428	500	7
Rutin	—[a]	—[a]	0

[a] No fluorescence could be observed.

Discussion

Fluorescence detection of flavonols after chelation with Al^{3+} is adequate to determine flavonols in body fluids. With the optimized postcolumn HPLC system, a limit of detection of 2 ng/ml can be achieved for quercetin in plasma. Due to the selectivity and sensitivity of this method, sample pretreatment is simple without the need for sample cleanup.

Structural Requirements for Complexation

Metal ions may bind to the 3-hydroxyl and 4-keto oxygen and to the 5-hydroxyl and 4-keto oxygen (Fig. 1). Two adjacent hydroxyls, such as the 3'- and 4'-hydroxyls in quercetin, may also be involved in chelation, but at low pH this complex is labile.[13] It was shown[8] in this postcolumn system that only flavonoids that contain both a 3-hydroxyl and a 4-keto oxygen had sufficient fluorescence intensity. Flavonoids without a 3-hydroxyl group, namely flavones and catechins, escape fluorescence detection. A free 3-hydroxyl group is essential as rutin, which contains a sugar bound to the 3-hydroxyl group, does not form a fluorescent chelate either (Table I).

Extraction of Flavonols from Plasma and Urine

Flavonols are present in plasma and urine either as glucuronides or as sulfates.[14] Glucuronic acid or sulfate may be bound to the 3-hydroxyl group, and thus prevent formation of the fluorescent aluminum complex. Deconjugation is required for detection as a fluorescent complex. Enzymatic methods using β-glucuronidase/sulfatase have been described for flavonols in plasma.[7] Acid hydrolysis with HCl offers a rapid alternative.[6] However, the long period of 5 to 8 hr needed to liberate quercetin completely from plasma and urine came as a surprise. Sulfates are hydrolyzed easily,[13] and glucuronides of quercetin in foods were completely hydrolyzed within 2 hr.[6] Binding of quercetin to proteins in plasma might explain this delay in extraction. Manach et al.[15] reported that circulating quercetin and added aglycone bind to plasma albumin. However, the identity of the constituents in urine that bind quercetin or quercetin conjugates remains to be clarified.

Gryglewski[16] found that quercetin is bound selectively to rabbit platelets *in vitro*. In our study, quercetin concentrations did not differ between platelet-rich and platelet-poor plasma (results not shown). Thus, we could not confirm that quercetin is bound to platelets after absorption from foods.

---

[13] K. R. Markham, "Techniques of Flavonoid Identification." Academic Press, London, 1982.

[14] C. Manach, O. Texier, F. Régérat, G. Agullo, C. Demigné, and C. Rémésy, *Nutr. Biochem.* **7,** 375 (1996).

[15] C. Manach, C. Morand, O. Texier, M.-L. Favier, G. Agullo, C. Demigné, F. Régérat, and C. Rémésy, *J. Nutr.* **125,** 1911 1995).

[16] R. Gryglewski, *Biochem. Pharmacol.* **36,** 317 (1987).

This postcolumn HPLC system has been used extensively for the determination of quercetin in body fluids.[9–12,17–19] Because of its high sensitivity and selectivity, minimal sample preparation is required, and 40 samples a day can be analyzed easily.

Acknowledgments

We thank Michel Buijsman and John van Trijp for excellent technical assistance. This work was supported by grants from the Foundation for Nutrition and Health Research and the Netherlands Heart Foundation (94.128).

[17] M. R. Olthof, P. C. H. Hollman, and M. B. Katan, *J. Nutr.* **130,** 1200 (2000).
[18] A. A. Aziz, C. A. Edwards, M. E. J. Lean, and A. Crozier, *Free Radic. Res.* **29,** 257 (1998).
[19] P. C. H. Hollman, M. S. van der Gaag, M. J. B. Mengelers, J. M. P. van Trijp, J. H. M. de Vries, and M. B. Katan, *Free Radic. Biol. Med.* **21,** 703 (1996).

[9] Determination of Quantity and Quality of Polyphenol Antioxidants in Foods and Beverages

By Joe A. Vinson, John Proch, and Pratima Bose

Introduction

More than 4000 phenol and polyphenol compounds have been identified in vascular plants.[1] Each year there are more than 1000 citations for these types of compounds. They are of interest primarily because of their ability to act as antioxidants, and for the fact that several epidemiological studies have linked their consumption in foods and beverages to a decreased risk of heart disease.[2–4] Two foods in particular, tea (especially green tea) and red wine, have been widely publicized in the lay press for their heart benefits. There is now a very large effort by many research groups in the private and public sector to identify these compounds in foods and beverages commonly consumed, and in those extracts used in the

[1] E. Middleton and C. Kandaswami, *in* "The Flavonoids: Advances in Research since 1986" (J. B. Harborn, ed.), p. 619. Chapman and Hall, London, 1994.
[2] M. G. L. Hertog, P. C. H. Hollman, and M. B. Katan, *Lancet* **342,** 1007 (1993).
[3] M. G. L. Hertog, D. Kromhout, C. Aravanis, H. Blackburn, R. Buzina, F. Fidanza, S. Giampaoli, A. Jansen, A. Menotti, S. Nedeljkovic, M. Pekkarinen, S. S. Bozida, H. Toshima, E. J. M. Feskens, P. C. H. Hollman, and M. B. Katan, *Ann. Intern. Med.* **155,** 381 (1995).
[4] P. Knekt, R. Järvinen, A. Reunanen, and J. Maatela, *Br. Med. J.* **312,** 478 (1996).

rapidly growing supplement and nutraceutical industry. Most of the published methods for food polyphenol analysis involve multiple and expensive reagents and instruments that are not widely available and do not determine separately the quantity and quality of the polyphenol antioxidants. This article describes some methods for determining these parameters using models relevant to heart disease.

Quantity of Polyphenols

Two methods have generated extensive data on the antioxidant activity of polyphenols alone or in foods: the Trolox equivalent antioxidant capacity (TEAC)[5] method available commercially from Randox Laboratories (San Diego, CA) and the automated oxygen radical absorbance capacity (ORAC) method.[6] In these methods, a free radical species is generated and there is a means of detecting its concentration directly by colorimetry (TEAC) or by its chemical damage to a target molecule by fluorescence (ORAC). The change in concentration is monitored with time in the ORAC method or in the TEAC method at a fixed time and compared with a standard, Trolox, which is a water-soluble form of vitamin E, and a blank. Both methods measure the quantity and quality of antioxidants in the water-soluble extracts of foods or in beverages, and the activity of the sample is designated by the millimole or micromole of Trolox equivalents per gram or liter. The TEAC procedure has been improved.[7] To obtain "total" antioxidants, the ORAC method requires the separate measurement of both aqueous and acetone extracts. Both procedures have been used to measure "free" polyphenols in foods and beverages. As a result they underestimate the concentration because in unprocessed foods and beverages there can be considerable phenolic groups conjugated to sugars. These groups can be hydrolyzed in the digestive tract prior to absorption in the blood and thus can contribute to the total antioxidant potential supplied by the food or beverage.

A simple method has been developed to determine free and total phenols and polyphenols in foods and beverages that uses the Folin–Ciocalteu reagent and colorimetry. The use of the Folin reagent in an alkali medium for the reaction has been reviewed.[8] The Folin method is an excellent colorimetric method as it operates by an oxidation–reduction mechanism and thus measures antioxidants. However, using this alkaline method it takes more than 2 hr to perform the two-step analysis and considerable time to prepare the reagent. We have developed a simpler, faster Folin procedure using a stable, commercially available reagent that contains phosphoric acid and is done under acidic conditions in which polyphenols are more stable.

[5] C. A. Rice-Evans and N. J. Miller, *Br. J. Food* **97**, 35 (1995).
[6] G. Cao and R. Prior, *Methods Enzymol.* **299**, 50 (1999).
[7] R. Roberta, N. Pellerini, A. Proteggente, A. Pannala, M. Yang, and C. Rice-Evans. *Free Radic. Biol. Med.* **26**, 1231 (1999).
[8] V. A. Singleton, R. Orthofer, and R. M. Lamuela-Raventós., *Methods. Enzymol.* **299**, 152 (1999).

Measuring Quantity of Polyphenols in Foods and Beverages

Sample Preparation

Edible portions of fruits or vegetables are weighed, cut into small portions, blended with liquid nitrogen, and then freeze dried overnight. Oils (10-g samples mixed with 10 ml 60% methanol/water) are extracted several times with 10 ml volumes of hexane to remove fats and lipophilic vitamins such as carotenoids and tocopherols. Previously freeze-dried chocolate samples are defatted by multiple extraction with 5 ml of hexane. Duplicate weighed portions (50 to 500 mg) of food, or an aliquot of the beverage, are put in plastic screw-capped test tubes and the polyphenols are extracted for 2 hr with 8 ml of 50% methanol/water at 95° with vortexing every 30 min. [*Note:* All water used in this and the following procedures has been distilled and passed through both carbon and ion-exchange columns in a commercial apparatus Nanopure II, Barnsted/Thermolyne, Dubuque, IA. Alternatively, HPLC grade water (Fisher Scientific Co., Pittsburgh, PA) can be used]. After low speed centrifugation for 10 min at room temperature this extract is diluted to 10 ml with methanol and is used for the measurement of free polyphenols. Total phenols are determined by extraction and hydrolysis of the weighed sample with 8 ml of 1.2 M hydrochloric acid in 50% (v/v) methanol/water, and treated as described previously. The extracts can be stored at $-20°$ after deaerating the solution with nitrogen.

Analysis

For the Folin analysis a 1000 μM solution of catechin is used as the standard. An aliquot of standard, blank, or the extract (up to 100 μl) is added to 1000 μl of the Folin–Ciocalteu reagent (Sigma Chemical Company, St. Louis, MO), which has been diluted previously 1 : 9 with water (stored in plastic bottles). The solution is then allowed to stand at 20–25° for 20 min before colorimetric measurement at 750 nm.

Interferences

Sulfite, present in wines and some dried fruits, reacts with the Folin reagent to give a color. The sulfite can be eliminated by adjusting 1 ml of the extract to pH 3 with 1 ml of an acetate buffer, adding 20 μl of acetaldehyde (99% reagent grade), and holding the sample for 30 min at room temperature before Folin analysis.

Ascorbic acid reacts positively with Folin with the same sensitivity as catechin in the acidic medium. It is not always possible to completely remove the interference by adding ascorbate oxidase in the food extract matrix. This may be due to the production of the oxidant hydrogen peroxide that reacts with the Folin reagent. Therefore the best method is to measure ascorbate concentration directly by high-performance liquid chromatography (HPLC), and then to subtract the ascorbate concentration from the total polyphenol concentration. Weigh out 50 mg of the freeze dried food and dissolve the ascorbate in the sample in 8 ml of 5% metaphosphoric acid in 50% methanol/water and incubate in a screw-capped test tube for 2 hr at 95°. Centrifuge for 10 min at room temperature. Then dilute the centrifugate to 10 ml with Nanopure water and place on the column. A 25-cm reversed phase 10-μm C_{18} column can be used with an isocratic solvent 90% (4% acetic acid in water) and 10% methanol at a flow rate of 2 ml/min. Ascorbate can be detected at 254 nm with a retention time of 2.7 min.

Sugars do not interfere with the acidic Folin reaction at concentrations found in high sugar foods such as chocolate,[9] but they interfere in the alkaline Folin method if present in high concentrations.[8] Any food extract can be tested for possible sugar interference by passing 1 ml of a <1000 μM sample and ≥90% water/methanol solution through 50 mg of a polyvinylpolypyrrolidone (Sigma) column (1 × 1 cm) and testing the eluate, which contains no polyphenols, for Folin activity. Alternatively, interferences can be removed using a more expensive Sephadex LH-20 column.[8]

Results

Using vegetables as an example of the method, 23 commonly consumed vegetables were assayed in duplicate for both free and total polyphenols. The average intraassay precision was 7.6% for free phenols and 8.6% for total phenols.[10] These extracts can also be used to identify and quantify the individual polyphenols in their free form, which is much simpler than in their glycosylated form where many different sugar links are possible. Table I shows results for free and total polyphenols in some representative foods and beverages. The Folin method may also be used to estimate the average daily per capita amount of polyphenols consumed in indivdual foods or food groups by assuming an average molecular weight of 290 (the anhydrous molecular weight of the catechin standard) and using government consumption data. This has been done for vegetables.[10] These calculations should be useful in epidemiological studies examining polyphenols in food and their effect on health and disease.

The Folin colorimetric method, which is simpler to perform, can be compared with the ORAC method, which, although automated, requires an expensive fluorometric centrifugal analyzer. The free polyphenols found in 16 vegetables using the Folin procedure were correlated with the ORAC[13] values using 2,2′-azobis(2-amindinopropane) dihydrochloride (AAPH), a peroxyl (ROO·) radical

[9] J. A. Vinson, L. Zubik, and J. proch, *J. Agric. Food Chem.* **47,** 4821 (1999).
[10] J. A. Vinson, Y. Hao, X. Su, and L. Zubik, *J. Agric. Food Chem.* **46,** 3630 (1998).
[11] J. A. Vinson, *in* "Flavonoids in the Living System" (E. Manthey and B. Buslig, eds.), p. 151. Plenum Press, New York, 1998.
[12] J. A. Vinson and B. A. Hontz, *J. Agric. Food Chem.* **43,** 401 (1995).
[13] G. Cao, E. Sofic, and R. Prior, *J. Agric. Food Chem.* **44,** 3426 (1996).

TABLE I
POLYPHENOL ANALYSIS IN FOODS AND BEVERAGES

Foodstuff	Free polyphenols[a]	Total polyphenols[a]	Refs.
Beets ($n = 2$)	45.2 ± 0.4	53.4 ± 7.6	[b]
Yellow onions ($n = 2$)	4.7 ± 3.8	22.9 ± 1.4	[b]
Kidney beans ($n = 2$)	31.9 ± 5.6	35.9 ± 8.2	[b]
Tomatoes ($n = 2$)	9.5 ± 7.0	18.9 ± 11.7	[b]
Oranges ($n = 2$)	8.1 ± 4.6	18.9 ± 10.7	[c]
Apples ($n = 2$)	16.5 ± 2.3	34.1 ± 4.8	[c]
Orange juices ($n = 5$)	0.77 ± 0.18	n.d.	[c]
Black teas ($n = 4$)	21.1 ± 6.8	n.d.	[c]
Red wines ($n = 5$)	6.74 ± 3.04	n.d.	[d]
Olive oils ($n = 2$)	0.12 ± 0.01	0.18 ± 0.00	[e]
Cinnamon ($n = 2$)	n.d.	408 ± 86	[f]
Milk chocolate ($n = 5$)	52.2 ± 20.4	52.2 ± 20.4	[g]

[a]Polyphenols are expressed as the mean ± SD in μmol of catechin equivalents per gram of dry weight food or per ml of beverage. n.d., not determined.
[b]From Vinson et al.[10]
[c]From Vinson.[11]
[d]From Vinson and Hontz.[12]
[e]From Vinson and Acharya, unpublished results, 1999.
[f]From Vinson and Adnan, unpublished results, 1998.
[g]From Vinson et al.[9]

generator (Fig. 1). The peroxyl radical is considered to be a biologically relevant oxidant. The linear equation is ORAC = 2.15 (Folin μmol catechin/g) + 15.1, with a Pearson correlation coefficient of 0.736, $p = 0.001$. The ORAC vegetable and fruit values give a significant positive linear correlation with electrochemical data obtained by coulometric array.[6] Thus the Folin assay is well correlated with both redox and kinetic antioxidant properties.

Quality of Polyphenols and Other Antioxidants

In addition to the quantity of polyphenol antioxidants in foods and beverages, the quality of the antioxidants is very important. Absorption studies done to date with such foods as apples,[14] onions,[14] tea,[15] wine,[16] and chocolate[17] have shown

[14] P. C. Hollman, J. M. van Trijp, M. N. Buysman, M. S. Van der Gaag, M. J. Mengelers, J. H. De Vries, and M. B. Katan, *FEBS Lett.* **418**, 152 (1997).
[15] C. S. Yang, L. Chen, M. J. Lee, D. Balentine, M. C. Kuo, and S. P. Schantz, *Cancer Epidemilol. Biomarkers Prev.* **7**, 351 (1998).
[16] J. L. Donovan, J. R. Bell, S. Kasim-Karakas, J. B. German, R. L. Walzem, R. J. Hansen, and A. L. Waterhouse, *J. Nutr.* **129**, 1662 (1999).
[17] M. Richelle, I. Tavazzi, M. Enslen, and E. A. Offord, *Eur. J. Clin. Nutr.* **53**, 22 (1999).

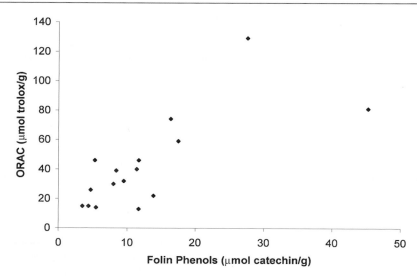

FIG. 1. Correlation of Folin assay of 16 polyphenols in vegetable extracts with oxygen radical antioxidant capacity (ORAC) against peroxyl radicals. Both assays are expressed on a dry matter basis.

that the maximum human plasma concentration of individual polyphenols is approximately 1 μM. Therefore, to be effective antioxidants *in vivo,* the polyphenols must have a high quality, i.e., be strong antioxidants. TEAC and ORAC methods are suitable for determining the quality of antioxidants with the caveat that the antioxidants are pure compounds. These methods use a nonbiological target molecule and determine the concentration of Trolox with the equivalent antioxidant potential to 1 mM solution of the compound analyzed.

Lipoprotein Oxidation Methods

The method described herein uses a biologically relevant target, the lower density lipoproteins LDL (low density lipoprotein) and VLDL (lower density lipoprotein). LDL and VLDL become atherogenic when oxidized *in vivo* below the surface of the artery.[18] LDL or other targets such as linoleic acid have often been used for *in vitro* comparison of antioxidants at one or several concentrations. Measurement after a fixed time is performed on thiobarbituric acid-reactive substances or alkanals such as hexanal.[19] This kind of comparison is fast but not entirely valid, as the inhibition of lipoprotein oxidation is a sigmoidal curve

[18] D. Steinberg, S. Parathasarathy, T. E. Carew, J. C. Khoo, and J. L. Witzum, *N. Engl. J. Med.* **320,** 914 (1989).
[19] E. N. Frankel, *Methods Enymol.* **299,** 190 (1999).

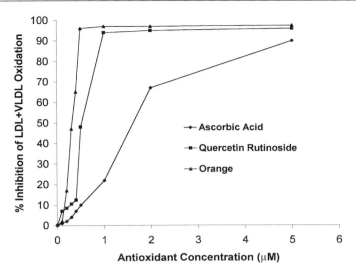

FIG. 2. Determination of the quality of antioxidant components of oranges and an orange extract by measurement of the inhibition of LDL + VLDL oxidation induced by cupric ions.

(see Fig. 2). Thus with a low or high percentage inhibition, the quality of the compounds or extracts is difficult to differentiate because the slope of concentration vs inhibition is low. The most accurate method for comparison of antioxidants is to perform a dose–response experiment to get several points on the steep slope near the middle of the sigmoidal plot and to calculate this 50% inhibition concentration value.

LDL is most often isolated by a lengthy ultracentrifugation in the presence of EDTA to prevent oxidation, followed by a slow dialysis to remove EDTA. Evidence shows that ultracentrifugation changes the physical properties of the lipoproteins because of the high artificial gravity produced. This oxidizes the lipoproteins, thereby decreasing the amount of the endogenous vitamin E bound to the lipoprotein.[20] An alternative procedure used a dextran–magnesium ion reagent to precipitate LDL + VLDL, which is then washed several times, followed by centrifugation, and finally resuspended prior to oxidation.[21]

The following method uses an affinity column for isolation of LDL + VLDL suitable for oxidation. This has been used for the determination of the quality of pure antioxidants and polyphenols,[22] wines,[12] vegetables,[10] chocolates,[9] and

---

[20] L. M. Scheek, S. A. Wiseman, L. B. M. Tijburg, and A. van Tol, *Atherosclerosis* **117**, 139 (1995).
[21] A. Zhang, J. Vertommen, L. Van Gaal, and I De Leeuw, *Clin. Chim. Acta* **227**, 159 (1994).
[22] J. A. Vinson, Y. A. Dabbagh, M. M. Serry, and J. Jang, *J. Agric. Food Chem.* **43**, 2800 (1995).

some fruits and beverages,[11] including teas and tea extracts.[23] The method has been critically evaluated.[24]

Procedure for Analysis of Quality of Antioxidants

A heparin-agarose affinity resin (Sigma Chemical Company) is used to adsorb the lipoproteins that are separated and desorbed by saline solutions. The reusable column (1 × 1.5 cm) has a porous plastic disk at the bottom and another placed above 1 ml of the resin. The upper disk is pushed down until it just contacts the top of the resin bed. The column is allowed to drain until the flow stops. Then 1000 μl of α-eluant (0.7% NaCl in Nanopure water) is added to equilibrate the column and the eluate is discarded.

With the column positioned over a 10-ml screw-capped plastic tube in an ice bath, 200 μl of human plasma (isolated in heparin or EDTA) is added directly onto the top disk and allowed to flow into the column. Then 50 μl of α-eluant is added. After 5 min, 2000 μl of α-eluant is added. The combined α-fractions containing HDL and albumin are discarded. A new tube is placed under the column, and 2500 μl of β-eluant (2.7% NaCl in Nanopure water) is added and the β-fraction containing LDL + VLDL is collected over ice. The solution is deaerated with nitrogen and mixed several times by gentle inversion. The LDL + VLDL can be used immediately or allocated into polypropylene microcentrifuge tubes under nitrogen and stored at $-80°$ or lower until use. Once thawed the samples must be used or discarded. (The columns can be reused several times by preconditioning with 1 ml of β-eluant, 2 ml of Nanopure water, followed by 2 ml of α-eluant and stored over α-eluant.) The LDL + VLDL fraction is measured for protein content using human albumin as the standard and a microcolorimetric procedure with Coomassie blue as the reagent (Sigma).

All samples for the oxidation are prepared in duplicate. For analysis, one needs samples to which a pure antioxidant or mixture is added, natives, blanks, and controls (see later). Antioxidants are added to the samples at five or more different concentrations (typically 0.005 to 5 μM of antioxidant), along with LDL + VLDL, cupric ion, and phosphate-buffered saline (PBS). Concentrations in the oxidation tubes for food extracts or beverage samples are expressed as Folin catechin equivalents. The antioxidant can be in methanol or water, but the methanol must not exceed 5% of the total volume in the oxidation tube or the proteins may precipitate. All oxidation samples have a total volume of 400 μl and are contained in 10-ml plastic screw-capped tubes. Cupric acetate from Sigma (1 mM) is prepared just before use in Nanopure water as was 0.67% 2-thiobarbituric acid (TBA, Sigma). Phosphate-buffered saline, pH 7.4 (Sigma), 10 mM EDTA, and 10% trichloroacetic acid

---

[23] J. A. Vinson, and Y. A. Dabbagh, *Nutr. Res.* **18**, 1067 (1998).
[24] I. V. Kaplan, G. A. Hobbs, and S. S. Levinson, *Clin. Chim. Acta* **283**, 89 (1999).

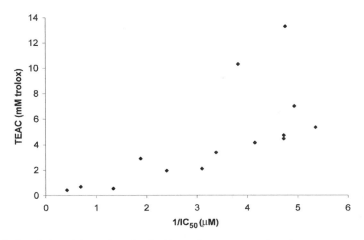

FIG. 3. Correlation of the quality of antioxidants of pure compounds and extracts as measured by $1/IC_{50}$ and Trolox equivalent antioxidant capacity (TEAC).

(TCA) are prepared in Nanopure water. To each tube enough LDL + VLDL is added (usually less than 100 μl) to give a final concentration of 70 μg/ml and is made up to 390 μl with PBS. Cupric acetate (10 μl) is added last to give 25 μM cupric ion. The native sample contains only LDL + VLDL and PBS. The blank has only PBS and the control has LDL + VLDL and cupric ion.

The blank, native, control, and oxidation samples are sealed and heated for 6 hr at 37°C, removed, and treated with 10 μl of EDTA and 1 ml each of TCA and TBA. They are heated at 95° for 10 min, removed, cooled with tap water, and filtered using a 3-ml syringe and 0.45-μm filters or centrifuged. The samples are transferred to disposable cuvettes and read vs the blank in a fluorometer at an excitation of 525 nm and an emission of 553 nm.

$$\%\text{Inhibition} = \frac{\text{Control fluorescence} - \text{native fluorescence} - \text{sample fluorescence}}{\text{Control fluorescence}} \times 100$$

This method has a precision of <10% for the determination of the concentration needed to inhibit the oxidation 50% (IC_{50}). The lower the value the better the quality of the antioxidant(s). An example of the graph is shown in Fig. 2 for an orange sample and orange components.

To illustrate the validity of this method for the determination of quality, the inverse of IC_{50} values for 12 polyphenols and 2 vitamins is correlated linearly with TEAC values[25] and are shown in Fig. 3. The equation is TEAC = 0.309 (1/IC_{50}) + 1.91. The Pearson correlation coefficient with the two aberrant gallates

[25] C. A. Rice-Evans, N. J. Miller, and G. Paganga, *Free Radic. Biol. Med.* **20,** 933 (1996).

(epigallocatechin and epigallocatechin gallate) was 0.695, $p < 0.01$, and without them 0.926, $p < 0.0001$.

IC_{50} values range from >16 to 0.075 μM for pure phenols with the tea gallates having the lowest values, i.e., being the most powerful antioxidants. Vitamins are poor antioxidants: vitamin C, 1.45 μM; tocopherol, 2.40 μM; β-carotene, 4.30 μM; and ubiquinol-10, 1.33 μM.[22] Many pure polyphenols have IC_{50} values <1 μM, indicating that they could provide antioxidant protection for lower density lipoproteins *in vivo* at physiological concentrations. The 23 vegetables studied ranged from 0.28 to 1.77 μM,[10] the 20 chocolates 0.20 to 0.45 μM,[9] and the 12 teas (green, oolong, and black) 0.23 to 1.00 μM.[23]

Lipoprotein-Bound Antioxidant Activity

The oxidation theory of atherogenesis states that the primary site of oxidation of lower density (LDL + VLDL) lipoproteins is beneath the surface of the aorta in the aortic subendothelium. Here the proteins are bound to the extracellular matrix proteins and are slowly oxidized in this extracellular milieu where there is a decrease in the concentration of large molecule protective antioxidants such as proteins and enzymes.[26] Tocopherol, the primary lipoprotein-bound antioxidant, is present at a concentration of 6–10 molecules/molecule of LDL.[27] Polyphenols, which have both water and lipid solubility, could provide protection below the aortic endothelium, if they bind to lipoproteins and protect them from oxidation. There is some evidence in the literature that this occurs. Tikkanen *et al.*[28] have shown that soy isoflavones bind strongly to LDL following human supplementation although they are present in low concentration in the LDL. Hayek *et al.*[29] have found that red wine polyphenols, quercetin and catechin, bonded strongly to LDL and reduced atherosclerosis in apo-E-deficient mice. In both these human and animal studies, the polyphenol supplementation decreased the susceptibility of lipoproteins to oxidation. Our group has shown that both black and green tea, given to normal and cholesterolfed hamsters, protected the LDL + VLDL from oxidation.[30] Five compounds, two identified as rutin and quercetin, have been detected in LDL isolated from unsupplemented individuals.[31] By our calculations, an estimated 0.2–40 molecules of polyphenols were present in the LDL molecule. This is the first evidence in unsupplemented subjects for LDL binding of polyphenols.

[26] B. Frei, *Crit. Rev. Food Sci. Nutr.* **35**, 83 (1998).
[27] H. Esterbauer, J. Gebicki, H. Puhl, and G. Jürgens, *Free Radic. Biol. Med.* **13**, 241 (1992).
[28] M. J. Tikkanen, K. Wahala, S. Ojala, V. Vihma, and H. Aldercreutz, *Proc. Natl. Acad. Sci. USA* **95**, 3106 (1998).
[29] T. Hayek, B. Fuhrman, J. Vaya, M. Rosenblat, P. Belinky, R. Coleman, A. Elis, and M. Aviram, *Arterioscler. Thromb. Vasc. Biol.* **17**, 2744 (1997).
[30] J. A. Vinson and Y. A. Dabbagh, *FEBS Lett.* **433**, 44 (1998).
[31] R. M. Lamuela-Raventós, M.-I. Covas, M. Fitós, J. Marrugat, and M. C. De la Torre-Boronat, *Clin. Chem.* **45**, 1870 (1999).

A simple *ex vivo* plasma model has been developed to demonstrate that polyphenols as pure compounds, beverages, or as extracts can enrich lipoproteins and protect them from oxidation.[32] The method (slightly modified) has been tested with 26 common pure dietary phenols, and it has been found that the lipoprotein-bound activity correlates with the ability of phenols to bind to protein.[33]

Procedure for Measurement of Lipoprotein-Bound Antioxidant Activity

Polyphenol solutions in plasma are prepared by combining 500 μl of plasma with a solution of the desired pure polyphenol, test substance, or extract. The volume added should be ≤50 μl and may be pure water, or ≤25 μl of methanol or ethanol mixed with water. A control is also prepared using plasma and the appropriate solvent. The plasma/antioxidant mixture is placed in a 2-ml polypropylene microcentrifuge tube, nitrogen is blown over the solution, which is then vortexed for 15 sec, and incubated at 37° for 2 hr. After cooling to room temperature, 200 μl is placed on a heparin-agarose column (see previous procedure). The protein concentration of the LDL + VLDL is measured and the volume is calculated to obtain a concentration of 70 μg/ml in 2 ml of total solution. Then the LDL + VLDL aliquot in 2.7% NaCl, a volume of Nanopure water to dilute it to 0.9% NaCl, and a volume of PBS to obtain a final volume of 1950 μl are combined and vortexed for 15 sec in a 4.0-ml, 1-cm path length quartz cuvette. The cuvette is placed in a kinetic UV spectrophotometer equipped with a constant temperature water circulator at 37° (Spectronic Genesys 5 with Application II SoftCard, Milton Roy, Rochester, NY). After 10 min of equilibration, 50 μl of 1 m*M* cupric acetate is added to the cuvette, the cuvette is removed, the contents of the cuvette are vortexed for 5 sec, and the covered cuvette is returned to the spectrophotometer. The 234-nm absorbance of the cuvette is measured at 5-min intervals until the rapid increase in absorbance is complete and the absorbance is constant or slightly decreasing. The lag time is calculated as the time when the lag phase (slow oxidation) line intersects the propagation phase (rapid oxidation rate) line.

Results

Lag times for LDL + VLDL samples from 10 different plasmas were determined (range of 76 to 105 min) with an average standard deviation of 4%. In Fig. 4, spiking of EGCG, the major ingredient of green tea, and green tea are shown to produce very similar lag times, 177 min vs 169 min, respectively. Both have significant lipoprotein antioxidant activity and thus much longer lag times compared with the control value of 67 min. This antioxidant activity has been found for pure antioxidant vitamins such as α-tocopherol (vitamin E), retinol (vitamin A), and

---

[32] J. A. Vinson, J. Jang, Y. A. Dabbagh, M. M. Serry, and S. Cai, *J. Agric. Food Chem.* **43**, 2798 (1995).
[33] W. Wang and M. T. Goodman, *Nutr. Res.* **19**, 191 (1999).

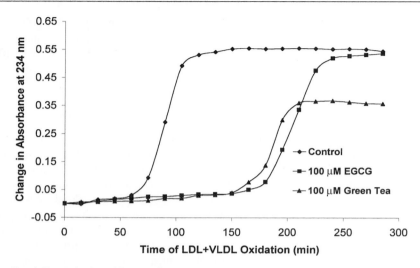

FIG. 4. Determination of lipoprotein-bound antioxidant activity by *ex vivo* spiking of plasma with pure epigallocatechin gallate (EGCG) and green tea and subsequent isolation and oxidation of the LDL + VLDL.

β-carotene (provitamin A) but not for ascorbate. Also, beverages that have been found in supplementation studies to decrease LDL oxidizability, such as wine and tea, have potent lipoprotein-bound antioxidant activity.[32,34]

Summary

The methods described in this article are quick, simple, and inexpensive to perform. The Folin quantitation method can determine both free and total polyphenol antioxidants in foods and beverages as described, as well as botanical extracts. This assay may also be used to estimate the daily per capita consumption of polyphenols in foods. The dose–response *in vitro* lower density lipoprotein antioxidant activity measurement (IC_{50}) can be employed to compare antioxidants as pure compounds, or in mixtures after quantitating the polyphenols. The *ex vivo* lipoprotein-binding antioxidant activity can be measured simply and rapidly to determine possible *in vivo* binding of pure compounds or extracts from foods. Supplementation and epidemiology studies can utilize the rapid and inexpensive affinity column isolation method of lower density lipoproteins for the determiniation of lipoprotein oxidative susceptibility.

[34] J. A. Vinson, J. Jang, J. Yang, Y. A. Dabbagh, X. Liang, M. Serry, J. Proch, and S. Cai, *J. Agric. Food Chem.* **47,** 2502 (1999).

[10] Preparation and Characterization of Flavonoid Metabolites Present in Biological Samples

By Christine Morand, Claudine Manach, Jennifer Donovan, and Christian Remesy

Introduction

The nature of flavonoids in biological samples is different from the forms that are found in plants. In plants, most flavonoids, except catechins, are present as glycosides, whereas in biological samples, flavonoids are present as glucuronide and sulfate conjugates. There are numerous methods to analyze the forms of flavonoids that exist in plants, however, these methods are not usually suitable for analysis of the derivatives present in biological samples. For example, methods for the hydrolysis of plant glycosides are not adapted for the hydrolysis of plasma metabolites. Furthermore, in most studies dealing with flavonoid bioavailability, the plasma has been analyzed after total hydrolysis by enzymes or acid and these procedures do not determine the nature of the circulating forms. This article describes methods to characterize directly the circulating metabolites as well as methods to synthesize the specific metabolites that may exist *in vivo*. It also describes methods to analyze and quantify the aglycone forms after enzymatic hydrolysis of the conjugates.

Extraction of Flavonoids in Biological Samples

The rat constitutes a good model to study the circulating metabolites of flavonoids and, unlike humans, rats can be fed controlled diets containing known quantities of purified flavonoids. In addition, it is possible to administer much higher doses of flavonoids, which are often necessary to characterize the circulating metabolites. However, regardless of the origin of the biological sample (humans or animals), the treatment and hydrolysis of samples after collection are similar.

Sampling Procedure

After anesthesia of the rats, blood is drawn from the abdominal aorta into heparinized tubes. To prevent losses in flavonoid content, plasma is acidified with 10 mM acetic acid. The resulting samples are used immediately for analysis or stored at $-80°$.

Urine can be collected from rats with metabolic cages fitted with urine/feces separators or sampled directly from the bladder using a needle. The latter procedure avoids hydrolysis of conjugated derivatives by residual bacterial contaminant activity. Samples are used immediately for analysis or stored at $-80°$.

Livers are perfused *in situ* with a thermostatted Krebs buffer (pH 7.4) to drain off the blood before excision. Then, livers are removed, dried on filter paper, frozen in liquid nitrogen, and stored at $-80°$ until use.

Extraction Procedure

Plasma or urine (195 μl) is spiked with 5 μl of an appropriate internal standard, and then flavonoids are extracted by adding 500 μl of methanol/HCl (200 mM). After centrifugation at 14,000g for 4 min at ambient temperature, 20 μl of the resulting supernatant is injected and analyzed by high-performance liquid chromatography (HPLC). For some polyphenols, it may be necessary to adapt the extraction procedure with another liquid/liquid or solid/liquid method, but the methanol/HCl extraction is quite suitable for the recovery of flavonols, flavones, isoflavones, flavonones, and flavanols, when present as aglycones, glycosylated, glucuronidated, and sulfated forms.

Livers are homogenized in 2 volumes of ice-cold 0.05 M sodium acetate buffer (pH 5.0) containing 1 g/liter ascorbate to prevent oxidation of flavonoids. The homogenate is centrifuged at ambient temperature for 5 min at 15,000g, and the resulting supernatant is spiked with an appropriate internal standard and extracted with methanol/HCl as described earlier.

When biological samples are analyzed using gas chromatography coupled to mass spectrometry (GC/MS), an internal standard must always be added at the beginning of analysis and samples must be extracted in a suitable organic solvent that can be dried before the flavonoids are derivatized. Catechin has been analyzed by this method in plasma and urine.[1] Plasma (500 μl) is mixed with 1 ml dichloromethane (to remove protein and lipids) and 500 μl of water. The samples are vortexed and then centrifuged at 4500g for 10 min at ambient temperature. The aqueous supernatant is removed, and the remaining portion is extracted a second time with 750 μl water. The combined aqueous extracts are then extracted twice with ethyl acetate (first with 2.0 ml and then with 1.5 ml). The extracts are combined and residual water is removed by passing the extract through a Pasteur pipette packed with anhydrous sodium sulfate. The extract is then evaporated to dryness under nitrogen gas, redissolved in 20 μl pyridine, and derivatized with 30 μl of BSTFA in an oven at 65–75° for 2 hr.

Enzymatic Hydrolysis of Flavonoid Conjugates

When rats are fed a diet supplemented with flavonoids, no free aglycones are present in the plasma, all the ciruclating forms are glucuronidated or sulfated. However, traces of free epicatechin have been detected in rat plasma after

---

[1] J. L. Donovan, J. R. Bell, S. Kasim-Karakas, J. B. German, R. L. Walzem, R. J. Hansen, and A. L. Waterhouse, *J. Nutr.* **129**, 1662 (1999).

intragastric administration of this flavanol.[2] This may result from a saturation of the conjugation activities because of the acute supply of epicatechin. Such saturation was observed in rats by Piskula and Terao[3] after the administration of two different doses of epicatechin. Sfakianos et al.[4] also showed a reduction of hepatic glucuronidation in rats, with increasing rates of genistein perfusion in the portal vein.

Total Hydrolysis for Quantitation

Aglycone forms of flavonoids can be quantified by HPLC. In contrast, the conjugated derivatives recovered in plasma represent a complex mixture, usually not completely identified, and are thus difficult to quantify. For this reason, the determination of the total levels of metabolites is performed using the aglycones released after enzymatic hydrolysis.

Before hydrolysis, plasma or urine samples (175 μl) are acidified to pH 4.9 with 20 μl of 0.58 M acetic acid solution, and spiked with 5 μl of internal standard. Samples are incubated at 37° in the presence of 10 μl of an *Helix pomatia* extract (Sigma, St. Louis, MO) containing 1×10^5 units/ml ß-glucuronidase and 5×10^3 units/ml sulfatase. The optimal incubation period for complete hydrolysis may vary with the nature of the flavonoid. For example, 30 min has been used for catechin and quercetin, 4 hr for naringenin, and 6 hr for daidzein and genistein.

A procedure using concentrated acid has also been described for hydrolysis of the conjugated forms in biological samples.[5] This procedure was adapted directly from a procedure for hydrolysis of flavonoid glycosides in plants described by Hertog et al.[6] However, this method is much less specific and seems more tedious than hydrolysis by enzymes because samples must be incubated at 90° for 5 hr (plasma) or 8 hr (urine) in aqueous methanol containing 2 M HCl. In addition, acid hydrolysis may produce compounds that do not necessarily exist in plasma. For example, anthocyanins can be formed from the hydrolysis of flavan-3-ol oligomers in similar conditions.

Differential Hydrolysis

In some cases it may be of interest to determine the levels of the different classes of conjugated derivatives. This can be achieved by differential enzymatic hydrolysis. Before hydrolysis, plasma or urine (175 μl) is acidified to pH 4.9 with 20 μl of 0.58 M acetic acid solution, and spiked with 5 μl of internal standard.

---

[2] E. Da Silva, M. Piskula, and J. Terao, *Free Radic. Biol. Med.* **24,** 1209 (1998).
[3] M. Piskula and J. Terao, *J. Nutr.* **128,** 1172 (1998).
[4] J. Sfakianos, L. Coward, M. Kirk, and S. Barnes, *J. Nutr.* **127,** 1260 (1997).
[5] P. C. H. Hollman, J. M. P. Van Trijp, M. N. C. P. Buysman, M. S. Gaag, M. J. B. Mengelers, J. H. M. deVries, and M. B. Katan, *FEBS Lett.* **418,** 152 (1997).
[6] M. G. L. Hertog, P. C. H. Hollman, and D. P. Venema, *J. Agric. Food Chem.* **40,** 1591 (1992).

For the hydrolysis of glucuronidated metabolites, plasma is incubated at 37° with 2000 units of bovine ß-glucuronidase type B3 (Sigma). For quercetin glucuronides, the incubation is performed during 120 min; however, the incubation time must be determined experimentally for each flavonoid. For the hydrolysis of sulfate conjugates, plasma is mixed with 50 units of arylsulfatase type VIII from *Abalone entrails* (Sigma) in the presence of D-saccharic acid 1,4-lactone (4 mg/ml) to inhibit contaminant ß-glucuronidase activity present in the sulfatase preparation. For complete hydrolysis of quercetin sulfates, the incubation is performed for 240 min at 25°. It is worthy to note that liver homogenates can be treated in the same way as biological fluids, except that the amounts of enzymes used should be three-fold higher.

The aglycone concentrations found after differential hydrolysis of samples can be used to estimate the percentages of the different types of conjugates (glucurono, sulfo, and glucurono–sulfo conjugates). Using this method, we have reported previously that the conjugated derivatives present in plasma of rats fed a diet containing 0.2% quercetin were constituted by 91.5% glucurono–sulfo conjugates and 8.5% glucuronides.[7] Using similar methodology, catechin was present as 68% glucurono–sulfo conjugates, 22% sulfates, and 8% glucuronides in human plasma 1 hr after red wine consumption.[1]

In Vitro Conjugation of Flavonoids

Flavonoids are intensively glucuronidated and sulfated by the intestinal mucosa, the liver and the kidney, and sometimes methylated in the liver and kidney. In some cases it may be of interest to study the respective role of these tissues in the conjugating process. Moreover, it may be useful to synthesize some conjugated derivatives for use as standards or to study their biological properties.

Flavonoid Glucuronidation from Microsomal Preparation

Intestinal and liver microsomes from rats are prepared by differential ultracentrifugation. Liver or mucosal scrapings are homogenized in an ice-cold buffer containing 50 mM Tris–HCl, pH 7.2, 100 mM sucrose, 10 mM EDTA, 2 mM dithiothreitol (DTT), and 1 μM leupeptin. For mucosal scrapings, 0.025% (w/v) of trypsin inhibitor type I-S (Sigma) is added to the homogenization buffer to prevent UDP-glucuronosyltransferase inactivation by pancreatic enzymes. The resulting homogenate is centrifuged at 105,000g at 4° for 1 hr, and then the microsomal pellet is resuspended in a buffer containing 100 mM HEPES, pH 7.2, and 100 mM sucrose (supplemented with trypsin inhibitor for intestinal fractions) and kept frozen at $-20°$C until use.

---

[7] C. Morand, V. Crespy, C. Manach, C. Besson, C. Demigné, and C. Rémésy, *Am. J. Physiol.* **275**, R212 (1998).

Before the glucuronidation reaction, the microsomal preparation is adjusted to a final protein concentration of about 5 mg/ml measured according to the Pierce B.C.A. protein reagent kit (Interchim, Montluçon, France). Incubations are carried out as follows: 100 μl of microsomal suspension (50 μg protein) activated *in situ* by 60 μl of a 0.2% solution of Triton X-100 is added to 540 μl of HEPES buffer (75 mM, pH 7.3, $MgCl_2$ 10 mM) containing 50 μl of an UDP-glucuronic acid solution at 67.5 mM (final concentration: 4.5 mM). The reaction is started by the addition of 2 μl of a concentrated solution of aglycone [18.75 mM in dimethyl sulfoxide (DMSO)] to obtain a final concentration of 50 μM. Incubations are performed at 37° (2 hr for flavonols), and the reaction mixture is extracted with methanol/HCl as described earlier.

Flavonoid Glucuronidation with Purified Enzyme

Aglycones (50 μM) are incubated at 37° (2 hr for flavonols) in HEPES buffer (25 mM, pH 7.4) in the presence of 10 mM $MgCl_2$, 4 mM UDP-glucuronic acid, 2 mM UDP-N-acetylglucosamine, and 0.12 U/l uridine 5′-diphosphoglucuronyltransferase from rabbit liver (EC 2.4.1.17, Sigma). When experiments are conducted with quercetin, the conjugation rate is about 95%.

Our studies with quercetin have shown that regardless of the procedure used for *in vitro* glucuronidation (from liver microsomes or from the purified enzyme), the resulting glucuronide conjugates are identical.[7]

Flavonoid Sulfation with a Cytosolic Liver Extract

For the enzyme preparation, fresh liver is homogenized in 3 volumes of ice-cold 0.1 M Tris–HCl, pH 7.2, 0.25 M sucrose. The homogenate is centrifuged for 20 min at 12,500g and the supernatant is further centrifuged at 105,000g for 1 hr. The final protein concentration in the cytosolic extract is adjusted to 25 mg/ml.

The *in vitro* sulfation of quercetin and isorhamnetin by cytosolic sulfotransferases of rat liver is performed using PAPS (3′-phosphoadenosine 5′-phosphosulfate) as the sulfate donor. The standard assay mixture consists of 20 μM of the flavonoid substrate dissolved in DMSO (final concentration in the assay 0.4%), 25 μM PAPS, 5 mM $MgCl_2$, and 0.5 mg of cytosolic protein in 25 mM Tris–HCl buffer, pH 7.2, in a total volume of 500 μl. The reaction is initiated by the addition of the cytosolic extract and incubated at 37° (60 min for flavonols), and then the reaction mixture is extracted with methanol/HCl as described previously.

Catechol O-Methyltransferase-Catalyzed O-Methylation of Flavonoids in Vitro

The reaction mixture consists of 0.01 M Tris–HCl buffer (pH 7.4), 1.2 mM $MgCl_2$, 1 mM DTT, 1.25 mM S-adenosyl-L-methionine, chloride salt, and 50 μM aglycones in a final volume of 1 ml. The reaction is started by the addition of

200 units of porcine liver catechol O-methyltransferase (COMT, Sigma). After an incubation for 3 hr at 37°, the reaction mixture is extracted immediately using methanol/HCl as described earlier. According to Shaw and Griffiths,[8] methylation of catechin by COMT takes place almost exclusively in the 3' position.

Chemical O-Methylation of Flavonoids in Vitro

Chemical methylation of flavonoids has been described for catechin and epicatechin. The reaction mixture consists of 250 mg catechin, 500 mg potassium carbonate, and 1 ml of methyliodide in 20 ml acetone and is performed for 2.5 hr in an ultrasonic bath. The two major products of the reaction are the 3'- and 4'-methyl ethers and these products can then be purified by reversed-phase semipreparative HPLC as described previously.[9] This method may be adapted for the methylation of other flavonoids; however, conditions should be optimized for each flavonoid studied.

Chromatographic Analysis

HPLC Coupled to UV Detection

The HPLC system is fitted with a 5-μm C_{18} Hypersil BDS analytical column (150 × 4.6 mm i.d.) (Life Sciences International, Cergy, France).

For quantitative determinations, HPLC analysis can be performed using isocratic conditions. The mobile phase consists of 73% H_2O/H_3PO_4 (99.5 : 0.5 v/v) and 27% acetonitrile at a flow rate of 1.5 ml/min, and the UV detector is set at 370 nm for the determination of flavonols, 350 nm for flavones, or at 320 nm for isoflavones and flavanones.

To visualize and separate the conjugated metabolites of flavonoids, a gradient elution procedure must be used. For flavonols, the percentage of acetonitrile in the mobile phase increases from 15 to 40% over a 20-min period (flow rate 1 ml/min) and is then returned to initial conditions for a 15-min equilibration period.

HPLC Coupled to Multielectrode Coulometric Detection

Multielectrode coulometric detection provides a precise and accurate method for the characterization and quantification of flavonoid metabolites. In this detection system, described extensively elsewhere,[8] each compound reacts on several electrodes according to their redox properties. The response profile on successive electrodes is characteristic of a given compound, and the identification of metabolites is based on their retention time and on their electrochemical behavior. Moreover, depending on the flavonoid studied, the sensitivity is about 50- to 100-fold higher using coulometric compared to UV detection.

---

[8] I. Shaw and A. Griffiths, *Xenobiotica* **10,** 905 (1980).
[9] J. L. Donovan, D. L. Luthria, P. Stremple, and A. L. Waterhouse, *J. Chromatogr. B* **726,** 277 (1999).

An 8-electrode CoulArray Model 5600 (Eurosep, Cergy, France) is fitted with a 5-μM C_{18} Hypersil BDS analytical column (150 × 4.6 mm i.d.) (Life Sciences International). The mobile phases consist of 30 mM NaH_2PO_4 buffer (pH 3) (A) and acetonitrile (B). The column temperature is 35° with a flow rate between 0.8 and 1 ml/min using gradient conditions and potentials adapted for each class of flavonoids. For example, analysis conditions of the circulating metabolites of catechin are as follows[11]: 0–5 min, 5% B; 5–15 min, linear gradient from 5 to 20% B; 15–30 min, 20% B; 30–33 min, linear gradient from 20 to 50% B; 33–38 min, 50% B; 38–41 min, gradient from 50 to 5% B; 41–60 min, 5% B. Settings for potentials are 5, 30, 110, 280, 310, 350, 400, and 500 mV.

Gas Chromatography with Mass Spectral Detection

The trimethylsilyl derivatives of flavonoids can be analyzed by GC/MS. Specific conditions have been described for catechin and epicatechin in plasma.[1] However, chromatographic conditions must be optimized for each flavonoid studied, although detection in selective ion monitoring mode is highly specific and reduces the chance of interferences. For quantitation of flavonoids in plasma by GC/MS, standard curves must be prepared in blank or control plasma containing an internal standard and the plasma should be prepared and analyzed exactly as the samples.

Conclusion

Numerous studies have shown the antioxidant activity and various cellular effects of aglycones, without considering the fact that the circulating metabolites are conjugated forms and undergo metabolism such as methylation in the liver. Normally, the conjugation of biologically active molecules serves to facilitate elimination and reduces biological activity. However, the conjugated derivatives of flavonoids, which constitute elimination forms, are the only ones that are present in plasma. Thus, the biological effects attributed to flavonoids should be investigated further using the metabolites that are present in biological fluids. We have obtained glucurono and sulfo conjugates of quercetin and methylated quercetin using the procedures described in this article. We showed that these compounds still display a significant protective effect on human or rat lipoprotein oxidation, even if it is lower than that of quercetin itself.[7,11]

Furthermore, the efficiency of absorption and elimination, as well as the rates of methylation and conjugation, certainly depends on the nature of the flavonoid, the dose ingested, and the species as well as the individual studied. These parameters must also be considered when evaluating the nature and the biological effects of the circulating metabolites of flavonoids.

[10] C. N. Svendsen, *Analyst* **118**, 123 (1993).
[11] C. Manach, C. Morand, V. Crespy, C. Demigné, O. Texier, F. Régérat, and C. Rémésy, *FEBS Lett.* **426**, 331 (1998).

[11] Caffeic Acid as Biomarker of Red Wine Intake

By PAOLO SIMONETTI, CLAUDIO GARDANA, and PIERGIORGIO PIETTA

Introduction

A number of experimental and epidemiological studies have indicated that low levels of endogenous antioxidants are correlated with higher risks for atherosclerosis development.[1] Vitamins C and E, carotenoids, and ubiquinone have been shown to exert a protective effect ascribed to their capacity to prevent low-density lipoprotein (LDL) oxidation.[2]

In addition to these antioxidants, other dietary constituents have been regarded as possible agents that may retard the development of atherosclerotic complications. Polyphenols belong to this group, as they are the most ubiquitous plant metabolites and, therefore, largely present in the human diet.[3] Polyphenols have received increasing attention, and some biological activities have been proved. Among these, the ability to protect circulating LDL against oxidation has been evaluated.[4,5] These experimental studies have ben confirmed by the observation that France's population has a lower incidence of coronary heart disease (CHD) than expected based on their high intake of saturated fat (14–15% of the energy).[6] This observation, known as the French paradox, has been explained by the regular moderate consumption of red wine. This beverage affords ethanol, which is known to exert potentially beneficial effects on fibrinolytic factors and high-density lipoprotein (HDL)-cholesterol.[7] However, the alcoholic component of red wine cannot be considered the unique cardioprotective factor, as red wine limits the risk for CHD more than other alcoholic beverages. Therefore, other components are implicated in the cardioprotective capacity of red wine. These are the polyphenols, which are present in red wines at levels averaging 1200 mg/liter,[8] whereas only one-sixth of that concentration is found in white wines. These polyphenols comprise flavonoids and nonflavonoids. The flavonoids include anthocyanins, flavan-3-ols (e.g., catechins and their polymers known as proanthocyanidins), and, in a lesser extent, some flavonols (e.g., rutin, quercetin, and myricetin). Nonflavonoids are

---

[1] M. N. Diaz, B. Frei, J. A. Vita, and J. F. Keaney, *N. Engl. J. Med.* **337,** 408 (1997).
[2] H. Esterbauer, G. Wag, and H. Puhl, *Br. Med. Bull.* **49,** 566 (1993).
[3] N. C. Cook and S. Samman, *J. Nutr. Biochem.* **7,** 66 (1996).
[4] E. N. Frankel, A. L. Waterhouse, and J. E. Kinsella, *Lancet* **341,** 454 (1993).
[5] G. I. Soleas, E. P. Diamandis, and D. M. Goldberg, *J. Clin. Lab. Anal.* **11,** 287 (1997).
[6] S. Renaud and M. de Lorgeril, *Lancet* **339,** 1523 (1992).
[7] P. Marques-Vidal, J. P. Cambou, V. Nicaud, G. Luc, A. Evans, D. Arveiler, A. Bingham, and F. Combien, *Atherosclerosis* **115,** 225 (1995).
[8] P. Simonetti, P. G. Pietta, and G. Testolin, *J. Agric. Food Chem.* **45,** 1152 (1997).

represented by hydroxybenzoates, hydroxycinnamates, and trihydroxystillbenes (resveratrol isomers).[9] All these polyphenols have free radical scavenging ability, depending on their reduction potential,[10] and are able to break the propagation of oxidizing lipids, depending on their lipophilicities.[11] However, the detection of red wine polyphenols in humans has scarcely been investigated. Total phenol plasma concentration has been measured with the Folin–Ciocalteau method.[12] Unfortunately, this method is rather unspecific, because it is sensitive to nearly all oxidable compounds. Anthocyanins have been detected in human urine after red wine intake,[13] whereas there is no report on the analysis in human plasma of the hydroxycinnamates caffeic acid, ferulic acid, and *p*-coumaric acid. These acids have been identified in the urine of subjects after a high intake of fruit.[14] Extending this work, the same authors have reported on the urinary excretion of ferulic acid after a single bolus of tomatoes.[15]

This article describes the pharmacokinetics of caffeic acid, which has been selected as a biomarker of hydroxycinnamates in plasma of subjects who received different amounts of red wine (RW) or dealcoholized red wine (DRW).

Analytical Approach to Detection of Caffeic Acid in Human Plasma

Volunteers

Ten healthy subjects (5 males, 5 female), mean age 25 ± 6 years, Body Mass Index (BMI) 24.1 ± 1.4 and 19.9 ± 1.9 kg/m^2, respectively, for men and women received either 200 ml RW or DRW. Volunteers participated on two occasions 1 week apart, consuming one of the beverages each time. Volunteers refrain from specific phenolic-rich foods and beverages for 3 days prior to the study, including bran cereals; vegetables such as onions, spinach, tomatoes, lettuce; fruits such as apples, apricot, blackberries, cranberries, grapes, grapefruit, strawberries; and beverages such as beer, coffee, fruit juices, tea, and wine.

Chemicals

The high-performance liquid chromatography (HPLC)-grade methanol and acetonitrile are obtained from BDH Laboratory Supplies (Poole, England). 2,2′-Diazobis(2-amidinopropane) dihydrochloride (ABAP) is from Wako Chemicals

---

[9] G. I. Soleas, E. P. Diamandis, and D. M. Goldberg, *Clin. Biochem.* **30**, 91 (1997).
[10] S. V. Jovanovic, S. Steenken, M. G. Simic, and Y. Haza, in "Flavonoids in Health and Disease" (C. A. Rice-Evans and L. Packer, eds.), p. 137. Dekker, New York, 1998.
[11] J. Brown, H. Khods, R. Hider, and C. A. Rice-Evans, *Biochem J.* **330**, 1173 (1998).
[12] S. V. Nigdikar, N. R. Williams, B. A. Griffin, and A. N. Noward, *Am. J. Clin. Nutr.* **68**, 258 (1998).
[13] T. Lapidot, S. Harel, R. Granit, and J. Kanner, *J. Agric. Food Chem.* **46**, 4297 (1998).
[14] L. C. Bourne and C. A. Rice-Evans, *Free Radic. Res.* **28**, 1429 (1998).
[15] L. C. Bourne and C. A. Rice-Evans, *Methods Enzymol.* **299**, 91 (1999).

(Richmond, VA). *R*-Phycoerythrin (R-PE) is from Molecular Probes (Eugene, OR). Caffeic acid and 6-hydroxy-2,5,7,8-tetramethylchroman-2-carboxylic acid (Trolox) are from Aldrich Chemical (Milwaukee, WI). All other chemicals are from Merck (Darmstadt, Germany). The red wine (Cabernet, 1998–Friuli, Italy) is from a local commercial winery.

Standards

A stock solution of standard caffeic acid is prepared by dissolving 20 mg of sample in 200 ml of methanol. This solution is stored at 0–4° and used within 4 weeks after preparation. Working solutions are prepared by diluting the stock solution with the mobile phase (1–100 ng/ml).

Wine Sample Preparation

Ethanol is removed from the sample by rotary evaporation (<40°) to 50% of the original volume and subsequently the DRW is normalized to the original volume with water. One milliliter of RW or DRW is diluted to 3 ml with water and the resulting solution is applied to a previously activated (by washing with 3 ml of methanol and 6 ml of water) Sep-Pak C_{18} cartridge (Water, Milford, MA). After washing with 6 ml of 10 mM HCl and 3 ml of 10 mM HCl–methanol (85 : 15, v/v), the caffeic acid-containing fraction is eluted with 3 ml of 10 mM HCl–methanol (50 : 50, v/v). The eluate is evaporated by dryness under nitrogen, and the residue is dissolved in 1 ml of 30 mM dihydrogen sodium phosphate (pH 3)–acetonitrile (88 : 12, v/v).

Plasma Sample Preparation

Fasting venous blood samples are taken in vacuotainer tubes containing sodium heparin before and at 20, 40, 60, 120, 180, 240, and 300 min after RW or DRW intake. Plasma is separated by centrifugation at 10,000g for 1 min at room temperature.

An aliquot of plasma (250 μl) is added to 2 ml of 10 mM HCl; the resulting solution is applied to a previously activated (3 ml of methanol followed by 6 ml of water) Sep-Pak C_{18} cartridge. After washing with 6 ml of 10 mM HCl and 3 ml of 10 mM HCl–methanol (85 : 15, v/v), the caffeic acid fraction is eluted with 3 ml of 10 mM HCl–methanol (50 : 50, v/v), dried under nitrogen, and dissolved in 120 μl of 30 mM dihydrogen sodium phosphate (pH 3)–acetonitrile (88 : 12, v/v).

Analysis of Caffeic Acid by HPLC

The HPLC system consists of a Model 510 pump (Waters) connected with a Coulochem II detector (ESA, Chelmsford, MA) and a Millenium workstation (Waters). The injection is by means of a Model 7125 Rheodyne injector (Waters), and the volume injected is 50 μl. A Symmetry C_{18} column (220 × 4.6 mm) with a

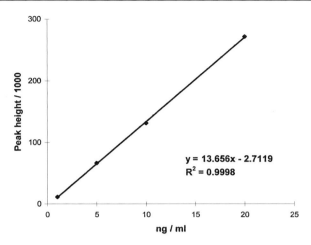

FIG. 1. Caffeic acid calibration plot.

5-μm particle size (Waters) is used. Elution (1.3 ml/min) is performed with a mobile phase composed of 30 mM dihydrogen sodium phosphate (pH 3)–acetonitrile (88:12, v/v). The chromatograms are obtained with detection under the following conditions: guard cell, −250 mV; control cell, −50 mV; and analytical cell, 350 mV. All injections are performed in duplicate. Quantification is carried out using a caffeic acid calibration plot (Fig. 1). This is obtained by calibration runs with caffeic acid (range 1–20 ng/ml). The limit of detection is around 0.3 ng/ml.

Percentage Recovery of Caffeic Acid

Blank plasma samples (1 ml) are added with increasing amounts of caffeic acid standard (1–10 ng). An aliquot (250 μl) is treated and analyzed by HPLC, as described earlier. The recovery is 96 ± 2%.

Determination of Plasma Total Radical-Trapping Antioxidant Parameter (TRAP)

Plasma TRAP is determined by the method of Ghiselli *et al.*[16]

Plasma peroxidation induced by ABAP is monitored by the loss of fluorescence of the protein R-phycoerythrin. The lag phase of the plasma is compared to that of a known concentration of the internal standard Trolox and is then related quantitatively to the antioxidant capacity of the plasma. One hundred microliters of plasma is diluted 10 times with a phosphate buffer (75 mM at pH 7), and 80 μl of this solution is added to 250 μl distilled water, 1.55 ml of phosphate buffer

[16] A. Ghiselli, M. Serafini, G. Maiani, E. Azzini, and A. Ferro-Luzzi, *Free Radic. Biol. Med.* **18,** 29 (1995).

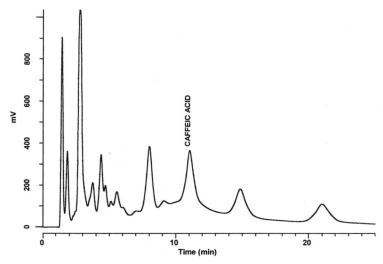

FIG. 2. Typical HPLC of a Cabernet red wine.

(75 mM at pH7), and 20 μl of 1.5 × 10^{-8} M R-PE in the same phosphate buffer. The resulting mixture is maintained at 37° for 5 min in a quartz fluorometer cell. After the mixture is stabilized, the oxidation reaction is started by adding 100 μl of ABAP (with a final concentration in the fluorescence spectrometer of 4.0 mM, which corresponds to 538 mg in 10 ml of phosphate buffer). After 25 min, 60 μl of 120 μM Trolox (7.51 mg in 250 ml water) is added. The decay of R-PE fluorescence is monitored by a Model LS50B fluorescence spectrometer (Perkin Elmer Limited, Beaconsfield, England). The detector is set at an excitation wavelength of 495 nm and an emission wavelength of 575 nm.

The TRAP value is calculated according to the following formula:

$$\text{TRAP}(\mu M \text{ Trolox}) = \frac{\text{Plasma lag phase} \times 250 \times 120 \times 2}{\text{Trolox lag phase} \times 34.333}$$

where 2 is the Trolox factor, 250 is the dilution in cuvette, 120 is the Trolox concentration, and 34.333 is the Trolox dilution in cuvette.

Results

The HPLC analysis of RW is shown in Fig. 2. Caffeic acid is identified with a retention time at around 11 min, and its identity is confirmed by spiking the sample with a standard and by mass spectrometry.[17] The content of caffeic acid in RW is 9.01 ± 0.29 mg/liter.

[17] P. G. Pietta, P. Simonetti, and C. Gardana, submitted for publication.

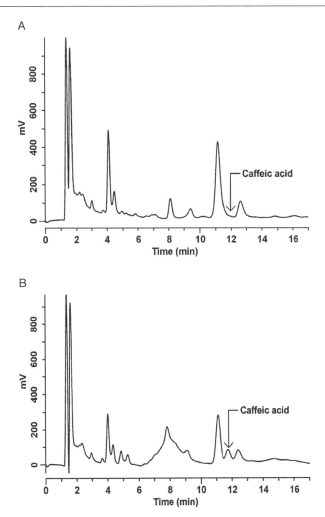

FIG. 3. (A) HPLC of human plasma before red wine intake. (B) HPLC of human plasma after red wine intake.

Figure 3 illustrates the HPLC analysis of the human plasma before and after red wine intake. Plasma caffeic acid concentrations increase after consumption of either RW or DRW. In men the mean concentration peak (10.8 ± 2.3 ng/ml) is reached approximately at 60 min, whereas in women this peak is attained quicker (20–40 min) and is lower (6.6 ± 1.1 ng/ml) (Fig. 4). Alcohol improves the absorption of caffeic acid in all volunteers, as suggested by the significantly different concentrations of plasma caffeic acid detected after RW or DRW intake (Table I).

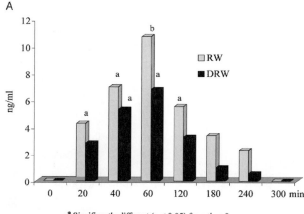
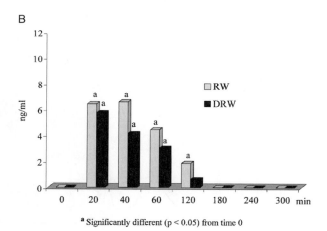

FIG. 4. (A) Time course of plasma caffeic acid concentration after ingestion of red wine (RW) and dealcoholized red wine (DRW) in men. (B) Time course of plasma caffeic acid concentration after ingestion of red wine (RW) and dealcoholized red wine (DRW) in women.

The caffeic acid plasma concentration postingestion of different amounts (100–300 ml) of red wine in men is shown in Table II. A dose–response relationship is established, thereby indicating that caffeic acid may be regarded as a suitable plasma biomarker for red wine.

The TRAP values measured after RW and DRW consumption are shown in Table III, and they are in accord with the time course of plasma caffeic acid concentration, as exemplified in Fig. 5.

TABLE I
PLASMA CAFFEIC ACID IN MEN AND WOMEN BEFORE AND AFTER DRINKING 200 ml OF RW AND DRW

Sample Time (min)	Plasma caffeic acid levels (ng/ml)[a] at time (min):							
	0	20	40	60	120	180	240	300
Women								
RW	0.0 ± 0.0	6.4 ± 1.1	6.6 ± 1.1[b]	4.4 ± 0.8[b]	1.8 ± 0.4[b]	0.0 ± 0.0	0.0 ± 0.0	0.0 ± 0.0
DRW	0.0 ± 0.0	5.7 ± 1.9	4.1 ± 1.3	3.0 ± 1.0	0.6 ± 0.3	0.0 ± 0.0	0.0 ± 0.0	0.0 ± 0.0
Men								
RW	0.0 ± 0.0	4.3 ± 1.5[b]	7.0 ± 2.2[c]	10.8 ± 2.3[b]	5.5 ± 2.1	3.4 ± 1.6	2.3 ± 1.3	0.0 ± 0.0
DRW	0.0 ± 0.0	2.8 ± 1.5	5.3 ± 2.4	6.8 ± 2.4	3.3 ± 1.4	1.0 ± 0.8	0.6 ± 0.5	0.0 ± 0.0

[a] Values are means ± SE, $n = 5$ for red wine (RW) and dealcoholized red wine (DRW).
[b] Significantly different ($p < 0.05$) from DRW.
[c] Significantly different ($p < 0.01$) from DRW.

TABLE II
CAFFEIC ACID PLASMA LEVELS IN MEN AFTER INTAKE OF DIFFERENT AMOUNTS OF RED WINE

Red wine (ml)	Caffeic acid plasma levels (ng/ml)[a] at time (min):					
	0	20	40	60	120	180
100	0.00 ± 0.00	0.49 ± 0.07	0.79 ± 0.20	1.19 ± 0.27	0.45 ± 0.06	0.00 ± 0.00
200	0.00 ± 0.00	0.69 ± 0.09	2.06 ± 0.23	3.15 ± 0.35	1.12 ± 0.10	0.00 ± 0.00
300	0.00 ± 0.00	2.48 ± 0.47	2.72 ± 1.47	5.15 ± 0.45	2.67 ± 0.61	0.00 ± 0.00

[a] Values are means ± SE.

TABLE III
TRAP LEVELS IN MEN AND WOMEN BEFORE AND AFTER DRINKING 200 ml OF RW AND DRW

Sample Time (min)	TRAP levels (μmol/liter)[a] at time (min):							
	0	20	40	60	120	180	240	300
Women								
RW	919 ± 82	934 ± 90	1023 ± 92	942 ± 54	941 ± 81	856 ± 76	861 ± 105	876 ± 139
DRW	741 ± 63	751 ± 51	811 ± 34	744 ± 67	744 ± 30	784 ± 74	754 ± 60	730 ± 72
Men								
RW	863 ± 10	926 ± 17	934 ± 21	1031 ± 25	922 ± 17	928 ± 26[b]	905 ± 16	908 ± 32
DRW	836 ± 12	866 ± 16	919 ± 23	977 ± 19	913 ± 25	863 ± 23	874 ± 49	877 ± 23

[a] Values are means ± SE, $n = 5$ for red wine (RW) and dealcoholized red wine (DRW).
[b] Significantly different ($p < 0.01$) from DRW.

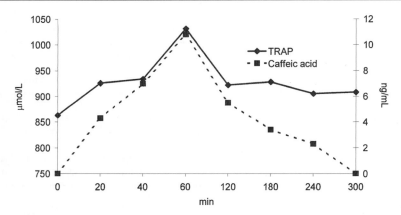

Fig. 5. Time course of plasma caffeic acid concentration and percentage variation of TRAP in men after red wine (RW) intake.

Based on these data, caffeic acid can be considered a representative biomarker of polyphenol uptake from red wine. In conclusion, the study described here is a further evidence that dietary hydroxycinnamates are bioavailable and, specifically, that caffeic acid is a valuable candidate for their pharmacokinetics.

[12] Measurement of *trans*-Resveratrol, (+)-Catechin, and Quercetin in Rat and Human Blood and Urine by Gas Chromatography with Mass Selective Detection

By GEORGE J. SOLEAS, JOE YAN, and DAVID M. GOLDBERG

Introduction

When consumed in moderation, beverage alcohol is associated with a reduced risk for coronary heart disease (CHD) (see Ref 1 for review). This conclusion has received overwhelming support from population-based analyses that are largely consistent in demonstrating a lower incidence of CHD mortality,[2-4] hospitalization,[5]

[1] D. M. Goldberg, S. E. Hahn, and J. G. Parkes, *Clin. Chim. Acta* **237,** 155 (1995).
[2] C. A. Camargo, Jr., M. J. Stampfer, R. J. Glynn, F. Grodstein, J. M. Gaziano, J. E. Manson, J. E. Buring, and C. H. Hennekens, *Ann. Intern. Med.* **126,** 372 (1997).
[3] A. L. Klatsky, M. A. Armstrong, and G. D. Friedman, *Ann. Intern. Med.* **117,** 646 (1992).
[4] W. B. Kannel and R. C. Ellison, *Clin. Chim. Acta* **246,** 59 (1996).
[5] A. L. Klatsky, M. A. Armstrong, and G. D. Friedman, *Am. J. Cardiol.* **80,** 416 (1986).

and early clinical symptoms[6] among moderate drinkers than among abstainers. Several biochemical features known to be a consequence of alcohol consumption and that can prevent or attenuate the development of atherosclerotic lesions, the pathologic basis for CHD, are believed to account for its cardioprotective properties. On the basis of both epidemiologic and experimental observations these include an increase in high-density lipoproteins (HDL),[7] inhibition of platelet aggregation,[8] enhanced fibrinolysis,[9] and prevention of smooth muscle cell proliferation and migration.[10,11]

Epidemiologic studies into the relative merits of specific beverages have yielded conflicting results. The hypothesis (the French paradox) was proposed that the lower incidence of CHD mortality in France than in most other developed countries, despite an extremely high consumption of dairy fat, was due to the protective effects of high wine consumption.[12] Earlier analyses of the relationship between national per capita data for wine consumption and CHD mortality derived from World Health Organization reports[13] had led to similar conclusions. Since then, population studies using questionnaire-based epidemiologic techniques have generated mixed reports, and a meta-analysis incorporating the results of 25 investigations concluded that wine was no better than other alcoholic beverages in lowering CHD risk.[14] Although wine, especially red wine, was reported to diminish the incidence of atherosclerosis in cholesterol-fed rabbits more effectively than beer, whiskey, or pure ethanol,[15] subsequent experiments utilizing similar protocols failed to reproduce these findings.[16,17]

These negative results are somewhat surprising in view of the fact that red wines are very rich in polyphenols, a class of compounds with important *in vitro* biological properties conducive to protection against atherosclerosis. Three, in particular (*trans*-resveratrol, (+)-catechin, and quercetin), are powerful antioxidants

---

[6] E. B. Rimm, E. L. Giovannucci, W. C. Willett, G. A. Colditz, A. Ascherio, B. Rosner, and M. J. Stampfer, *Lancet* **338**, 464 (1991).

[7] J. M. Gaziano, J. E. Buring, J. L. Breslow, S. Z. Goldhaber, B. Rosner, M. VanDenburgh, W. Willett, and C. H. Hennekens, *N. Engl. J. Med.* **329**, 1829 (1993).

[8] R. Rubin and M. L. Rand, *Alcohol Clin. Exp. Res.* **18**, 105 (1994).

[9] M. L. Aikens, H. E. Grenett, R. L. Benza, E. M. Tabengwa, G. C. Davis, and F. M. Booyse, *Alcohol Clin. Exp. Res.* **22**, 375 (1998).

[10] R. Locher, P. M. Suter, and W. Vetter, *Am. J. Clin. Nutr.* **67**, 941 (1999).

[11] R. J. Hendrickson, S. S. Okuda, P. A. Cahill, E. Yankah, J. V. Sitzmann, and E. M. Redmond, *J. Surg. Res.* **84**, 64 (1999).

[12] S. Renaud and M. De Lorgeril, *Lancet* **339**, 1523 (1992).

[13] A. S. St. Leger, A. L. Cochrane, and F. Moore, *Lancet* **1**, 1017 (1979).

[14] E. B. Rimm, A. Klatsky, D. Grobbee, and M. J. Stampfer, *Br. Med. J.* **312**, 731 (1996).

[15] D. M. Klurfeld and D. Kritchevsky, *Exp. Mol. Pathol.* **34**, 62 (1981).

[16] J. S. Munday, K. G. Thompson, K. A. James, and B. W. Manktelow, *Coron. Artery Dis.* **10**, 97 (1999).

[17] Y. Wakabayashi, *J. Agric. Food Chem.* **47**, 4724 (1999).

that can prevent the oxidation of low-density lipoproteins (LDL) *in vitro*,[18,19] an important step in the initiation and progression of atherosclerotic lesions.[20] *trans*-Resveratrol and quercetin can also inhibit platelet aggregation, as well as the synthesis of procoagulant and proinflammatory eicosanoids *in vitro*.[21]

The failure of red wine to live up to the expectations raised by its favorable *in vitro* properties and by the results of epidemiologic population-based investigations could arise from the unsuitability of this approach for the demonstration of direct causal relationships, given the existence of so many confounding factors, not least being the fact that ethanol alone possesses (with the exception of antioxidant activity) virtually all of the antiatherosclerotic properties of wine polyphenols.[22] However, it manifests quite definite prooxidant effects *in vivo,* and this property could mask or even reverse the antioxidant benefits of polyphenols alone. In this scenario, direct intervention studies in animals and humans are required to test the *in vivo* efficacy of wine polyphenols. A number of limited investigations based on this approach have been described, but these have used red wine as the source of polyphenols. Red wine given to human volunteers increases blood antioxidant activity, although not as much as small doses of vitamin C.[23,24]

About half of this increase is attributable to raised blood uric acid concentrations, a known *in vivo* effect of ethanol.[25] Many reports have described resistance of LDL drawn from these subjects to *in vitro* copper-mediated oxidation,[26–28] but others found that prior wine consumption enhanced sensitivity to oxidation by reducing the length of the lag phase preceding the oxidative propagation phase,[29] and in at least two no effect whatsoever could be detected.[30,31] Controversial findings have also been reported for *in vivo* investigations examining the effects of red wine on blood lipids and coagulation. In humans, medium term consumption of red and white wine led to equivalent increases in HDL and its main apolipoproteins, apoAI

---

[18] J. Kanner, E. Frankel, R. Granit, B. German, and J. E. Kinsella, *J. Agric. Food Chem.* **42,** 64 (1993).

[19] G. J. Soleas, E. P. Diamandis, and D. M. Goldberg, *J. Clin. Lab. Analysis* **11,** 287 (1997).

[20] D. Steinberg, S. Parsatharathy, T. E. Carew, J. C. Khoo, and J. L. Witztum, *N. Engl. J. Med.* **320,** 915 (1989).

[21] C. R. Pace-Asciak, S. Hahn, E. Diamandis, G. Soleas, and D. M. Goldberg, *Clin. Chim. Acta* **235,** 207 (1995).

[22] D. M. Goldberg, G. J. Soleas, and M. Levesque, *Clin. Biochem.* **32,** 505 (1999).

[23] S. Maxwell, A. Cruickshank, and G. Thrope, *Lancet* **344,** 193 (1994).

[24] T. P. Whitehead, D. Robinson, S. Allaway, J. Syms, and A. Hale, *Clin. Chem.* **41,** 32 (1995).

[25] S. R. J. Maxwell, *in* "Wine: Nutritional and Therapeutic Benefits" (T. R. Watkins, ed.), p. 150. American Chemical Society, Washington, DC, 1997.

[26] B. Fuhrman, A. Lavy, and M Aviram, *J. Clin. Nutr.* **61,** 549 (1995).

[27] Y. Miyagi, K. Miwa, and H. Inoue, *Am. J. Cardiol.* **80,** 1627 (1997).

[28] S. V. Nigdikar, N. R. Williams, B. A. Griffin, and A. N. Howard, *Am. J. Clin. Nutr.* **68,** 258 (1998).

[29] P. H. Van Golde, L. M. Sloots, W. P. Vermeulen, J. P. Wielders, H. C. Hart, B. N. Bouma, and A. van de Wiel, *Atherosclerosis* 365 (1999).

[30] R. A. Caccetta, K. D. Croft, L. J. Beilin, and I. B. Puddey, *Am. J. Clin. Nutr.* **71,** 67 (2000).

[31] Y. B. de Rijke, P. N. Demacker, N. A. Assen, L. M. Sloots, M. B. Katan, and A. F. Stalenhoef, *Am. J. Clin. Nutr.* **63,** 329 (1996).

and apoAII,[32] but both red and white wines were alleged to reduce serum HDL in mice fed an atherogenic diet.[16] Comparable effects of red and white wines on coagulation parameters were observed in human subjects,[33] but in rats, a rebound hypercoagulability phenomenon following alcohol consumption was not observed with red wine or when grape phenolics were added to alcohol.[34]

Clearly, a welter of information derived from indirect experimental approaches has led to confusion about whether wine polyphenols have any *in vivo* benefits that might add to the protection against CHD provided by ethanol. Many of these compounds, with the exception of *trans*-resveratrol,[35] are quite widely distributed among fruits and vegetables, and it is pertinent to enquire whether these dietary sources can also enhance the antioxidant, anti-inflammatory, and anticoagulant status of consumers. In other words, a number of important questions need to be answered to assess the health potential of dietary antioxidants whether in wine or other food sources: (1) Are they absorbed in the human intestinal tract? (2) Do they persist in human blood and tissues at concentrations compatible with biological activity? (3) How are they stored and/or excreted? (4) Is bioavailability in wine greater than in solid matrices, and is absorption from ethanolic solutions more efficient than from aqueous media?

Answers to these questions require the administration of individual polyphenols to experimental animals or humans under standardized conditions followed by measurement of their *in vivo* biological effects on lipid metabolism, coagulation, fibrinolysis, eicosanoid synthesis, lipid and membrane oxidation, and so forth. Such experiments, however, are predicated on first demonstrating that their intestinal absorption is adequately efficient to generate pharmacologically meaningful concentrations at their sites of action. Knowledge of their pharmacokinetic and pharmacodynamic behavior and their metabolism is also necessary. These investigations depend, in turn, on the availability of methods that allow the assay in blood, other body fluids, and tissues of the small quantities likely to accumulate after the intake of natural foods and beverages as well as in artificially enriched dietary sources or substitutes.

Method

Sample Preparation and Derivatization

One hundred microliters of sample (serum, urine, or homogenates containing a known number of red or white blood cells, or platelets, previously separated

---

[32] D. M. Goldberg, V. Garovic-Kocic, E. P. Diamandis, and C. R. Pace-Asciak, *Clin. Chim. Acta* **246**, 183 (1996).
[33] C. R. Pace-Asciak, O. Rounova, S. E. Hahn, E. P. Diamandis, and D. M. Goldberg, *Clin. Chim Acta* **246**, 163 (1996).
[34] J. C. Ruf, J.-L. Berger, and S. Renaud, *Arterioscler. Thromb. Vasc. Biol.* **15**, 140 (1995).
[35] G. J. Soleas, E. P. Diamandis, and D. M. Goldberg, *Clin. Biochem.* **30**, 91 (1997).

TABLE I
SOME CHARACTERISTICS OF GC-MS METHOD

Characteristic	trans-Resveratrol	(+)-Catechin	Quercetin
Mass spectral features			
Target ion	444	368	647
Qualifier ions	445,446	369,370	648,649
Retention time (min)	10.03	10.65	11.46
LOD (µg/liter)	0.1	0.2	0.2
LOQ (µg/liter)	1.0	2.0	2.0

on a dextran sulfate gradient)[36] is extracted twice with 0.5 ml of ethyl acetate by vortexing for 1 min in a 1-ml vial. The mixture is centrifuged for 5 min at 1500g, and the top layers (ethyl acetate) are removed, pooled in another tube, and evaporated to dryness under nitrogen. One hundred microliters of ethyl acetate and 100 µl of bis(trimethylsilyl)trifluoroacetamide (BSTFA) are added, and after vortexing the mixture for 30 sec it is heated for 2 hr at 70°.

Gas Chromatography–Mass Spectrometry Analysis

The derivatized sample is analyzed by Gas Chromatography–Mass Spectrometry (GC-MS) using a 5890 GC interfaced to a 5970 MSD, both from Hewlett-Packard (Mississauga, ON, Canada). One microliter is injected into a DB-5 column 15 m long, 0.25 mm id., and 0.25 µm film thickness (J & W, Folsom, CA). The injector and detector are set at 260° and 300°, respectively. The temperature program (oven) comprises two phases: in the first, temperature is ramped from 150° to 215° at a rate of 8°/min and held for 5 min; in the second, temperature is ramped to 300° at a rate of 30°/min and held for 6 min. Ultrahigh purity helium with an in-line Supelpure moisture trap and hydrocarbon trap (Supelco Canada, Mississauga, ON, Canada) is used as carrier gas. The carrier gas line is set at 48 psi, column head pressure at 8 psi, and the septum purge at 1.2 ml/min. The total run time is 22 min. The elution times and SIM parameters (one target ion and two qualifying ions, dwell time per ion 100 msec) for each compound are presented in Table I. A typical chromatogram for rat whole blood is shown in Fig. 1; a comparable resolution is achieved for rat serum and urine and for human whole blood, serum, and urine.

Calibration and Quality Control

Quantitation is accomplished by comparing the peak abundance with those of authentic pure standards purchased from Sigma-Aldrich Canada Ltd. (Mississauga,

---

[36] W. J. Williams, E. Beutler, and A. J. Erslev (eds.), "Haematology," 4th Ed. McGraw-Hill, New York. 1990.

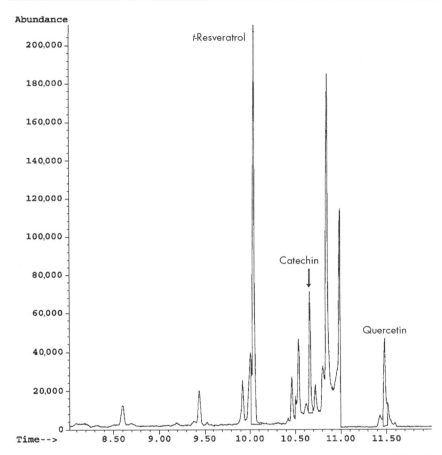

FIG. 1. Chromatogram (total ion abundance) of three polyphenolic antioxidants in whole blood of rat. Each was present at a concentration of 10 μg/liter.

Ontario, Canada). Stock concentrated standards at 1 g/liter in 95% (v/v) ethanol are stored at 4° under nitrogen and wrapped in foil, and are stable for 6 months. From these, six dilutions in 95% (v/v) ethanol covering the range of 1–100 μg/liter are prepared, stored as described earlier, and are stable for 2 weeks. A full calibration curve with four assays at each concentration is prepared weekly and limits for standard deviation (1 SD and 2 SD) are established. One high and one low standard are run at the beginning and the end of each day on which unknown samples are analyzed. Quality control (QC) criteria require that not more than three of the four samples are above or below the calibration line, not more than two outside the limit of 1 SD and <2 SD on either side of the line, and none beyond the 2 SD limits. A fresh calibration curve with newly defined SDs as described earlier is constructed on the rare occasion when QC criteria are not met, and the GC-MS analysis of the unknown samples is repeated.

Results

Analytical Sensitivity

Based on standard criteria,[37] the limits of detection (LOD), defined as the lowest concentration to exceed the mean baseline value by >3 SD, and the limits of quantitation (LOQ), defined as the lowest concentration of each compound that gave a signal-to-noise ratio of 10, were determined for standards and for whole blood, serum, and urine from rats and humans by spiking each matrix with 100 μg/liter of each compound and analyzing each sample 10 times. The results are shown in Table I. Good sensitivity was obtained with this method; LOD did not exceed 0.2 μg/liter and LOQ did not surpass 2.0 μg/liter for any of the three compounds in the matrices analyzed (Table I).

Recovery

This was examined by adding each of the three compounds to each matrix, independently and combined, at two different final concentrations. Each sample was analyzed, beginning with the extraction procedure, six times. The results where each compound was added separately are presented in Table II as the mean recovery (amount found as a percentage of the amount added) and its relative standard deviation (RSD) as a percentage of the mean. Analysis of variance[38] revealed that the recovery of quercetin (80.5–92.1%) was significantly less ($P < 0.01$) than that of *trans*-resveratrol (88.8–96.6%) or (+)-catechin (90.4–99.2%), but the latter two were not significantly different from each other. The recovery values when all three compounds were added simultaneously showed slightly lower percentages in most but not all cases (data not presented); the differences averaged out to around 1.5% overall (a little more in the case of quercetin than with the other two polyphenols) and the RSD values were higher by a similar amount, but analysis of variance of data for each individual polyphenol under both sets of conditions revealed no significant differences. No further action appears necessary for *trans*-resveratrol or (+)-catechin, but in the case of quercetin, accuracy is enhanced by use of an internal standard, for which purposes fisetin is recommended. The poorer recovery of quercetin is probably due to the fact that it is much less soluble in organic solvents than the other two polyphenols.

---

[37] R. O. Kringle, in "Tietz Textbook of Clinical Chemistry" (C. A. Burtis and E. R. Ashwood, eds.), p. 384. Saunders, Philadelphia, 1994.
[38] G. G. Snedecor and W. G. Cochran, "Statistical Methods." Iowa State Univ. Press, Ames, IA, 1967.

TABLE II
RECOVERY OF THREE POLYPHENOLS ADDED TO WHOLE BLOOD, BLOOD SERUM, AND URINE OF RATS AND HUMANS[a]

Species and compound	Concentration (μg/liter)	Mean recovery (%) ± RSD (%)		
		Whole blood	Serum	Urine
Rat				
trans-Resveratrol	10	98.3 ± 12.1	93.0 ± 4.3	93.2 ± 3.4
(+)-Catechin	10	93.7 ± 8.4	92.1 ± 4.7	96.1 ± 2.0
Quercetin	10	81.4 ± 16.3	83.5 ± 5.1	92.1 ± 8.6
trans-Resveratrol	100	90.2 ± 3.8	94.2 ± 5.2	91.9 ± 8.2
(+)-Catechin	100	87.6 ± 5.1	95.3 ± 4.8	99.2 ± 7.9
Quercetin	100	85.9 ± 7.6	88.6 ± 6.7	83.7 ± 15.8
Human				
trans-Resveratrol	10	94.0 ± 7.8	96.6 ± 4.4	101.6 ± 9.6
(+)-Catechin	10	91.5 ± 6.8	90.4 ± 3.3	94.5 ± 8.3
Quercetin	10	83.1 ± 11.7	82.7 ± 8.1	86.0 ± 12.9
trans-Resveratrol	100	88.8 ± 1.5	95.5 ± 2.2	96.3 ± 8.5
(+)-Catechin	100	97.2 ± 5.3	92.7 ± 5.9	105.1 ± 7.3
Quercetin	100	80.5 ± 4.3	85.4 ± 6.3	93.5 ± 10.5

[a] At two concentrations. All data are means of six independent assays.

Linearity

This was determined by adding independently to each of the six matrices sufficient of each polyphenol to generate six different concentrations over the range of 1–100 μg/liter for resveratrol and 1–150 μg/liter for catechin and quercetin. The linearity of all polyphenols in all matrices was excellent, and the quantitative parameters based on correlation and regression analyses[38] are listed in Table III. The closeness of fit is documented by the fact that the lowest r value recorded was 0.993 (catechin in human urine). In two-thirds of instances the line deviated from zero (y intercept) by <1% and in a further one-sixth by <2%. In only two instances, (+)-catechin and quercetin in human urine, did this deviation reach statistical significance ($P < 0.05$).

In human biological fluids, the slope for each polyphenol (also a measure of sensitivity of the detector response combined with the efficiency of extraction and derivatization) was virtually independent of the matrix, but based on the Pearson test utilizing the derived SE values for the slopes,[38] sensitivity for quercetin was significantly less than that for trans-resveratrol and (+)-catechin ($P < 0.02$), which did not differ from each other. In rat fluids, more variability in response was noted, although the values for each polyphenol did not differ significantly among the three matrices. However, the response of quercetin was consistently and significantly lower than that of trans-resveratrol ($P < 0.05$), whereas quercetin and (+)-catechin

TABLE III
LINEARITY OF POLYPHENOL ASSAYS IN WHOLE BLOOD, BLOOD SERUM, AND URINE OF RATS AND HUMANS[a]

Matrix and compound	Correlation coefficient (r)	Slope (R) (mv per μg)	y intercept[b] (mv)
Rat whole blood			
trans-Resveratrol	0.999	5.76×10^4	0.69×10^3 (1.2)
(+)-Catechin	0.997	8.86×10^3	0.34×10^2 (0.4)
Quercetin	0.998	4.70×10^3	0.64×10^2 (1.4)
Rat serum			
trans-Resveratrol	0.997	3.82×10^4	1.05×10^2 (0.3)
(+)-Catechin	0.997	8.78×10^3	0.34×10^2 (0.4)
Quercetin	0.998	1.94×10^3	0.16×10^2 (0.8)
Rat urine			
trans-Resveratrol	0.998	5.48×10^4	0.73×10^3 (1.3)
(+)-Catechin	0.998	8.42×10^3	0.25×10^2 (0.3)
Quercetin	0.997	7.03×10^3	0.43×10^2 (0.6)
Human whole blood			
trans-Resveratrol	0.996	1.09×10^5	0.92×10^3 (0.8)
(+)-Catechin	0.997	1.45×10^5	0.34×10^4 (2.3)
Quercetin	0.998	3.24×10^4	1.23×10^2 (0.4)
Human serum			
trans-Resveratrol	0.997	1.06×10^5	0.47×10^3 (0.4)
(+)-Catechin	0.997	1.37×10^5	0.36×10^4 (0.3)
Quercetin	0.999	2.81×10^4	1.42×10^2 (0.5)
Human urine			
trans-Resveratrol	0.998	1.09×10^5	0.46×10^3 (0.4)
(+)-Catechin	0.993	1.33×10^5	0.38×10^4 (2.9)
Quercetin	0.998	2.89×10^4	1.04×10^3 (3.6)

[a] Evaluated by correlation and regression analysis.[36]

[b] Data in parentheses represent the percentage deviation from zero, which, when evaluated by the Pearson test utilizing the derived SE,[38] was significant ($P < 0.05$) only in the case of catechin and quercetin in human urine, indicating minor bias possibly due to the presence of low amounts of these compounds in this matrix.

showed no significant differences. The response of resveratrol was higher than that of (+)-catechin in all matrices by 6.5-fold (whole blood and urine) and 4.4-fold (serum), with the first two being statistically significant at a level of $P < 0.001$ and the last at a level of $P < 0.02$.

Imprecision

Table IV presents the results of experiments in which the imprecision of the entire method (extraction plus derivatization plus chromatography and detection) was assessed in all matrices at low, medium, and high concentrations of all three polyphenols added simultaneously to the matrix. Even at 2 μg/liter (approximately,

TABLE IV
IMPRECISION OF THREE POLYPHENOL ASSAYS IN WHOLE BLOOD, BLOOD SERUM, AND URINE OF RATS AND HUMANS[a]

Matrix and compound	Approximate concentration		
	2 μg/liter	10 μg/liter	100 μg/liter
Human whole blood			
trans-Resveratrol	9.8	3.5	1.2
(+)-Catechin	7.8	4.4	4.3
Quercetin	12.2	6.6	3.8
Human serum			
trans-Resveratrol	8.8	4.0	1.8
(+)-Catechin	10.2	3.1	4.2
Quercetin	12.5	6.8	4.9
Human urine			
trans-Resveratrol	8.3	3.7	1.6
(+)-Catechin	9.4	3.3	3.2
Quercetin	13.8	7.4	4.7
Rat whole blood			
trans-Resveratrol	9.9	6.6	5.2
(+)-Catechin	10.0	9.6	4.8
Quercetin	11.3	8.2	6.7
Rat serum			
trans-Resveratrol	8.2	5.6	4.3
(+)-Catechin	9.1	8.4	4.7
Quercetin	14.0	8.8	5.1
Rat urine			
trans-Resveratrol	8.2	3.4	2.0
(+)-Catechin	7.9	2.0	2.3
Quercetin	15.8	8.6	6.9

[a] At three concentrations. Data are RSD (%) based on six replicate analyses of the same samples spiked simultaneously with all three polyphenols approximating the stated concentration (±15%).

LOQ), imprecision for *trans*-resveratrol and (+)-catechin did not exceed 9.8 and 10.2%, respectively. At the two higher concentrations, both of these polyphenols could be measured with excellent reproducibility in human whole blood and blood serum and in the urine of both humans and rats, with imprecision in rat whole blood and serum being somewhat higher. The quercetin assays were the least precise of the three, but apart from the lowest concentration, the RSD values were <10% in all matrices.

Further experiments were undertaken to define the sources of imprecision. With *trans*-resveratrol and (+)-catechin, the imprecision was not diminished significantly when replicate analyses were performed on ethyl acetate extracts, indicating that the extraction procedure was highly reproducible. This was not so for quercetin, where the RSD was reduced by 30% on analyzing the ethyl acetate extracts.

Comparing replicate analyses performed on the dried ethyl acetate residues before and after BSTFA treatment, it appeared that the RSD values for *trans*-resveratrol and (+)-catechin were on average only about 15% lower after derivatization, whereas with quercetin this reduction was 20–25%. Variance due to the chromatographic separation and MSD procedures accounted for <50% of the total analytical variance for quercetin and >70% for *trans*-resveratrol and (+)-catechin.

Distribution of Polyphenols among Blood Components

In previous experiments utilizing a reversed-phase high-performance liquid chromatography (HPLC) procedure with isocratic elution,[39,40] we noted that *trans*-resveratrol, when added to blood, becomes tightly bound to serum proteins and to the membranes of red and white blood cells and platelets, with the further possibility that part of the resveratrol is internalized. Of four solvents tested (ethanol, methanol, acetonitrile, and acetone), only acetone was able to affect complete extraction of *trans*-resveratrol together with protein precipitation, allowing complete phase separation on centrifugation. Even then, three extractions with a total of 17 volumes of solvent were required. We did not test ethyl acetate in these experiments. The distribution of added *trans*-resveratrol was calculated for the serum and various cellular fractions of the blood using the present method, and the results are shown diagrammatically in Fig. 2. Serum accounted for 54.8% of added *trans*-resveratrol in human and 49.8% in rat blood. However, taking protein content into account, the combined leukocyte/platelet fraction (buffy coat) was the most highly enriched in *trans*-resveratrol. In view of the large amounts present in cellular fractions, as subsequently confirmed by others,[41] assays performed on serum alone would seriously underestimate the uptake of *trans*-resveratrol following its oral administration to humans and animals.

Using serum and cell suspensions in a manner analogous with our previous studies, we have found complete extraction of added polyphenols with the present method, with recoveries around 95% for *trans*-resveratrol and (+)-catechin in all fractions; the recovery of quercetin was a little lower, but in every instance the amount of polyphenol obtained with a third ethyl acetate extraction did not exceed 1.5% of the total dose. At variance with our previous studies,[39,40] we did not find evidence for specific binding of any of these polyphenols; rather, the amount associated with any cell fraction was essentially a function of its total protein

---

[39] L. Tham, D. M. Goldberg, E. P. Diamandis, A. Karumanchiri, and G. J. Soleas, *Clin. Biochem.* **28**, 339 (1995).

[40] D. M. Goldberg, L. Tham, E. P. Diamandis, A. Karumanchiri, and G. J. Soleas, *Clin. Chem.* **41**, S115 (1995).

[41] A. Blache, I. Rustan, P. Durand, G. Lesgards, and N. Loreau, *J. Chromotogr. B* **702**, 103 (1997).

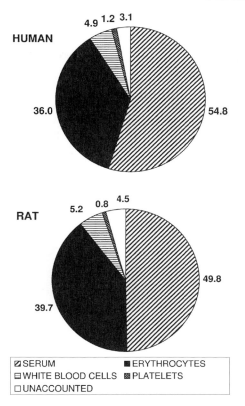

FIG. 2. Distribution (percentage of dose) among blood components of *trans*-resveratrol added to whole human blood (top) and whole rat blood (bottom) after separation of fractions[36] and extraction as described in the text. Blank space represents percentage not recovered.

content or volume. However, the distribution of *trans*-resveratrol in blood serum was in good agreement with our previous work: $56.6 \pm 2.6\%$ in human blood and $48.4 \pm 3.4\%$ in rat blood (mean $\pm$ SD, $n = 4$). As with our earlier results, this difference is consistent with the higher hematocrit in the rat compared with human blood. For (+)-catechin and quercetin, the distribution in human blood greatly favored the serum, 72.0 ± 4.2 and $86.5 \pm 6.1\%$, respectively ($n = 4$); comparable values for rat blood were of a similar order but about 10% lower, again showing consistency with a higher hematocrit.

Preliminary Demonstration of trans-Resveratrol Absorption in Humans

Blood is withdrawn at 30-min intervals from two healthy male subjects who, after an 18-hr fast, drank 100 ml of white wine containing 25 mg of added

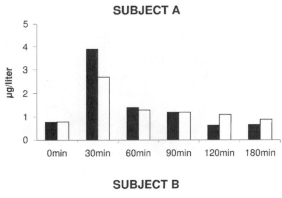
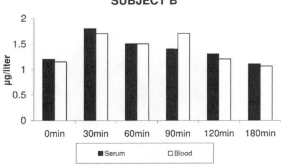

FIG. 3. Concentration of *trans*-resveratrol in whole blood and serum of two healthy males at various times after oral consumption of 25 mg of the compound in 100 ml of white wine.

trans-resveratrol over a period <5 min. The results (Fig. 3) demonstrate the appearance of *trans*-resveratrol in the whole blood and serum at concentrations consistent with the absorption of at least 10–15% of the administered dose, calculated as an approximation based on the area under the curve and total body water. A much smaller amount (<1% of the dose) is excreted in the urine of these subjects over a 24-hr period. These data can be contrasted with our observations in rats given an oral dose of [^{3}H]resveratrol and in whom around 70% of the isotope appears in the urine over the next 24 hr, despite only trace amounts in the blood and tissues.[42] Taken together, these findings are consistent with the notion that *trans*-resveratrol is absorbed rapidly by the mammalian intestine and may undergo extensive metabolism prior to its excretion in the urine.

[42] G. J. Soleas, M. Angelini, L. Grass, E. P. Diamandis, and D. M. Goldberg, *Methods Enzymol.* **335** [13] 2001 (this volume).

Discussion

Several decades ago, the absorption and urinary excretion of (+)-catechin in human subjects was demonstrated indirectly by the measurement of total phenols[43] and tracer studies with (+)-[^{14}C]catechin.[44] Absorption exceeded 50% of the dose administered, and blood levels, which peaked around 3 hr, persisted for at least 12 hr.[44] More recent investigations using HPLC methods were consistent with these observations.[45,46] Other methods utilizing UV[47] and fluorescence[48] detection following HPLC have also been presented. These two detection systems were evaluated,[46] and fluorescence proved to be superior; the method finally developed by these authors was very precise [coefficient of variation $(CV) = 3.9$–6.5% at 50 μg/liter], was linear from 45 to 1.35 mg/liter, and had LOD and LOQ of 5 and 40 μg/liter, respectively; however, recovery was only 85%, possibly because of a membrane filtration step during sample preparation. Assay of (+)-catechin by the present method offers superior sensitivity and recovery.

Gugler and colleagues[49] studied the disposition of quercetin in human subjects after oral and intravenous (IV) administration using a method with LOQ = 0.1 mg/liter. Even after an oral dose of 4 g, quercetin and its main metabolite were undetectable in urine. Subsequent experiments performed on human ileostomy patients revealed that only 48–83% of a single oral dose of quercetin and two of its derivatives given as onions reached the terminal ileum;[50] the difference was attributed to absorption, although blood levels were not measured, and only 0.5% of the amount presumed to be absorbed was recovered in the urine over the next 13 hr. Nevertheless, reports have described the absorption from the human intestine of anthocyanins, flavonoid derivatives that are no less chemically complex than quercetin and its conjugates,[51,52] and the presence of quercetin glycosides in plasma from nonsupplemented human subjects has been documented.[53] Moreover,

---

[43] N. P. Das, *Biochem. Pharmacol.* **20**, 3435 (1971).
[44] A. M. Hackett and L. A. Griffiths, *Xenobiotica* **13**, 279 (1983).
[45] A. L. Waterhouse, R. M. Walzem, P. L. Teissedre, J. B. German, R. J. Hansen, and E. N. Frankel, in "Polyphenols Communication" (J. Vercauteren, C. Cheze, M. C. Dumon, and J.F. Weber, eds.), p. 407. Groupe Polyphenols, Bordeaux, 1996.
[46] S. Carando, P.-L. Teissedre, and J.-C. Cabanis, *J. Chromatogr. B* **707**, 195 (1998).
[47] H. Y. Pan, D. L. Liu, P. P. Xu, and M. L. Lu, *Acta. Pharmacol. Sin.* **26**, 371 (1991).
[48] Y. Ho, Y. L. Lee, and K. Y. Hsu, *J. Chromatogr. B* **665**, 383 (1995).
[49] R. Gugler, M. Leschik, and H. J. Dengler, *Eur. J. Clin. Pharmacol.* **9**, 229 (1975).
[50] P. C. H. Hollman, J. H. M. de Vries, S. D. van Leeuwen, M. J. B. Mengelers, and M. B. Katan, *Am. J. Clin. Nutr.* **62**, 1276 (1995).
[51] T. Lapidot, S. Harel, R. Granit, and J. Kanner, *J. Agric. Food Chem.* **46**, 4297 (1998).
[52] T. Miyazawa, K. Nakagawa, M. Kudo, K. Muraishi, and K. Someya, *J. Agric. Food Chem.* **47**, 1083 (1999).
[53] G. Paganga and C. A. Rice-Evans, *FEBS Lett.* **401**, 78 (1997).

Hollman et al.[54] supplemented their earlier report[50] by recording plasma concentrations of 200 μg/liter quercetin after an oral dose of 64 mg, a result consistent with the absorption of at least 10–25% of the dose. The presence of mean concentrations of 126 μg/liter in blood of humans given a large oral dose has also been described.[55] This would be consistent with the absorption of around 5–7.5% of the dose given. The analysis employed HPLC with coulometric detection and LOD of 1.7 μg/liter, but no other analytical features were described. Because quercetin has shown promising anticancer activity when given IV to human patients,[56] its potential absorption needs to be unequivocally ascertained, and the present method should aid that endeavor.

Absorption of *trans*-resveratrol has been demonstrated in the rat.[57,58] These authors used a liquid chromatography–mass spectrometry (LC-MS) system employing two HPLC columns in tandem, preceded by a fairly lengthy extraction step. The method is incompletely described and no analytical performance data were presented apart from a statement that the LOQ is 1 μg/liter. The method of Blache et al.[41] utilizes GC with flame-ionization detection and clearly lacks the positive identification by ion spectrum inherent in our method. Linearity and recovery were excellent, imprecision at 3.24 mg/liter was 3.3% (RSD), but LOD was 50 μg/liter (LOQ not indicated). Two methods for the measurement of plasma *trans*-resveratrol have been described with detection limits of 20[59] and 5 μg/liter,[60] respectively. Only the former has actually been applied in whole animals.

In conclusion, the present method is versatile, offers excellent analytical features, including high sensitivity and positive peak identification, and can be applied to whole blood, blood serum, blood cell fractions, and urine. It is suitable for pharmacokinetic and pharmacodynamic investigations on these three biologically important polyphenolic antioxidants and has been used in a preliminary study to demonstrate for the first time that some absorption of *trans*-resveratrol does occur

---

[54] P. C. H. Hollman, M. V. D. Gaag, M. J. B. Mengelers, J. M. P. Van Trijp, J. H. M. DeVries, and M. B. Katan, *Free Radic. Biol. Med.* **21,** 703 (1996).

[55] C. Manach, C. Morand, V. Crespy, C. Demigne, O. Texier, F. Regerat, and C. Remesy, *FEBS Lett.* **426,** 331 (1998).

[56] D. R. Ferry, A. Smith, J. Malkhandi, D. W. Fyfe, P. G. deTakats, D. Anderson, J. Baker, and D. J. Kerr, *Clin. Cancer Res.* **2,** 659 (1996).

[57] A. A. E. Bertelli, L. Giovannini, R. Stradi, A. Bertelli, and J.-P. Tillement, *Int. J. Tissue Reac.* **17,** 67 (1996).

[58] A. A. E. Bertelli, L. Giovannini, R. Stradi, S. Urien, J.-P. Tillement, and A. Bertelli, *Int. J. Clin. Pharm. Res.* **16,** 77 (1996).

[59] M. E. Juan, R. M. Lamuela-Raventós, M. C. de la Torre-Boronat, and J. M. Planas, *Anal. Chem.* **71,** 747 (1999).

[60] Z. Zhu, G. Klironomos, A. Vachereau, L. Neirinck, and D. W. Goodman, *J. Chromatogr. B* **724,** 389 (1999).

in humans. No method for these matrices published to date with adequate documentation can match its analytical performance for these three important polyphenolic antioxidants.

Acknowledgments

This work was performed by Mr. George Soleas in partial fulfillment of the requirements of a Ph.D. degree at the University of Toronto. The authors thank Mrs. Patricia Machado for her skillful preparation of this manuscript. Initial technical help at the outset of this study was provided by Lucy Tham and Dr. Jan Palaty. We are grateful to the National Research Council of Canada (IRAP Program), the Canadian Wine Institute, and the Wine Institute (San Francisco) for their generous support.

[13] Absorption of *trans*-Resveratrol in Rats

By GEORGE J. SOLEAS, MARK ANGELINI, LINDA GRASS, ELEFTHERIOS P. DIAMANDIS, and DAVID M. GOLDBERG

Introduction

Discrepancies between the *in vitro* biological activities of antioxidant polyphenols present in red wine, such as *trans*-resveratrol, (+)-catechin and quercetin, and the *in vivo* effects of red wine oral consumption or that of the individual polyphenols by animals and humans have been described.[1] This paradox is especially striking for *trans*-resveratrol. Despite its ability to inhibit the synthesis and secretion of apolipoprotein B and cholesterol esters[2,3] and to block platelet aggregation and thromboxane synthesis,[4,5] daily administration of 375 ml of red wine enriched in this compound to healthy human volunteers for 4 weeks did not alter any of these functions (platelet aggregation, eicosanoid synthesis, or lipoprotein metabolism)[6,7] *In vitro, trans*-resveratrol manifests significant estrogen-like

---

[1] G. J. Soleas, J. Yan, and D. M. Goldberg, *Methods in Enzymol.* **335**, [12] 2001 (this volume).
[2] D. M. Goldberg, S. E. Hahn, and J. G. Parkes, *Clin. Chim. Acta* **237**, 155 (1995).
[3] D. M. Goldberg, G. J. Soleas, S. E. Hahn, E. P. Diamandis, and A. Karumanchiri, *in* " Wine Composition and Health Benefits" (T. R. Watkins, ed.), p. 24. American Chemical Society, Washington, DC, 1997.
[4] M.-I. Chung, C.-M. Teng, K.-L. Cheng, F.-N. Ko, and C.-N. Lin, *Planta Med.* **58**, 274 (1992).
[5] C. R. Pace-Asciak, S. Hahn, E. Diamandis, G. Soleas, and D. M. Goldberg, *Clin. Chim. Acta* **235**, 207 (1995).
[6] C. R. Pace-Asciak, O. Rounova, S. E. Hahn, E. P. Diamandis, and D. M. Goldberg, *Clin. Chim. Acta* **246**, 163 (1996).
[7] D. M. Goldberg, V. Garovic-Kocic, E. P. Diamandis, and C. R. Pace-Asciak, *Clin. Chim. Acta* **246**, 183 (1996).

activity,[8,9] but after oral administration it has been very difficult to demonstrate consistent *in vivo* estrogenic effects.[10] Other examples of this paradox have been outlined in another comprehensive review,[11] raising the issue of failure of absorption as a reason for the lack response.

In 1996, Bertelli and colleagues[12,13] reported the absorption of *trans*-resveratrol by rats after a very small intragastric dose (25 μg) and went on to describe the results of pharmacokinetic studies involving sequential assays on blood, urine, and various tissues. Peak concentrations in plasma (around 30 μg/liter) occurred at 60 min, according to data presented. These papers provide little methodological detail, and no explanation was proposed for the attainment of such a high plasma concentration relative to the dose administered. Three years later, it was reported anecdotally that the *trans*-resveratrol concentration of rat plasma 15 min after an oral dose of approximately 0.5 mg was 175 μg/liter.[14] No experimental details (e.g., number of animals) were given, nor were data for other time periods described, although the authors stated that blood was also drawn at 30 and 45 min after the dose.

For several years we have been grappling with the problem of measuring *trans*-resveratrol concentrations in the blood at levels occurring after oral administration, and have developed a suitable method for doing so.[1] Before this was finally achieved, we were able to perform a series of experiments using *trans*-resveratrol radiolabeled with [^{3}H] in a stable nonexchangeable position within the first benzene ring. These observations were amplified by use of the new method[1] and are described in this article.

Materials and Methods

trans-Resveratrol, tritiated in the 4-position of the first benzene ring and containing 170 μCi/ml, is from Sibtech, Inc. (Newington, CT) by special requisition, with purity >99% guaranteed. Of the original methanolic solution, 350 μl is diluted to 1 ml with 30% (v/v) ethanol to yield a stock solution of 60 μCi/ml. For intragastric administration, 2 μl (120 nCi) of this solution is added to 1 ml of one

---

[8] B. D. Gehm, J. M. McAndrews, P. Y. Chien, and J. L. Jameson, *Proc. Natl. Acad. Sci. U.S.A.* **94**, 14138 (1997).
[9] S. Stahl, T. Y. Chun, and W. G. Gray, *Toxicol. Appl. Pharmacol.* **152**, 41 (1998).
[10] R. T. Turner, G. L. Evans, M. Zhang, A. Maran, and J. D. Sibonga, *Endocrinology* **140**, 50 (1999).
[11] G. J. Soleas, E. P. Diamandis, and D. M. Goldberg, *in* "Nutrition and Cancer Prevention: New Insights into the Role of Phytochemicals" (American Institute for Cancer Research, eds.). Kluwer Academic/Plenum Publishers, New York, 2001, in press.
[12] A. A. E. Bertelli, L. Giovannini, R. Stradi, A. Bertelli, and J.-P. Tillement, *Int. J. Tissue Reac.* **17**, 67 (1996).
[13] A. A. E. Bertelli, L. Giovannini, R. Stradi, S. Urien, J.-P. Tillement, and A. Bertelli, *Int. J. Clin. Pharm. Res.* **16**, 77 (1996).
[14] M. E. Juan, R. M. Lamuela-Raventós, M. C. de la Torre-Boronat, and J. M. Planas, *Anal. Chem.* **71**, 747 (1999).

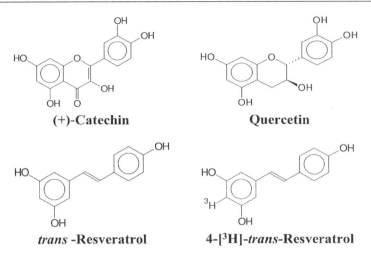

FIG. 1. Structures of unlabeled and [^{3}H]-labeled *trans*-resveratrol, (+)-catechin, and quercetin.

of the following three matrices: 10% (v/v) ethanol, V-8 homogenized vegetable cocktail (Campbell Foods), and white grape juice (Welch Food Co.) The matrices have values of pH 4.5–6.0. In some experiments, unlabeled *trans*-resveratrol, (+)-catechin, or quercetin (all from Sigma, St. Louis, MO) are coadministered together with the labeled *trans*-resveratrol. Their structures are illustrated in Fig. 1.

Animal Procedures

Male Wistar rats averaging 350 g are fed water and chow *ad libitum* and are acclimatized for 7 days. They are fasted overnight, lightly stunned with CO_2, and given one of the three matrix solutions with or without antioxidant polyphenols, as described earlier, by gavage: 1 ml initially and a further 1 ml of the matrix containing no additive by the same syringe to ensure washing out of all the material. The animals are placed individually in metabolic cages and observed carefully for the first 20 min to ensure that vomiting or regurgitation does not occur. On a few occasions the animal is rejected for this reason. Where appropriate, urine and feces are collected independently for a period of 24 hr. Two hours after gavage, chow and water are restored to the cages *ad libitum* except for short-term experiments where the animals are sacrificed within the first 2 hr to examine blood concentrations over this period. They are sacrificed by cardiac puncture and blood withdrawal; some is set aside for analysis as whole blood and the rest is centrifuged after clotting to produce serum.

Liver, kidneys, heart, and spleen are removed, washed in ice-cold isotonic saline, trimmed free of fat, finely chopped, and homogenized in a PRO-200 blender (Pro Scientific, CT) in 5 volumes of 95% (v/v) ethanol for three 1-min periods separated by 30-sec intervals. The homogenate is centrifuged (1,500g for 15 min at 4°), the supernatant is transferred, and the residue is returned to the blender for two further extractions with the same volume of ethanol. The supernatants are pooled, and the volume is measured and recorded. Whole blood and blood serum are similarly extracted.

The bladder is carefully dissected, the contents are washed into the blender with 95% (v/v) ethanol, the bladder is finely chopped and added together with the 24-hr urine collection, and the mixture is extracted as described earlier. The final volume of the pooled supernatants is carefully measured. This material is referred to as urine.

The entire colon from the ileocecal junction to the anus is dissected free, briefly washed in isotonic saline, placed in a flat dish, and sliced longitudinally, and the contents are carefully scraped free with a spatula assisted by rinsing with 95% (v/v) ethanol. The contents are quantitatively transferred to the blender, together with the 24-hr fecal collection. The entire colon is finely chopped and added to the contents. Homogenization is performed following the procedure used for tissues, but with one additional extraction step. This material is referred to as stool.

Preliminary experiments demonstrated virtually complete recovery of radioactivity from all sources using the techniques just described; in no instance did a further extraction yield >1.5% of the value obtained with the standard procedures. During tissue preparation and extraction, all materials are kept on ice or at 4° and protected against light by metal foil. The receptacles used to collect the 24-hr urine and fecal samples are similarly protected against light and temperature. The full protocol is approved by the Animal Experimentation Committee of the University of Toronto.

Assays

Radioactivity is determined as disintegrations per minute (dpm) after color quench corrections by adding 500 μl of the ethanol extract to 5 ml of scintillation fluid (Ready-Value, Beckman Instruments, Fullerton, CA) in a Beckman 600 scintillation counter. By reference to the total volume of extract, the total dpm attributable to each sample is calculated. Counts are measured over a 5-min period. Duplicates are run on each extract, and the results are averaged; if, as with very low values, the duplicates differed by >10%, two further replicates are analyzed and the results of all four assays are averaged.

Results

Overall Absorption and Matrix Effect

The percentage of tritiated *trans*-resveratrol absorbed over a 24-hr period is calculated according to the liquid matrix in which it was administered. Only trace amounts (<1% of the dose administered) are detectable in the liver, kidneys, heart, or spleen, with the aggregate value for these tissues being <2%. Values for blood and plasma at that time are scarcely above background. Around 75% of the dose administered with each matrix is accounted for by adding the radioactivity of stool and urine. Taking as the measure of absorption the difference between the amount of radioactivity given and the amount recovered in the stool (24-hr feces plus colonic contents and colon), it appears that 77–80% of *trans*-resveratrol may be absorbed in the rat intestine, there being no differences among the three liquid matrices (Fig. 2). Clearly, all of the tritiated label present in the urine must have been absorbed and subsequently excreted, the values ranging from 49 to 61% with, again, no differences among the three matrices (Fig. 3). It therefore appears that, by any criteria, at least 50% of *trans*-resveratrol is absorbed by the rat and that alcohol up to 10% by volume does not enhance absorption. Therefore, the compound is no more bioavailable in wine than in aqueous beverages, lending credence to the notion that it can, if considered desirable, be administered as a food or beverage additive. At this time, we cannot explain the difference (approximately 25%) between the sum of stool plus urine radioactivity and the amount actually

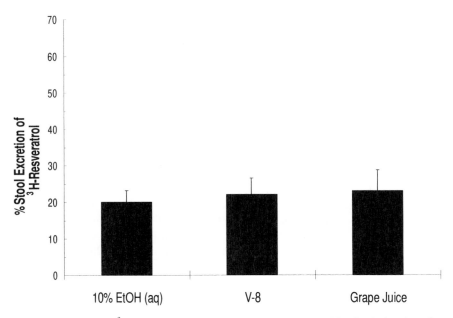

FIG. 2. Recovery of [^{3}H] label as percentage of dose given in the rat stool fraction (colon plus colon contents plus feces) 24 hr after intragastric administration of [^{3}H]-labeled *trans*-resveratrol in three different liquid matrices (mean ± SEM, $n = 8$). All assays are in duplicate.

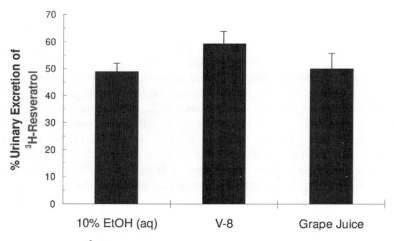

FIG. 3. Recovery of [^{3}H] label as percentage of dose given in the rat urine fraction (bladder plus contents plus urine) 24 hr after intragastric administration of [^{3}H]-labeled *trans*-resveratrol in three different liquid matrices (mean ± SEM, $n = 8$). All assays are in duplicate.

administered. Some could be accounted for by excretion via sweat and respiratory water as well as by metabolism to CO_2. Keeping in mind its high lipid solubility, *trans*-resveratrol might be deposited over a 24-hr period in adipose tissue and other tissues with high lipid content, such as the brain and nervous system. Skeletal muscle is another possible source that has not been examined. The amount of unaccounted radioisotope is not large (approximately 30 nCi), and the concentrations in such tissues, if any, would be difficult to measure accurately.

Competition Experiments

The 24-hr urinary excretion of radioactivity was measured in rats in whom unlabeled *trans*-resveratrol, (+)-catechin, and quercetin were coadministered in a matrix of 10% (v/v) ethanol in concentrations ranging from 10 nM to 1 mM. No significant inhibition by these compounds was observed (Fig. 4). There would not appear to be competition among these three antioxidant polyphenols for a common absorptive mechanism, and the failure of unlabeled *trans*-resveratrol to reduce urinary excretion suggests that its absorption is not saturable over the range of concentrations used.

Time Course of trans-*Resveratrol Absorption*

Values for the radioactivity in whole blood and serum of two rats after the administration of tritiated *trans*-resveratrol over the first 2-hr period are presented in Fig. 5. Significant radioactivity was present at 30 min after administration, but

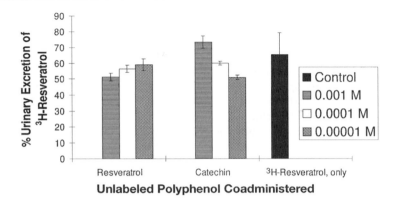

FIG. 4. Recovery of [^{3}H] label (120 nCi tritiated *trans*-resveratrol) as percentage of dose given together with one of three concentrations of *trans*-resveratrol, (+)-catechin, or quercetin in 10% (v/v) ethanol by gavage in the rat urine fraction 24 hr after administration (mean ± SEM, $n = 8$). All assays are in duplicate.

remained around the same concentrations over the next 90 min. These observations are not consistent with the rapid rise and fall of *trans*-resveratrol in the blood of rats, peaking at 1 hr previously reported,[12] or as observed by us in human experiments.[1] However, it is invidious to compare experiments in which only the parent compound is analyzed with others in which this, as well as metabolites, is included in the estimate.

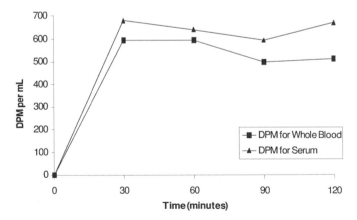

FIG. 5. Radioactivity (dpm per ml) of whole blood and blood serum in rats given 120 nCi of [^{3}H]-labeled *trans*-resveratrol at four intervals after intragastric administration in 10% (v/v) ethanol. Mean of results for two rats, all assays are in duplicate.

Assays for Unlabeled trans-Resveratrol

When the methods for assay of *trans*-resveratrol in serum and urine became available,[1] they were applied to its determination in all of the 24-hr urine samples (as well as in a few 24-hr serum samples) collected in experiments where cold and tritiated *trans*-resveratrol were administered simultaneously. These had been stored at $-20°C$ for up to 4 months. The mean percentage of the dose administered measurable in the 24-hr urine ranged from 2.5% (10 nM) to 7.4% (100 nM) to 14.7% (1 mM). These results suggest that whereas the absorption of *trans*-resveratrol as indicated by the excretion of radioactivity is not affected by the dose administered over this range, the metabolic conversion of the parent compound is saturable so that as the dose is increased, a higher percentage is excreted unchanged in the urine. As expected, the *trans*-resveratrol concentrations of the 24-hr serum samples assayed ranged between the limit of detection (0.1 µg/liter) and the limit of quantitation (1 µg/liter) of the method.

A new set of experiments was conducted, in the first of which (A) rats were given 0.5, 1.5, and 2.5 mg of *trans*-resveratrol in 1 ml of 20% (v/v) ethanol as described earlier, and blood was withdrawn by cardiac puncture 60 min later. The mean concentrations in serum were 2.5 µg/liter after 0.5 mg, 3.6 µg/liter after 1.5 mg, and 5.7 µg/liter 2.5 mg; mean blood concentrations were, respectively, 2.2, 2.5, and 4.5 µg/liter. In the second set (B), the amount given was 5 mg and for two rats, blood was withdrawn at 15, 30, and 60 min. The results are illustrated in Fig. 6. High concentrations of *trans*-resveratrol were already evident in serum by 15 min after gavage, peaking at 30 min, and falling precipitously over the next 30 min. A similar time course was seen in whole blood, although the concentrations were consistently lower than those of serum.

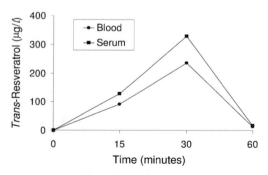

FIG. 6. *trans*-Resveratrol concentrations in whole blood and blood serum of rats at three intervals after intragastric administration in 20% (v/v) ethanol of 5 mg unlabeled compound. Data are means for two rats with all assays in duplicate.

Discussion

These experiments extend the informations provided by previous investigations[12–14] in several directions. First, peak concentrations of *trans*-resveratrol occur in blood and serum very rapidly, around 15 min from the time of administration. The difference between this result and that of Bertelli *et al.*[12] may, at least in part, be attributable to the much greater dose used in the present work. Second, while the concentration of parent compound falls sharply after this peak, radioactive metabolites decline far more slowly. Third, taking account of excretion of the tritium label in feces and urine, 50–75% of orally administered *trans*-resveratrol appears to be absorbed.

This last estimate may be conservative. (+)-Catechin has been shown to undergo biliary excretion,[15] whereas quercetin, which does not seem to be excreted in rat bile, can be secreted by rat intestine after conjugation and absorption.[16] It is therefore conceivable that some of the stool radioactivity after gastric *trans*-resveratrol administration may represent excretion of part of the initial dose after absorption and metabolism.

Following its absorption, quercetin, which is highly hydrophobic, is converted rapidly to more water-soluble conjugates (glucuronide and sulfate) and glycosides.[17–19] Further, these conjugates are potent antioxidants, being fourfold more effective than Trolox in blocking the oxidation of low-density lipoproteins.[20] A similar metabolic pattern may prevail for *trans*-resveratrol. In support of this notion, we found that radioactivity was best extracted from all tissues and matrices by 95% (v/v) ethanol rather than by ethyl acetate, which was the most efficient solvent for extracting *trans*-resveratrol from these same preparations.[1] This suggests that the radioactivity was in large measure associated with metabolites more water soluble than the parent compound. It should also be noted that Casper *et al.*[21] found that when *trans*-resveratrol was administered to rats it was much more potent in blocking the function of the aryl hydrocarbon receptor (required for the conversion of proximal carcinogens to metabolites) than when incubated with cell

---

[15] N. P. Das and S. P. Sothy, *Biochem, J.* **125,** 417 (1971).

[16] V. Crespy, C. Morand, C. Manach, C. Besson, C. Demigne, and C. Remesy, *Am. J. Physiol.* **277,** G120 (1999).

[17] G. Paganga and C. A. Rice-Evans, *FEBS Lett.* **401,** 78 (1997).

[18] C. Manach, C. Morand, V. Crespy, C. Demigne, O. Texier, F. Regerat, and C. Remesy, *FEBS Lett.* **426,** 331 (1998).

[19] C. Manach, O. Texier, C. Morand, V. Crespy, F. Regerat, C. Demigne, and C. Remesy, *Free Radic. Biol. Med.* **27,** 1259 (1999).

[20] C. Morand, V. Crespy, C. Manach, C. Besson, C. Demigne, and C. Remesy, *Am. J. Physiol.* **275,** R212 (1998).

[21] R. F. Casper, M. Quesne, I. M. Rogers, T. Shirota, A. Jolivet, E. Milgrom, and J. F. Savouret, *Mol. Pharmacol.* **56,** 784 (1999).

cultures *in vitro*; they speculated that hepatic biotransformation generated one or more metabolites with much greater activity than the parent compound.

In certain red wines, substantial amounts of resveratrol glucosides (polydatins) are present, which, on hydrolysis with glucosidases, yield the free stilbene.[22] Since Hollman et al.[23] have shown that flavonoid glycosides are absorbed by humans much more effectively than free flavonoids,[23] it is reasonable to speculate that polydatins are likewise highly bioavailable and capable of supplementing the favorable biological effects attributable to *trans*-resveratrol.

On a final technical note, it is intriguing that when added to whole blood *in vitro*,[1] a much higher proportion of *trans*-resveratrol is associated with the cells than is the case when *in vivo* absorption into the blood takes place, as in the present work. This calls for considerable caution in extrapolating the results of *in vitro* distribution experiments to the whole organism.

Acknowledgments

This work was performed by Mr. George Soleas in partial fulfillment of the requirements for the degree of Ph.D. at the University of Toronto. We thank Mrs. Patricia Machado for her skillful preparation of this manuscript and the Wine Institute (San Francisco) for its financial support, especially Mrs. Elizabeth Holmgren of the latter organization.

---

[22] D. M. Goldberg, E. Ng, A. Karumanchiri, E. P. Diamandis, and G. J. Soleas, *Am. J. Enol. Vitic.* **47**, 415 (1996).

[23] P. C. Hollman, M. N. Bijsman, Y. van Gameren, E. P. Cnossen, J. H. de Vries, and M. B. Katan, *Free Radic. Res.* **31**, 569 (1999).

Section III

Antioxidant Action

[14] Galvinoxyl Method for Standardizing Electron and Proton Donation Activity

By Honglian Shi, Noriko Noguchi, and Etsuo Niki

Introduction

Flavonoids are diphenylpropanes that commonly occur in plants and are frequently components of human diet. It has been found that flavonoids and other polyphenols possess antitumoral,[1–5] antiallergic,[6,7] antiplatelet,[8–10] anti-ischemic,[11–13] and anti-inflammatory[14–19] activities. Most of these biological effects are believed to come from their antioxidant properties. Flavonoids can exert their antioxidant activity by inhibiting the activities of enzymes, including lipoxygenase and cyclooxygenase,[20–22] by chelating metal ions,[20,23–27] and, most importantly, by scavenging free radicals.[28–33] The free radical-scavenging ability of flavonoids or other phenols is ascribed to the active phenolic hydroxy groups attached to ring structures. Phenolic acids, polyphenols, and in particular flavonoids quench free radicals by donating hydrogen in the active hydroxyl

---

[1] J. Brown, *Mutat. Res.* **75,** 243 (1980).
[2] V. Elangovan, N. Sekar, and S. Govindasamy, *Cancer Lett.* **87,** 107 (1994).
[3] E. E. Deschner, J. Ruperto, G. Wong, and H. L. Newmark, *Carcinogenesis* **12,** 1193 (1991).
[4] J. K. Lin, Y. C. Liang, and S. Y. Lin-Shiau, *Biochem. Pharmacol.* **58,** 911 (1999).
[5] T. L. Wadsworth and D. R. Koop, *Biochem. Pharmacol.* **57,** 941 (1999).
[6] H. Cheong, S. Y. Ryu, M. H. Oak, S. H. Cheon, G. S. Yoo, and K. M. Kim, *Arch. Pharm. Res.* **21,** 478 (1998).
[7] T. Kanda, H. Akiyama, A. Yanagida, M. Tanabe, Y. Goda, M. Toyoda, R. Teshima, and Y. Saito, *Biosci. Biotechnol. Biochem.* **62,** 1284 (1998).
[8] S. H. Tzeng, W. C. Ko, F. N. Ko, and C. M. Teng, *Thromb. Res.* **64,** 91 (1991).
[9] P. L. da Luz, C. V. Serrano Junior, A. P. Chacra, H. P. Monteiro, V. M. Yoshida, M. Furtado, S. Ferreira, P. Gutierrez, and F. Pileggi, *Exp. Mol. Pathol.* **65,** 150 (1999).
[10] H. E. Osman, N. Maalej, D. Shanmuganayagam, and J. D. Folts, *J. Nutr.* **128,** 2307 (1998).
[11] A. F. Rump, R. Schussler, D. Acar, A. Cordes, R. Ratke, M. Theisohn, R. Rosen, W. Klaus, and U. Fricke, *Gen. Pharmacol.* **26,** 603 (1995).
[12] H. van Jaarsveld, J. M. Kuyl, D. H. Schulenburg, and N. M. Wiid, *Res. Commun. Mol. Pathol. Pharm.* **91,** 65 (1996).
[13] S. al Makdessi, H. Sweidan, K. Dietz, and R. Jacob, *Basic Res. Cardiol.* **94,** 71 (1999).
[14] B. H. Juurlink and P. G. Paterson, *J. Spinal Cord Med.* **21,** 309 (1998).
[15] S. H. Tsai, S. Y. Lin-Shiau, and J. K. Lin, *Br. J. Pharmacol.* **126,** 673 (1999).
[16] B. Gil, M. J. Sanz, M. C. Terencio, M. L. Ferrandiz, G. Bustos, M. Paya, R. Gunasegaran, and M. J. Alcaraz, *Life Sci.* **54,** 333 (1994).
[17] E. Middleton Jr., and C. Kandaswami, *Biochem. Pharmacol.* **43,** 1167 (1992).
[18] F. Yang, W. J. de Villiers, C. J. McClain, and G. W. Varilek, *J. Nutri.* **128,** 2334 (1998).
[19] J. C. Monboisse, P. Braquet, A. Randoux, and J. P. Borel, *Biochem. Pharmacol.* **32,** 53 (1983).

groups to form resonance-stabilized phenoxyl radicals, thus they are usually called hydrogen-donating antioxidants. The capacities of the polyphenolic compounds to donate hydrogens depend on the redox properties of their phenolic hydroxy groups and the structural relationships between the different parts of the chemical structures.[33,34] Although there are many ways to test the antioxidant activity of polyphenols, it is of great interest to determine their hydrogen-donating efficacies in a simple and like manner.

For this we have introduced the stable free radical galvinoxyl to measure and compare the efficacy of the hydrogen-donating ability of phenols.[35] Based on its structure, galvinoxyl is a rather stable molecule and, while it accepts an electron or hydrogen radical to become a stable and diamagnetic molecule, it can be reduced irreversibly. Because of its odd electron, galvinoxyl shows a strong absorption band at 428 nm (in ethanol); at low concentration a solution of galvinoxyl appears yellow. As the electron is paired off, the absorption vanishes and the resulting decolorization is stoichiometric with respect to the number of electrons taken up. Taking advantage of the color change of galvinoxyl in the presence of an antioxidant, dynamics of the antioxidant activity, hydrogen-donating activity, can be measured easily.

To test whether a certain substance acts as a radical scavenger, reaction rate constants with radicals must be determined. Although stoichiometric number determines the duration of the inhibition period or lag time, the rate constant can give the extent of inhibition in oxidation. This article describes methods for measuring not only the stoichiometric number of active phenolic hydrogens of a substance,

---

[20] M. J. Laughton, P. J. Evans, M. A. Moroney, J. R. C. Hoult, and B. Halliwell, *Biochem. Pharmacol.* **42,** 1673 (1991).

[21] J. R. S. Hoult, M. A. Moroney, and M. Paya, *Methods Enzymol.* **234,** 443 (1994).

[22] M. A. Moroney, M. J. Alcaraz, R. A. Forder, F. Carey, and J. R. Hoult, *J. Pharm. Pharmacol.* **40,** 787 (1988).

[23] J. E. Brown, H. Khodr, R. C. Hider, and C. A. Rice-Evans, *Biochem. J.* **330,** 1173 (1998).

[24] J. F. Moran, R. V. Klucas, R. J. Grayer, J. Abian, and M. Becana, *Free Radic. Biol. Med.* **22,** 861 (1997).

[25] S. A. van Acker, D. J. van den Berg, M. N. Tromp, D. H. Griffioen, W. P. van Bennekom, W. J. van der Vijgh, and A. Bast, *Free Radic. Biol. Med.* **20,** 331 (1996).

[26] I. Morel, G. Lescoat, P. Cillard, and J. Cillard, *Methods Enzymol.* **243,** 437 (1994).

[27] Q. Guo, B. Zhao, M. Li, S. Shen, and W. Xin, *Biochim. Biophys. Acta* **1304,** 210 (1996).

[28] L. Packer, G. Rimbach, and F. Virgili, *Free Radic. Biol. Med.* **27,** 704 (1999).

[29] N. Cotelle, J. L. Bernier, J. P. Cattear, J. Pommery, J. C. Wallet, and E. M. Gaydou, *Free Radic. Biol. Med.* **20,** 35 (1996).

[30] T. Nakayama, M. Yamada, T. Osawsa, and S. Kawakishi, *Biochem. Pharmacol.* **45,** 265 (1993).

[31] C. A. Rice-Evans, N. J. Miller, P. G. Bolwell, P. M. Bramley, and J. B. Pridham, *Free Radic. Res.* **22,** 375 (1995).

[32] G. Sichel, C. Corsaro, calia, A. J. D. Bilio, and R. P. Bonomo, *Free Radic. Biol. Med.* **11,** 1 (1991).

[33] C. A. Rice-Evans, N. J. Miller, and G. Paganga, *Free Radic. Biol. Med.* **20,** 933 (1996).

[34] W. Bors, W. Heller, C. Michel, and M. Saran, *Methods Enzymol.* **186,** 343 (1990).

[35] H. Shi and E. Niki, *Lipids* **33,** 365 (1998).

but also the rate constant for related antioxidants in the reaction with galvinoxyl using a stopped-flow technique. The method can be used to determine and compare the antioxidative activity of hydrogen-donating compounds, either as pure substances or as mixtures.

Reagents

Galvinoxyl, quercetin, and *n*-propyl gallate are from Wako Pure Chemical Co. Ltd. (Osaka, Japan). Kaempferol, myricetin, and catechin are from Aldrich (Milwaukee, WI). *Ginkgo biloba* extract (GBE) is provided from Takehaya Co., Ltd (Tokyo, Japan). α-Tocopherol is kindly supplied by Eisai Co., Ltd. (Tokyo, Japan). All of the substances are dissolved in analytical reagent grade ethanol and used as received. Experiments are always performed on freshly prepared solutions. (Although galvinoxyl is considered as a stable free radical, its solutions deteriorate slowly, so freshly prepared galvinoxyl solution is also used throughout the experiments.) The structures of substances used are shown in Fig. 1.

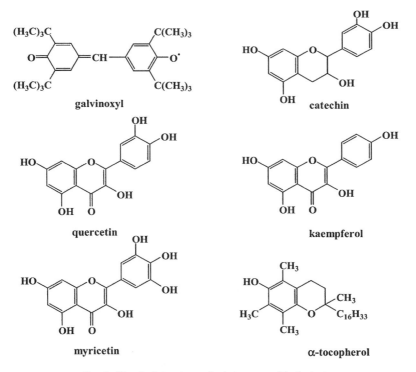

FIG. 1. Chemical structures of substances used in the text.

Stoichiometric Study

Galvinoxyl is reduced by hydrogen-donating free radical scavengers as shown in reaction 1:

$$G\cdot + IH \longrightarrow GH + I\cdot \qquad (1)$$

where G· stands for galvinoxyl; GH, reduced galvinoxyl; IH, hydrogen-donating free radical scavenger; and I·, corresponding radical from IH.

On the basis that one galvinoxyl molecule reacts with one active hydroxyl group,[35,36] we can determine the quantity of active phenolic hydrogens in the reaction with galvinoxyl by the absorbance decrease of galvinoxyl in the reaction solution under the condition of [galvinoxyl] > [IH], which allows all of the IH to take part in the reaction with galvinoxyl. The stoichiometric number may be calculated as follows from the fall of absorbance of galvinoxyl and the concentration of IH:

$$n = \Delta A/(\epsilon l[IH]) \qquad (2)$$

where n, is the stoichiometric number of IH in the reaction with galvinoxyl; ΔA, the absorbance difference of galvinoxyl at 428 nm between the initial and the end reaction; l, cell length; ϵ, molar extinction coefficient of galvinoxyl ($\lambda_{428nm} = 1.50 \times 10^5 M^{-1} cm^{-1}$ in ethanol); and [IH], the concentration of an antioxidant. In the case of pure substance, the unit of [IH] is mol/liter; in the case of a mixture such as *G. biloba* extract, the unit g/liter could be used.

Generally, the concentration of galvinoxyl can be set at 5–8 μM (final concentration). For pure compounds, 0.1–2 μM concentrations can be selected (final concentration). For mixtures, preliminary experiments should be carried out to determine the concentration range to be used. The reaction is started by mixing the galvinoxyl ethanol solution with a hydrogen-donating compound ethanol solution. The fall of absorbance of galvinoxyl at 428 nm is traced until completion of the reaction by using a spectrophotometer Beckman DU 640 at 25°.

Mean absorbance and standard deviation (SD) are measured for each sample and the corresponding blank. The difference between the absorbance of the sample and the blank is the actual absorbance of the sample under investigation.

Kinetic Study

The kinetics of the hydrogen-donating reaction between tested substances and galvinoxyl is carried out by following the decrease in absorbance of galvinoxyl at 428 nm under pseudo first-order conditions, [IH] ≫ [galvinoxyl]. The concentration of galvinoxyl radical used is always more than five times less than that

---

[36] J. Tsuchiya, T. Yamada, E. Niki, and Y. Kamiya, *Bull. Chem. Soc. Jpn.* **58**, 326 (1985).

of the tested substances to obtain the pseudo first-order rate constant, k_1. The second-order rate constant, k_2, is calculated from the k_1 values using Eq. (3).

$$-d[\text{galvinoxyl}]/dt = k_1[\text{galvinoxyl}] = k_2[\text{IH}][\text{galvinoxyl}] \quad (3)$$

Taking into consideration the stoichiometric number, we can change reaction (1) to the following form:

$$\text{G}\cdot + (1/n)\text{IH} \longrightarrow \text{GH} + (1/n)\,\text{I}\cdot \quad (4)$$

and Eq. (3) can be rewritten as Eq. (5).

$$-d[\text{galvinoxyl}]/dt = k_{2'}[\text{IH}][\text{galvinoxyl}] \quad (5)$$

where $k_{2'} = k_2/n$ is introduced. $k_{2'}$ may be defined as the average rate constant of active hydroxyl groups of a hydrogen-donating compound. The rate constants, k_2 and k_2', may be considered as the total reactive potential and the average reactive potential of active hydroxyl groups in the substances studied, respectively.

A stopped-flow spectrophotometer (RA-2000, Photal Otsuka Electronics, Osaka, Japan) equipped with an advanced data acquisition and processing system is used for the fast reaction kinetics. Appropriate volumes of the reactants are placed in two cells: cell A contains galvinoxyl in ethanol (20 μM) and cell B contains test substances in ethanol. All the kinetic runs are performed with a sweep time longer than 10 half-lives of the corresponding reaction going to completion. The data acquisition system records absorbance versus time data and computes the rate constants from $\log(A_t - A_\infty)$ versus time plots, where A_t and A_∞ correspond to absorbance at time t and infinite time (after completion of the reaction), respectively.

Each kinetic run is performed five times, and the mean k_1 and standard deviation (SD) are measured for each sample and the corresponding blank. The difference between the k_1 value of the sample and the blank is the actual k_1 value of the sample under investigation.

Applications

Stoichiometric Studies

The stoichiometric numbers for myricetin, quercetin, kaempferol, catechin, propyl gallate, α-tocopherol, and GBE with galvinoxyl were determined. The concentration range of GBE used was selected as 0–0.02 g/liter so that considerable galvinoxyl remains at the end of reaction. Figure 2 shows a time-dependent decrease in absorbance at 428 nm in the presence of quercetin at different concentrations. The decrease in absorbance was proportional to the concentrations of antioxidants. A representative plot of data for quercetin and α-tocopherol is shown in Fig. 3. Excellent linear correlation was obtained for the plots of all the

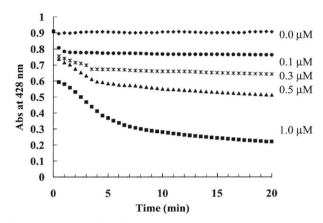

FIG. 2. Time-dependent absorbance decrease of galvinoxyl in the presence of quercetin at different concentrations.

antioxidants studied. By using Eq. (3), n values could be determined. The value of ϵ for galvinoxyl at 428 nm in Eq. (5) was $1.50 \times 10^5 M^{-1} cm^{-1}$, with cell length, l being 1 cm. The slopes of the lines in Fig. 3 correspond to $n/(\epsilon l)$, from which n was calculated as summarized in Table I. The n value for GBE was determined as 1.1×10^{-4} mol/g, which means that 1 g GBE could react with 1.1×10^{-4} mol of galvinoxyl. In other words, there are 1.1×10^{-4} mol (6.62×10^{19} molecules)

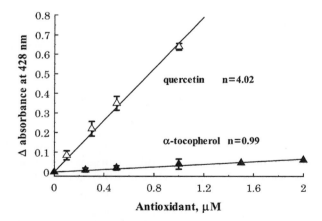

FIG. 3. Representative plots of decrease in absorbance of galvinoxyl at 428 nm against the concentrations of hydrogen-donating antioxidants. Data points represent means ± SD of five individual experiments.

TABLE I
STOICHIOMETRIC NUMBER AND SECOND-ORDER RATE CONSTANT

Antioxidant	n	$k_2{}^b$	$k_{2'}{}^b$
Myricetin	4.5	1.1×10^6	2.4×10^5
Quercetin	4.0	5.9×10^3	1.5×10^3
Kaempferol	1.9	2.1×10^3	1.1×10^3
Catechin	3.0	1.5×10^3	5.0×10^2
Propyl gallate	3.1	1.2×10^4	3.9×10^3
α-Tocopherol	1.0	2.4×10^3	2.4×10^3
GBE	$1.1 \times 10^{-4\,a}$	0.13^c	1.2×10^3

[a] Moles of active hydroxyl groups/g of antioxidant.
[b] Second-order rate constant for molecule (k_2) and per active hydrogen ($k_{2'}$) in M^{-1} sec^{-1}.
[c] (g/liter)$^{-1}$ sec^{-1}.

active hydrogens in 1 g GBE. The number of active hydrogens from myricetin, quercetin, kaempferol, catechin, propyl gallate, and α-tocopherol are 4.5, 4.0, 1.9, 3.0, 3.1, and 1.0, respectively.

Kinetic Studies

Figure 4 is a typical time course showing the decrease in absorbance at 428 nm measured by the stopped-flow spectrophotometer when an ethanol solution of galvinoxyl was mixed with that of an antioxidant under pseudo first-order conditions. The theoretical curve b fits very well with the experimentally obtained curve a. A typical log plot versus time is also shown in Fig. 4 (curve c), which gives a pseudo first-order rate constant k_1. The first-order rate constants observed at 428 nm were linearly dependent on the concentrations of the substances tested. Examples are shown in Fig. 5. Excellent linear correlation was obtained for the plots of all the antioxidants studied. The plots of k_1 against the concentrations of tested substances according to Eq. (3) yielded the rate constants k_2 for myricetin, quercetin, kaempferol, propyl gallate, α-tocopherol, and GBE in the reaction with galvinoxyl, respectively. These data are also summarized in Table I. The k_2 value obtained for GBE was 0.13 (g/liter)$^{-1}$ sec^{-1}. For the pure substances, the values decreased in the order of myricetin > propyl gallate > quercetin > α-tocopherol > kaempferol. However, because it is a mixture, it is difficult to compare the scavenging capacity of GBE with other antioxidants. By introducing the other second-order rate constant, $k_{2'}(=k_2/n)$, which gives the average reactive rate constant for each active hydroxyl group, the relative reactivities of active hydroxyl groups in different pure substances or mixtures like GBE were determined and compared. The results in Table I show that the reactivity of the active phenolic hydrogen of

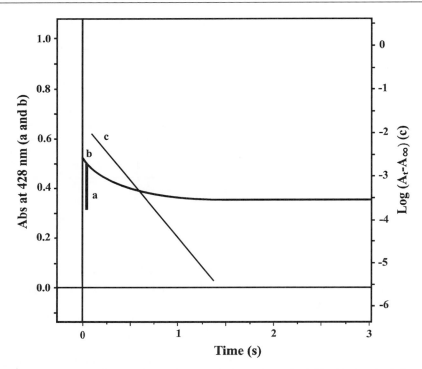

FIG. 4. A representative experiment of the pseudo first-order study. A 20 μM ethanol solution of galvinoxyl was mixed with a 0.4 mM ethanol solution of propyl gallate (curve a). Curve b is a theoretical fit to curve a. A typical log plot versus time is also shown (curve c), which gives a pseudo first-order rate constant of 2.4 sec^{-1}.

GBE is greater than that of catechin, similar to those of quercetin and kaempferol, slightly smaller than those of α-tocopherol and propyl gallate, and much lower than myricetin. The average reactive potential of active hydroxyl groups studied decreased in the order myricetin > propyl gallate > α-tocopherol > quercetin > GBE = kaempferol > catechin.

The structure–activity relationships of flavonoids have been elucidated by several groups.[31,33,34,37] There are mainly three structure groups for determining the free radical scavenging and/or antioxidative potential of flavonoids[31,34]: a catechol moiety of the B ring, the 2,3 double bond in conjugation with a 4-oxo function of a carbonyl group in the C ring, and the additional presence of hydroxyl groups at the 3 and 5 positions in the A ring. The results presented here are in accordance with

[37] N. Cotelle, J. L. Bernier, J. P. Henichart, J. P. Cattear, E. Gaydou, and J. C. Wallet, *Free Radic. Biol. Med.* **13,** 211 (1992).

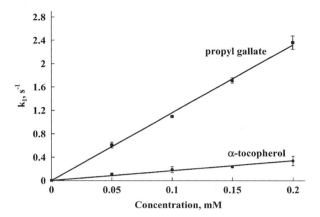

FIG. 5. Representative plots of first-order rate constants (k_1) as a function of concentrations of the antioxidant tested. Data points represent means ± SD of five individual experiments. The measurements were carried out at 25°.

the just-described explanation of the structure–activity relationships. Quercetin possesses all three structure groups and thus gives higher total reactive potential and average reactive potential than kaempferol, which has no catechol moiety in the B ring. The presence of a hydroxyl group at 5′ in the B ring remarkably increases the antioxidant potential. The stoichiometric number of myricetin is only 0.5 greater than that of quercetin. However, the total reactive potential (k_2) and the average reactive potential ($k_{2'}$) are two orders higher than quercetin. This indicates that the more hydroxyl groups, the higher the reactivity. This agrees with other reports indicating that the antioxidant activity of myricetin is more effective than quercetin in emulsions,[38,39] oils,[40,41] liposomes,[42] and low-density lipoproteins (LDL).[43,44] The relative activities of quercetin and α-tocopherol are reversed when compared in terms of the total reactive potential (k_2) and the average reactive potential ($k_{2'}$). The total activity of the scavenging galvinoxyl of α-tocopherol is lower than that of quercetin, whereas the activity of the phenolic hydroxyl group of α-tocopherol is higher than that of the active hydroxyl groups of quercetin.

[38] M. S. Taya, E. E. Miller, and D. E. Pratt, *J. Am. Oil. Chem. Soc.* **61,** 928 (1984).
[39] N. P. Das and T. A. Pereira, *J. Am. Oil. Chem. Soc.* **67,** 255 (1990).
[40] Z. Y. Chen, P. T. Chan, K. Y. Ho, K. P. Fund, and J. Wang, *Chem. Phys. Lipids* **79,** 157 (1996).
[41] C. Mehta and T. R. Seshadri, *J. Sci. Ind. Res.* **18b,** 24 (1959).
[42] M. H. Gordon and A. Roedig-Penman, *Chem. Phys. Lipids* **97,** 79 (1998).
[43] E. N. Frankel, A. L. Waterhouse, and P. L. Teissedre, *J. Agric. Food Chem.* **43,** 890 (1995).
[44] P. L. Teissedre, E. N. Frankel, A. L. Waterhouse, H. Pele, and J. B. German, *J. Agric. Food Chem.* **70,** 55 (1996).

Limitations

The antioxidative potentials compared here are the reactivities of substances studied with galvinoxyl radical. To understand the whole activity of polyphenols as an antioxidant is much more complex because polyphenols may act not only by scavenging radicals, but also (1) by chelating metal ions via the phenolic o-dihydroxy structure of the flavonoids, although it has been suggested that the antioxidative and lipid peroxidation-inhibiting potential of flavonoid predominantly resides in the radical-scavenging capacity rather than the chelation of metals,[45,46] or (2) by regenerating α-tocopherol through reduction of the α-tocopheroxyl radical. To act as a hydrogen-donating antioxidant is one of the mechanisms by which flavonoids and other phenolic compounds can be effective antioxidants. Furthermore, the total antioxidant efficiency of radical-scavenging antioxidants is determined not only by their reactivities toward radicals, but also by physical factors such as location and mobility at the microenvironment, fate of antioxidant-derived radicals, and interactions with other antioxidants.[47]

Acknowledgments

This study was supported in part by a Grant-in-Aid for Scientific Research and COE Research from the Ministry of Education, Science, Sports and Culture, Japan.

---

[45] C. G. Iraga, V. S. Martino, G. E. Ferraro, J. F. Coussio, and A. Boveris, *Biochem. Pharmacol.* **36**, 717 (1987).
[46] A. K. Ratty and N. P. Das, *Biochem. Med. Metab. Biol.* **39**, 69 (1988).
[47] E. Niki, N. Noguchi, H. Tsuchihashi, and N. Gotoh, *Am. J. Clin. Nutr.* **62**(Suppl.), 1322S (1995).

[15] Structure–Activity Relationships Governing Antioxidant Capacities of Plant Polyphenols

By WOLF BORS, CHRISTA MICHEL, and KURT STETTMAIER

Introduction

As members of a large family of plant secondary metabolites, flavonoids represent various structural classes that can be distinguished by their biosynthetic pathways and chemical behavior.[1,2] In the context of the antioxidative function of the flavonoids, the latter aspect is of particular relevance.

---

[1] J. B. Harborne, "The Flavonoids: Advances in Research since 1986." Chapman & Hall, London, 1993.

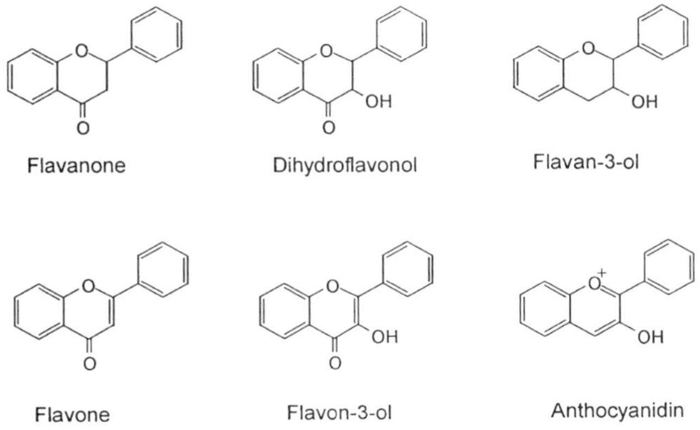

SCHEME 1. Structures of flavonoids.

Not only the reactivity with various types of radicals but also the fate and subsequent reactions of the flavonoid radicals are ultimately determined by their structures. In Scheme 1, the flavonoids (excluding isoflavones, neoflavones, or biflavonoids), which are derived from six basic structures, are arranged vertically according to their biosynthetic pathways.[3] The structural class of the flavan-3-ols, derived from either (+)-catechin or (−)-epicatechin, is also known as proanthocyanidins (the often used term procyanidins only refers to compounds with B-ring catechol structures and, in consequence, "prodelphinidins" would denote compounds with B-ring pyrogallol structures).[4] A clear distinction exists for the proanthocyanidins, as they are the only ones that can form gallate esters at the 3-hydroxy group and various oligomeric structures depending on the plant sources[5] (Scheme 2).

Various types of covalent binding exist for the oligomers; those listed are but a few examples.[5] The nomenclature for the dimers is quite detailed, as it denotes the enantiomeric structures and the binding sites (the numbers correspond with those in Scheme 1), whereas starting with the trimers, the different stereoisomers are considered together.

---

[2] W. Bors, W. Heller, C. Michel, and K. Stettmaier, in "Handbook of Antioxidants" (E. Cadenas and L. Packer, eds.), p. 409. Dekker, New York, 1996.

[3] W. Heller and G. Forkmann, in "The Flavonoids: Advances in Research since 1986" (J. B. Harborne, ed.), p. 499. Chapman & Hall, London, 1994.

[4] H. A. Stafford, in "Flavonoid Metabolism" (H. A. Stafford, ed.), p. 63. CRC Press, Boca Raton, FL, 1990.

[5] R. Kaul, *Pharm. Zeit* **25**, 175 (1996).

	R_1	R_2	R_3
Cat	H	OH	OH
EC	H	OH	OH
ECG	H	OH	OG
EGC	OH	OH	OH
EGCG	OH	OH	OG

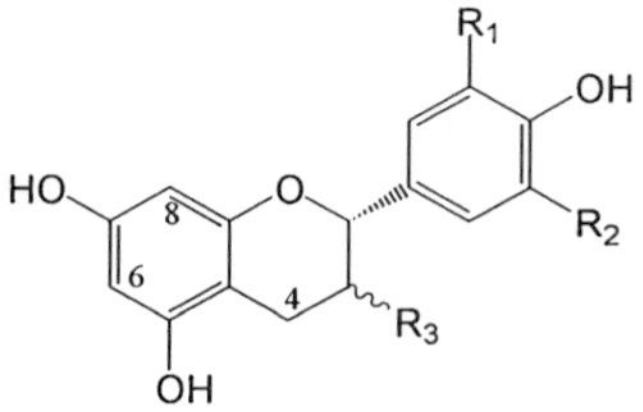

SCHEME 2. Structures of proanthocyanidin monomers.

Dimers: A series with C–C bond *and* ether bond
PC-A$_2$ EC 4β → 8, 2β → O → 7 EC

Dimers: B series with C–C bond
PC-B$_2$ EC 4β → 8 EC
PC-B$_5$ EC 4β → 6 EC

Trimers: with C–C bonds
PC-C$_2$ EC → EC → Cat

All other flavonoids exist in plants mainly as glycosides. As outlined later, this distinction is crucial for the respective function of the flavonoids.

Aside from the proanthocyanidins with a flavonoid skeleton, another group of important polyphenolic structures exists, the so-called hydrolyzable tannins in contrast to the condensed tannins, which is still another term for the proanthocyanidins. Hydrolyzable tannins are gallate esters of glucose, such as the structures depicted in Scheme 3 (where OG denotes the gallate ester), but they may also exist

β–glucogallin	R_2 = OG
pentagalloyl glucose	R_2-R_6 = OG
Chinese tannin (8xOG)	R_2,R_3,R_5 = OG
	R_4 = OG-OG
	R_6 = OG-OG-OG
tannic acid (10xOG)	undefined structure

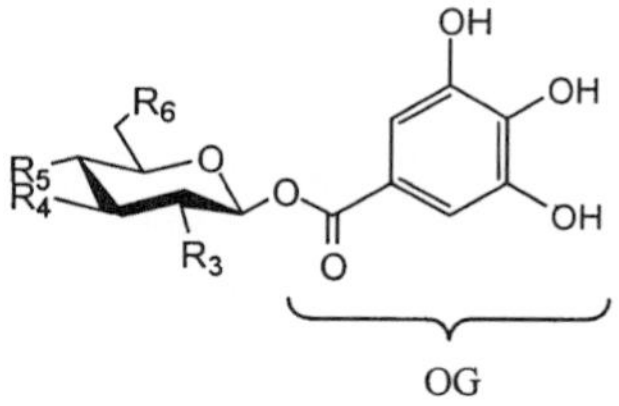

hamameli-tannin

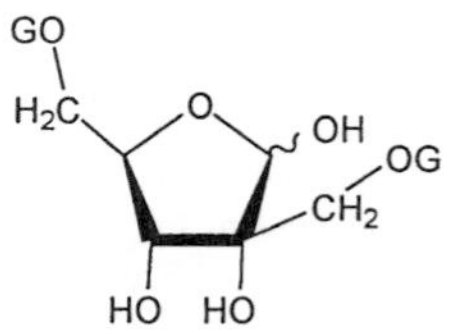

SCHEME 3. Structures of hydrolyzable tannins

as ellagitannins with gallate ester dimers, the hexahydroxydiphenic acid moieties, and derivatives thereof.[6,7]

Antioxidant vs Prooxidant (Biocidal) Activities

Structure–activity relationships have first been correlated for the the inhibition of lipid peroxidation,[8,9] mutagenicity,[10,11] and cytotoxicity[12,13] as end points. Structural criteria for these opposing effects are for all practical purposes identical, which suggests that all of these properties depend on the ease of the individual compounds to enter into radical reactions.[2] In the antioxidant mode, electron transfer from a phenolate anion or hydrogen donation to an oxidizing radical (cf. the 1,1-diphenyl-2-picrylhydrazyl radical, DPPH•[14]) forms the organic radical, whereas futile redox cycling involving the antioxidant radicals and their oxidized forms (in the case of polyphenols semiquinones and quinones) is most likely the basis for the prooxidant or biocidal mode.[12,13,15,16] The cytotoxicity of flavanols, in contrast, seems to be due to the formation of ternary complexes with copper ions and DNA leading to Fenton-type reactions.[17,18]

Radical-Scavenging Activity of Polyphenols

Flavonoids have been widely accepted as so-called chain-breaking antioxidants because they react very effectively as radical scavengers and their aroxyl radicals

---

[6] K. T. Chung, T. Y. Wong, C. I. Wei, Y. W. Huang, and Y. Lin, *Crit. Rev. Food Sci. Nutr.* **38,** 421 (1998).
[7] T. Yokozawa, C. P. Chen, E. Dong, T. Tanaka, G. I. Nonaka, and I. Nishioka, *Biochem. Pharmacol.* **56,** 213 (1998).
[8] J. Terao and M. K. Piskula, in "Flavonoids in Health and Disease" (C. Rice-Evans and L. Packer, eds.), p. 227. Dekker, New York, 1997.
[9] D. S. Leake, in "Flavonoids in Health and Disease" (C. Rice-Evans and L. Packer, eds.), p. 253. Dekker, New York, 1997.
[10] M. Nagao, N. Morita, T. Yahagi, M. Shimizu, M. Kuroyanagi, M. Fukuoka, K. Yoshihira, S. Natori, T. Fujino, and T. Sugimura, *Environ. Mutagen.* **3,** 401 (1981).
[11] J. T. MacGregor, in "Plant Flavonoids in Biology and Medicine" (V. Cody, E. Middleton, and J. B. Harborne, eds.), p. 411. R. Liss, New York, 1986.
[12] G. H. Cao, E. Sofic, and R. L. Prior, *Free Radic. Biol. Med.* **22,** 749 (1997).
[13] J. Gaspar, in "Natural Antioxidants and Food Quality in Atherosclerosis and Cancer Prevention" (J. T. Kumpulainen and J. T. Salonen JT, eds.), p. 290. Royal Chem. Soc., Cambridge, UK, 1996.
[14] N. D. Yordanov and A. G. Christova, *Fresenius J. Anal. Chem.* **358,** 610 (1997).
[15] Y. H. Miura, I. Tomita, T. Watanabe, T. Hirayama, and S. Fukui, *Biol. Pharm. Bull.* **21,** 93 (1998).
[16] D. Metodiewa, A. K. Jaiswal, N. Cenas, E. Dickancaité, and J. Segura-Aguilar, *Free Radic. Biol. Med.* **26,** 107 (1999).
[17] L. A. Chrisey, G. H. S. Bonjar, and S. M. Hecht, *J. Am. Chem. Soc.* **110,** 644 (1988).
[18] S. Shirahata, H. Murakami, K. Nishiyama, K. Yamada, G. Nonaka, I. Nishioka, and H. Omura, *J. Agric. Food Chem.* **37,** 299 (1989).

are sufficiently stable to avoid chain-propagating reactions.[2] Rate constants for the strongly electrophilic hydroxyl (•OH) and azide (•N$_3$) radicals are in the diffusion-controlled range for practically all investigated substances.[19,20] Peroxyl radical reactions are in the intermediary range, and rate constants with superoxide anions ($O_2^{•-}$) are of the order of 10^4 M^{-1} sec^{-1}.[21] Very little distinction is apparent, if one relates the rate constants to the individual structural classes.

Quite distinct, however, is the behavior of the proanthocyanidins in which an increase in the reactivity correlates linearly with the number of reactive sites, defined as the catechol and/or pyrogallol groups existing in the individual molecules. Using pulse radiolysis, this correlation was observed with •OH and $O_2^{•-}$, but surprisingly not with •N$_3$ radicals.[22] The fact that the rate constants with •OH radicals ($k_{•OH}$) for pentagalloyl glucose (PGG) and tannic acid (TA) clearly exceed diffusion-controlled limits is further proof that multiple target sites exist in these molecules.[23] A similar linear correlation is also obtained using either the Trolox-equivalent antioxidant capacity (TEAC) assay[24] or the inverse values of the DPPH assay[7] to determine the antioxidant potential (Fig. 1). Extending the studies to procyanidin oligomers, Plumb and co-workers[25] showed that this correlations holds true in aqueous solution up to trimers with the TEAC assay. The decrease observed with higher oligomers is probably due to their insolubility and precipitation in aqueous solution. The study using the DPPH assay,[7] in fact, was the most comprehensive ever done with plant polyphenolic antioxidants and likewise showed an increase in the hydrogen-donating capacity to the DPPH radical, the higher the number of reactive hydroxy groups in the molecule. It is interesting to note that TA fits quite well into the linear relationship of the inverse DPPH plot (inset of Fig. 1), whereas the $k_{•OH}$ value in the pulse-radiolytic experiments is actually lower than that of PGG with 15 hydroxy groups as compared to 25 for TA. We consider this to be due to sterical hindrance.[22]

Fate and Reactivity of Polyphenol Aroxyl Radicals

An obligatory constraint for the effectivity of antioxidants is the relative stability or low reactivity of the secondary antioxidant radicals.[26] In the case of polyphenolic flavonoids, the presence of B-ring catechol structures and the respective

---

[19] W. Bors, W. Heller, C. Michel, and M. Saran, *Methods in Enzymol.* **186,** 343 (1990).

[20] W. Bors, W. Heller, C. Michel, and M. Saran, in "Free Radicals and the Liver" (G. Csomos and J. Feher, eds.), p. 77. Springer, Berlin, 1992.

[21] W. Bors, C. Michel, and M. Saran, *Methods in Enzymol.* **234,** 420 (1994).

[22] W. Bors and C. Michel, *Free Radic. Biol. Med.* **27,** 1413 (1999).

[23] L. M. Dorfman and G. E. Adams, *NSRDS-NBS Report,* No. 46, (1973).

[24] N. Salah, N. J. Miller, G. Paganga, L. Tijburg, G. P. Bolwell, and C. Rice-Evans, *Arch. Biochem. Biophys.* **322,** 339 (1995).

[25] G. W. Plumb, S. de Pascual-Teresa, C. Santos-Buelga, V. Cheynier, and G. Williamson, *Free Radic. Res.* **29,** 351 (1998).

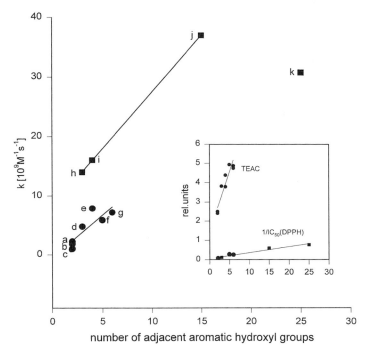

FIG. 1. Correlation of rate constants with hydroxyl radicals vs the number of reactive hydroxy groups. (●) *Proanthocyanidins:* a, Catechin; b, epicatechin; c, Pycnogenol; d, epigallocatechin; e, proanthocyanidin A2; f, epicatechin gallate; g, epigallocatechin gallate. (■) *Hydrolyzable tannins:* h, Propyl gallate; i, β-glucogallin; j, pentagalloylglucose; k, tannic acid. (Inset) Correlation of the number of reactive hydroxy groups vs TEAC values[24,25] or $1/IC_{50}$ values of the reaction with the DPPH radical[7] (the symbols are the same). Numbers in the abscissae correspond to those in column 3 (*n* values) of Table I.

o-semiquinones renders these intermediates more stable than those of aroxyl radicals without these moieties, especially in alkaline medium.[19,20] The semiquinone radicals of flavones and flavonols with 2,3-double bonds are further stabilized due to the extensive electron delocalization over all three ring systems.[19]

Transient spectra observed after the oxidation of individual flavonoids by various oxidizing radicals are quite similar, and structural dependencies are only apparent with regard to the presence or absence of 2,3-double bonds.[20] This rather trivial distinction is only broken by the anthocyanidins, or flavylium salts, which due to the oxonium functionality in the heterocyclic C ring are intensely colored. In

---

[26] W. Bors, C. Michel, W. Heller, and H. Sandermann, Jr., in "Free Radicals, Oxidative Stress, and Antioxidants. Pathological and Physiological Significance" (T. Özben, ed.), p. 85. Plenum Press, New York, 1998.

aqueous solution, however, only pelargonidin chloride with one hydroxy group in the B ring shows a strong visible absorption, as cyanidin chloride (B-ring catechol) or delphinidin chloride (B-ring pyrogallol) readily form colorless leuco adducts with water.[19] In nature, their intense color in fruits, berries, and flower petals is only retained by complexation with flavonols[27] or flavanols.[28,29]

Because aroxyl radicals of flavonoids generally decay in second-order reactions, the obvious disproportionation reaction has to result in half of the aroxyl radicals being reduced back to the parent polyphenol and half being oxidized to a quinoid structure. None of the quinones derived from flavon(ol)s have ever been isolated, yet kinetic evidence supports their formation[30] and the prooxidant role of quercetin has been ascribed to futile redox cycling of the elusive quinone.[15,16]

This behavior contrasts quite sharply with that of proanthocyanidin-derived aroxyl radicals. They also decay in second-order reactions and may even be capable of intramolecular dismutation if adjacent semiquinones interact,[31] but the quinones formed are more likely to combine with the polyhydroxylated parent compound in a second-order nuclear substitution reaction (S_N2).[32] This reaction is alternatively called phenolic coupling and its prime feature is the retention of the number of hydroxyl groups, which results in a doubling of the reactive sites after the dimerization process.[33-35] We could actually follow this reaction by electron paramagnetic resonance (EPR) spectroscopy in the case of (−)-epigallocatechin gallate (EGCG),[36] and Scheme 4 shows the steps involved in the dimerization of this compound. It is interesting to note that the biological oligomers of the B type involve mainly 4–6 or 4–8 coupling of the (+)-catechin and/or (−)-epicatechin monomers,[4,5] yet not one of the various synthetic approaches ever yielded these specific bond types. The EGCG coupling in Scheme 4 depicts a 2′,2″-bond formation, a structure that has actually been found in fermented tea and is called theasinensin,[37] after oxidation under mildly alkaline conditions.[38]

Because there is sufficient evidence that phenolic coupling prevails during the oxidation of proanthocyanidins,[33,34] in contrast to potential futile redox cycling

---

[27] J. M. Baranac, N. A. Petranovic, and J. M. Dimitric-Markovic, *J. Agric. Food Chem.* **44**, 1333 (1996).

[28] E. M. Francia-Aricha, M. T. Guerra, J. C. Rivas-Gonzalo, and C. Santos-Buelga, *J. Agric. Food Chem.* **45**, 2262 (1997).

[29] N. E. Es-Safi, H. Fulcrand, V. Cheynier, and M. Moutounet, *J. Agric. Food Chem.* **47**, 2096 (1999).

[30] W. Bors, C. Michel, and S. Schikora, *Free Radic. Biol. Med.* **19**, 45 (1995).

[31] F. Ursini, I. Rapuzzi, R. Toniolo, F. Tubaro, and G. Bontempelli., *Methods in Enzymol.*, **335** [29] (2001) (this volume).

[32] S. Quideau, K. S. Feldman, and H. M. Appel, *J. Org. Chem.* **60**, 4982 (1995).

[33] D. Ferreira and R. Bekker, *Nat. Prod. Rep.* **13**, 411 (1996).

[34] D. Ferreira, E. V. Brandt, J. Coetzee, and E. Malan, *Progr. Chem. Organ. Natural Prod.* **77**, 21 (1999).

[35] T. Escribano-Bailon, O. Dangles, and R. Brouillard, *Phytochemistry* **41**, 1583 (1996).

[36] W. Bors, C. Michel, and K. Stettmaier, *Arch. Biochem. Biophys.* **374**, 347 (2000).

[37] F. Hashimoto, G. Nonaka, and I. Nishioka, *Chem. Pharm. Bull.* **36**, 1676 (1988).

[38] K. Yoshino, M. Suzuki, K. Sasaki, T. Miyase, and M. Sano, *J. Nutr. Biochem.* **10**, 223 (1999).

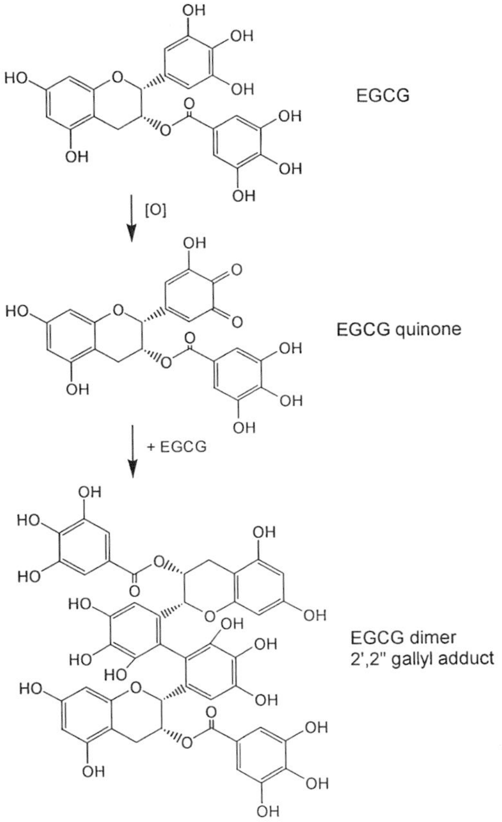

SCHEME 4. Oxidation and phenolic coupling of EGCG.

of flavon(ol) quinones,[15,16] we propose this to be the most pronounced distinction for the various types of polyphenols to act as antioxidants. Thus the initial scavenging of radicals is governed less by structural criteria except among the tannins themselves. A strong structure–activity relationship exists for the fate of quinones derived either from flavonoids proper or from proanthocyanidins. In effect, proanthocyanidins might actually enhance their antioxidant capacity after the initial radical scavenging whenever the quinones formed are kinetically capable of phenolic coupling reactions with the parent compound.

As is the case for other phenolic antioxidants, polyphenols could also be subject to redox cycling in the presence of reducing substances such as ascorbate[30,39] and

---

[39] S. V. Jovanovic, S. Steenken, M. G. Simic, and Y. Hara, in "Flavonoids in Health and Disease" (C. Rice-Evans and L. Packer, eds.), p. 137. Dekker, New York, 1997.

thus participate in the so-called "antioxidant network."[40,41] While this would avoid the formation of quinoid structures or the transfer of reducing equivalents from the semiquinone to oxygen with the formation of superoxide anions ($O_2^{\bullet-}$),[15,16] data available are insufficient to allow any structural correlation.

Theoretical Aspects

The mostly empirical QSAR correlations, which in effect compared a sufficient number of compounds with distinct structural features for various activities,[42] have been superseded by more genuine "quantitative" correlations using various computer software programs.[43–45] An early example is the CASE method (computer-automated structure evaluation) developed by Klopman and employed with flavonoids functioning as glyoxalase I inhibitors.[46] Aside from a multivariate statistical analysis,[47] later emphasis was based on molecular properties such as three-dimensional structures,[43,44,48–50] hydrogen bonding,[44] and aromaticity[48,51] to explain the inhibitory effects on several enzymes. An example is the evaluation of bond dissociation energies,[52] showing for some proanthocyanidins a rather labile C_2–H bond, which would result in a different type of radical than a catechol or pyrogallol semiquinone. Quantum chemical estimations were first attempted by Ferrel and co-workers as early as 1979,[48] yet, for *ab initio* calculations (e.g., using Gaussian software), large computers with parallel processors are still needed.[53–55]

[40] M. Lopez-Torres, J. J. Thiele, Y. Shindo, D. Han, and L. Packer, *Br. J. Dermatol.* **138,** 207 (1998).
[41] N. Haramaki, D. B. Stewart, S. Aggarwal, H. Ikeda, A. Z. Reznick, and L. Packer, *Free Radic. Biol. Med.* **25,** 329 (1998).
[42] W. Bors, W. Heller, and C. Michel, *in* "Flavonoids in Health and Disease" (C. Rice-Evans and L. Packer, eds.), p. 111. Dekker, New York, 1997.
[43] L. Costantino, G. Rastelli, and A. Albasini, *Eur. J. Med. Chem.* **31,** 693 (1996).
[44] Z. Nikolovska-Coleska, L. Suturkova, K. Dorevski, A. Krbavcic, and T. Solmajer, *Quant. Struct. Act. Relat.* **17,** 7 (1998).
[45] H. Y. Zhang, *Sci. China B* **42,** 106 (1999).
[46] G. Klopman and M. L. Dimayuga, *Mol. Pharmacol.* **34,** 218 (1988).
[47] B. Limasset, C. le Doucen, J. C. Doré, T. Ojasoo, M. Damon, and A. Crastes de Paulet, *Biochem. Pharmacol.* **46,** 1257 (1993).
[48] J. E. Ferrell, P. D. G. Chang Sing, G. Loew, R. King, J. M. Mansour, and T. E. Mansour, *Mol. Pharmacol.* **16,** 556 (1979).
[49] K. C. S. Chen Liu, S. S. Lee, M. T. Lin, C. W. Chang, C. L. Liu, J. Y. Lin, F. L. Hsu, S. Ren, and E. J. Lien, *Med. Chem. Res.* **7,** 168 (1997).
[50] D. Amic, D. Davidovic-Amic, D. Beslo, B. Lucic, and N. Trinajstic, *Croat. Chem. Acta* **70,** 905 (1997).
[51] D. Amic and N. Trinajstic, *J. Chem. Soc. Perkin Trans. II* 891 (1991).
[52] K. Kondo, M. Kurihara, N. Miyata, T. Suzuki, and M. Toyoda, *Free Radic. Biol. Med.* **27,** 855 (1999).
[53] P. J. O'Malley, *J. Phys. Chem. A* **101,** 6334 (1997).
[54] L. A. Eriksson, *in* "Density-Functional Methods in Chemistry and Materials Sciences" (M. Springborg, ed.), p. 125. Wiley, Chichester, UK, 1997.
[55] D. E. Wheeler, J. H. Rodriguez, and J. K. McCusker, *J. Phys. Chem. A* **103,** 4101 (1999).

In fact, using the simpler self-consistent field calculations to approximate experimental coupling constants of polyphenol semiquinones obtained by EPR spectroscopy proved inadequate for these complex structures. Nevertheless, simple quantum mechanical studies of flavonoids acting as inhibitors of protein tyrosine kinase inhibitors[44] and xanthine oxidase[56] or with regard to their antioxidative activity[57] have already been performed.

Experimental Techniques

General Aspects

According to a recent review,[58] we can distinguish between relative and absolute methods to determine antioxidant activities of individual compounds with the total radical-scavenging antioxidant potential (TRAP) method,[59] the preferred technique for compound mixtures in biological fluids.

The TEAC assay method has been described in detail,[59] but doubts have risen concerning its accuracy and reliability[60,61] and it has since been improved.[62,63] The DPPH assay was first used to determine the antioxidative potential of vitamin E[64] and can be applied using either visible light or EPR spectroscopy.[14] The most comprehensive study of the antioxidative potential of plant polyphenols has been done with this method[7] after prior confirmation of the experimental results with nuclear magnetic resonance (NMR) spectroscopy.[65]

Pulse Radiolysis

This method has been described with regard to flavonoid antioxidants.[19,21] In the study of plant polyphenol reactions with •OH, •N$_3$, and O$_2^{•-}$ care must be taken with catechins, as they are in general more susceptible to autoxidation than flavon(ol)s.[66] For the reaction with •OH, solutions are saturated with N$_2$O

[56] G. Rastelli, L. Costantino, and A. Albasini, *Eur. J. Med. Chem.* **30,** 141 (1995).
[57] S. A. B. E. van Acker, M. J. de Groot, D. J. van den Berg, M. N. J. L. Tromp, G. Donn-Op den Kelder, W. J. F. van der Vijgh, and A. Bast, *Chem. Res. Toxicol.* **9,** 1305 (1996).
[58] W. Bors, C. Michel, M. Saran, and K. Stettmaier, *in* "Different Pathways through Life: Biochemical Pathways of Plant Biology and Medicine" (A. Denke, K. Dornisch, F. Fleischmann, J. Graßmann, I. Heiser, S. Hippeli, W. Oβwald, and H. Schempp, eds.), p. 9. Lincom Europe, Munich, 1999.
[59] C. A. Rice-Evans and N. J. Miller, *Methods in Enzymol.* **234,** 279 (1994).
[60] M. Strube, G. R. M. M. Haenen, H. van den Berg, and A. Bast, *Free Radic. Res.* **26,** 515 (1997).
[61] L. Mira, M. Silva, R. Rocha, and C. F. Manso, *Redox Rep.* **4,** 69 (1999).
[62] R. Re, N. Pellegrini, A. Proteggente, A. Pannala, M. Yang, and C. Rice-Evans, *Free Radic. Biol. Med.* **26,** 1231 (1999).
[63] R. van den Berg, G. R. M. M. Haenen, H. van den Berg, and A. Bast, *Food Chem.* **66,** 511 (1999).
[64] M. P. Imam and R. O. Recknagel, *Toxicol. Appl. Pharmacol.* **42,** 455 (1977).
[65] Y. Sawai and K. Sakata, *J. Agric. Food Chem.* **46,** 111 (1998).
[66] Z. Y. Chen, Q. Y. Zhu, Y. F. Wong, Z. S. Zhang, and H. Y. Chung, *J. Agric. Food Chem.* **46,** 2512 (1998).

and the pH is adjusted to pH 8.0–9.0, depending on the individual pK values in whose neighborhood strong spectral changes might cause artifacts during the kinetic evaluations. Azide radicals are generated in N_2O-saturated solutions with the additional presence of 10 mM sodium azide. Superoxide anions are produced in oxygenated solutions with sodium formate (10 mM).

Delivering an intense pulse of energy (about 1.8 MeV) within a very short time (100 nsec)—these values pertain to the "Febetron 708" type of accelerator, while higher pulse energies and variable pulse lengths can be obtained with linear accelerators[67]—generates in aqueous solutions about equal amounts of hydrated electrons (e_{aq}^-), •OH radicals, and, with 10% of the total yield, hydrogen atoms (H•) at a yield of 0.28, 0.27, and 0.06 μmol/J, respectively.[68]

The choice of the added solutes or gases then controls the amount and type of the invidual radicals present, e.g., N_2O scavenges e_{aq}^- very rapidly and converts it to •OH, in effect doubling its yield. The additional presence of sodium azide then causes rapid conversion of •OH into •N_3, another strongly electrophilic species, which is known to react selectively with phenolates to form the phenoxyl radicals.[69] In contrast, •OH can also add to the aromatic ring with the intermediary formation of hydroxycyclohexadienyl radicals[70–72] (Fig. 2a shows an example of such a biphasic buildup kinetic). The exclusive presence of $O_2^{•-}$ in O_2-saturated solutions containing sodium formate is due to the rapid conversion of both e_{aq}^- and H• into $O_2^{•-}$ and $HCOO^-$ converting •OH into the formate radicals ($CO_2^{•-}$), which transfers its reducing equivalent to O_2.[73]

All these reactions occur at subnanosecond time scales, whereas the reactions with substrates are generally slower. With optical spectroscopy in the nanosecond-to-second time range, three technical approaches are in use: (i) monochromators,[74] (ii) spectrograph-based monitoring systems, both using UV-sensitive photomultipliers,[75] and (iii) photodiode arrays.[76,77] While one experiment (or repetitive pulses

---

[67] M. C. Sauer, *in* "The Studies of Fast Processes and Transient Species by Electron Pulse Radiolysis" (J. H. Baxendale and F. Busi, eds.), p. 35. Reidel, Dordrecht, NL, 1982.

[68] C. von Sonntag and H. P. Schuchmann, *Methods in Enzymol.* **233,** 3 (1994).

[69] Z. B. Alfassi and R. H. Schuler, *J. Phys. Chem.* **89,** 3359 (1985).

[70] A. Mantaka, D. G. Marketos, and G. Stein, *J. Phys. Chem.* **75,** 3886 (1971).

[71] P. Neta and R. W. Fessenden, *J. Phys. Chem.* **78,** 523 (1974).

[72] H. Taniguchi and R. H. Schuler, *J. Phys. Chem.* **89,** 3095 (1985).

[73] W. Bors, M. Saran, C. Michel, and D. Tait, *in* "Advances on Oxygen Radicals and Radioprotectors." (A. Breccia, C. L. Greenstock, and M. Tamba, eds.), p. 13. Lo Scarabeo, Bologna, 1984.

[74] L. K. Patterson, *in* "Radiation Chemistry: Principles and Applications" (Farhataziz and M. A. J. Rodgers, eds.), p. 65. VCH Verlag, Weinheim, 1987.

[75] M. Saran, G. Vetter, M. Erben-Russ, R. Winter, A. Kruse, C. Michel, and W. Bors, *Rev. Sci. Instrum.* **58,** 363 (1987).

[76] G. Roffi, *in* "The Studies of Fast Processes and Transient Species by Electron Pulse Radiolysis" (J. H. Baxendale and F. Busi, eds.), p. 63. Reidel, Dordrecht, NL, 1982.

[77] E. P. L. Hunter, M. G. Simic, and B. D. Michael, *Rev. Sci. Instrum.* **56,** 2199 (1985).

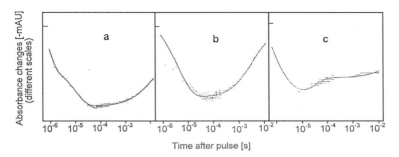

FIG. 2. Kinetic traces (log scale) for catechin, β-glucogallin, and pentagalloyl-β-D-glucose. Abscissa: Time after pulse [sec]; ordinate: negative changes of absorption (arbitrary mAU at different scales). Examples of typical buildup and decay kinetics: a, intermediary formation of hydroxycyclohexadienyl radicals (catechin = 95 μM, pH 9.3, [•OH] = 11.2 μM, 310 nm, maximal absorption 84 mAU); b, decay in the absence of absorbing quinone product (β-glucogallin = 36 μM, pH 9.3, [•OH] = 12.3 μM, 275 nm, maximal absorption 52 mAU); c, fast first-order decay and subsequent second-order decay (pentagalloylglucose = 13.4 μM, pH 9.2, [•OH] = 15.9 μM, 405 nm, maximal absorption 23 mAU).

under identical conditions to enhance signal-to-noise ratios) only provides spectral data, concentration-dependent changes in the kinetics of the buildup or decay of individual absorption peaks can be evaluated for second-order scavenging rate constants or first- and second-order decay rate constants.

A special feature of pulse-radiolytic investigations of plant polyphenols pertains to their poor solubility in slightly alkaline solutions and their strong molar absorptivities (aside from the limited amounts available). While preferentially a large excess of substrate over radical concentrations is used to ensure pseudo first-order conditions, with the polyphenols these ratios may sometimes even be inverted. The assumption that one still operates under pseudo first-order condition, can be corroborated by kinetic modeling studies in which all possible reactions are compiled in a kinetic scheme and iterative optimization of hypothetic rate constants is achieved by computerized comparison with digitized experimental data,[30,78,79] an approach that has also been called numerical modeling.[80] The reaction responsible for retaining pseudo first-order conditions at low substrate vs radical concentrations is a very rapid dimerization of the primary radicals (2 •OH → H_2O_2 or 2 •N_3 → $3N_2$), resulting in substoichiometric scavenging of these radicals. As an example, Table I depicts a comparison of rate constants for tannins, obtained either by linear regression analysis or by kinetic modeling,

---

[78] W. Bors and C. Michel, in "Lipoic Acid in Health and Disease" (J. Fuchs, L. Packer, and G. Zimmer, eds.), p. 33. Dekker, New York, 1997.
[79] R. Goldman, W. Bors, C. Michel, B. W. Day, and V. E. Kagan, Environ. Nutr. Interact. 1, 97 (1997).
[80] P. Wardman, in "S-Centered Radicals" (Z. B. Alfassi, ed.), p. 289. Wiley, Chichester, 1999.

TABLE I
RATE CONSTANTS OF FLAVANOLS AND GALLATES OBTAINED BY KINETIC MODELING OR REGRESSION ANALYSIS AFTER ATTACK BY •OH RADICALS

Substance	λ (nm)	n^a	$k_{•OH}$ ($\times 10^9 M^{-1} sec^{-1}$)^b		$2k_{PhO•}$ ($\times 10^6 M^{-1} sec^{-1}$)		$\varepsilon(M^{-1} cm^{-1})$	
			Regression^c	Model^d	Regression^c	Model^d	PhO•	Ph = O
(−)-Epicatechin	320	2	1.01	2.06	13.5	4.15	3800	3500
(+)-Catechin	310	2	2.18	2.19	7.84	4.42	7050	6000
Pycnogenol	320^e	2	0.52	1.35	15.4	5.73	2400	3000
(−)-Epigallocatechin	290^f	3	4.7	4.85	4.80	2.72	2800	3850
	550		—	4.28	215.25	278	1300	800
Procyanidin A2^g	355	4	7.8	n.d.	—	—	—	—
Procyanidin B2^h	355	4	2.3	n.d.	—	—	—	—
(−)-Epicatechin gallate	400	5	5.8	5.8	(I.O.)	29.0	1950	2100
(−)-Epigallocatechin gallate	400	6	7.1	10.7	29.3	10.1	2150	2550
Propyl gallate	405	3	13.7	17.1	(I.O.)	173	2500	2350
Glucogallin	405	3	16.2	17.4	29.4	15.2	1850	1250
Pentagalloylglucose^i	405	15	37.2	45.5	0.75	0.43	2050	1500
Tannic acid^i	420	25	31.4	48.2	1.85	1.12	2950	2550

^{a} Number of reactive hydroxy groups (see text).
^{b} Pulse-radiolytic experiments at pH 8.9–9.3.
^{c} Linear regression analysis of first- and second-order plots.
^{d} Kinetic modeling with extrapolation to final product absorption (see last column).
^{e} Pulse-radiolytic experiments at pH 9.3; molecular weight based on catechin content.
^{f} Superposition of parent substance absorption.
^{g} Weak signal, decay kinetics could not be evaluated.
^{h} Oily syrup, no accurate concentration could be determined.
^{i} Biphasic kinetics, kinetic evaluation only possible for late phase whereas results from kinetic modeling are shown in more detail in Table II; values in last column thus represent ε of secondarily formed PhO•, #2, not the final product, Ph = O.

demonstrating the close similarity of data obtained by these diverse methods. For the gallate esters, only the values at the higher wavelengths representing the ester bond are listed, as they are less prone to kinetic distortions.[22]

This kinetic modeling approach allows the resolution of more complex kinetic behavior, e.g., the biphasic buildup kinetics due to the intermediary formation of hydroxycyclohexadienyl radicals after reaction of •OH with aromatic compounds or successive first- and second-order decay kinetics as observed for the hydrolyzable tannins pentagalloylglucose and tannic acid.[22] Tables II and Table III list respective data, while in Fig. 2 the kinetic traces for these biphasic behaviors are shown.

TABLE II
KINETIC MODELS FOR BIPHASIC BUILDUP KINETICS OF (+)-CATECHIN, (−)-EPICATECHIN, AND PYCNOGENOL

Parameters	Catechin ($\lambda = 310$ nm)	Epicatechin ($\lambda = 320$ nm)	Pycnogenol ($\lambda = 320$ nm)
Regression analysis data			
$k_{\bullet OH}(\times 10^9\ M^{-1}\ \text{sec}^{-1})$	2.18	1.01	0.52
$2k_{PhO\bullet}(\times 10^6\ M^{-1}\ \text{sec}^{-1})$	7.84	13.5	15.4
$\varepsilon(PhO^\bullet)(M^{-1}\text{cm}^{-1})$	10.900	6365	4725
Kinetic modeling data			
$k_{(PhOH \to PhO^-)}(\times 10^6\ \text{sec}^{-1})$	3.15	5.0	3.15
$k_{(\bullet OH+PhO^- \to PhO\bullet)}(\times 10^9\ M^{-1}\ \text{sec}^{-1})$	17.1	20.8	12.0
$k_{(\bullet OH+PhOH \to HCHD\bullet)}(\times 10^9\ M^{-1}\ \text{sec}^{-1})$	25.9	26.5	22.7
$k_{(HCHD\bullet \to PhO\bullet)}(\times 10^4\ \text{sec}^{-1})$	7.89	4.99	3.36
$2k(PhO^\bullet)(\times 10^6 M^{-1}\ \text{sec}^{-1})$	4.42	4.15	5.73
$\varepsilon(PhO^\bullet)(M^{-1}\ \text{cm}^{-1})$	7060	3790	2425
$\varepsilon(Ph=O)(M^{-1}\ \text{cm}^{-1})$	6000	3500	3000

EPR Spectroscopy

Transient spectra after pulse radiolytic generation of the organic radicals in the case of the gallate esters (both hydrolyzable and condensed tannins) provided evidence for semiquinone generation and the participation of the gallate ester bond in the overall reactions.[22] EPR spectroscopy yields much more structural information, yet due to its much slower time resolution, it is sometimes unlikely that the same radicals as in pulse radiolysis are observed.[36] A particular feature that

TABLE III
KINETIC MODELS FOR BIPHASIC DECAY KINETICS OF PENTAGALLOYLGLUCOSE (P66) AND TANNIC ACID (TA)

Parameter	PGG ($\lambda = 405$ nm)	TA ($\lambda = 420$ nm)
Regression analysis data		
$k_{\bullet OH}(\times 10^9\ M^{-1}\ \text{sec}^{-1})$	37.2	31.4
$2k_{PhO\bullet,\#2}(\times 10^6\ M^{-1}\ \text{sec}^{-1})$	0.75	1.85
$\varepsilon(PhO^\bullet, \#2)(M^{-1}\ \text{cm}^{-1})$	1800	2800
Kinetic modeling data		
$k_{\bullet OH}(\times 10^9\ M^{-1}\ \text{sec}^{-1})$	45.5	48.2
$k_{(PhO^\bullet,\#1 \to PhO^\bullet,\#2)}(\times 10^4\ \text{sec}^{-1})$	1.60	0.54
$2k_{PhO\bullet,\#2}(\times 10^6\ M^{-1}\ \text{sec}^{-1})$	0.43	1.12
$\varepsilon(PhO^\bullet, \#1)$	2050	2950
$\varepsilon(PhO^\bullet, \#2)$	1500	2550

could only be observed with EPR was the fact that the signal of EGCG changed from a single radical species to that of an oligomeric species.[36]

To obtain the radicals for EPR measurements, horseradish peroxidase/hydrogen peroxide oxidation of the respective phenols was carried out *in situ* in the EPR cuvette (tyrosinase, 5 U/ml, yielded identical EPR signals). In detail, 1 mM N_2-saturated substrate solutions were mixed with 6.7 U/ml horseradish peroxidase, and the reaction was started by the addition of 4 mM H_2O_2 at pH 8.5–9.0. To compare the results with literature data,[81–83] radicals can also be generated *in situ* by autoxidation of 1 mM substrate in a slightly alkaline solution (0.05 M NaOH, pH > 10). EPR spectra are recorded in the X-band mode at a modulation amplitude of 0.5 G, a scan speed of 0.5 G/sec, a 20-mW power setting, and a gain of 10^6.

Conclusions

While the relative assay methods (TEAC and/or DPPH• bleaching) enable a rapid screening of antioxidant potential at low substrate concentrations, this information lacks mechanistic details. Only pulse radiolysis combined with kinetic spectroscopy is capable of providing an insight into the very first moments after a radical-initiated reaction was started. With EPR spectroscopy yielding more detailed information on the structures of radical intermediates, these data can then be combined with analytical data obtained by modern methods such as mass spectrometry combined with high-performance liquid chromatography to obtain a comprehensive picture of the function of polyphenolic antioxidants. Because in general these methods are limited to *in vitro* systems, the results can only be utilized indirectly to provide information on the *in vivo* behavior of these compounds. In fact, only for the anticarcinogenic potential of polypenols have SAR studies been performed in cellular model systems.[84,85]

Acknowledgment

We appreciate the intense discussions with Manfred Saran throughout the experimental period and during the preparation of the manuscript.

[81] O. N. Jensen and J. A. Pedersen, *Tetrahedron* **39,** 1609 (1983).
[82] H. Yoshioka, K. Sugiura, R. Kawahara, T. Fujita, M. Makino, M. Kamiya, and S. Tsuyumu, *Agric. Biol. Chem.* **55,** 2717 (1991).
[83] Q. N. Guo, B. L. Zhao, M. F. Li, S. R. Shen, and W. J. Xin, *Biochim. Biophys. Acta* **1304,** 210 (1996).
[84] I. K. Wang, S. Y. Lin-Shiau, and J. K. Lin, *Eur. J. Cancer* **35,** 1517 (1999).
[85] M. H. Pan, J. H. Lin, S. Y. Lin-Shiau, and J. K. Lin, *Eur. J. Pharmacol.* **381,** 171 (1999).

[16] Antioxidant and Prooxidant Abilities of Foods and Beverages

By LEE HUA LONG and BARRY HALLIWELL

Introduction

There is considerable and growing interest in the beneficial health effects of certain foods and beverages, especially wines and green and black teas. Many of these proposed health benefits have been attributed to the antioxidants present in these foods and beverages; usually the focus has been on phenolic compounds, such as tocopherols, ferulic acid, catechins, and other flavonoids.[1–7] For example, red wines have high total antioxidant activity (TAA) *in vitro*, of which at least 25% can be accounted for by the major phenolic constituents.[7] White wines have, in general, lower TAA than red wines.[7] Similarly, ~78% of the antioxidant activity of green tea was accounted for by the catechins and catechin gallate esters.[7] Even beer has been examined for its antioxidant activity.[5,8] Multiple methods are available to assess the TAA of food extracts or beverages.[9–11] One of the first methods to be widely used employed 1,1-diphenyl-2-picrylhydrazyl (DPPH).[12] DPPH is a stable free radical with a distinctive electron spin resonance (ESR) signal. Its reactions with antioxidants can be followed by loss of the ESR signal, or the fall in absorbance at 517 nm.[13] For example, this assay has been used to relate the TAA of sake to its "sensory maturation" during storage.[14] Another stable free radical is galvinoxyl. For example, the total antioxidant activity of *Gingko biloba* extracts has been measured using this radical.[15]

Because food extracts and beverages are often colored, turbid, and contain components insoluble in water, an assay that can cope with these problems is

---

[1] S. A. Wiseman, D. A. Balentine, and B. Frei, *Crit. Rev. Food Sci. Nutr.* **37**, 705 (1997).
[2] J. Constant, *Clin. Cardiol.* **20**, 420 (1997).
[3] H. Fujiki, M. Suganuma, S. Okabe, *et al.*, *Mutat. Res.* **402**, 307 (1998).
[4] E. N. Frankel, A. L. Waterhouse, and P. L. Teissedre, *J. Agric. Food Chem.* **43**, 890 (1995).
[5] G. Paganga, N. Miller, and C. A. Rice-Evans, *Free Radic. Res.* **30**, 153 (1999).
[6] C. Rice-Evans, N. J. Miller, and G. Paganga, *Free Radic. Biol. Med.* **20**, 933 (1996).
[7] C. Rice-Evans, N. J. Miller, P. G. Bolwell, P. M. Bramley, and J. B. Pridham, *Free Radic. Res.* **22**, 375 (1995).
[8] L. Bourne, G. Paganga, D. Baxter, P. Hughes, and C. Rice-Evans, *Free Radic. Res.* **32**, 273 (2000).
[9] C. Rice-Evans, *Free Radic. Res.*, in press.
[10] H. Wang, G. Cao, and R. L. Prior, *J. Agric. Food Chem.* **44**, 701 (1996).
[11] E. E. Robinson, S. R. Maxwell, and G. H. Thorpe, *Free Radic. Res.* **26**, 291 (1997).
[12] A. K. Ratty, J. Sunatomo, and N. P. Das, *Biochem. Pharmacol.* **37**, 989 (1988).
[13] T. Yamaguchi, H. Takamura, T. Matoba, and J. Terao, *Biosci. Biotechnol. Biochem.* **62**, 1201 (1998).
[14] H. Kitagaki and M. Tsugawa, *J. Biosci. Bioeng.* **87**, 328 (1999).

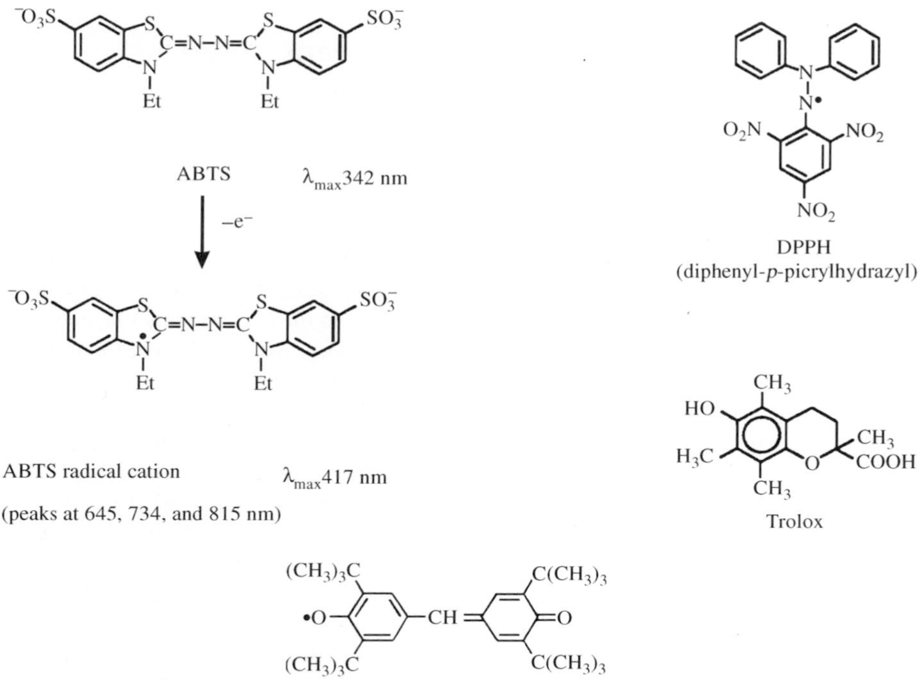

FIG. 1. Some of the compounds used in "total antioxidant assays." ABTS$^{•+}$ and DPPH are nitrogen-centerd radicals, whereas galvinoxyl (bottom structure) is an aromatic alkoxyl (phenoxyl) radical that has a strong absorbance at 428 nm in ethanol. Trolox, used to calibrate many assays, is a water-soluble form of α-tocopherol: the hydrophobic side chain is replaced by a –COOH group. One mole of Trolox can scavenge two $RO_2^•$ radicals. DPPH is dissolved in ethanol, whereas ABTS is soluble in aqueous solutions but can also be used in organic media.[16]

beneficial. One such assay, applicable to both aqueous and organic phases, is the ABTS assay.[9,16] Essentially, ABTS [2,2′-azinobis(3-ethylbenzothiazoline 6-sulfonate)] is oxidized to the colored nitrogen-centered radical cation ABTS$^{•+}$, which has absorption maxima at 645, 734, and 815 nm. Incidentally, DPPH is also a nitrogen-centered radical, whereas galvinoxyl is a phenoxyl (aromatic alkoxyl) radical (Fig. 1). In earlier versions of the ABTS assay, a mixture of myoglobin (or other heme protein) and H_2O_2 was used to oxidize ABTS to ABTS$^{•+}$. If all reagents (including the food/beverage to be tested) are added simultaneously to

---

[15] H. Shi and E. Niki, *Lipids* **33**, 365 (1998).
[16] N. J. Miller, J. Sampson, L. P. Candeias, P. M. Bramley, and C. A. Rice-Evans, *FEBS Lett.* **384**, 240 (1996).

such a reaction mixture, it is possible that decreases in the level of $ABTS^{•+}$ could be produced not only by scavenging of this radical, but also by inhibiting its formation, by removing H_2O_2, and/or by interacting with the higher oxidation states of myoglobin[17] that convert ABTS to $ABTS^{•+}$. Indeed, a wide range of phenolic compounds and other antioxidants are substances for the peroxidase activity of heme proteins and thus could potentially interfere.[18] Although such interference does not appear to have been a major artifact in most studies,[19,20] the problem can be avoided easily. For example, myoglobin/H_2O_2 can be used to preform $ABTS^{•+}$, followed by addition of the putative antioxidant once $ABTS^{•+}$ formation is complete. Alternatively, a chemical oxidant can be used, such as MnO_2, peroxyl radicals, or persulfate.[19–21] The protocol described by Re et al.[20] has been found to work well in our laboratory. The water-soluble vitamin E analog Trolox (Fig. 1) is used to provide a reference standard.

Materials and Methods

Chemicals

All chemicals are from Sigma-Aldrich Pte Ltd., Singapore, except for ethanol, KCN, and $HgCl_2$ (Merck, Germany), methanol, Na_2SO_3, and KOH (Fluka, Switzerland), and H_2SO_4 (J. T. Baker, Phillipsburg, NJ).

ABTS Assay

This is carried out essentially as described in Re et al.[20] using Trolox as a reference standard. ABTS is dissolved in water to a final concentration of 7 mM and potassium persulfate is added to a final concentration of 2.45 mM (i.e., ABTS is in excess). The mixture is allowed to stand at room temperature overnight (>12 hr) in the dark. The $ABTS^{•+}$ solution is diluted to an absorbance of 0.70 at 734 nm in phosphate-buffered saline (PBS) and 10 μl of antioxidant compound or Trolox standard added to 1 ml of $ABTS^{•+}$ solution. Absorbance is measured 1 min after initial mixing unless otherwise stated. Controls without $ABTS^{•+}$ are used to allow for any absorbance of the test compounds. One milliliter of PBS (instead of the $ABTS^{•+}$ solution) is mixed with the test compound at the same dilution and the absorbance at 734 nm is read after 1 min. Even with dark soy sauce preparations, the absorbances of these controls are small (∼0.03 at 734 nm for the dilution used).

---

[17] M. Strube, G. R. Haenen, H. Van Den Berg, and A. Bast, *Free Radic. Res.* **26,** 515 (1997).
[18] B. Halliwell, *Adv. Pharmacol.* **38,** 3 (1996).
[19] N. J. Miller and C. A. Rice-Evans, *Free Radic. Res.* **26,** 195 (1997).
[20] R. Re, N. Pellegrini, A. Proteggente, A. Pannala, M. Yang, and C. Rice-Evans, *Free Radic. Biol. Med.* **26,** 1231 (1999).
[21] E. A. Lissi, B. Modak, R. Torres, J. Escobar, and A. Urzua, *Free Radic. Res.* **30,** 471 (1999).

For beverages insoluble in water (e.g., sesame oil), the ABTS$^{\bullet+}$ solution is diluted to an absorbance of 0.70 at 734 nm in ethanol instead of PBS. Dilutions of the sesame oil and of the Trolox standard are also in ethanol.

Sulfite Determination

Sulfite is measured by the fuchsin–formaldehyde method.[22] Fuchsin reagent is prepared fresh before use by adding 1.6 ml of a 3% (w/v) basic fuchsin solution in ethanol (left stirring overnight at room temperature) to 93.6 ml deionized water containing 4.4 ml concentrated H_2SO_4. After thorough mixing, 0.43 ml of 37% (v/v) formaldehyde is added. The solution is decolorized by adding 400 mg of activated charcoal. After repeated shaking over a period of 15 min, the solution is filtered.

To a 2.5-ml sample of food extract or beverage, 0.5 ml of 1% (w/v) KOH (in ethanol) is added and mixed thoroughly. Of a saturated aqueous solution of $HgCl_2$, 1.0 ml is then added and mixed. After centrifuging at 3000 rpm (2000g) for 10 min, 1.0 ml of the clear supernatant is mixed with 4.0 ml fuchsin reagent. After 15 min at room temperature, the absorbance of each sample is measured at 580 nm. A standard curve is prepared using Na_2SO_3 solution (80 to 480 nmol).

Hydrogen Peroxide

This is measured by the Fox assay.[23,24] The catalase used is Sigma type C40, specific activity 25,000 units per milligram protein. Reagent 1 is 4.4 mM butylated hydroxytoluene (BHT) in HPLC-grade methanol and reagent 2 is 1 mM xylenol orange plus 2.56 mM ammonium ferrous sulfate in 250 mM H_2SO_4. One volume of reagent 2 is added to 9 volumes of reagent 1 to make the "working" FOX reagent. Beverage (0.33 ml), diluted as needed, is made up to 1.00 ml with deionized water and 90 μl is pipetted into each of two Eppendorf tubes. Methanol (10 μl) is added to each tube and the contents are vortexed for 5 sec and then incubated at room temperature for 30 min. The working FOX reagent (0.9 ml) is added to each tube, followed by vortexing and 30-min incubation as described earlier. Solutions are then centrifuged at 15,000g for 10 min at room temperature, and the absorbance at 560 nm is read against a methanol blank containing the appropriate amount of beverage to correct for any absorbance of the beverage itself. The FOX assay is calibrated using standard H_2O_2, diluted from stock, and its concentration is assessed using a molar extinction coefficient of 43 M^{-1} cm^{-1} at 240 nm. Calibration plots are linear in the concentration range 0–50 μM and all absorbances obtained from beverages are read within this range. The addition of 10 units of catalase is sufficient

---

[22] F. J. Leinweber and K. J. Monty, *Methods Enzymol.* **143**, 15 (1987).
[23] J. Nourooz-Zadeh, J. Tajaddini-Sarmadi, and S. P. Woolf, *Anal. Biochem.* **220**, 403 (1994).
[24] L. H. Long, A. N. B. Lan, F. T. Y. Hsuan, and B. Halliwell, *Free Radic. Res.* **31**, 67 (1999).

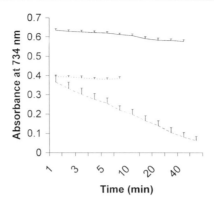

FIG. 2. Time course of ABTS$^{\bullet+}$ depletion in the presence of (dotted line) Trolox, (solid line) red wine (Turning Leaf Cabernet Sauvignon, California) or (dashed line dark soy sauce (Tai Hua Brand, Singapore). Error bars indicate mean ± SD, $n \geq 3$. Reaction with ascorbate, like that of Trolox, was complete within 1 min (data not shown).

to destroy all the H_2O_2 immediately and this level can therefore be used as a control in studies with the beverages to check for any absorbance due to constituents of the beverages themselves.

Antioxidant Activity Using ABTS/Persulfate System

We found that this system worked well with most food extracts or beverages tested, and amounts of ABTS$^{\bullet+}$ used up were proportional to the amount of extract/beverage added over a reasonable working range. In agreement with Re et al.,[20] the reaction of ABTS$^{\bullet+}$ with Trolox or with ascorbate was complete by the 1-min time point usually used (see the methods section). However, for several food extracts/beverages, this was not the case, in that depletion of ABTS$^{\bullet+}$ continued for substantial periods (Fig. 2). These foods and beverages must therefore contain mixtures of antioxidants that react with ABTS$^{\bullet+}$ at different rates. For example, several phenols need up to 4 min for complete reaction with ABTS$^{\bullet+}$, whereas luteolin and naringenin react more slowly.[20] The even longer time courses for red wine, and especially for dark soy sauce (Fig. 2), suggest that complex mixtures of antioxidants are present, including perhaps polymers of phenols.

We have applied the ABTS method to examine the antioxidant activity of seasonings commonly used in Singapore.[25] The Trolox equivalent antioxidant capacities (TEAC values) were, in general, substantial (Table I). Several different amounts of each sauce were tested, with appropriate controls to allow for any

[25] L. H. Long, D. C. T. Kwee, and B. Halliwell, *Free Radic. Res.* **32**, 181 (2000).

TABLE I
ANTIOXIDANT CAPACITIES (TEAC) OF SEASONINGS AS EVALUATED
BY ABTS ASSAY

Seasoning type	TEAC (mM)[a]
HP sauce	9.80 ± 0.53
Tomato sauce	3.23 ± 0.06
Kung Bo sauce	9.17 ± 0.91
Black vinegar	10.37 ± 1.00
Chinese cooking wine	6.17 ± 0.35
Chinese rice wine	0.36 ± 0.13
Chili sauce	11.10 ± 1.51
Dark soy sauce (Tiger brand)	147.33 ± 9.45
Dark soy sauce (Tai Hua brand)	127.33 ± 4.93
Dark soy sauce (Woh Hup brand)	47.1 ± 1.93
Light soy sauce (Woh Hup brand)	16.27 ± 1.08
Sweet dark soy sauce (Zara)	27.73 ± 2.54
Sweet soy sauce	35.43 ± 1.60
Oyster sauce	5.58 ± 2.75
Plum sauce	3.53 ± 1.27
Hoisin sauce	13.60 ± 2.16
Sweet flour sauce	10.40 ± 0.53
Soba sauce	9.27 ± 0.64
Sesame oil	3.27 ± 1.08

[a] Results are expressed as Trolox equivalent antioxidant capacity (TEAC) and were determined at 1 min after addition to ABTS. This may be an underestimate of TEAC (Fig. 2 and text): (mean ± SD, $n = 3$).

absorbance due to the sauce itself, and TEAC values were calculated from ranges in which the fall in absorbance was proportional to the amount of sauce added. Of special note was the fact that all the dark soy sauces tested had extremely high TEAC values (Table I). In fact, these values are underestimates, as data in Table I are based on measurements of ABTS$^{•+}$ depletion at 1 min, whereas ABTS$^{•+}$ depletion continued for long periods with all the dark soy sauces (Fig. 2). For example, in one experiment the TEAC of the Tai Hua brand dark soy sauce increased to a value of 200 at 5 min and 318 at 50 min. The agents responsible for such powerful *in vitro* antioxidant activity remain to be identified, but might include some contribution from isoflavones.[26,27]

---

[26] M. B. Ruiz-Larrea, A. R. Motian, G. Paganga, N. J. Miller, G. P. Bolwell, and C. A. Rice-Evans, *Free Radic. Res.* **26**, 63 (1997).

[27] A. Arora, M. G. Nair, and G. M. Strasburg, *Arch. Biochem. Biophys.* **356**, 133 (1998).

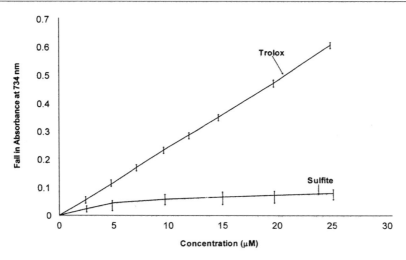

FIG. 3. Antioxidant activity of sodium sulfite and of Trolox in the ABTS assay. Reagents were added at the concentrations stated, and ability to scavenge ABTS$^{\bullet+}$ was determined as described in the methods section. Data are mean ± SD, $n \geq 3$. Reaction with Na_2SO_3 was complete within 1 min.

Effects of Preservatives

Several preservatives are added by manufacturers to foods and beverages, and these could potentially contribute to the measured TAA activity. Sodium benzoate is a widely used preservative, but it had no effect on the ABTS$^{\bullet+}$ radical when tested at concentrations up to 1 mM. This is not surprising because benzoic acid does not have a phenolic −OH group. In contrast, sulfite, used frequently as a food and beverage preservative,[28,29] was weakly reactive in the ABTS assay (Fig. 3), although at concentrations above 5 μM the reactivity with ABTS$^{\bullet+}$ did not maintain the same linear relationship with concentration as at lower concentrations, and the standard deviations were larger than those obtained with Trolox or ascorbate (Fig. 3). The reaction of SO_3^{2-} with ABTS$^{\bullet+}$ was complete by 1 min. This limited antioxidant activity of SO_3^{2-} may be sufficient to artificially enhance the TAA values of some beverages. For example, we found that the sulfite content of some white wines could contribute significantly to their TEAC.[25] Red wines, which have a lower sulfite content[29,30] and a generally greater content of antioxidant phenolics than white wines,[6] are unlikely to be affected by this artifact.[25] Commercially

---

[28] R. Walker, *Food Addit. Contam.* **2**, 5 (1985).
[29] J. P. Mareschi, M. Francois-Collange, and M. Suschetel, *Food Addit. Contam.* **9**, 541 (1992).
[30] J. J. Sullivan, T. A. Hollingworth, M. M. Wekell, V. A. Meo, A. Etemad-Moghadam, J. G. Phillips, and B. H. Gump, *J. Assoc. Official Anal. Chem.* **73**, 223 (1990).

purchased fruit juices often have added vitamin C, which could contribute to their TAA.

Prooxidant Effects

Many antioxidants can be made to exert prooxidant effects *in vitro* under certain assay conditions.[11,31] This often involves their interaction with transition metal ions. For example, ascorbate, flavonoids, and α-tocopherol can reduce Fe^{3+} and Cu^{2+} ions to Fe^{2+} and Cu^+, respectively, which can stimulate oxidative damage under certain assay conditions.[18,31–36] Any oxidized phenols (as quinones and semiquinones) formed from phenolics during such redox reactions could conceivably also be deleterious.[37–39] In addition, oxidation of phenols and other constituents in foods and beverages can generate H_2O_2, which can cause direct damage to food constituents and human tissues (e.g., by oxidizing SH groups on proteins) and might also become converted[18,40] into highly reactive radicals (OH•) by reaction with Fe^{2+} or Cu^+. H_2O_2 appears to account for much, if not all, of the reported mutagenicity of coffees, teas, and wines in bacterial test systems.[41,42] The Fox assay[23,24] can conveniently be used to measure H_2O_2 production, although other methods are available.[41–46] Figure 4 shows representative data on the levels of H_2O_2 that can be achieved in commonly prepared beverages. Instant coffees generate particularly large amounts of H_2O_2. In contrast, no H_2O_2 could be detected in milk (fresh pasteurized, evaporated, or condensed), which instead appears to be able to scavenge H_2O_2.[24] The cocoa preparations tested appeared inefficient at generating H_2O_2 by comparison with the other beverages (Fig. 4). Dark soy

---

[31] B. Halliwell, *Food Sci. Agric. Chem.* **1**, 67 (1999).
[32] N. Yamanaka, O. Oda, and S. Nagao, *FEBS Lett.* **401**, 230 (1997).
[33] M. Maiorino, A. Zamburlini, A. Roveri, and F. Ursini, *FEBS Lett.* **330**, 174 (1993).
[34] P. Otero, M. Viana, E. Herrera, and B. Bonet, *Free Radic. Res.* **27**, 619 (1997).
[35] S. A. Kang, Y. J. Gang, and M. Park, *Free Radic. Res.* **28**, 93 (1998).
[36] M. J. Laughton, B. Halliwell, P. J. Evans, and J. R. Hoult, *Biochem. Pharmacol.* **38**, 2859 (1989).
[37] D. Metodiewa, A. K. Jaiswal, N. Cenas, E. Dickancaite, and J. Segura-Agilar, *Free Radic. Biol. Med.* **26**, 107 (1999).
[38] A. T. Canada, E. Giannella, T. D. Nguyen, and R. P. Mason, *Free Radic. Biol. Med.* **9**, 441 (1990).
[39] G. Galati, T. Chan, B. Wu, and P. J. O'Brien, *Chem. Res. Toxicol.* **12**, 521 (1999).
[40] R. H. Stadler, J. Richoz, R. J. Turesky, D. H. Wetti, and L. B. Fay, *Free Radic. Res.* **24**, 225 (1996).
[41] E. Alejandre-Duran, A. Alonso-Moraga, and C. Pueyo, *Mutat. Res.* **188**, 251 (1987).
[42] R. R. Ariza and C. Pueyo, *Mutat. Res.* **251**, 115 (1991).
[43] H. U. Aeschbacher, U. Wolleb, J. Loliger, J. C. Spadone, and R. Liardon, *Food Chem. Toxicol.* **27**, 227 (1989).
[44] R. H. Stadler, R. J. Turesky, O. Muller, J. Markovic, and P. M. Leong-Morgenthaler, *Mutat. Res.* **308**, 177 (1994).
[45] Y. Fujita, K. Wakabayashi, M. Nagao, and T. Sugimura, *Mutat. Res.* **144**, 227 (1985).
[46] J. Ruiz-Laguna and C. Pueyo, *Mutagenesis* **14**, 95 (1999).

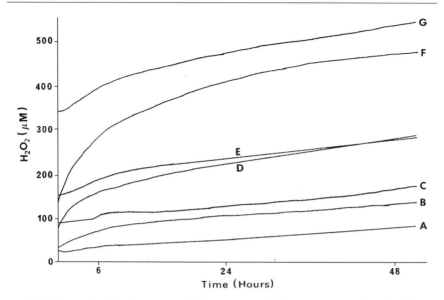

FIG. 4. Generation of hydrogen peroxide by beverages. A, Pure cocoa powder from Van Houten (Germany); B, oolong tea from Nam Wan tea (Malaysia); C, green tea from Ujinotsuyuseicha (Japan); D, black tea from Lipton (London); E, ground Arabica coffee (Gourmet Bean, Washington, DC); F, tea granules from Lipton (London); G, instant coffee from Nestle (Singapore). 0.25 g of the beverages was weighed, and 25 ml of deionized water (at 90°) was added and stirred on a hot plate to maintain the temperature for 15 min. This was taken as zero time. At every 30 min (up to 6 hr), and at 24 and 48 hr, 330 µl was taken and made up to 1.0 ml with deionized water. Ninety microliters was taken for the Fox assay to determine the amount of H_2O_2 generated. Data abstracted from Long et al.[24] courtesy of Harwood Academic Publishers.

sauces at room temperature also contained little H_2O_2.[25] Phenols can interact with cell-culture media to produce H_2O_2.[47]

Conclusion

Many foods and beverages contain a complex mixture of antioxidant and prooxidant activities, which can be detected by various assays *in vitro*, including the simple assays described in this article. Of course, demonstrating that these antioxidant or prooxidant effects are relevant *in vivo* raises a whole new set of methodological questions.[31,48] Antioxidants can exert protective effects both within the gastrointestinal tract (prior to their absorption or excretion) and after uptake into the

[47] L. H. Long, M. V. Clement, and B. Halliwell, *Biochem. Biophys. Res. Commun.* **273,** 50 (2000).
[48] B. Halliwell, *Nutr. Rev.* **57,** 104 (1999).

human body.[31,48,49] The use of "biomarkers" of oxidative damage to proteins, lipids, and DNA is now becoming widely used to examine the *in vivo* antioxidant effects of foods and beverages; recent examples may be found elsewhere.[49–54]

[49] B. Halliwell, *Free Radic. Res.* **29,** 469 (1998).
[50] H. Verhagen, H. E. Poulsen, S. Loft, G. van Poppel, M. I. Willems, and P. J. van Bladeren, *Carcinogenesis* **16,** 969 (1995).
[51] X. S. Deng, J. Tuo, H. E. Poulsen, and S. Loft, *Free Radic. Res.* **28,** 323 (1998).
[52] A. Rehman, L. C. Bourne, B. Halliwell, and C. A. Rice-Evans, *Biochem. Biophys. Res. Commun.* **262,** 828 (1999).
[53] J. E. Klaunig, Y. Xu, C. Han, L. M. Kamendulin, J. Chen, C. Heiser, M. S. Gordon, and E. R. Mohler III, *Proc. Soc. Exp. Biol. Med.* **220,** 249 (1999).
[54] J. F. Young, S. E. Nielsen, J. Haraldsdottir, *et al.*, *Am. J. Clin. Nutr.* **69,** 87 (1999).

[17] Metal Chelation of Polyphenols

By ROBERT C. HIDER, ZU D. LIU, and HICHAM H. KHODR

Chelation Process

The protonated phenolic group is not a particularly good ligand for metal cations, but once deprotonated, an oxygen center is generated that possesses a high charge density, a so-called "Hard" ligand [Eq. (1)]. Although the pK_a value of most phenols is in the region of 9.0–10.0, in the presence of suitable cations for instance iron(III) or copper(II), the proton is displaced at much lower pH values, e.g., 5.0–8.0. Thus metal chelation by phenols can occur at physiological pH values. Aliphatic alcohols do not share this property, as the resulting oxygen anion is not stabilized by the mesomeric effect typical of phenols [Eq. (2)].[1] In principle, a pyrone oxygen can also bind metal cations due to the partial delocalization of the lone pairs associated with the heteroatom [Eq. (3)]. Such delocalization is more prominent in nonfused rings such as maltol (**1**) than in compounds with fused rings, e.g., the bicyclic compound (**2**).[2]

$$\text{PhOH} + M^{n+} \rightleftharpoons \text{PhO}^-M^{n+} + H^+ \quad (1)$$

[1] R. C. Hider and A. D. Hall, *Progress in Med. Chem.* **28,** 41 (1991).
[2] P. S. Dobbin, R. C. Hider, L. Venkatramani, J. Siripitayananon, and D. Vanderhelm, *J. Heterocycl. Chem.* **30,** 723 (1993).

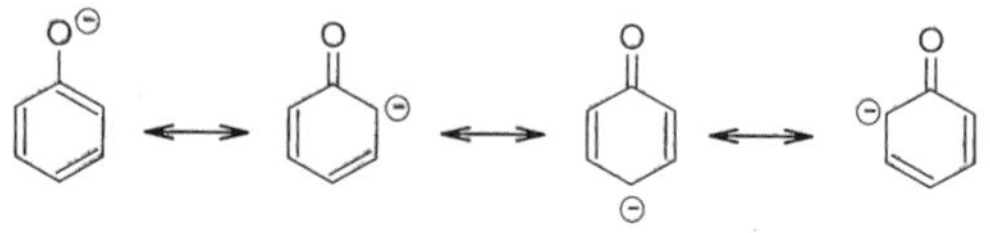

(2)

(3)

For chelation, bidentate ligands are much more powerful scavengers of metal cations than monodentate ligands, thus catechol (**3**) binds iron(III) tightly at pH 7.0, whereas phenol does not. Clearly 1,3- (**4**) and 1,4-dihydroxybenzene (**5**) behave like phenol as they cannot chelate metals in bidentate mode. The *peri* positions of two fused rings may also provide a chelation center, e.g., 1,8-dihydroxynaphthalene (**6**) and juglone (**7**).[3]

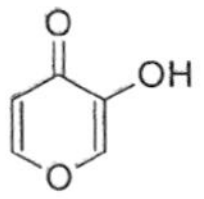

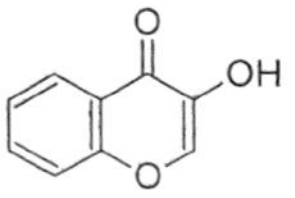

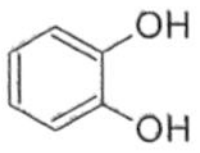

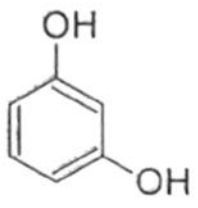

Flavonoids have been widely reported to chelate metals[4–9] and, using the basic rules just outlined, potential metal-binding sites are readily identified on polyphenol compounds, e.g., quercetin (**8**) and kaempferol (**9**). Quercetin has three potential metal-binding sites, A, B, and C, with the affinity of site A being greater than sites B and C at pH 7.0. With kaempferol there are two potential binding sites,

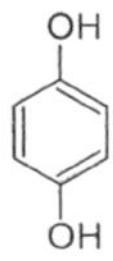

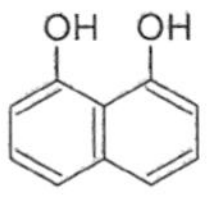

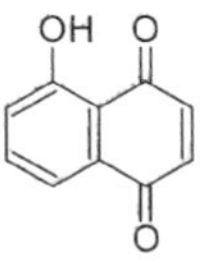

---

[3] R. C. Hider, *in* "Iron, Siderophores, and Plant Diseases" (T. R. Swinburne, ed.), p. 49. Plenum Press, New York, 1986.

namely D and E, but only one or the other will be used for chelation, both cannot be used simultaneously. When a phenol group is conjugated with a carbohydrate moiety, it can no longer bind metals, as the dissociatable proton is lost, thus rutin (**10**) only possesses two metal-binding sites, F and G.

Because the phenoxide group is a "Hard" ligand, it favors interaction with cations of high charge density, e.g., iron(III) (ferric iron), iron(II) (ferrous iron), copper(II), and zinc(II). In general, tribasic cations are preferred to dibasic cations[1] and therefore polyphenols will also bind aluminum(III). They do not chelate alkali and alkaline earth cations, e.g., sodium, potassium, and calcium. There may be some weak interaction with magnesium, but a catechol function will lose magnesium ions in competition with phosphate.

Polyphenol Metal Complexes

Stoichiometry

Bidentate ligands can form 1:1, 2:1, and 3:1 complexes with cations, thus iron(III) forms three complexes with catechol-containing moieties, e.g.,

---

[4] A. K. Ratty and N. P. Das, *Oncology* **39**, 69 (1988).
[5] G. Sichel, C. Corsaro, M. Scalia, A. J. Di Bilio, and R. P. Bonomo, *Free Radic. Biol. Med.* **11**, 1 (1991).
[6] W. F. Hodnick, E. B. Milosavljevic, J. H. Nelson, and R. S. Pardini, *Biochem. Pharmacol.* **37**, 2607 (1988).
[7] N. Suzuki, *Spectrum* **6**, 21 (1993).
[8] S. V. Jovanovic, S. Steenken, M. Tosic, B. Marjanovic, and M. G. Simic, *J. Am. Chem. Soc.* **116**, 4846 (1994).
[9] J. E. Brown, H. Khodr, R. C. Hider, and C. A. Rice-Evans, *Biochem. J.* **330**, 1173 (1998).

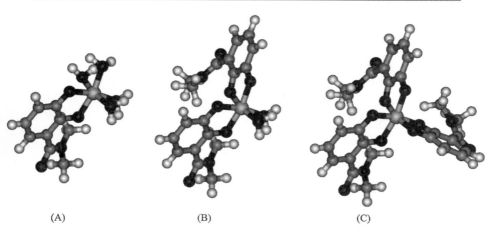

(A) (B) (C)

FIG. 1. Iron(III) complexes of N,N-dimethyl-2,3-dihydroxybenzamide: (A) 1 : 1 iron(III) complex, with four coordinated water molecules; (B) 2 : 1 iron(III) complex, with two coordinated water molecules; and (C) 3 : 1 iron(III) complex.

N,N-dimethyl-2,3-dihydroxybenzamide (Fig. 1). The relative proportions of each form depends on the concentration of both metal and ligand and the pH of the solution (Fig. 2). Thus at iron concentrations in the range 10^{-6}–10^{-4} M, the 2 : 1 species predominates at pH 7.0. The 1 : 1 and 2 : 1 species possess hydrated iron(III) surfaces (Fig. 1) and as such are susceptible to hydroxyl radical production either by redox cycling or fission of hydrogen peroxide.

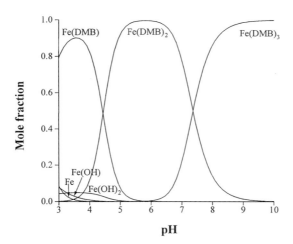

FIG. 2. Speciation plot of iron(III) in the presence of N,N-dimethyl-2,3-dihydroxybenzamide (DMB); $[Fe^{3+}]_{total} = 10^{-5} M$; $[DMB] = 10^{-4} M$.

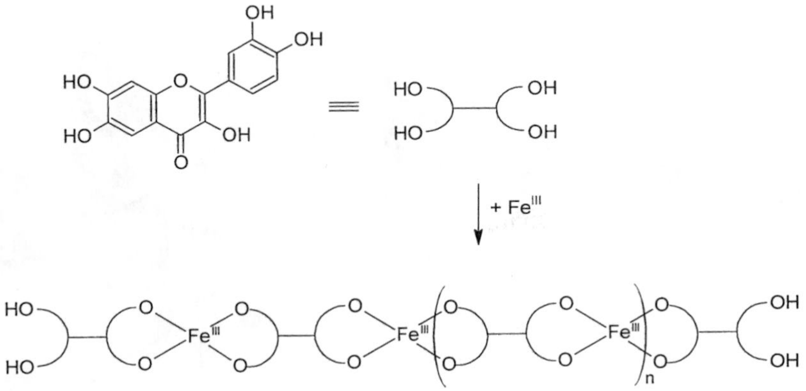

FIG. 3. Iron(III)-induced polymerization of flavonoid structure that possesses two metal-chelating centers.

Many polyphenols possess more than one metal-binding site, e.g., (**8**) and (**10**), and therefore are capable of polymerization and oligomerization as indicated in Fig. 3. This aggregation effect will reduce the ability of polyphenols to partition in membranes and therefore to gain access to intracellular compartments.

Internal Redox Reaction

In contrast to hydroxypyranones (e.g., **1** and **2**), catechols are susceptible to oxidation [Eq. (4)]. The oxidation products, semiquinone and quinone, are also able to coordinate cations, but generally with reduced affinity.[10–13] Assignment of a formal oxidation state to the coordinated metal is not always unambiguous as the orbitals involved are delocalized over the metal and at least one ring. The electronic structure of the metal quinone chelate ring can be viewed in terms of three isoelectronic forms [Eq. (5)]. These various complex types can be induced electrochemically by the addition or removal of electrons via an electrode surface or a chemical agent. Metal–catechol complexes (Fe and Cu) are able to undergo intramolecular electron transfer reactions. The positions of equilibrium of such reactions are influenced by the solution pH value and the presence of other ligands. Catechol complexes of copper show such a ligand dependence. Nitrogen donors favor copper(II) catecholate complexes (**11**),[14] whereas phosphorous donors favor

---

[10] D. R. Eaton, *Inorg. Chem.* **3**, 1268 (1964).
[11] F. Rohrscheid, A. L. Blach, and R. H. Holm, *Inorg. Chem.* **5**, 1542 (1966).
[12] C. Floriani, R. Henzi, and F. Calderazzo, *J. Chem. Soc. Dalton* 2640 (1972).
[13] C. G. Pierpont and R. M. Buchanan, *Coord. Chem. Rev.* **38**, 45 (1981).
[14] D. G. Brown, J. T. Rinprecht, and G. C. Vogel, *Inorg. Chem. Lett.* **12**, 399 (1976).

the copper(I) semiquinone form (**12**).[15] A similar phenomenon occurs with iron(III) catechol complexes, where an internal redox reaction is triggered by a change in pH.[16] Over the pH range 2.0 to 4.5, the green iron(II) semiquinone complex is the dominant species as determined by Mössbauer spectroscopy[16] and magnetic moment measurements.[17] Negatively charged ligands tend to stabilize iron(III), as electrostatic repulsion will favor removal of an electron from iron(II). However, neutral unsaturated ligands, like the monoprotonated semiquinone, will stabilize the divalent state [Eq. (6)]. This inversion is largely due to the greater crystal field stabilization of a d^6 system (Fe^{II}) compared to that of a d^5 system (Fe^{III}) in strong fields.[18] In this pH range the coordinated iron(II) prevents the semiquinone undergoing disproportionation to catechol and benzoquinone. However, at pH values below 2.0, disproportionation does occur.[19]

$$\text{Catechol} \rightleftharpoons \text{Semiquinone} \rightleftharpoons \text{Quinone} \quad (4)$$

$$\quad (5)$$

$$[(H_2O)_4 Fe^{III}\cdots]^{2+} \rightleftharpoons [(H_2O)_4 Fe^{II}\cdots]^{2+} \quad (6)$$

(a)   (b)

Thus the redox state of iron coordinated to catechol is dependent on the pH of the solution and can be cycled between iron(II) and iron(III) by manipulation of pH.[20] The Fe^{III}/Fe^{II} ratio for various iron–polyphenol solutions is presented in Fig. 4 (structures **13–15**). Within the group of compounds reported, epicatechin (**13**) is the most effective at reducing iron(III) to iron(II) via an internal redox mechanism.[21]

---

[15] G. A. Razuvaev, V. K. Cherksov, and G. A. Abamukov, *J. Organomet. Chem.* **160**, 361 (1978).
[16] R. C. Hider, A. R. Mohd-Nor, J. Silver, I. E. G. Morrison, and L. V. C. Rees, *J. Chem. Soc. Dalton* 609 (1981).
[17] R. C. Hider, B. Howlin, J. R. Miller, A. R. Mohr-Nor, and J. Silver, *Inorg. Chim. Acta* **80**, 51 (1983).
[18] F. Basolo and R. G. Pearson, Mechanism of Inorganic Reactions, a Study of Metal Complexes in Solution, 2nd Ed., p. 76. Wiley, New York, 1967.
[19] E. Mentasti and E. Pelizzetti, *J. Chem. Soc. Dalton* 2605 (1973).
[20] H. K. J. Powell and M. C. Taylor, *Aust. J. Chem.* **35**, 735 (1982).
[21] J. A. Kennedy and H. K. J. Powell, *Aust. J. Chem.* **38**, 879 (1985).

Autoxidation

Ligands with appreciable affinity for iron(III) facilitate the autoxidation of iron(II) at neutral pH values.[22] Thus the half-life of 4×10^{-4} M iron at pH 7.0 in the presence of phosphate (1.2×10^{-3} M) is 210 sec and in the presence of citrate (1.5×10^{-3} M) is 60 sec. Similar rates of autoxidation will occur in the presence of catechol-containing molecules [Eq. (7)]. Thus the interactions of polyphenols

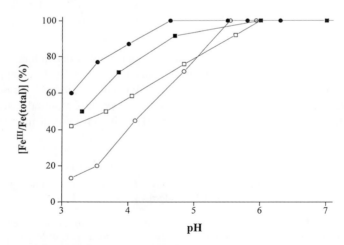

FIG. 4. The pH dependence of the percentage of iron(III) in the presence of various ligands; $[Fe]_{total} = 3.5 \times 10^{-4} M$; [Ligand] $= 2.2 \times 10^{-3}$ M. (○) Epicatechin (**13**); (□) catechol (**3**); (●) gallic acid (**15**); and (■) pyrocatechnic acid (**14**).

with iron(II) should be investigated in the absence of O_2.

$$\left[\underset{O}{\overset{O}{\bigcirc}} Fe^{II}(H_2O)_4 \right]^0 + O_2 \rightleftharpoons \left[\underset{O}{\overset{O}{\bigcirc}} Fe^{III}(H_2O)_4 \right]^+ + O_2^- \quad (7)$$

Methods to Characterize Speciation of Iron and Copper in Presence of Polyphenols

Techniques for Minimization of Flavonoid Autoxidation in Presence of Iron(III)

Rigorous exclusion of O_2 in aqueous solutions (pH ≥ 7.0) is essential for any analytical work using polyphenols.[20,23] A convenient method that uses sonification has been reported by van Acker et al.[24] To prevent the flavonoids from oxidizing in the presence of the iron, oxygen can be removed from the nanopure water by sonification for 10 min followed by purging with O_2-free nitrogen for at least 1 hr. Buffers should be prepared from such water and be purged continuously with oxygen-free nitrogen or argon. During measurement, cuvettes should be sealed with a cap. Stock solutions of the flavonoids can be prepared in 5% dimethyl sulfoxide (DMSO) in oxygen-free nanopure water up to a concentration of 500 μM. Flavonoids with a very low solubility can be dissolved in DMSO and absolute ethanol (1 : 1). This solution can be mixed with oxygen-free nanopure water to achieve concentrations of 500 μM. Nitrogen gas should be scrubbed with NaOH (6 M) to remove traces of CO_2 and with an acidic vanadium(II) solution over a zinc amalgam. Solutions should not be used with oxygen concentrations ≥0.01 ppm, which can be monitored using a standard oxygen electrode.

Spectrophotometric Analysis

As there is a growing interest in iron chelators with therapeutic potential,[25,26] there is an increasing requirement for a user-friendly analytical system. A relatively simple system that is capable of reliably determining ionization constants (pK_a) and affinity constants of ligands and their metal complexes has been introduced.[27]

---

[22] D. C. Harris and P. Aisen, *Biochim. Biophys. Acta* **329**, 156 (1973).
[23] J. A. Kennedy, M. H. G. Munro, H. K. J. Powell, L. J. Porter, and F. L. Yeap, *Aust. J. Chem.* **37**, 885 (1984).
[24] S. A. B. E. van Acker, D. J. van den Berg, M. N. J. L. Tromp, D. H. Griffioen, W. P. van Bennekom, W. J. F. van der Vijgh, and A. Bast, *Free Radic. Biol. Med.* **20**, 331 (1996).
[25] R. J. Bergeron, J. Wiegand, M. Wollenweber, J. S. McManis, S. E. Algee, and K. Ratliff-Thompson, *J. Med. Chem.* **39**, 1575 (1996).
[26] G. S. Tilbrook and R. C. Hider, in "Metal Irons in Biological Systems" (A. Sigel and H. Sigel eds.), Vol. 35, p. 691. Dekker, New York, 1998.
[27] H. H. Khodr, Z. D. Liu, G. S. Tilbrook, and R. C. Hider, submitted for publication.

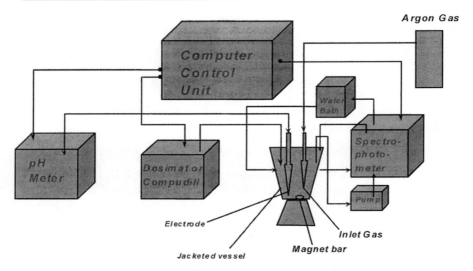

FIG. 5. Schematic presentation of titration system.

This system has the advantage of determining pK_a values and affinity constants by synchronous operation of spectrophotometric and potentiometric techniques. In addition, a microtitration mode has been developed that enables spectrophotometric measurement of compounds possessing limited aqueous solubility ($\leq 10^{-5}\ M$). Data obtained by autotitration are processed to provide pK_a and β values using the Gauss–Newton–Marquardt algorithm.[28] The unit is operated by a personal microcomputer programmed in BASIC. The system contains four major components (Fig. 5): (1) Metrohm 665 dosimat (Metrohm Ltd., CH-9100 Herisau, Switzerland) and/or COMPUDILL (Hook and Tucker Instruments, Vulcan Way, Croydon, England); (2) Perkin-Elmer lambda 5 UV/VIS spectrophotometer (Bodenseewerk Perkin-Elmer & Co. GmbH, Postfach 1120, D-7770 Uberlingen, Bindesrepublik Deutschland); (3) Corning ion analyzer 255 pH meter (Corning Science Products, Corning Glass Works, Corning, NY); and (4) 286 PC Opus technology, which controls the integrated system. Solutions are typically monitored by a Sirius combined electrode (Sirius Analytical Instruments Limited, Riverside, Forest Row, Business Park, East Sussex RH18 5DW, UK) in a jacketed vessel maintained at $25 \pm 1°$ under an argon atmosphere and are acidified by 0.15-ml aliquots of $0.2\ M$ HCl. Blank titrations are carried out against $0.2\ M$ KOH and then repeated in the presence of the ligand or metal–ligand complex. The solutions are also pumped continuously through a UV flow cell (HELLMA PRAZISIONS-KUVETTEN aus Quarzglas

---

[28] W. H. Press, S. A. Teukolsky, W. T. Vetterling, and B. P. Flannery, "Numerical Recipes in Fortran, the Art of Scientific Computing," 2nd Ed., Chap. 15. Cambridge Univ. Press, 1992.

SUPRASIL) using a peristaltic pump. The β values represent the overall stability constant between the metal cation M and the bidentate ligand L as indicated in Eqs. (8)–(12). Spectrophotometric monitoring of competition with EDTA can also provide an independent value of β_3, a method particularly useful for high-affinity ligands.[27,30]

$$[M] + [L] \xrightleftharpoons{K_1} [ML] \qquad K_1 = \frac{[ML]}{[M][L]} \qquad (8)$$

$$[ML] + [L] \xrightleftharpoons{K_2} [ML_2] \qquad K_2 = \frac{[ML_2]}{[ML][L]} \qquad (9)$$

$$[ML_2] + [L] \xrightleftharpoons{K_3} [ML_3] \qquad K_3 = \frac{[ML_3]}{[ML_2][L]} \qquad (10)$$

$$[M] + 2[L] \xrightleftharpoons{\beta_2} [ML_2] \qquad \beta_2 = K_1 K_2 \qquad (11)$$

$$[M] + 3[L] \xrightleftharpoons{\beta_3} [ML_3] \qquad \beta_3 = K_1 K_2 K_3 \qquad (12)$$

The pM value, which is the negative logarithm of the free concentration of metal ion (M), is a useful parameter for comparing the chelating abilities of different ligands under physiological conditions. The value of pM depends on the pK_a, pH, and β values together with the hydrolysis values of the metal ion and the concentrations of metal and ligand.[26,29,30]

*Characterization of Interaction of Monocatechol, Aminochelin (**16**), with Iron(III)*

In order to fully comprehend the interaction of polyphenol compounds with metals under physiological conditions, it is essential to also characterize their interaction with protons. The spectrophotometric titration of aminochelin (**16**) (Fig. 6B) generates spectral changes over the pH range 3 to 11. Potentiometric and spectrophotometric titration data obtained in Figs. 6A and 6C, respectively, were processed by the *NONLINWI* program to provide pK_a values.[27] Three pK_a values were identified: two possessing a strong absorption at 320 nm, namely 7.05 and 12.08, which were attributed to the 2-hydroxyl and 3-hydroxyl functions, respectively. The other, 10.22, was UV-silent and assigned to the terminal amine function [Eq. (13)]. The resulting speciation plot (Fig. 6D) demonstrates that the major species at pH 7.4 are (LH_3^+) and (LH_2).

---

[29] Z. D. Liu, H. H. Khodr, D. Y. Liu, S. L. Lu, and R. C. Hider, *J. Med. Chem.* **42**, 4814 (1999).
[30] K. N. Raymond, G. Muller, and B. F. Matzanke, *Top. Curr. Chem.* **58**, 49 (1984).

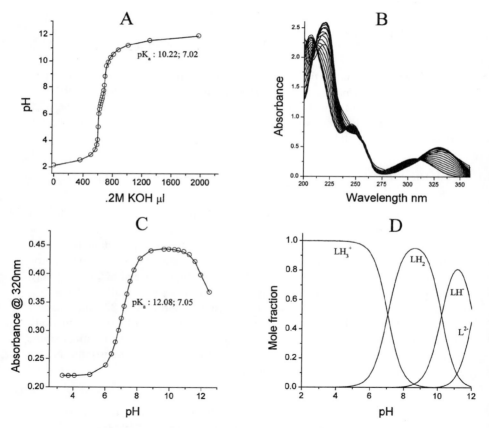

The interaction of aminochelin with iron(III) was monitored spectrophotometrically (Fig. 7A) and the absorbance values at 620 nm were used to obtain β_1, β_2, and β_3 of iron(III)–aminochelin complexes (Fig. 7B). The iron(III) complexes of the aminochelin are prepared in a 5:1 (2.8×10^{-4} M) molar ratio in acidic

FIG. 6. Determination of pK_a values of aminochelin (**16**): (A) potentiometric titration, (B) spectrophotometric titration, (C) spectrophotometric titration monitored at 320 nm, and (D) speciation plot.

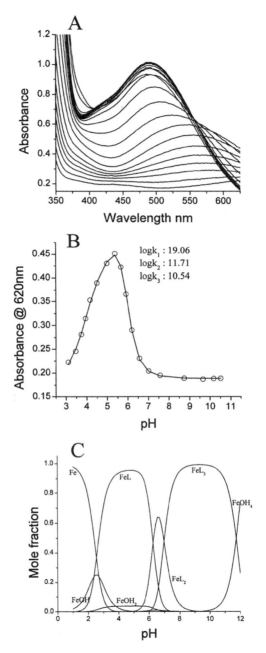

FIG. 7. Determination of aminochelin (**16**) affinity constants for iron(III). (A) Spectrophotometric titration of aminochelin (**16**) in the presence of iron(III) over the pH range 2.87–10.32, (B) spectrophotometric titration monitored at 620 nm, and (C) speciation plot of iron(III) in the presence of aminochelin (**16**), $[Fe^{3+}]_{total} = 10^{-5} M; [(16)] = 10^{-4} M$.

medium by adding 80 ml of 37% HCl. The computed line fitted to experimental data (Fig. 7B) is obtained by optimizing these β values using the *STABOPT* program.[27] These β values correspond to K_1, K_2, and K_3 as described in Eqs. (8)–(12). The corresponding speciation plot (Fig. 7C) demonstrates that the 1 : 1 complex dominates over the pH range 3–6 and that the 3 : 1 complex begins to dominate at pH values above 6.0.

(16)   (17)

*Characterization of Interaction of Tricatechol, Protochelin (**17**), with Iron(III)*

The three lower protonation constants of protochelin (**17**) were determined by a simultaneous spectrophotometric and potentiometric titration and found to be $pK_{a4} = 9.68$, $pK_{a5} = 8.09$, and $pK_{a6} = 6.67$. The proton-dependent stability constant $K^*_{Fe-(17)} = [Fe(\mathbf{17}^{3-})][\mathbf{H}^+]^3/([Fe^{3+}][(\mathbf{17})])$ for the iron protochelin complex was determined spectrophotometrically by means of competitive complex formation with EDTA[31] in aqueous solution at pH 6.9 as $10^{8.3}$, using the known stability constant for iron(III)–EDTA of $\log K_{FeEDTA} = 25.1$.[31] To obtain the conventional (proton-independent) formation constant $K_{Fe-(17)} = [Fe(\mathbf{17}^{6-})]/([Fe^{3+}][(\mathbf{17}^{6-})])$, an average pK_a value[31] of 12.1 was assumed for the experimentally inaccessible three higher protonation constants of protochelin. Based on this value, a $\log K_{Fe-(17)}$ value of 44.6 has been estimated for ferric protochelin. Thus the stability of ferric protochelin is comparable with the stability of iron(III) complexes of other tris(catecholamide) ligands, such as linear enterobactin ($\log K_{FeL} = 43$),[32] and the enterobactin analog mecam ($\log K_{FeL} = 43$).[33]

Technique for Determination of Redox State of Iron in Solution

Iron(III) can be determined quantitatively by spectrophotometric measurement as the $Fe(NCS)-(H_2O)_5^{2+}$ complex. Samples of the test solution (1 ml) are withdrawn by syringe from the titration cell; the syringe contains 1 ml of 1 M HCl and the combined solution is added to a 10-ml volumetric flask. Acidification

---

[31] W. R. Harris, C. J. Carrano, S. R. Cooper, S. R. Sofen, A. E. Avdeef, J. V. McArdle, and K. N. Raymond, *J. Am. Chem. Soc.* **101,** 6097 (1979).
[32] R. C. Scarrow, D. J. Ecker, C. Ng, S. Liu, and K. N. Raymond, *Inorg. Chem.* **30,** 900 (1991).
[33] W. R. Harris and K. N. Raymond, *J. Am. Chem. Soc.* **101,** 6534 (1979).

effectively quenches any redox process. Ammonium thiocyanate (1 ml; 2 M) is added and distilled water is added to the mark. The absorbance of the solution is measured immediately at 480 nm. Ferrous iron was estimated as the difference between known total iron and $[\text{iron(III)}]^{20}$ (Fig. 4). It is also possible to determine iron(II) levels directly as $\text{Fe(bipyridyl)}_3^{2+}$ in the presence of both iron(III) and phenolic-reducing agents by complexing the iron(III) with nitrilotriacetic acid.[34]

[34] H. Fadrus and J. Maly, *Analyst* **100**, 549 (1975).

[18] Mechanism of Antioxidant Effect of Catechins

By KAZUNARI KONDO, MASAAKI KURIHARA, and KIYOSHI FUKUHARA

Introduction

Catechins are an important class of polyphenols that are found in abundance in green tea. It is well known that catechins can be protective against cancer and inflammatory and cardiovascular diseases.[1-6] These protective effects have been largely due to their antioxidant activity. Therefore, many researchers have reported the effectiveness of catechins in suppressing the formation of lipid peroxidation in biological tissues and subcellular fractions such as microsomes and low-density lipoprotein (LDL).[7-11] However, a negative result on the inhibitory effect of catechins has been reported by van het Hof *et al.*[12] Some antioxidants can act as prooxidants, such as ascorbic acid and quercetin. It has been reported that

[1] S. Renaud and M. de Lorgeril, *Lancet* **339**, 1523 (1992).
[2] M. G. L. Hertog, E. J. M. Feskens, P. C. H. Hollman, M. B. Katan, and D. Kromhout, *Lancet* **342**, 1007 (1993).
[3] J. P. A. Brown, *Mutat. Res.* **75**, 243 (1980).
[4] E. Middeton and C. Kondasmami, *Biochem. Pharmacol.* **43**, 1167 (1992).
[5] T. P. Whitehead, D. Robinson, S. Allaway, J. Syms, and A. Hale, *Clin. Chem.* **41**, 32 (1995).
[6] B. Fuhrman, A. Lavy, and M. Aviram, *Am. J. Clin. Nutr.* **61**, 549 (1995).
[7] S. Miura, J. Watanabe, T. Tomita, M. Sano, and I. Tomita, *Biol. Pharm. Bull.* **17**, 1567 (1995).
[8] N. Salah, N. J. Miller, G. Paganga, L. Tijburg, G. P. Bolwell, and C. Rice-Evans, *Arch. Biochem. Biophys.* **322**, 339 (1995).
[9] C. Desmarchelier, S. Barros, M. Repetto, L. R. Latorre, M. Kato, J. Coussio, and G. Ciccia, *Planta Med.* **63**, 561 (1997).
[10] A. S. Meyer, M. Heinonen, and E. N. Frankel, *Food Chem.* **61**, 71 (1998).
[11] D. A. Pearson, E. N. Frankel, R. Aeschbach, and J. B. German, *J. Agric. Food Chem.* **46**, 1445 (1998).
[12] K. H. van het Hof, H. SM. de Boer, S. A. Wiseman, N. Lien, J. A. Weststrate, and L. BM. Tijburg, *Am. J. Clin. Nutr.* **66**, 1125 (1997).

quercetin may act as a cytotoxic prooxidant.[13] This suggests that the antioxidant action of flavonoids, including catechins on scavenging free radicals, may not be very simple. In addition, little is known about the antioxidant mechanisms of catechins, despite the fact that much attention has been paid to the effect of catechins on lipid peroxidation.

To clarify the potential of flavonoids in living systems, a detailed investigation of the antioxidant effect of flavonoids and their mechanisms is required. Numerous studies on free radical research using electron spin resonance (ESR) and pulse radiolysis studies have provided important results about antioxidants.[14–19] The aim of this article is to elucidate the antioxidant mechanisms of catechins by investigating the products produced during the first stages of radical oxidation using liquid chromatography/mass spectrometry (LC/MS), photodiode array detector (PDA), and semiempirical molecular orbital (MO) calculations in addition to chemiluminescence and electrochemical analyses.

Methods

Mass Spectrometric Analyses

An LCQ mass spectrometer (Thermoquest, Manchester, UK) equipped with an electrospray ionization (ESI) or an atmospheric pressure chemical ionization (APCI) source and coupled to a L-7000 series HPLC is used in a positive ionization mode for the mass spectral confirmation of the reaction intermediates of catechins. The column is TSK-GEL ODS-80Ts (2.0 × 150 mm, Toso, Tokyo, Japan). A mixture of methanol and water containing 1% (v/v) acetic acid (2:8) is used as an eluent, and the flow rate is 0.2 ml/min. To the catechin solution (50 μM), 2,2'-azobis(2-aminopropane)hydrochloride (AAPH) (2 mM) is added, and aliquots of the reaction mixture (10 μl) are withdrawn at specific intervals and then injected onto the high-performance liquid chromatography (HPLC) column. The structure of the products formed during the first stages is analyzed by LC/MS and PDA. MS/MS analyses are achieved using helium gas at a pressure of 0.1 Pa (10^{-3} Torr).

Spectrophotometric Study

Spectrophotomeric data are obtained using a Toso HPLC system (Tokyo, Japan) with a photodiode array detector (MCPD-3600, Otsuka Electronics Co.,

---

[13] D. Metodiewa, A. K. Jaiswal, N. Cenas, E. Dickancaite, and J. Segura-Aguilar, *Free Radic. Biol. Med.* **26**, 107 (1999).
[14] S. V. Jovanovic, Y. Hara, E. Steenken, and M. G. Simic, *J. Am. Chem. Soc.* **117**, 9881 (1995).
[15] J. Ueda, N. Saito, Y. Shimazu, and T. Ozawa, *Arch. Biochem. Biophys.* **333**, 377 (1996).
[16] T. Yoshida, K. Mori, T. Hatano, T. Okuhara, I. Uehara, K. Komagoe, Y. Fujita, and T. Okuda, *Chem. Pharm. Bull.* **37**, 1919 (1989).
[17] Q. Guo, B. Zhao, M. Li, S. Shen, and W. Xin, *Biochim. Biophys. Acta* **1304**, 210 (1996).
[18] P. T. Gardner, D. B. McPhail, and G. G. Duthie, *J. Sci. Food Agric.* **76**, 257 (1998).
[19] F. Nanjo, K. Goto, R. Seto, M. Suzuki, M. Sakai, and Y. Hara, *Free Radic. Biol. Med.* **21**, 895 (1996).

Osaka, Japan) and a Shimadzu HPLC system (Kyoto, Japan) with a PDA (Shimadzu SPD-M10Avp). The column is TSK-GEL ODS-80Ts (4.6 × 150 mm). Aliquots of the reaction mixture (10 μl) are withdrawn and injected onto the column.

Formation of Superoxide

Superoxide formation from catechins during the inhibitory action is investigated in the absence and presence of superoxide dismutase (SOD) by the luminol-dependent chemiluminescence method using a Bio-Orbit Luminometer 1251 (Turku, Finland). Catechins (1 μM) and Cu,Mn-SOD (70 U/ml) or buffer (without SOD) is preincubated in luminol (130 μM) at 37° for 5 min, and then AAPH (2 mM) is added to the mixtures. Chemiluminescence is measured in a light-tight housing controlled thermodynamically at 37° for 15 min.

Semiempirical Molecular Orbital Calculation

The program SPARTAN (version 5.0, Wavefunction Inc., Irvine, CA) is used for MO calculations. First, the catechin structures are optimized by a conformational search (MacroModel version 6.5, Schrödinger Inc., Jersey City, NJ; MonteCalro method, MM2* force field) and then the bond dissociation enthalpies (BDEs) are calculated using the semiempirical AM1 method. In the BDE calculations, global minimum energy conformations are used. BDEs are calculated as described in Eq. (1).

Bond dissociation enthalpy (BDE) = heat of formation (HF) of (**2**) + 52.102 − HF of (**1**) (kcal/mol)

$$\text{(1)} \longrightarrow \text{(2)} + \text{H}^\bullet \quad (52.102 \text{ kcal/mol}) \tag{1}$$

Electrochemical Analysis

Cyclic voltammetry is performed at room temperature in 50 mM (final concentration) sodium cacodylate buffer, pH 7.2, using a BAS100B electrochemical analyzer (Bioanalytical Systems Inc., IN). A three-electrode system is used. The glassy carbon working electrode (i.d. 3.0 mm) is polished before each experiment with aluminum powder. The counter electrode is a platinum wire. The reference is Ag|AgCl (RE-1B) and is separated from the sample solution by a salt bridge containing saturated KCl. To remove oxygen in the sample solution, argon gas is used before each experiment. Catechins are dissolved in dimethyl sulfoxide, and then diluted with the buffer to a final concentration of 2 mM. The cyclic voltammogram is recorded at a scan speed of 200 mV/sec.

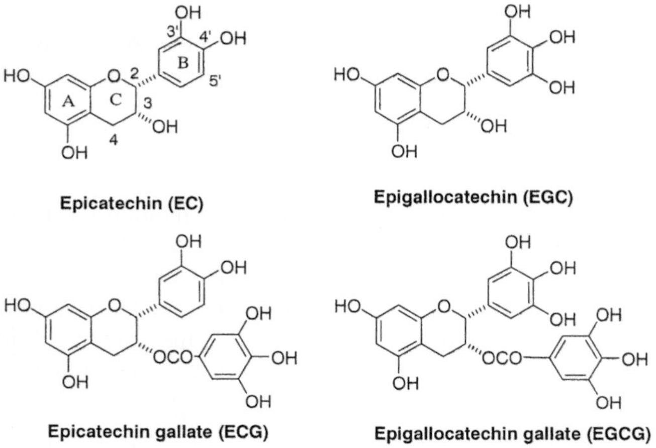

FIG. 1. Structures of four catechins.

Result

Effect of Catechins on Liposomal Phospholipid Peroxidation

The structures of catechins used are listed in Fig. 1. EC and ECG are compounds with the catechol structure in the B ring that have two hydroxyl groups at the C-3′ and C-4′ positions. However, EGC and ECGC are compounds with the pyrogallol structure that have three hydroxyl groups at the C-3′, C-4′, and C-5′ positions. The effect of catechins on lipid peroxidation in phosphatidylcholine (PC) liposomes initiated by 2,2′-azobis(2-aminopropane)hydrochloride (AAPH) are shown in Fig. 2. Apparently, EGC was the least effective among four catechins. EGC increased PC peroxides rapidly after an initial inhibition period (57 min). In the case of EGCG, a marked increase was also observed after an inhibition period (83 min). These results indicate that catechins with the pyrogallol structure in the B ring may act as prooxidants.

The kinetic parameters were calculated according to Ioku *et al.*,[20] showing that the inhibition period of EGC and EGCG on lipid peroxidation was shorter (t_{inh} = 57 and 83 min, respectively) than that of EC (156 min), although the former two catechins could quickly scavenge free radicals (k_{inh}/k_p = 232 and 628, respectively) (Table I).[21]

---

[20] K. Ioku, T. Tsushida, Y. Takei, N. Nakatani, and J. Terao, *Biochim. Biophys. Acta* **1234**, 99 (1995).
[21] K. Kondo, M. Kuruhara, N. Miyata, T. Suzuki, and M. Toyoda, *Arch. Biochem. Biophys.* **362**, 79 (1999).

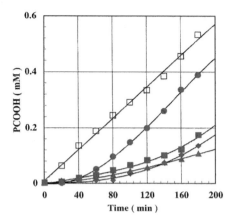

FIG. 2. Inhibitory effect of catechins on the peroxidation of soybean PC liposomes. The reaction system consisted of soybean PC (5 mM) and catechins (5 μM) in 10 mM Tris–HCl buffer (pH 7.4) containing 0.5 mM DTPA. Control (□), EC (■), EGC (●), ECG (▲), and EGCG (◆). Reprinted from Kondo et al., Arch. Biochem. Biophys. **362,** 81 (1999).

Scavenging Mechanism of Catechins

The reaction products generated from each catechin were investigated using LC/MS equipped with the ESI source for EC and EGC and the APCI source for ECG and EGCG, and using LC with a PDA detector. The total ion chromatograms of the reaction mixtures (EC or EGC and AAPH) are shown in Fig. 3. Two peaks (A and B) were observed from EC approximately 2 hr after the AAPH-induced

TABLE I
INHIBITORY EFFECT OF CATECHINS ON PEROXIDATION OF SOYBEAN PC LIPOSOMES

Sample	t_{inh} (min)[a]	R_{inh} (M sec^{-1})[a]	$R'_{\text{p}}/R_{\text{p}}$[b]	$k_{\text{inh}}/k_{\text{p}}$[c]	Inhibition (%)[d]	n[e]
EC	156	13.3×10^{-9}	0.95	41	68	3.1
EGC	57	6.3×10^{-9}	1.06	232	26	1.1
ECG	—	—	0.21	—	79	—
EGCG	83	1.6×10^{-9}	0.43	628	75	1.7

[a] t_{inh} and R_{inh}, the inhibition period and the inhibition rates in the presence of catechins, respectively.
[b] R_{p} and R'_{p}, the radical propagation rates in the absence and presence of catechins, respectively.
[c] $k_{\text{inh}}/k_{\text{p}}$, the ratio of the rate constant for inhibition to that for chain propagation.
[d] Inhibition ratio (%) when control is 0%.
[e] Stoichiometric number of radicals trapped by antioxidant.

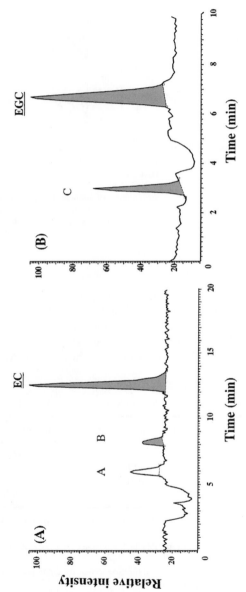

FIG. 3. Mass chromatograms of reaction mixtures (EC or EGC, and AAPH). Oxidation was initiated by the addition of AAPH at 37°. Reaction mixtures containing (A) EC or (B) EGC were withdrawn at specific intervals and then injected onto a column of LC/MS. Reprinted from Kondo et al., Arch. Biochem. Biophys. **362**, 82 (1999).

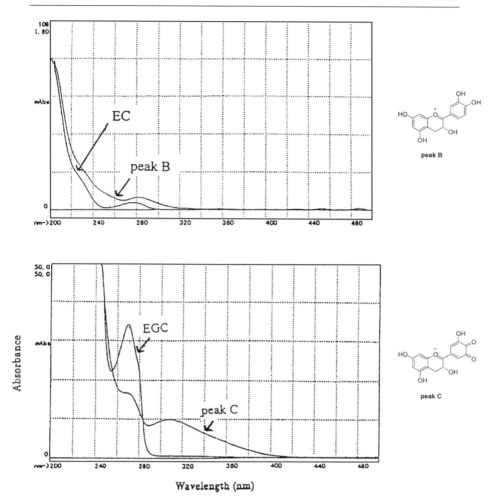

FIG. 4. Spectra of peak B (top) formed from EC and peak C (bottom) formed from EGC, and their structures. Spectra were measured by HPLC with photo diode array detector. Reprinted from Kondo et al., Arch. Biochem. Biophys. **362,** 85 (1999).

radical reaction was started. The UV spectrum of peak B was almost the same as that of EC (λ_{max} = 277 nm) except for a bathochromic shift, showing that peak B has the catechol structure in the B ring. In addition, the result of the MS/MS analysis for peak B (m/z 163, 271) suggested the structure shown in Fig. 4. In EGC, however, only one peak (peak C) was observed. The UV spectrum was different from that of the intact EGC. The absorption at 270 nm of EGC decreased and

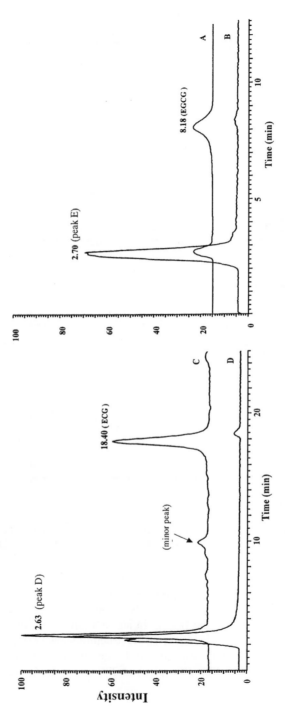

FIG. 5. Chromatograms monitored using the UV detector (280 nm) (A and C) and total ion chromatograms (TIC) (B and D) of the reaction mixtures produced from EGCG (left) and ECG (right). Oxidation was initiated by the addition of AAPH at 37°. Reaction mixtures containing EGCG or ECG were withdrawn at specific intervals and then injected onto a LC/MS column. Reprinted from Kondo et al., *Free Radic. Biol. Med.* **27**, 858 (1999) with permission from Elsevier Science.

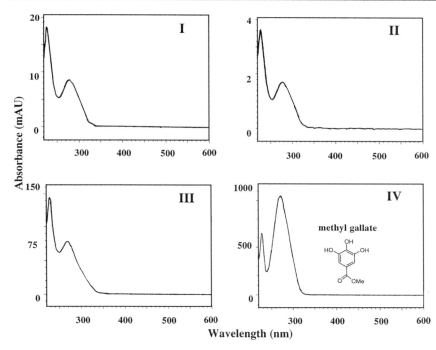

FIG. 6. Spectra of peaks D, E, F, and methyl gallate. All spectra were measured by HPLC with a photo diode array detector (PDA). I, peak D; II, peak E; III, peak F; IV, methyl gallate. Reprinted from Kondo et al., *Free Radic. Biol. Med.* **27,** 859 (1999) with permission from Elsevier Science.

another absorption at 308 nm was formed, indicating that peak C is quite different from EGC in the B ring. The result of the MS/MS analysis for peak C showed two major fragments (m/z 154, 233), thus suggesting the structure in Fig. 4.

A similar investigation for ECG and EGCG was carried out as shown in Figs. 5 and 6. One major peak (peak D or E, Fig. 5) was produced in the case of ECG or EGCG, respectively. Each peak has the same retention time. The MS analysis for each peak suggests that peaks D and E are just the same. The structure of the product would be derived from the common partial structure of EGCG and ECG. Namely, it was probably from the gallate moiety. The same compound (peak F) was also produced from methyl gallate (MG, Fig. 6, IV). UV spectra of peaks D, E, and F (Fig. 6, I–III) were definitely the same (Fig. 6). The spectrum of MG showed a large absorption at 272 nm based on the benzene structure coupled with the carbonyl group. The peak for the counterpart of the gallate moiety was observed in the case of EGCG and ECG and their structures were analyzed as described previously.[22]

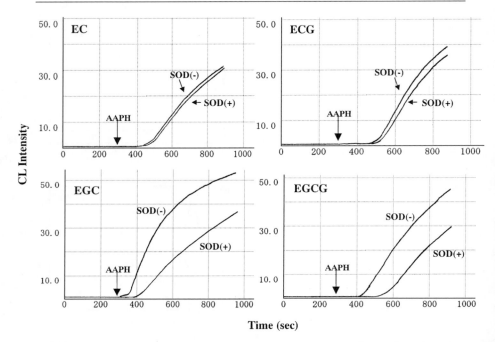

FIG. 7. Effect of SOD on catechins during the antioxidative action. The inhibitory effect of catechins on radical oxidation was examined in the absence [SOD (−)] and the presence [SOD (+)] of Cu,Zn-SOD. Chemiluminescence was monitored at 37° for 15 min. Oxidation was initiated by AAPH after preincubation. Reprinted from Kondo et al., Free Radic. Biol. Med. **27**, 861 (1999) with permission from Elsevier Science.

Formation of Superoxide

Superoxide formation from catechins during the antioxidant action was investigated in the absence and presence of SOD using the luminol-dependent chemiluminescence method. As shown in Fig. 7, SOD reduced the sharp increase in the chemiluminescence in EGC and EGCG, although SOD has little effect on the chemiluminescence of EC and ECG. This result indicates the involvement of superoxide during the inhibitory action in the case of EGC and EGCG, which have the pyrogallol structure in the B ring.

Bond Dissociation Enthalpies for Catechins

To clarify the antioxidative mechanisms of catechins, BDEs were evaluated using AM1 semiempirical (MO) calculations. All MO calculations for catechins

---

[22] K. Kondo, M. Kuruhara, N. Miyata, T. Suzuki, and M. Toyoda, *Free Radic. Biol. Med.* **27**, 855 (1999).

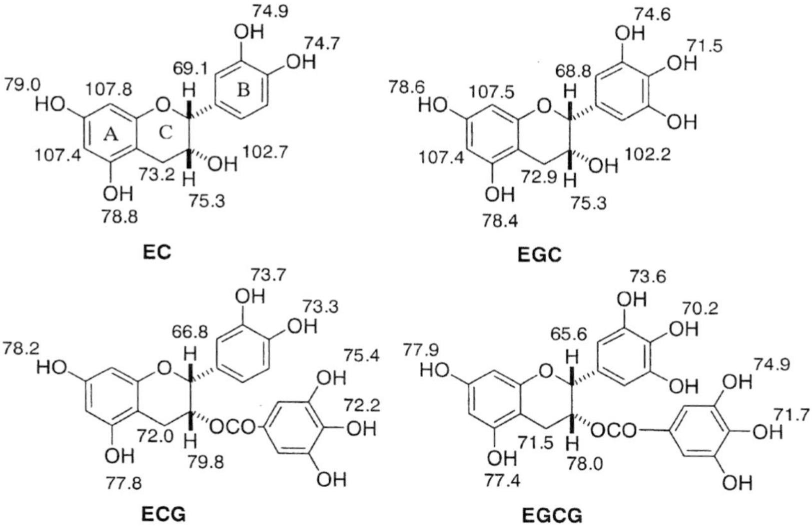

FIG. 8. Calculated bond dissociation enthalpies for EC, EGC, ECG, and EGCG. All MO calculations were carried out using the semiempirical AM1 method for global minimum energy conformations.

and catechin radicals were performed after the structural optimization by the conformational search. BDEs for the C-2 position of four catechins were lower (66–69 kcal/mol) than that for the phenolic OH (72–75 kcal/mol) as shown in Fig. 8.

Electrochemical Analysis of Catechins

Results of the cyclic voltammetry for catechins are shown in Fig. 9. The first oxidation potentials for EGC (159 mV) and EGCG (270 mV) with the pyrogallol structure are obviously lower than that of EC (307 mV) and ECG (362 mV) with the catechol structure, indicating that EGC and EGCG are subjected to a one electronoxidation.

The ratio of the rate constant for the inhibition to that for the chain propagation (k_{inh}/k_p) showed that EGC (232) and EGCG (628) can scavenge peroxyl radicals much more quickly than EC (41), suggesting that the rapid scavenging ability for free radicals increases with the increasing number of phenolic OHs. In other words, there is a significant relationship between the rapid scavenging ability and the first oxidation potential. However, catechins without the gallate group, such as EC and EGC, would be easily subjected to a two-electron oxidation, leading to the anthocyanin-like compounds shown in Fig. 10.

The results described in this article provided sufficient evidence to consider the antioxidant mechanisms of four catechins (Figs. 10 and 11).

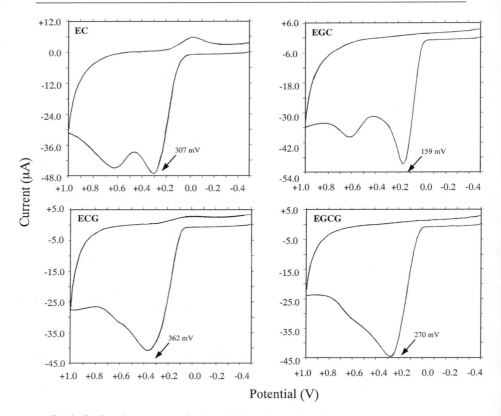

FIG. 9. Cyclic voltammograms for the oxidation of catechins. Catechins (2 mM) were analyzed using a glassy carbon electrode as a working electrode in 50 mM sodium cacodylate buffer (pH 7.2) at room temperature.

Discussion

The antioxidant mechanisms of EC, EGC, ECG, and EGCG were studied using LC/MS, spectrophotometry, chemiluminescence and electrochemical analyses, and semiempirical MO calculations. These results showed that the antioxidant effect, of catechins and their mechanisms were complicated. EGC has a lower oxidation potential (159 mV) than EC (307 mV) and, therefore, can scavenge peroxyl radicals more quickly ($k_{inh}/k_p = 232$) than EC ($k_{inh}/k_p = 41$). This indicates that the pyrogallol structure in the B ring play an important role in the rapid scavenging ability. EGCG having the gallate group at the C-3 position shows much more rapid scavenging ability ($k_{inh}/k_p = 628$) than EC and EGC, suggesting that as the number of phenolic OHs in catechins increases, the more rapidly they can

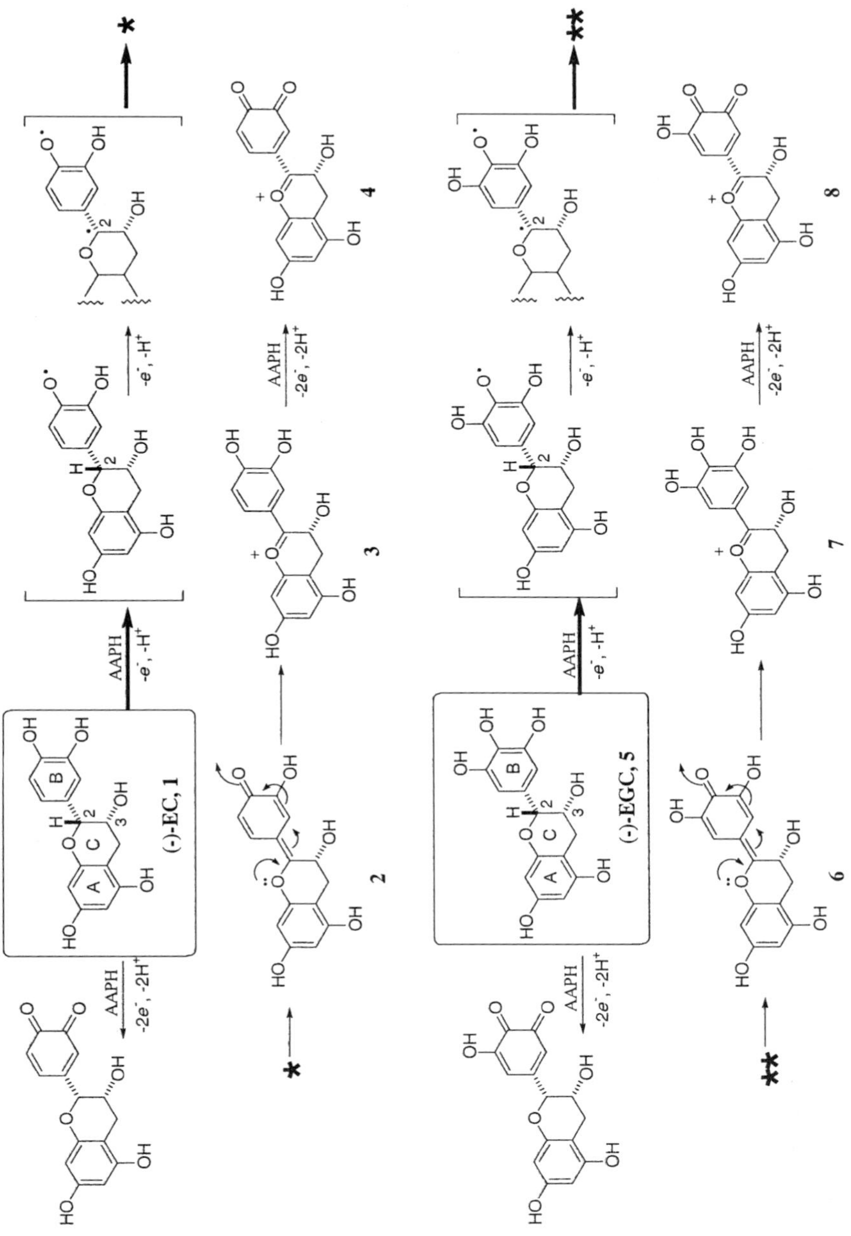

FIG. 10. Proposed mechanisms of EC and EGC during radical oxidation.

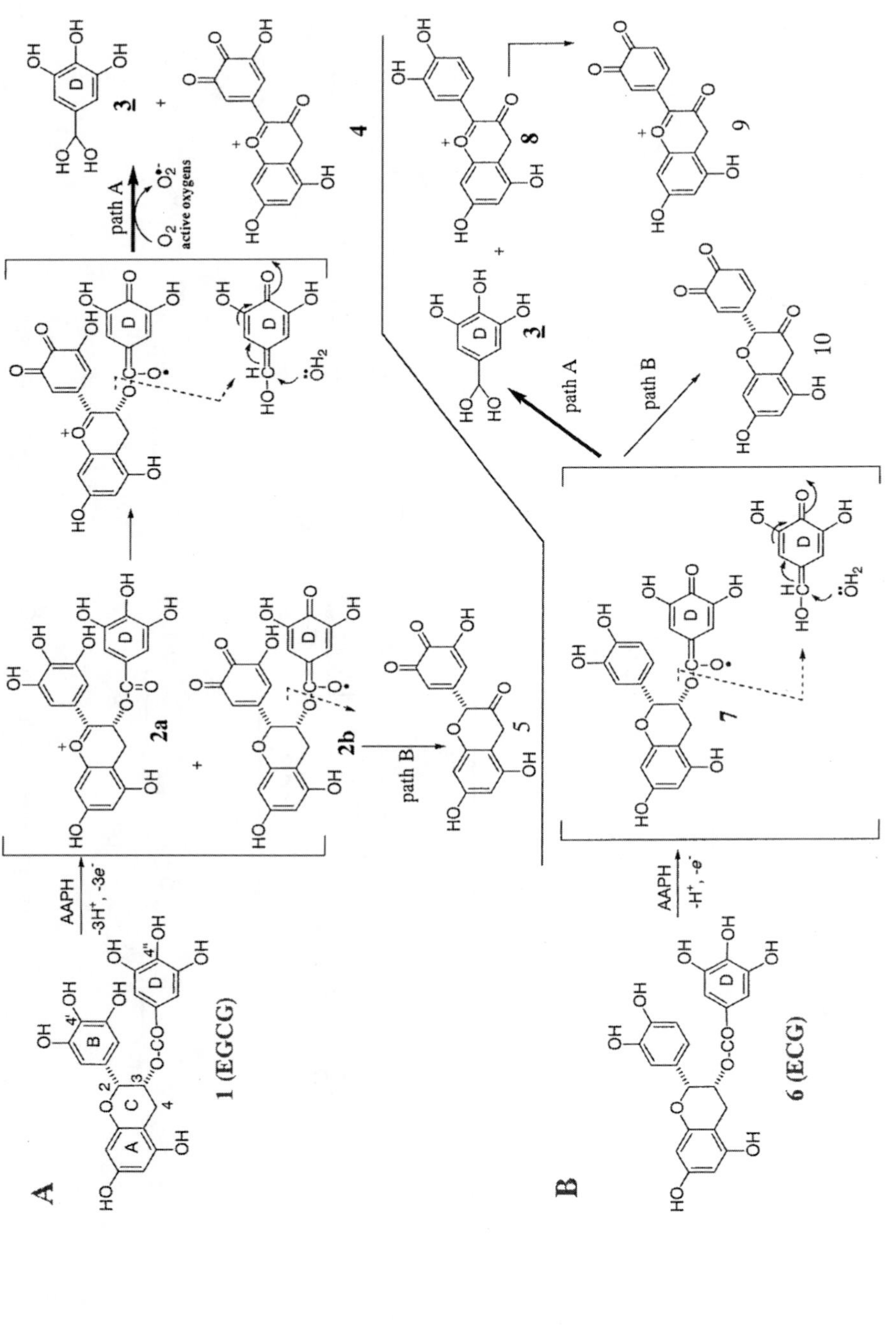

FIG. 11. Proposed mechanisms of ECG and EGCG during radical oxidation.

scavenge peroxyl radicals. However, EGC and EGCG with the pyrogallol structure have a negative aspect. They may generate superoxide and the antioxidant effect of them does not last for a long time. Catechins have both aspects as antioxidants and prooxidants as do other flavonoids. Catechin, with the lowest oxidation potential, does not exert the strongest antioxidant effect. Also, the effect may change based on the experimental conditions such as the solvent system or radical species being used as an initiator.

[19] Free Radical Scavenging by Green Tea Polyphenols

By BAOLU ZHAO, QIONG GUO, and WENJUAN XIN

Introduction

Tea is one of the most popular beverages all over the world. Epidemiological studies have shown that drinking tea is beneficial to people's health by protecting the body from cancer, heart disease, and aging. It is thought that antitumor and antimutagenic materials in tea are polyphenols. Green tea polyphenols isolated from green tea were shown to inhibit the backward mutation induced by aflatoxin B_1 (AFB1) and benzo[α]pyrene (BαP) in *Salmonella typhimurium*. Green tea polyphenols were also found to inhibit gene mutation in V_{79} (Chinese hamster) cells treated with AFB1 and BαP[1-4] and could prevent cancer by inhibiting urokinase (uPA) and angiogenesis[5,6]. Experiment indicate that the beneficial effects of green tea polyphenols are connected with their free radical scavenging property.[7-10] Green tea extract contains four types of polyphenols—(−)-epicatechin(EC),(−)-epicatechin gallate(ECG), (−)-epigallocatechin(EGC), and (−)-epigallocatechin gallate(EGCG)—and their epimers (Fig. 1). They all can scavenge free radicals,

---

[1] S. K. Katiyar, R. R. Mohan, R. Agarwal, and H. Mukhtar, *Carcinogesis* **18,** 497 (1997).
[2] R. D. Ley and V. E. Reeve, *Environ. Health Pespect.* **105** (*Suppl.* 981) (1997).
[3] I. E. Dreosti, M. J. Wargovich, and C. S. Yang, *Crit. Rev. Food Sci. Nutr.* **37,** 761 (1997).
[4] S. J. Cheng, in "Proceeding of the International Symposium on Natural Antioxidants: Molecular Mechanisms and Health Effects (L. Packer, M. G. Traber, and W. Xin, eds.). AOCS Press, Champaign, IL, 1995.
[5] J. Jankun, S. H. Selman, R. Swierez, and E. Skrzyperak-Jankun, *Nature* **387,** 561 (1997).
[6] Y. Cao and R. Cao, *Nature* **398,** 381 (1999).
[7] B.-L. Zhao, X.-J. Li, S.-J. Cheng, and W.-J. Xin, in "Medical, Biochemical and Chemical Aspects of Free Radicals" (Hayaishi *et al.,* eds.). Amsterdam,1989.
[8] B.-L. Zhao, X.-J. Li, and W.-J. Xin, *Cell Biophys.* **14,** 175 (1989).
[9] S.-R. Shen, F.-J. Yang, B.-L. Zhao, and W.-J. Xin, *J. Tea Sci.* **13,** 141 (1993).
[10] B.-L. Zhao, J.-C. Wang, and W.-J. Xin, *Chin. Sci. Bul.* **41,** 923 (1996).

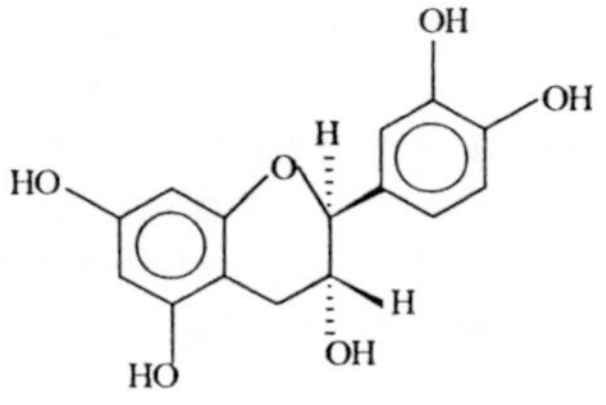

FIG. 1. Chemical structures of green tea polyphenols and their epimers.

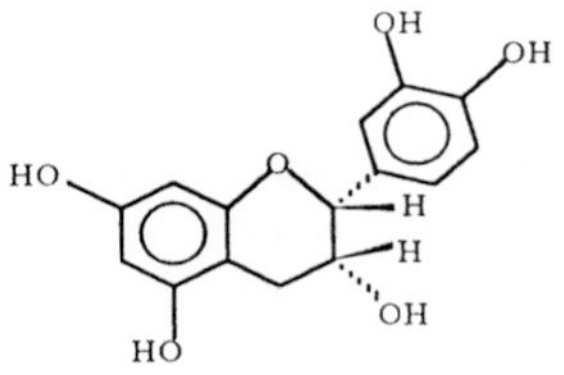

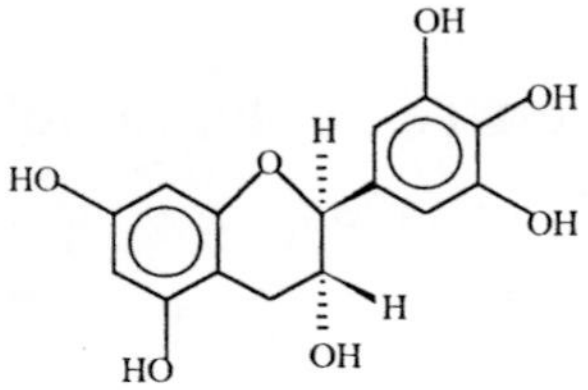

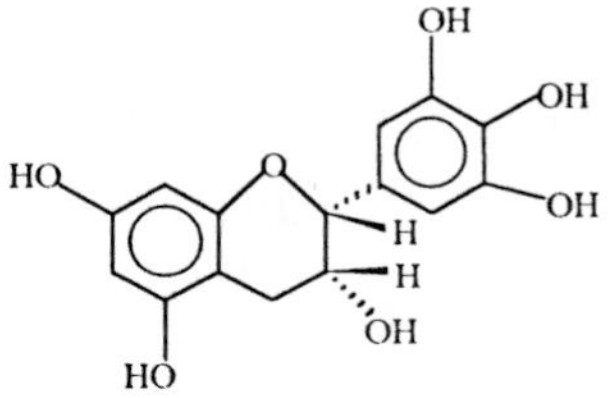

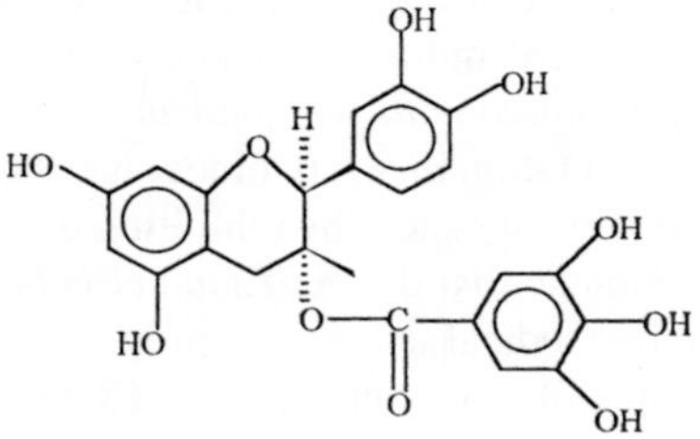

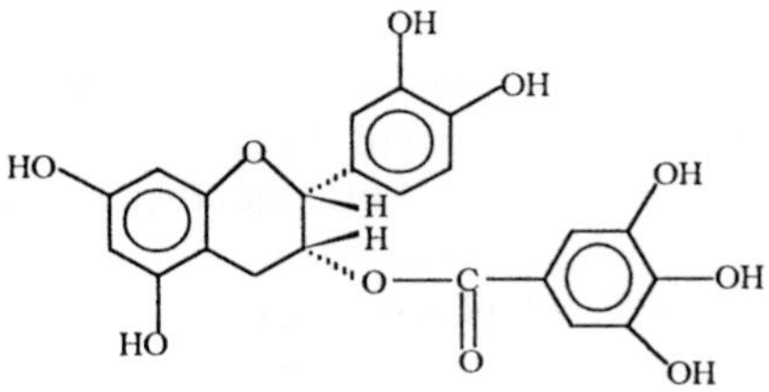

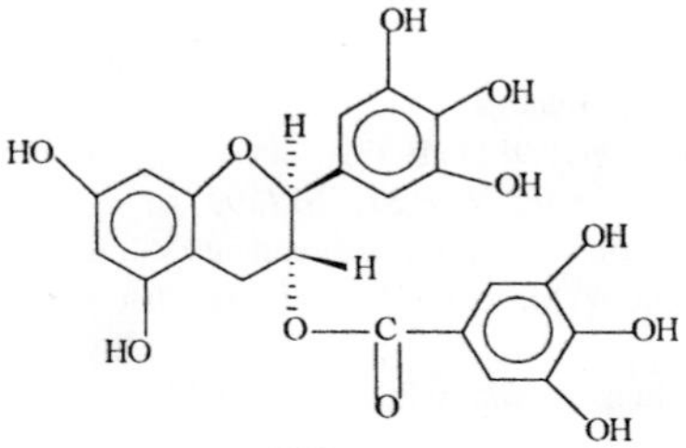

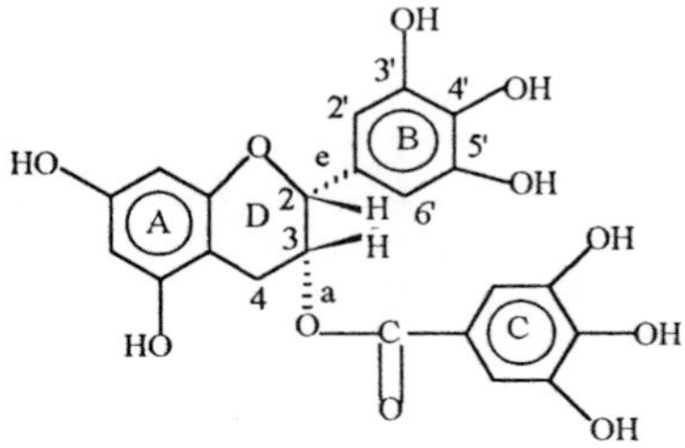

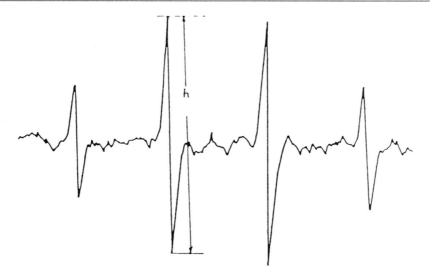

FIG. 2. ESR spectrum of the hydroxyl-free radical generated from the Fenton reaction (0.1% H_2O_2 + Fe^{2+}) and trapped by DMPO (50 mM). ESR condition: The spectrum was recorded at Varian E109 ESR spectrometer, X-band, 100 kHz modulation with amplitude 1 G, microwave power 1 mW, time constant 0.128 sec, central magnetic field 3250 G, swept width 200 G, room temperature.

especially oxygen-free radicals, but their scavenging effects are different because they have different structures. These structures are very important for their pharmacological usage.

Electron spin resonance (ESR) is the most effective and direct method for detecting free radicals and their scavenging. The life of oxygen-free radicals is very short, e.g., the life of the hydroxyl radical is about 10^{-6} sec. A spin trapping technique has to be used with ESR to detect the short-lived free radicals and their scavenging in this case.[11–13] The methods and results regarding free radical scavenging of green tea polyphenols are presented in this article.

Methods and Results

Scavenging Effect of Green Tea Polyphenols on Hydroxyl Radicals Generated from Fenton Reaction

The Fenton reaction system can generate hydroxyl radicals, which can be trapped by DMPO (5,5-dimethyl-1-pyrroline 1-oxide) and give a typical four-line ESR spectrum with intensity 1 : 2 : 2 : 1 ($a_N = a_H = 14.9$G) (Fig. 2).[1]

---

[11] Q. Guo, B.-L. Zhao, M.-F. Li, S.-R. Shen, and W.-J. Xin, *Biochim. Biophys. Acta* **1304**, 210 (1996).

Reagents. DMPO (purified by active charcoal before use), H_2O_2 ferrous ammonium sulfate, green tea polyphenols, deionized water, vitamin E, rosemary, curcumin, and vitamin C.

Procedure. Mix 50 mM DMPO, 0.1% H_2O_2, and 100 μM ferrous ammonium sulfate in deionized water and transfer to a quartz capillary. Measure ESR spectra 3 min after mixing. The conditions of ESR measurement are as follows: X band, 100-kHz modulation with amplitude 1 G, microwave power 10 mW, time constant 0.128 sec, and room temperature. The relative ESR intensity is taken as the amount of hydroxyl radical in control (h_0).

Mix 50 mM DMPO, 0.1% H_2O_2, 100 μM ferrous ammonium sulfate, and different concentrations of green tea polyphenols in deionized water and transfer to a quartz capillary. Measure ESR spectrum 3 min after mixing (the same conditions as the control). Relative ESR intensity is taken as the amount of hydroxyl radical of sample (h_x).

The scavenging effect of green tea polyphenols on hydroxyl radicals is calculated from Eq. (1):

$$E = [(h_o - h_x)/h_o]100\% \tag{1}$$

Results. The scavenging effects of green tea polyphenols and the other antioxidants, vitamin E, rosemary, curcumin, and vitamin C, on hydroxyl radical generated from the system were 12 ± 3.6, 35 ± 13.5, 16 ± 9.0, 69 ± 11.7, and $56 \pm 5.9\%$, respectively. It indicates that green tea polyphenol shows a weak scavenging effect on the hydroxyl radical in the Fenton reaction.[1,2]

Scavenging Effect of Green Tea Polyphenols on Hydroxyl Radical Generated from Photolysis of H_2O_2

In order to avoid the interference of iron ions, the irradiation of H_2O_2 can also be used to generate a hydroxyl radical, which can be trapped by DMPO, and gives similar ESR spectrum to that in Fig. 2.[1,2]

Reagents. H_2O_2, DMPO (purified by active charcoal before use), phosphate-buffered saline (PBS), green tea polyphenols, and their epimers.

Procedure. Irradiate a mixture containing 0.04% H_2O_2 and 0.08 M DMPO in 0.05 M PBS with an UV lamp set at 254 nm for 1 min. Transfer the reaction mixture into a quartz capillary and measure ESR spectrum 3 min after irradiation. The ESR measurement conditions are the same as those described earlier. The relative ESR intensity is taken as the amount of hydroxyl radical in the control (h_o).

---

[12] B.-L. Zhao, Q. Guo, and W.-J. Xin, *In* "Modern Applications of EPR/ESR from Biophysics to Materials Science" (C. Z. Rudowicz, K. N. Yu, and H. Hiraoka, eds.), p. 74, Springer, New York, 1998.

[13] Q. Guo, B.-L. Zhao, J.-W. Hou, and W.-J. Xin, *Biochim. Biophys. Acta* **1427**, 13 (1999).

Irradiate 0.04% H_2O_2, 0.08 M DMPO, and different concentrations of green tea polyphenols in 0.05 M PBS by an UV lamp set at 254 nm for 1 min. Transfer the reaction mixture into a quartz capillary and measure ESR spectra 3 min after irradiation. The ESR measurement conditions are the same as the control. The relative ESR intensity is taken as the amount of hydroxyl radical of sample (h_x).

The scavenging effect of green tea polyphenols on hydroxyl radicals generated from irradiation of H_2O_2 is calculated from Eq. (1).

Results. EGC (0.75 mM) was sufficient to reduce the amount of hydroxyl radical by 46.5%. EC at the same concentration reduced the amount of hydroxyl radical by 19.1%. However, EGCG and EGC at the same concentration did not exhibit any hydroxyl radical scavenging property. When the concentration of EGCG was increased to 1.5, 3.75, and 7.5 mM, it reduced the amount of hydroxyl radical by 61.1, 83.3, and 87.9%, respectively, whereas when the concentration of EGC was increased to 7.5 mM, it still did not exhibit any hydroxyl radical scavenging activity. Their hydroxyl radical scavenging activity can be classified as follows: ECG > EC > EGCG > EGC.[11]

Scavenging Effect of Green Tea Polyphenols on Superoxide Radicals Generated from Irradiation of Riboflavin/EDTA

The superoxide radical can be generated from the irradiation of the riboflavin/EDTA system. The free radical can be trapped by DMPO, and its 12-line ESR spectrum ($a_N = 14.3$ G, $a_H^\beta = 11.3$ G, $a_H^\gamma = 1.25$ G) is shown in Fig. 3.[1,2]

Reagents. Riboflavin, EDTA, DMPO (purified by active charcoal before use), green tea polyphenols and their epimers, vitamin E, rosemary, curcumin, and vitamin C.

Procedure. Mix 0.3 mM riboflavin, 5 mM EDTA, and 0.1 M DMPO transfer to a quartz capillary, and put into the cavity of the ESR spectrometer. Irradiate the sample with a xenon lamp for 30 sec. Measure the ESR spectrum 3 min after irradiation. The ESR measurement conditions are the same as described earlier. The relative ESR intensity is taken as the amount of superoxide radical in the control (h_o).

Mix 0.3 mM riboflavin, 5 mM EDTA, 0.1 M DMPO, and different concentrations of green tea polyphenols transfer to a quartz capillary, and put into the cavity of the ESR spectrometer. Irradiate the sample with a xenon lamp for 30 sec. Measure ESR spectra 3 min after irradiation. The ESR measurement conditions are the same as the control. The relative ESR intensity is taken as the amount of superoxide radical of sample (h_x).

The scavenging effect of green tea polyphenols on superoxide radicals generated from irradiation of riboflavin/EDTA is calculated from Eq. (1).

Results. The scavenging effects of green tea polyphenols and other antioxidants, such as vitamin E, rosemary, curcumin, and vitamin C on superoxide radicals

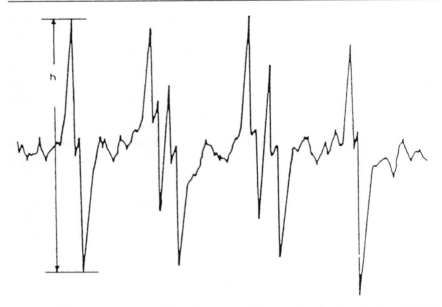

FIG. 3. ESR spectrum of superoxide radical generated from irradiation of the riboflavin (0.5 M)/EDTA(50 mM) system and trapped by DMPO(0.1 M). The ESR spectrum was recorded 1 min after irradiation. The ESR condition was the same as that in Fig. 2.

generated from the system, were 72 ± 7.4, 23 ± 6.1, 23 ± 5.1, 17 ± 5.5, and 96 ± 6.4%, respectively. It was found that the scavenging effect of green tea polyphenols on superoxide radicals was stronger than that of vitamin E, rosemary, and curcumin, but weaker than that of vitamin C.[1,2]

We also studied the scavenging effects of green tea polyphenols and their epimers on superoxide radicals in this system. The superoxide radical scavenging activities of those catechins increased as follows: EC < (+)-C < EGC < GC < EGCG <GCG. In addition, it was found that the differences between the steric structures of catechins (EGCG, EGC, and EC) and their corresponding epimers [GCG, GC, and (+)-C] at low concentration had a significant influence on their scavenging effects. The lower the concentrations of the polyphenols, the higher the differences of their scavenging effects. The scavenging activities of GCG, GC, and (+)-C were higher than those of their corresponding epimers, respectively.[13]

Scavenging Effect of Green Tea Polyphenols on Oxygen-Free Radicals Generated from PMA-Stimulated PMN

Polymorphonuclear leukocytes (PMN) are one of the main sources for the production of active oxygen radicals in the human body. When stimulated by phorbol myristate acetate (PMA), the PMN will be activated to have a respiratory burst and to generate oxygen-free radicals, which can be trapped by DMPO and give a similar ESR spectrum to that in Fig. 3.[1,2]

Reagents. PMN (separated according to the literature[14]), PMA, DMPO (purified by active charcoal before use), green tea polyphenols, DETAPAC (diethylenetriaminepentaacetic acid), vitamin E, rosemary, curcumin, and vitamin C.

Procedure. Incubate 10^7/ml PMN, 0.1 mM DETAPAC, and 100 ng/ml PMA for 2 min in a water bath (37°), Add 0.1 M DMPO and mix homogeneously before transferring the mixture to a quartz capillary. Measure the ESR spectrum 3 min after mixing. The ESR measurement conditions are the same as described earlier. The relative ESR intensity is taken as the amount of oxygen radical in control (h_o).

Mix 10^7/ml PMN, 0.1 mM DETAPAC, 200 μg/ml of green tea polyphenols and other antioxidants, and 100 ng/ml PMA and incubate for 2 min in a water bath (37°). Add 0.1 M DMPO and mix homogeneously before transferring to a quartz capillary. Measure the ESR spectrum 3 min after mixing. The ESR measurement conditions are the same as the control. The relative ESR intensity is taken as the amount of oxygen radical of the sample (h_x).

The scavenging effect of green tea polyphenols on oxygen radicals generated from PMA-stimulated PMN is calculated from Eq. (1).

Results. At the 200-mg/ml concentration, green tea polyphenols scavenged completely the oxygen-free radicals generated from PMA-stimulated PMN, whereas vitamin C, vitamin E, rosemary, and curcumin scavenged about 73, 27, 71, and 55% of the free radical, respectively.[1,2]

Scavenging Effect of Green Tea Polyphenols on Lipid-Free Radical Generated from Lipid Peroxidation of Lecithin Catalyzed by Lipoxidase

Lecithin can be peroxidized in the presence of lipoxidase. The lipid-free radical generated from this system can be trapped by the spin trap agent 4-POBN [α-(4-pyridyl-1-oxide)-N-tert-butylnitrone]. The 2 × 3 line ESR spectrum of lipid free radical trapped by 4-POBN is shown in Fig. 4 (a_N = 15.5G, a_H = 2.7G).[11]

Reagents. Lecithin, 4-POBN, lipoxidase, PBS, and components of green tea polyphenols.

Procedure. Mix 20 mg/ml lecithin, 0.2 M 4-POBN, and 50 mg/ml lipoxidase in 0.05 M PBS, transfer into a quartz capillary, and measure ESR spectra after incubating at room temperature for 30 min. The ESR measurement conditions are the same as described earlier. The relative ESR intensity is taken as the amount of lipid radical in the control (h_o).

Mix 20 mg/ml lecithin, 0.2 M 4-POBN, 50 mg/ml lipoxidase, and different concentrations of green tea polyphenol in 0.05 M PBS, transfer into a quartz capillary, and measure ESR spectra after incubating at room temperature for 30 min. The ESR measurement conditions are the same as control. The relative ESR intensity is taken as the amount of lipid radicals of the sample (h_x).

[14] J. L. Mark, *Science* **253**, 529 (1987).

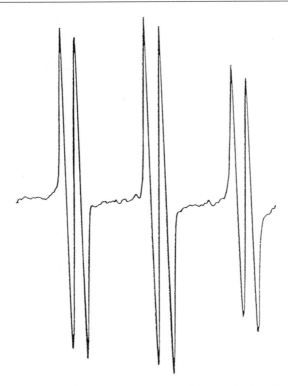

FIG. 4. ESR spectrum of lipid-free radical generated from lipid peroxidation of lecithin (20 mg/ml) catalyzed by lipoxidase (50 mM) and trapped by 4-POPN(0.2 M). The ESR condition was the same as that in Fig. 2.

The scavenging effect of green tea polyphenols on lipid radicals generated from lipid peroxidation of lecithin catalyzed by lipoxidase is calculated from Eq. (1).

Results. The concentrations of the scavenging effect of ECG, EGCG, EGC, and EC on the lipid free radical for 50% were 0.37, 0.46, 2.7, and 3.0 mM, respectively. The free radical scavenging activity was classified as follows: ECG > EGCG $\gg$ EGC > EC.[11]

Scavenging Effect of Green Tea Polyphenols on Lipid-Free Radicals Generated from Iron-Induced Lipid Peroxidation of Synaptosome

Synaptosome can be oxidized in the presence of iron ions. The lipid-free radical generated from this system can be trapped by the spin trapping agent 4-POBN. The ESR spectrum is similar to that in Fig. 4.[11]

Reagents. Synaptosome (purified according to the literature[11]), 4-POBN, Fe(NH$_4$)$_2$(SO$_4$)$_2$, FeCl$_3$. PBS, and components of green tea polyphenols.

Procedure. Mix 2.85 mg/ml synaptosome, 0.03 M 4-POBN, 0.385 mM Fe(NH$_4$)$_2$(SO$_4$)$_2$, and 1.15 mM FeCl$_3$ in 0.05 M PBS, transfer into a quartz capillary, and measure ESR spectra after incubating at room temperature for 30 min. The ESR measurement conditions are the same as described earlier. The relative ESR intensity is taken as the amount of lipid free radical in the control (h_o).

Mix 2.85 mg/ml synaptosome, 0.03 M 4-POBN, 0.385 mM Fe(NH$_4$)$_2$(SO$_4$)$_2$, 1.15 mM FeCl$_3$, and different concentrations, of green tea polyphenols in 0.05 M PBS, transfer into a quartz capillary, and measure ESR spectrum after incubating at room temperature for 30 min. The ESR measurement conditions are the same as control. The relative ESR intensity is taken as the amount of lipid radical of the sample (h_x).

The scavenging effect of green tea polyphenols on lipid radicals generated from the iron-induced lipid peroxidation of synaptosome is calculated from Eq. (1).

Result. Preincubation of synaptosome with EGCG, ECG, EGC, or EC before exposure to Fe^{2+}/Fe^{3+} treatment resulted in a decrease in the amount of lipid free radical and was dose dependent. It was found that IC$_{50}$ of EGCG, ECG, EGC, and EC on lipid-free radicals was 0.1, 0.19, 0.24, and 0.35 mM, respectively. Their lipid-free radical scavenging activities could be classified as follows: EGCG > ECG > EGC > EC.[11]

Scavenging Effect of Green Tea Polyphenols on Carbon-Centered Free Radicals from AAPH

When AAPH [2,2'-azobis(2-amidinopropane) hydrochloride] is heated at 37°, it decomposes to generate AAPH-free radicals, which can be trapped by 4-POBN. The 2 × 3 (a_N = 12.2G, a_H = 2.4G)ESR spectrum of AAPH free radical is similar to that in Fig. 4.[13]

Reagents. AAPH, 4-POBN, components of green tea polyphenols, and their epimers.

Procedure. Mix 10 mM AAPH and 0.1 M 4-POBN, transfer into a quartz capillary, and measure ESR spectra after incubating at 37° for 30 min. The ESR measurement conditions are the same as described earlier. The relative ESR intensity is taken as the amount of AAPH-free radical of the control (h_o).

Mix 10 mM AAPH, 0.1 M 4-POBN, and different concentrations of the green tea polyphenol component, transfer into a quartz capillary, and measure ESR spectra after incubating at 37° for 30 min. The ESR measurement conditions are the same as the control. The relative ESR intensity is taken as the amount of AAPH-free radical of the sample (h_x).

The scavenging effect of green tea polyphenols on AAPH-free radicals generated from the decomposition of AAPH is calculated from Eq. (1).

Results. A dose-dependent scavenging effect of green tea polyphenols components and their epimers on the carbon-centered free radicals generated from AAPH was observed. Results showed that the ability to scavenge the carbon-centered free

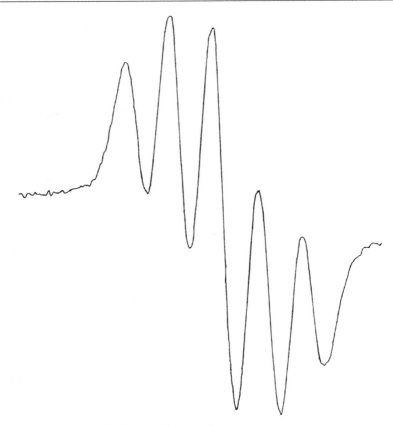

FIG. 5. ESR spectrum of DPPH (100 μM)-free radical in ethanol. The ESR condition was the same as that in Fig. 2.

radical differed among the catechins, which increased as follows: EC < (+)-C < EGC < GC < EGCG < GCG. In addition, it was found that their scavenging effects were dependent on their steric structures. The scavenging abilities of GCG, GC, and (+)-C were stronger than those of their corresponding epimers, respectively. These differences of their scavenging abilities were amplified at the lower concentration.[13]

Scavenging Effect of Green Tea Polyphenols on DPPH-Free Radical

DPPH (1,1-diphenyl-2-picrylhydrazyl) is a stable free radical. The five-line ESR spectrum of DPPH in ethanol is shown in Fig. 5.[13]

Reagents. DPPH, ethanol, components of green tea polyphenols, and their epimers.

Procedure. Transfer 100 μ*M* DPPH in ethanol into a quartz capillary and measure ESR spectra. The ESR measurement conditions are the same as described earlier. The relative ESR intensity is taken as the amount of DPPH-free radical of the control (h_o).

Mix 100 μ*M* DPPH and different concentrations of green tea polyphenol components in ethanol, transfer into a quartz capillary, and measure ESR spectra. The ESR measurement conditions are the same as control. The relative ESR intensity is taken as the amount of DPPH-free radical of the sample (h_x).

The scavenging effect of green tea polyphenols and their epimers on the DPPH-free radical is calculated from Eq. (1).

Results. It was found that the scavenging rates of green tea polyphenol components and their epimers increased as their concentrations increased from 12.5 to 25 μ*M*. At the two indicated concentrations, their DPPH radical scavenging activities were classified as follows: EC < (+)-C < EGC < GC < EGCG < GCG. The scavenging abilities of GCG, GC, and (+)-C were stronger than those of their corresponding epimers, respectively. The difference between the scavenging effects of the catechins and their corresponding epimers was greater at 12.5 μ*M* than that at 25 μ*M*. This suggested that the difference between the steric structures of the catechins and their epimers played a more important role in the ability of EGCG and GCG to scavenge DPPH than in the ability of GC, EGC, EC, and (+)-C because of their bulkier steric structure.[13]

Scavenging Effect of Green Tea Polyphenols on Singlet Oxygen

The photoradiation-hemoporphyrin system can generate a singlet oxygen, which can be trapped by TEMPONE (2,2,6,6-tetramethylpicrylhydrazyl). A three-line spectrum of the stable nitroxide radical generated from the reaction of singlet oxygen with TEMPONE is shown in Fig. 6. Its hyperfine splitting constant is $a_N = 15.6$ G.[13]

Reagents. Hemoporphyrin, TEMPONE, PBS, components of green tea polyphenols, and their epimers.

Procedure. Transfer a mixture containing 20 m*M* hemoporphyrin and 0.3 *M* TEMPONE in 0.05 *M* PBS to a quartz capillary and put into the cavity of the ESR spectrometer. Irradiate the mixture by a Xe lamp (power 1 kW) at room temperature for 20 min and record ESR spectrum every 30 sec. The ESR measurement conditions are the same as described earlier. Different concentrations of green tea polyphenols are added to the system, and the maximum of the ESR signals and the time to reach the maximum are measured.

Results. In the absence of components of green tea polyphenols and their epimers, the $^1O_2$ production increased with the time of irradiation and reached a maximum at about 12 min of irradiation and then declined rapidly. When the GC concentration was increased from 0.1 to 1.0 m*M*, the rate of $^1O_2$ production decelerated, the time to reach the maximal amount of $^1O_2$ production became

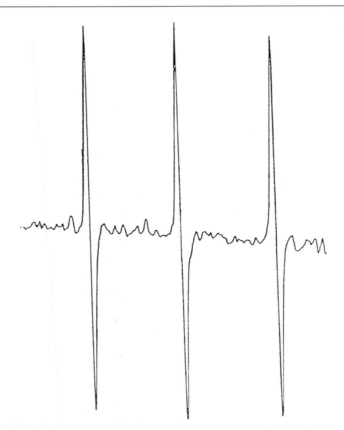

FIG. 6. ESR spectrum of TEMPON-free radical formed by the reaction of TEMPON (0.3 M) with singlet oxygen generated from the irradiation of hemoporphyrin (20 mM). The ESR condition was the same as that in Fig. 2.

significantly shorter, and the maximal amount became much less as well. When the concentration increased from 0.3 to 0.5 mM, there was no significant difference between the effect of EGCG and its epimer GCG on $^1O_2$ produced in this system. However, at a low concentration of 0.1 mM, a significant difference was observed in the time to reach maximum of $^1O_2$ production between GCG and EGCG, which was 4 and 6 min, respectively. It was found that the scavenging effect of GCG on $^1O_2$ was stronger than that of its epimer EGCG. The effects of either EGC and GC or EC and (+)-C were similar to those of EGCG and GCG. Also, the scavenging effects of GC and (+)-C were stronger than those of their epimers EGC and EC, respectively. So, the scavenging activity of the catechins was in the following order: EC [(+)-C] < EGC (GC) < EGCG (GCG).[13]

Scavenging Effect of Green Tea Polyphenols on Oxygen-Free Radicals Generated from Ischemic–Reperfusion Myocardium

Myocardial ischemia–reperfusion injury is a model widely used for studying the mechanisms of heart disease, which shows a close relationship between the oxygen-free radical paradox and the physiological function with myocardial ischemia–reperfusion injury.[16–19]

Reagents. Pentobarbital, heparin, standard physiological solution, Krebs bicarbonate (KB) buffer (NaCl, 124 mmol liter; KCl, 4.7 mmol/liter; $MgCl_2$ 1.2 mmol/liter; $NaHCO_3$, 19.5 mmol/liter; KH_2PO_4, 1.2 mmol/liter; $CaCl_2$, 2.5 mmol/liter; 0.5 mmol/liter EDTA-Na, glucose, 10 mmol/liter), 95% O_2, 5% CO_2, and green tea polyphenols.

Procedure. Prepare the ischemia–reperfusion of rat heart according to the modified Langendoorff method.[20] Anesthetize Wistar rats (250–300 g) with 3% pentobarbital (10 ml/kg). Excised the hearts and put immediately into standard physiological solution (4°) after heparinization (5000 U/kg). Cannulate the hearts through the artery and perfuse with oxygenated Krebs bicarbonate buffer at 37° under pressure of 80 mmHg [95% O_2, 5% CO_2 (v/v), pH 7.4] for 10 min. The rhythmically beating hearts are taken as the normal heart. Subject the hearts to ischemia for 30 min, which are taken as ischemic hearts. Ischemic hearts reperfused for 15 sec with oxygenated KB buffer are taken as the group of ischemia-reperfused hearts. Hearts reperfused with oxygenated KB buffer containing different concentration of green tea polyphenols are taken as the group of ischemia–reperfusion hearts with green tea polyphenols. Cut the myocardium into small cylinders of 2.5 mm diameter with a cold scissors (liquid nitrogen) and put immediately into a quartz tube of 3 mm diameter, packed 3 cm in height, and place in liquid nitrogen.

EPR measurement conditions: X-band, 100 kHz modulation with amplitude 8 G, microwave power 1 mW, time constant 0.128 sec, central magnetic field 3250 G, swept width 500 G, and temperature 123 K.

Results. Results showed that the isolated reperfused rat heart subjected to 30 min of ischemia followed by 15 sec of reperfusion induced a significant increase in the production of oxygen-free radicals. The concentration of oxygen-free radicals in the ischemic-reperfused myocardium was about 188% of that in the normal myocardium. After pretreatment of the heart with 0.1 mg/ml green tea polyphenols, oxygen-free radicals were reduced about 71% compared with those in ischemia–reperfusion ($p < 0.01$). This result indicates that the natural

---

[15] B.-L. Hao, N.-N. Huang, J.-Z. Zhang, and W.-J. Xin, *Chin. Sci. Bul.* **31**, 1139 (1986).
[16] J. L. Zweier, J. T. Flahery, and M. L. Weisfeldt, *Proc. Natl. Acad. Sci. U.S.A.* **84**, 1404 (1987).
[17] B.-L. Zhao, W.-D. Yang, H.-L. Zhu, and W.-J. Xin, *Chin. Sci. Bul.* **35**, 56 (1990).
[18] B.-L. Zhao, J.-G. Shen, M. Li, M.-F. Li, and W.-J. Xin, *Biochim. Biophys. Acta* **1315**, 131 (1996).
[19] X.-L. Zou, Q. Wan, M.-F. Li, B.-L. Zhao, and W.-J. Xin, *Chin. J. Magn. Res.* **12**, 237 (1995).
[20] M. J. Shlafer, *Thorac. Cardiovasc. Surg.* **83**, 830 (1982).

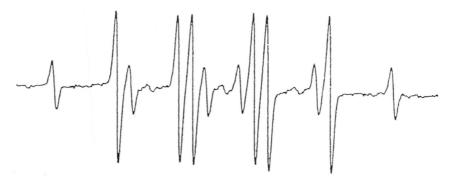

FIG. 7. ESR spectrum of methyl-free radical generated from the reaction of DMSO(1.4 mM) with peroxynitrite(4.8 mM) and trapped by MNP (80 mM). The ESR condition was the same as that in Fig. 2.

antioxidant green tea polyphenols may protect the heart from ischemic-reperfused injury by scavenging the oxygen-free radicals.[21]

Scavenging Effects of Green Tea Polyphenols on Methyl Radical Generated from Dimethyl Sulfate Oxidized by Peroxynitrite

Superoxide radical reacts with nitric oxide in many pathological cases to form peroxynitrite.[21] It has a pK_a of 6.6 at 0° and decays rapidly once protonated to give species such as hydroxyl radical-like and nitric dioxide, which can oxidize membrane lipids.[22] It can also oxidize dimethyl sulfoxide (DMSO) to form a methyl radical.[23] Using spin-trapping agent MNP (2-methyl-2-nitrosopropane), a free radical from the reaction of DMSO with peroxynitrite can be trapped. It was identified as the methyl-free radical spin adduct of MNP by the analysis of its spectrum ($a_N = 17.2$ G, $a_H = 14.2$ G) (Fig. 7).[23]

Reagents. Peroxynitrite (synthesized according to the literature[23]), MNP, DMSO, green tea polyphenols quercertin, and vitamin C.

Procedure. Mix 4.8 mM peroxynitrite, 80 mM MNP, and 1.4 mM DMSO, transfer to a quartz capillary, and measured ESR spectra 30 min after mixing. The ESR measurement conditions are the same as those described earlier. The relative ESR intensity is taken as the amount of methyl-free radical of the control (h_o).

Mix 4.8 mM peroxynitrite, 80 mM MNP, 1.4 mM DMSO, and different concentrations of green tea polyphenols, transfer into a quartz capillary, and measure ESR spectra 30 min after mixing. The ESR measurement conditions are the same as in

---

[21] Y. Henry, M. Lepoivre, J.-C. Dripier, C. Ducrocq, J. L. Boucher, and A. Guissani, *FASEB J.* **7**, 1124 (1993).

[22] R. Radi, J. S. Beckman, K. M. Bush, and B. A. Freeman, *Arch. Biochem. Biophys.* **288**, 481 (1992).

[23] B.-L. Zhao, W. Jiang, Y. Zhao, J.-W. Hou, and W.-J. Xin, *Biochem. Mol Biol. Int.* **38**, 17 (1996).

the control. The relative ESR intensity is taken as the amount of the methyl-free radical of the sample (h_x).

The scavenging effect of green tea polyphenols on methyl-free radicals is calculated from Eq. (1). This is a useful index for studying the oxidation property of peroxynitrite.

Results. It was found that the yield of methyl radical was dose dependent on both of the concentrations of peroxynitrite and DMSO. The scavenging effects of green tea polyphenols on methyl-free radicals generated from this system were studied and compared with the scavenging effects of quercertin and vitamin C. Although they could all effectively scavenge the methyl-free radical in this system, the scavenging effect of green tea polyphenols was the strongest whereas that of vitamin C was the weakest.[23]

Comments

1. The free radical scavenging of green tea polyphenols can be tested systematically and quantitatively with the methods described in this article. It can be studied by the ESR technique in water solution (Fenton reaction system), irradiation (riboflavin/EDTA), enzyme (lipoxidase/lecithin), cell (PMN), and tissue (ischemia–reperfusion heart) systems. Green tea polyphenols were tested to scavenge hydroxyl radical, superoxide radical, lipid radical, methyl radical, carbon-centered AAPH radical, and DPPH radical, as well as singlet oxygen.

2. When the scavenging effects of green tea polyphenols on hydroxyl radical and lipid radical are studied in the presence of Fe^{2+}/Fe^{3+}, the chelation of green tea polyphenols with Fe^{2+}/Fe^{3+} should be considered. Different components of green tea polyphenols have different chelating ability with Fe^{2+}/Fe^{3+}, which will affect the scavenging effects of green tea polyphenols on the free radicals.[13]

3. Green tea polyphenols will be changed to semiquinone-free radical and quinone after scavenging free radicals. The stability of the semiquinone-free radical and quinone is an important property that affects green tea polyphenol-free radical scavenging. The more stable the semiquinone-free radical formed from the reaction of GTP with oxygen-free radicals, the stronger of the green tea polyphenols free-radical scavenging will be.[13]

4. Green tea polyphenols are water soluble and suitable for studying methods described in this article. If a lipid-soluble-free radical scavenger is being studied, other more suitable methods should be found for the lipid-soluble-free radical scavenger.

5. Usually, the free radical scavenging ability of a polyphenol depends on the number of hydroxyl groups in the structure, but it also depends on the environment around the free radicals and scavengers.

Acknowledgment

This work was supported by a grant from the National Natural Science Foundation of China.

[20] Polyphenol Protection of DNA against Damage

By GUANGJUN NIE, TAOTAO WEI, SHENGRONG SHEN, and BAOLU ZHAO

Introduction

DNA is one of the major cellular components that are prone to oxidative attack. Excessive generation of reactive oxygen species (ROS) *in vivo* may result in DNA damage. At least two major human problems, aging and carcinogenesis, involve DNA damage.[1,2] As a model for investigating oxidative DNA damage, the copper 1,10-phenanthroline complex, a complex of the chelating agent 1,10-phenanthroline (OP) with Cu^{2+} ion [$(OP)_2Cu^{2+}$], has been widely used as a chemical nuclease for structural studies on DNA.[3,4] In the presence of hydrogen peroxide and ascorbic acid, it can induce several kinds of DNA damage, such as strand breaks, base modification, nonbasic sites, and DNA–protein cross-links.[5]

Green tea polyphenols have aroused considerable attention in recent years due to their diverse pharmacological functions, e.g., antimutagenic and anticarcinogenic effects.[6–8] The major compounds in green tea polyphenols are (−)-epigallocatechin gallate (EGCG), (−)-epicatechin gallate (ECG), (−)-epigallocatechin (EGC), and (−)-epicatechin (EC) (Fig. 1). A possible mechanism behind the antimutagenic and anticarcinogenic effects of green tea polyphenols might be that they act as potent antioxidants to scavenge many kinds of ROS, such as lipid free radicals, superoxide radicals, singlet oxygen ($^1O_2$), and hydroxyl radicals.[9–12] There are several reports concerning the protective effects of green tea polyphenols on oxidative DNA damage.[13–15] In most of the studies, green tea polyphenols are

---

[1] P. A. Cerutti, *Science* **227,** 375 (1985).
[2] H. Wiseman, H. Kaur, and B. Halliwell, *Cancer Lett.* **93,** 113 (1995).
[3] D. S. Sigman, *Acc. Chem. Res.* **19,** 180 (1986).
[4] D. S. Sigman, D. R. Graham, V. D'Aurora, and A. M. Stern, *J. Biol. Chem.* **254,** 12269 (1979).
[5] R. Rodriguez, R. Drouis, G. P. Holmguist, T. R. O'Connor, S. Boiteux, J. Laval, J. H. Doroshow, and S. A. Akman, *J. Biol. Chem.* **270,** 17633 (1995).
[6] H. Mukhtar, Z. Y. Wang, and S. K. Katiyar, *Prev. Med.* **21,** 351 (1992).
[7] Z. Y. Wang, S. J. Cheng, Z. C. Zhou, M. Athar, W. A. Khan, D. R. Bickers, and H. Mukhtar, *Mutat. Res.* **223,** 273 (1989).
[8] Z. G. Dong, W. Y. Ma, C. S. Huang, and C. S. Yang, *Cancer Res.* **57,** 4414 (1997).
[9] W. Bors and M. Saran, *Free Radic. Res. Commun.* **2,** 289 (1987).
[10] W. Bors, W. Heller, C. Michel, and M. Saran, *Methods Enzymol.* **186,** 343 (1990).
[11] Q. Guo, B. Zhao, M. Li, S. Shen, and W. Xin, *Biochim. Biophys. Acta* **1304,** 210 (1996).
[12] Q. Guo, B. Zhao, S. Shen, J. Hou, H. Hu, and W. Xin, *Biochim. Biophys. Acta* **1427,** 13 (1999).
[13] H. Wei, X. Zhang, J. Zhao, Z. Wang, D. Bickers, and M. Lebwohl, *Free Radic. Biol. Med.* **26,** 1427 (1999).
[14] P. Leanderson, Å. O. Faresjö, and C. Tagesson, *Free Radic. Biol. Med.* **23,** 235 (1996).
[15] J. Zhang, J. Qin, E. Cao, Z. Zhang, and Y. Zhang, *Acta Biophys. Sinica* **12,** 691 (1996).

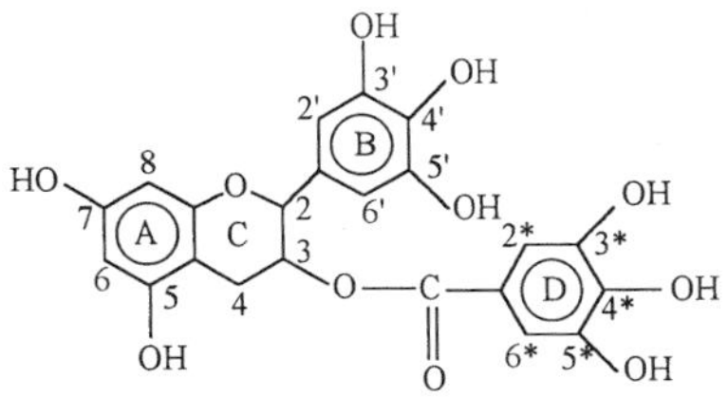

Epigallocatechin gallate (EGCG)

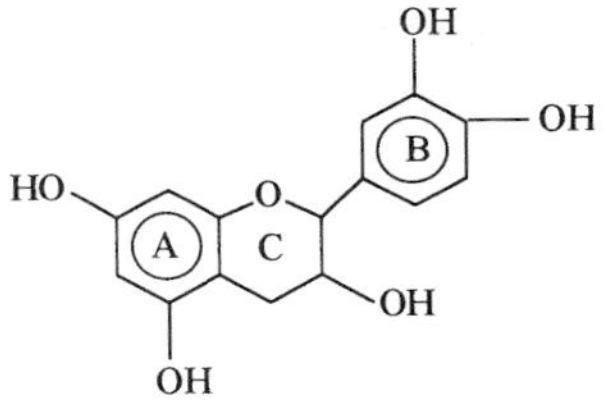

Epicatechin (EC)

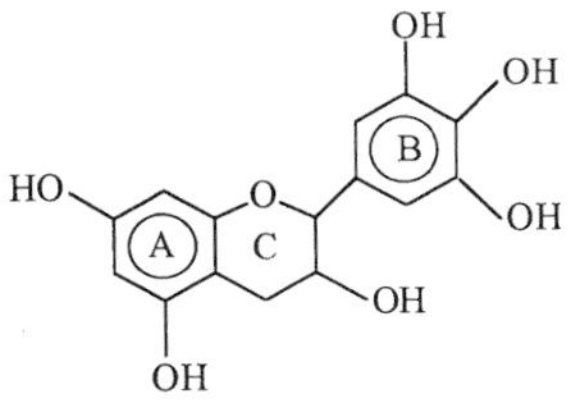

Epigallocatechin (EGC)

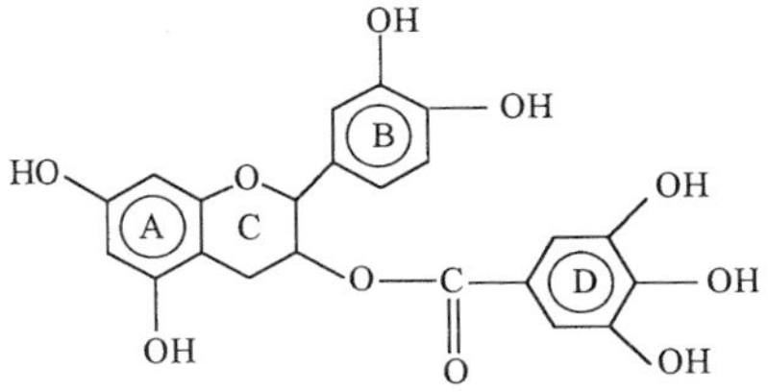

Epicatechin gallate (ECG)

FIG. 1. Chemical structures of green tea polyphenols.

used as a mixture or extractable fraction so that structure–activity relationships for the four major components have not yet been studied. Up to now, high-performance liquid chromatography coupled with electrochemical detection (HPLC-ECD) and gas chromatography/mass spectrometry with selected-ion monitoring (GC/MS-SIM) are the major methods for the detection of DNA damage. Oxidized bases, especially 8-oxo-7,8-dihydro-2′-deoxyguanosine (8-oxodGuo), are considered appropriate biomarkers to evaluate oxidative DNA damage *in vivo* and *in vitro* by both techniques.[16,17] Instead of providing information on the reaction process, these methods could only reveal the final oxidative products induced by ROS.[18] Chemiluminescence seems to be one of the earliest responses to oxidative stress in studies of oxidative DNA damage, increasing before any other changes are detected. The electron spin resonance (ESR) technique can directly provide the information about the free radicals that cause oxidative DNA damage.

In this article, we have investigated $(OP)_2Cu^{2+}$-mediated oxidative DNA damage using chemiluminescence and ESR methods and have studied the protective

[16] B. A. Ames and L. S. Gold, *Mutat. Res.* **250,** 3 (1991).
[17] R. A. Floyd, *Carcinogenesis* **11,** 1447 (1990).
[18] W. Ma, E. Cao, J. Zhang, and J. Qin, *J. Photochem. Photobiol.* **44,** 63 (1998).

effects of four components of green tea polyphenols. The different protective efficiency of these compounds is also discussed.

Methods

Materials

Reagents

0.1 M acetic acid–sodium acetate buffer, pH 5.2

1 μg/ml calf thymus DNA (Sino-American Biotechnology Company, Beijing, China): prepared from stock solution of 100 μg/ml DNA in acetic acid–sodium acetate buffer

150 μM $CuSO_4$ stock solution in acetic acid–sodium acetate buffer

1.05 mM OP stock solution in acetic acid–sodium acetate buffer: dissolve 38.5 mg 1,10-phenanthroline crystals in 50 μl of absolute ethanol prior to the addition of 200 μl acetic acid–sodium acetate buffer to make up the stock solution

525 μM/75μM 1,10-phenanthroline–Cu^{2+} solution: combine 1.05 mM OP stock solution and 150 mM $CuSO_4$ stock solution in a ratio of 1:1 (v/v)

3mM ascorbic acid in acetic acid–sodium acetate buffer

3% H_2O_2: dilute 10 ml of 30% H_2O_2 in 90 ml of doubly distilled water

0.5 M 5,5'-dimethyl-1-pyrroline 1-oxide (DMPO)(Sigma Chemical Co.) in doubly distilled water (purified with active charcoal before use)

0.2 mM EGCG, ECG, EC, and EGC (purified by HPLC as pure compounds) in acetic acid–sodium acetate buffer, respectively

Stability of Solutions. DNA stock solution is quite stable even at room temperature if microbial contamination is avoided. OP–$CuSO_4$ and DMPO solutions must be kept in the dark and DMPO also needs to be kept frozen. H_2O_2 and green tea polyphenol solutions should be prepared before use. Ascorbic acid is quite stable even in air-saturated buffer in the absence of catalytic metals. Storage in a relatively air-tight flask helps increase the shelf life. The appearance of a yellow color indicates significant ascorbic acid oxidation.

Chemiluminescence Assay

Principle of Assay. The 1,10-phenanthroline/Cu^{2+} complex reacts with H_2O_2 and ascorbic acid and then emits light (Fig. 2a). The maximal emission wavelength is around 460–480 nm.[18] The $O_2^{\bullet-}$ and hydroxyl radicals are considered the

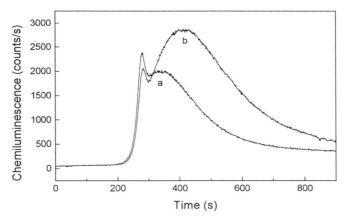

FIG. 2. Chemiluminescence curves in the absence of DNA (curve a) and in the presence of DNA (curve b) in the $(OP)_2Cu^{2+}$/ascorbic acid/H_2O_2 system at 16°. Final concentrations of the compounds in the assay mixture were $CuSO_4$, 50 μM; OP, 350 μM; DNA, 1 μg/ml; and H_2O_2, 130 μM, respectively.

contributors of the chemiluminescence.[19] When DNA is added to the OP/Cu^{2+}/ascorbic acid/H_2O_2 system, a markedly delayed peak of chemiluminescence is observed (Fig. 2b). The special light emission could be attributed to oxidative DNA damage, and guanine modification is likely to be the cause of the chemiluminescence.[15,18,20] The maximal emission wavelength is around 400–420 nm,[18] and the chemiluminescence species in oxidative DNA damage may be excited carbonyl species.[18] Because of the involvement of the oxidative reaction, antioxidants can affect the light emission in the chemiluminescence system. Because natural antioxidants such as flavonoids and polyphenols can inhibit the oxidative reaction and decrease light emission,[18] the method could be used to evaluate antioxidative properties.

Procedure. One milliliter (525 μM/75 μM) OP/Cu^{2+} and 150 μl (1 μg/ml) DNA are premixed and then 150 μl (3 mM) ascorbic acid and 200 μl (3%) H_2O_2 are added to initiate the luminescence reaction. Emission intensity is recorded with a BPCL-4 ultra weak chemiluminescence analyzer with a computerized high-sensitivity single-photon counter (SPC) at 16°. The voltage in the photomultiplier is kept at 1000 V, and the spectral range of chemiluminescence recorded is 340–800 nm. Before measuring the effects of green tea polyphenols on the chemiluminesence from DNA oxidation, different concentrations of EGCG, ECG, EC, and EGC are premixed with the $Cu^{2+}(OP)_2$/DNA solution, respectively. The chemiluminescent measurement is recorded according to the description just given.

[19] O. S. Fedorova, S. E. Olkin, and V. M. Berdnikof, *Z. Phys. Chem.* **263,** 529 (1982).
[20] J. Tie, L. Li, and B. Ru, *Acta Biophys. Sinica* **11,** 276 (1995).

Factors Influencing Chemiluminescence from DNA Oxidation. The kinetic curve of chemiluminescence is affected significantly by the concentrations of DNA, ascorbic acid, H_2O_2, and Cu^{2+}. The light emission is proportional to the concentration of DNA (0.2–1.2 μg/ml).[18] (The light emission intensity here refers to the difference between the total intensity of the measuring period and the background emission for no DNA.) The total intensity of chemiluminescence is also increased if the concentration of H_2O_2 is increased and reaches a plateau when the final concentration of H_2O_2 is 1%.[15]

The concentration of ascorbic acid affects the time course and the intensity of chemiluminescence from DNA oxidation (Fig. 3 and the inset of Fig. 3). When the concentration of ascorbic acid is low (<300 μM), the total light emission of the measuring period is increased by increasing the ascorbic acid concentration. When the ascorbic acid concentration is high (above 300 μM), the total emission intensity is decreased and the onset of light emission is accelerated.

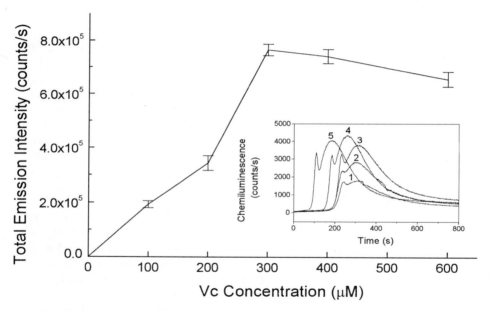

FIG. 3. The relationship between emission intensity of DNA oxidation (total intensity over 800 sec of the measuring period) and ascorbic acid concentrations at 16°. For each concentration of ascorbic acid, the emission intensity was acquired by subtracting the intensity for no DNA from the total intensity. Each data point was the mean of three independent tests, and the bar represents the standard deviation. Final concentrations of the compounds in the assay mixture were $CuSO_4$, 50 μM; OP, 350 μM; DNA, 1.2 μg/ml; and H_2O_2, 130 μM, respectively. (Inset) Chemiluminescence curves of different ascorbic acid concentrations. The ascorbic acid concentrations of curves 1–5 were 100, 200, 300, 400, and 600 μM, respectively.

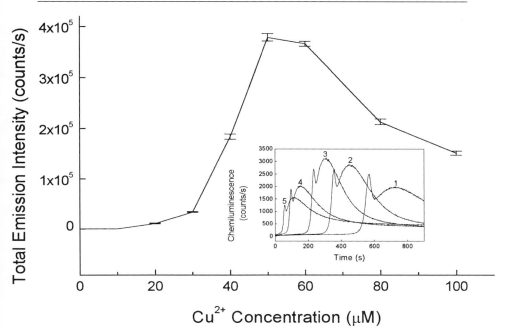

FIG. 4. The relationship between emission intensity of DNA oxidation (total intensity over 900 sec of the measuring period) and CuSO$_4$ concentrations. The final concentration of DNA was 0.5 μg/ml. Other conditions were the same as those in Fig. 3. (Inset) Chemiluminescence curves of different Cu^{2+} concentrations. The Cu^{2+} concentrations of curves 1–5 were 40, 50, 60, 80, and 100 μM, respectively.

Figure 4 shows that the time course and the acceleration of chemiluminescence are quite sensitive to the Cu^{2+} concentration. Generally, the onset of light emission is delayed by decreasing the concentration of Cu^{2+}. The maximal total intensity occurs when Cu^{2+} concentration is at 50 μM.

pH and temperature are other important factors influencing chemiluminescence from DNA oxidation. With an increase in temperature, the chemiluminescence peak appears earlier but the total intensity of the chemiluminescence decreases.[15] pH values influence the time course and acceleration of chemiluminescence significantly (Fig. 5). The onset of light emission is accelerated when the pH value increases from pH 3.6 to pH 5.6. The maximal total emission intensity peaks at pH 5.2 and declines thereafter.

Calculations. The inhibition effects of the four green tea polyphenol components on the emission intensity from DNA oxidation are calculated from Eq. (1):

$$I\% = (C_D - C_I)/(C_D - C_0) \times 100\% \qquad (1)$$

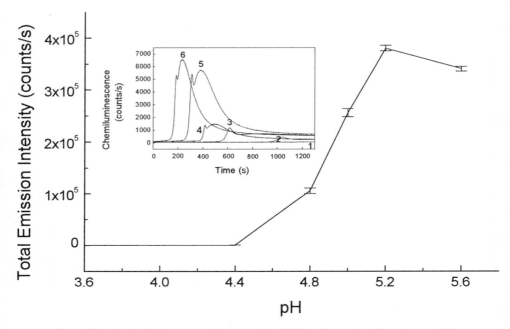

FIG. 5. The relationship between emission intensity of DNA oxidation (total intensity over 1200 sec of the measuring period) and pH value (3.6–5.6). The final concentration of DNA was 0.5 μg/ml. Other conditions were the same as those in Fig. 3. (Inset) Chemiluminescence curves of different pH value. The pH values of curves 1–6 were 3.6, 4.4, 4.8, 5.0, 5.2, and 5.6, respectively.

where C_D and C_0 are the total emission intensity over 900 sec with and without DNA, respectively, and C_I is the total emission intensity over 900 sec with green tea polyphenols in the OP/Cu^{2+}/ascorbic acid/H_2O_2/DNA system.

ESR Spin–Trapping Assay

Procedure. Twenty microliters (525 μM/75 μM) OP/Cu^{2+}, 3 μl (1 μg/ml) DNA (or 3 μl acetic acid–sodium acetate buffer), and 10 μl (0.5 M) DMPO are premixed and then 3 μl (3 mM) ascorbic acid and 4 μl (3%) H_2O_2 are added to initiate the reaction. When inhibition effects are measured, green tea polyphenols are premixed with OP/Cu^{2+}/DNA and then ascorbic acid and H_2O_2 are added. The concentrations of EGCG, EGC, EC, and ECG are 10 μM when used. The reaction mixture is transferred to a quartz capillary, which is fitted into the cavity of a Bruker ER-200D-SRC ESR spectrometer. The conditions of ESR measurement are as follows: X band, microwave power 10 mW, 100 kHz modulation with amplitude 1 G, scan width 200 G, time constant 200 msec, and room temperature (298 K).

Calculations. The scavenging capacities of the four components of green tea polyphenols are measured by comparing the hydroxyl radical spin adduct of DMPO with and without green tea polyphenols, respectively, 4 min after adding ascorbic acid and H_2O_2 to initiate the reaction. The second peak height of the spectrum of DMPO-OH in the control experiment (with and without DNA, respectively) represents 100% DMPO-OH formation. The scavenging effect of green tea polyphenols on the hydroxyl radicals can be calculated from Eq. (2):

$$E\% = (H_0 - H_X)/H_0 \times 100\% \qquad (2)$$

where H_X and H_0 are the ESR signal intensities of samples with and without green tea polyphenols, respectively.

Results

Inhibiting Effects of Green Tea Polyphenols on Chemiluminescence Concomitant with Oxidative DNA Damage. Preincubation of $Cu^{2+}(OP)_2$/DNA with EGCG, ECG, EC, and EGC, respectively, before adding ascorbic acid and H_2O_2 resulted in a marked decrease in the emission intensity of chemiluminescence, especially the second peak, i.e., the DNA-related oxidative luminescence. Figure 6 (inset) shows the typical dose-dependent relationship between the concentrations of EGC and the emission intensity from DNA oxidation. Apart from inhibiting emission intensity, the second peak of chemiluminescence is also delayed. Figure 6 shows the different inhibiting effects of the four components EGCG, ECG, EC, and EGC at 6.67 μM. All four compounds show chemiluminescence–inhibition capacity with a potency order of ECG > EC > EGCG > EGC (Fig. 7). The concentrations of these compounds for inhibiting 50% chemiluminescence from DNA oxidation (IC$_{50}$) were calculated from the concentration–activity curves (Fig. 7), and the IC$_{50}$ values of EGCG, EGC, ECG, and EC were 6.0, 11.5, 2.6, and 3.8 μM, respectively.

In order to investigate the effects of green tea polyphenols on the chemiluminescence kinetic curve of DNA oxidation directly, EGCG was added when the DNA oxidative chemiluminescence reached the maximal value. Figure 8 shows that EGCG at a concentration of 5.34 μM has a significant inhibitory effect on the DNA oxidative chemiluminescence. However, the same amount of doubly distilled water does not have an inhibitory effect.

Capacity of Green Tea Polyphenols to Scavenge Hydroxyl Radical Generated in $Cu^{2+}(OP)_2$/DNA/Ascorbic Acid/H_2O_2 System. Typical ESR spectra of the spin adduct DMPO-OH are detected in both the $Cu^{2+}(OP)_2$/ascorbic acid/H_2O_2 system and the $Cu^{2+}(OP)_2$/DNA/ascorbic acid/ H_2O_2 system. When DNA was added to the $Cu^{2+}(OP)_2$/ascorbic acid/H_2O_2 system, the intensity of the ESR signal was decreased slightly (Figs. 9a and 9b; $a^N = a^H = 14.9$ G). In order to confirm the generation of hydroxyl radicals in the system, absolute methanol was added to the reaction mixture and the spin adduct DMPO-CH$_2$OH was also detected (Fig. 9c; $a^N = 15.4$ G, $a^H = 18.56$ G). The amount of DMPO-OH increases during the

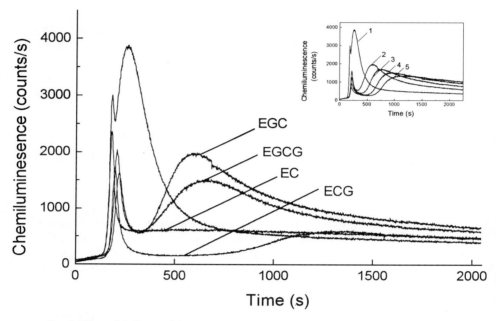

FIG. 6. Effect of 6.67 μM of four components of green tea polyphenols on chemiluminescence from DNA oxidation. The conditions were the same as those in Fig. 3. (Inset) Effect of EGC on chemiluminescence from DNA oxidation in a dose-dependent manner. The concentrations of curves 1–5 were 0, 6.67, 13.32, 26.64, and 39.96 μM of EGC, respectively. The conditions were the same as those in Fig. 3.

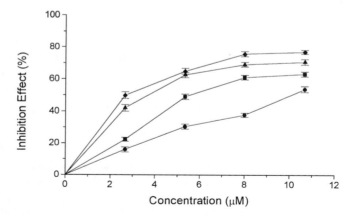

FIG. 7. The inhibition effect of (■) EGCG, (●) EGC, (▲) EC, and (◆) ECG on chemiluminescence from DNA oxidation in the $(OP)_2Cu^{2+}$/ascorbic acid/H_2O_2 system. The conditions were the same as those in Fig. 3.

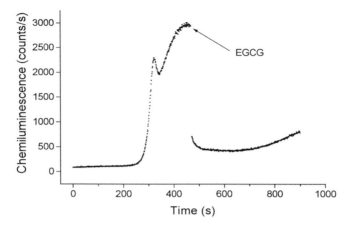

FIG. 8. Effect of EGCG (5.34 μM) on chemiluminescence from DNA oxidation. EGCG was added at the time indicated by the arrow. Other conditions were the same as those in Fig. 3.

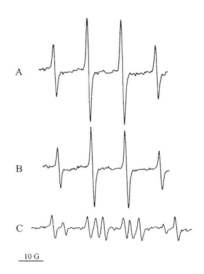

FIG. 9. (A) ESR spectrum of the hydroxyl radical spin adduct of DMPO produced by a solution of 0.1 M DMPO, 50 μM Cu^{2+}, 350 μM OP, 300 μM ascorbic acid, and 130 μM H_2O_2. (B) ESR spectrum of the hydroxyl radical spin adduct of DMPO produced by a solution of 0.1 M DMPO, 50 μM Cu^{2+}, 350 μM OP, 300 μM ascorbic acid, 130 μM H_2O_2, and 1 μg/ml DNA. (C) ESR spectrum of the hydroxyl radical and absolute methanol radical spin adduct of DMPO produced by a solution of 0.1 M DMPO, 50 μM Cu^{2+}, 350 μM OP, 300 μM ascorbic acid, 130 μM H_2O_2, and 1 μg/ml DNA. The amount of absolute methanol was 10 μl. ESR measurement conditions were described under the procedure of ESR.

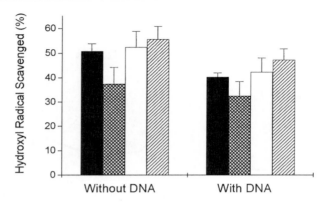

FIG. 10. The scavenging effects of EGCG (solid bar), EGC (cross-hatched bar), EC (open bar), and ECG (hatched bar) on the hydroxyl radical produced in the OP/Cu^{2+}/ascorbic acid/H_2O_2 and OP/Cu^{2+}/ascorbic acid/H_2O_2/DNA reaction systems. The concentration of the four green tea polyphenols was 10 μM. ESR measurement conditions were described under the procedure of ESR. Data are means ±SD ($n = 3$).

10-min period of the reaction and after 4 min the scavenging capacity of four green tea polyphenol components on hydroxyl radical with and without DNA is measured. When the four compounds at the same concentration (10 μM) were added to the $Cu^{2+}(OP)_2$/DNA/ascorbic acid/H_2O_2 system and the $Cu^{2+}(OP)_2$/ascorbic acid/H_2O_2 system, respectively, they all showed scavenging capacity of hydroxyl radicals with a potency order of ECG > EC > EGCG > EGC (Fig. 10).

Discussion

In this article, a rapid and convenient chemiluminescent method was developed for measuring the protective effects of green tea polyhenols against oxidative DNA damage. Chemiluminescence concomitant with oxidative DNA damage induced by OP/Cu^{2+}/ascorbic acid/H_2O_2 system was observed and it was found that green tea polyphenols could inhibit chemiluminescence greatly. DNA damage by the $(OP)_2Cu^{2+}$ complex was measured previously by several authors.[3,4,20,21] The 1,10-phenanthroline–copper complex, a tetrahedral complex, binds in the minor groove of DNA where the Cu^{2+} ions are reduced to Cu^+ by reducing agents such as ascorbic acid or mercaptoethanol. When hydrogen peroxide (H_2O_2) is added, the complex is oxidized to form a copper–oxo species; DNA damage is also noted.[20] The key species for DNA damage may be the hydroxyl radical (•OH) formed on the DNA molecular structure.[22]

[21] J. M. C. Gutteridge and B. Halliwell, *Biochem. Pharmacol.* **31,** 2801 (1982).
[22] M. Dizdaroglu, O. I. Aruoma, and B. Halliwell, *Biochemistry* **29,** 8447 (1990).

The chemiluminescence of oxidative DNA damage in the OP/Cu^{2+}H$_2$O$_2$/ascorbic acid system is specific for DNA damage. This can be found by comparing the chemiluminescence curves in the absence of DNA and in the presence of DNA (Fig. 2). When adding DNA, the maximal peak of chemiluminescence appears delayed and the maximal emission wavelength of chemiluminescence shifts from 460–480 to 400–420 nm.[18] Apart from these changes, the intensity of the light emission also increases greatly after adding DNA. Another significant advantage of the chemiluminescence method is its exceptional sensitivity. With a high-sensitivity single photon counter, the chemiluminescence analyzer seems to be able to measure and record one of the earliest changes during oxidative DNA damage. Most previous work monitored the end products of oxidative reactions, however, light emission is a marker at the beginning of oxidative DNA damage. Chemiluminescence is also a noninvasive method to follow the kinetic change in oxidative DNA damage.

Despite the high sensitivity and specificity of this method, several factors limit the application of this technique: (1) photomultipliers are not equally sensitive to the light emitted at all wavelengths and (2) the low intensity of emitted light also limits the use of interference filters to better characterize the emission wavelength. Despite these drawbacks, substantial information could be obtained in terms of the protective effects on oxidative DNA damage.

Mechanisms for the protective effects of green tea polyphenols on oxidative DNA damage may occur in two ways. First, green tea polyphenols could scavenge H$_2$O$_2$ directly. In the chemiluminescence assay, catalase was observed to inhibit chemiluminescence greatly, and the amount of H$_2$O$_2$ was also a major factor influencing the yield of chemiluminescence. Wei *et al.*[13] showed that green tea extracts potently scavenge H$_2$O$_2$ in a dose-dependent manner *in vitro*. Thus green tea polyphenols could decrease the formation of hydroxyl radicals. Second, green tea polyphenols also scavenge hydroxyl radicals and decrease the damage to DNA from hydroxyl radicals. With the ESR method, we observed that green tea polyphenols could effectively decrease the DMPO-OH signal.

In our experiments, the protective effects of green tea polyphenols were evaluated and it was found that all four compounds had inhibiting effects on chemiluminescence intensity and DMPO-OH signal intensity in oxidative DNA damage. The four compounds showed chemiluminescence inhibition capacity with a potency order of ECG > EC > EGCG > EGC. Compared with previous results [11] about the scavenging effects of green tea polyphenols on hydroxyl radicals, it is interesting that we obtained the same potency order of the four components on protection of oxidative DNA damage. To summarize these experimental results, the presence of a catechol group on the B ring and a gallate group at the 3-position play an important role in their free radical-scavenging activities against hydroxyl radical and protective effects on DNA damage.

In conclusion, with chemiluminescence and ESR methods, we compared the structure–activity relationship of four green tea polyphenols in inhibiting oxidative DNA damage. Because all four green tea polyphenols have a remarkably protective effect on the oxidative DNA damage induced by ·OH *in vitro,* these effects could, at least in part, explain the mechanisms of action by which green tea counteracts the carcinogenic processes.

Acknowledgment

This work was supported by a grant from the National Natural Science Foundation of China.

[21] Markers for Low-Density Lipoprotein Oxidation

By MICHAEL AVIRAM and JACOB VAYA

Low-Density Lipoprotein (LDL) Oxidation and Atherosclerosis

It is widely accepted that lipid peroxidation plays a central role in the development of cardiovascular diseases and that low-density lipoprotein oxidation is considered to be the hallmark of early atherosclerosis.[1,2] This disease, the major cause of morbidity and mortality in the Western world, is characterized by the accumulation of cholesterol and oxidized lipids in the arterial wall.[3] Oxidized LDL (Ox-LDL) is atherogenic, as it causes arterial cell death, and increases arterial infiltration of LDL and the accumulation of inflammatory cells in the arterial wall. It stimulates the release of the growth factors monocyte chemoattractant protein 1 (MCP-1), colony-stimulating factor (CSF), endothelium-derived relaxation factors (EDRF, NO), vascular endothelial growth factor (VEGF), and interleukins (IL). In addition, Ox-LDL contributes to platelet aggregation, smooth muscle cell proliferation, and thrombotic and inflammatory processes.[4] Oxidation of LDL is a free radical-driven lipid peroxidation process. The initial step in this process is the abstraction of a hydrogen atom from the allylic position (next to a carbon–carbon double bond) of polyunsaturated fatty acids (PUFA) to form an alkyl radical (L·).

---

[1] D. Steinberg, S. Parthasarathy, T. E. Carew, J. C. Khoo, and J. L. Witztum, *N. Engl. J. Med.* **320**, 915 (1989).

[2] M. Aviram, *Curr. Intervent. Cardiol. Rep.* **1**, 77 (1999).

[3] S. Yla-Herttuala, W. Palinski, M. E. Rosenfeld, S. Parthasarathy, T. E. Carew, S. Butler, J. L. Witztum, and D. Steinberg, *J. Clin. Invest.* **84**, 1086 (1989).

[4] J. A. Berliner, M. Navab, A. M. Fogelman, J. S. Frank, L. L. Demer, P. A. Edwards, A. D. Watson, and A. J. Lusis, *Circulation* **91**, 2488 (1995).

This is followed by a shift of one of the double bonds into a *trans* configuration, leading to conjugated diene formation (–C=C–C=C–). In the presence of oxygen, L· reacts with molecular oxygen (O_2) to form a peroxyl radical (LOO·), which can then reabstract a hydrogen atom from another molecule of PUFA to form lipid hydroperoxide (LOOH) and a new L·, initiating a chain reaction. Several catalysts catalyze the LDL autoxidation process *in vitro*, but the initiator *in vivo* is still unknown.

Various LDL oxidation pathways result in the production of several lipid peroxidation products, such as lipid hydroperoxides and aldehydes, oxysterols derivatives of esterified and unesterified cholesterol, hydroxy fatty acids, and isoprostanes. Analysis of the oxidative status of LDL, as well as the susceptibility to oxidation, can provide functional markers for the integrity of the lipoprotein. One single assay, however, is probably not sufficient to serve as a sole marker for lipid peroxidation in relation to atherosclerotic risk.

This article presents several procedures for the quantitative measurement of LDL lipid peroxidation products. Some of these compounds are not totally defined and are considered as indirect markers of oxidized LDL. These procedures include the formation of conjugated dienes (obtained from PUFA in the early stage of oxidation), the analysis of total lipid peroxides (PD), and the assay of aldehyde reactive products (the TBARS assay). Other more sensitive methods include analysis of specific constituents of Ox-LDL compounds with a known molecular structure, such as cholesterol oxides (oxysterols), cholesteryl linoleate hydroperoxide (CL-OOH), and hydroxide (CL-OH). In addition, hydroperoxide and hydroxide derivatives of linoleic acid (L-OOH and L-OH) and the prostaglandin-like structure, isoprostanes, are also specific markers for LDL oxidation. All of these markers result from the oxidation of PUFA, whereas two other markers of LDL oxidation are related to changes that occur in the protein moiety of the LDL particle, e.g., changes in the electrophoretic mobility of LDL, or the reaction of lipoprotein with specific antibodies against Ox-LDL. This latter method, together with the isoprostanes assay, is used in detecting the oxidative status of plasma or LDL *ex vivo*.

In plasma, LDL is protected from oxidation by potent antioxidants, whereas in the arterial wall the antioxidant environment is poor, and LDL can be oxidized easily by cells of the arterial wall.[5] Measurements of LDL susceptibility to oxidation are used in relation to cardiovascular diseases (hypercholesterolemia, diabetes, hypertension) and in evaluating the effects of dietary or pharmacological antioxidants on LDL oxidation. LDL is isolated from plasma by discontinuous density gradient ultracentrifugation,[6] and its susceptibility to oxidation is then tested under various inducers of oxidation, such as the free radical generator AAPH [2,2′-azobis

---

[5] A. Hatta and B. Frei, *J. Lipid Res.* **36**, 2383 (1995).
[6] M. Aviram, *Biochem. Med.* **30**, 111 (1983).

(2-amidinopropane) dihydrochloride], transition metal ions (Cu^{2+} or Fe^{2+}), and cultured arterial cells (macrophages, endothelial cells, smooth muscle cells).

The procedure for copper ion-induced LDL oxidation is as follows.[7]

a. Incubate human LDL (1 ml of 100 mg protein/liter, predialyzed overnight against EDTA-free phosphate-buffered saline, PBS) with $CuSO_4$ (1–10 μM) in PBS for 1–5 hr under air in a shaking water bath at 37°. If ethanol is used as a solvent for the antioxidant, its final concentration should not exceed 0.1%.

b. If the effect of a compound (e.g., antioxidant activity) is under investigation, then add the studied compound (usually 0–100 μM) in the appropriate solvent. The optimal incubation time varies for each marker, but in most cases about 1–2 hr of incubation with copper ions is required for the initiation of the oxidation process.

c. If a kinetic measurement is required, take samples from the incubation bath for analysis at regular times.

d. Test LDL samples for the markers described in the following section.

General Markers for LDL Oxidation

General markers, which are determined quantitatively in the measurement of LDL oxidation, are conjugated dienes (CD), lipid peroxides (PD), and thiobarbituric acid-reactive substances (TBARS).

Conjugated Dienes (CD) Assay

This method[7] allows a dynamic quantification of conjugated dienes (CD) formed as a result of LDL-associated PUFA oxidation by measuring UV absorbance at $\lambda = 234$ nm. The principle of this assay is that, during LDL oxidation, the C=C double bonds of PUFA are converted into conjugated double bonds, which are characterized by strong UV absorption at 234 nm. LDL (100 mg protein/liter) is placed in PBS and incubated in a spectrophotometer cuvette with 5 μM $CuSO_4$ at 25°. Absorbance is monitored at 10-min intervals. The initial background of the samples ranges between 0.1 and 0.2 OD. After initial absorbance has been recorded, the spectrophotometer is set to zero against the blank. A typical curve showing the lag phase required for the initiation of LDL oxidation, the propagation phase, and the decomposition phase is presented in Fig. 1.

Lipid Peroxides (PD) Assay

During LDL oxidation, polyunsaturated fatty acids, such as linoleic and arachidonic acids (assigned as LH), are oxidized in their allylic position in a nonspecific

---

[7] H. Esterbauer, G. Striegl, H. Puhl, and M. Rotheneder, *Free Radic. Res. Commun.* **6,** 67 (1989).

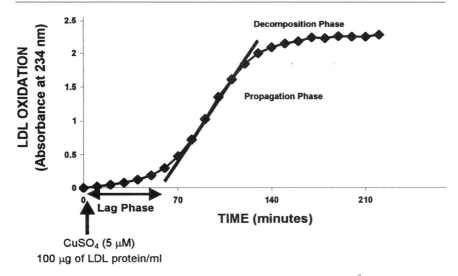

FIG. 1. Typical formation curve for conjugated dienes (H. Esterbauer et al.[7]). One hundred microgram of LDL protein/ml incubated with copper ion (5 μM, CuSO₄) and absorbance at 234 nm is measured at different time intervals. The curve shows the lag phase required for the initiation of LDL oxidation, the propagation phase, and the decomposition phase.

reaction to form an unstable mixture of lipid peroxides. The total amount of lipid peroxides can be detected iodometrically using the method originally developed by El-Saadani et al.[8] The amount of lipid peroxides accumulated in LDL during its oxidation reaches a maximum after about 5 hr incubation (600–1000 nmol of peroxides/mg LDL protein), thereafter declining (decomposition phase), yielding the formation of aldehydes. The procedure for the determination of lipid peroxides is as follows.

1. Add 1 ml color reagent [0.2 M potassium phosphate, pH 6.2, 0.12 M potassium iodide, 0.15 mM sodium azide, 2 g/liter Igepal CA-630 (Sigma, St. Louis, MO), 0.1 g/liter alkylbenzyldimethylammonium chloride and 10 μM ammonium molybdate] to 100 μl LDL (100 mg protein/liter).
2. Vortex the reaction mixture and place for 30 min in the dark at 25°.
3. Measure the absorbance at 365 nm against the color reagent as a blank.
4. Calculate lipid peroxide concentration, using a molar coefficient of I_3^- at 365 nm ($\varepsilon = 2.46 \pm 0.25 \times 10^4\ M^{-1}\ cm^{-1}$).

[8] M. El-Saadani, H. Esterbauer, M. El-Sayed, M. Goher, A. Y. Nassar, and G. Jurgens, J. Lipid Res. **30,** 627 (1989).

Thiobarbituric Acid Reactive Substances (TBARS) Assay

During lipoprotein oxidation, peroxides are formed, with subsequent formation of peroxyl radicals, followed by a decomposition phase to such aldehydes as hexanal, malondialdehyde (MDA), and 4-hydroxynonenal. Because MDA is a hydrophilic compound, most of it diffuses out of the LDL particle into the aqueous solution, unlike the lipophilic aldehydes, which remain associated with the LDL particle. The assay[9] is based on the detection of a stable product, which is formed between aldehydes and thiobarbituric acid (TBA) in the aqueous phase.

1. Dilute 100–500 μl of LDL (100 mg protein/liter) with water to a final volume of 500 μl. Add 1 ml TBA solution (0.375 g thiobarbituric acid, 2.5 ml concentrated HCl, 15 ml trichloroacetic acid. Fill with water up to 100 ml).
2. Vortex the mixture and heat for 20 min at 100°, followed by centrifugation (at 500g) for 10 min.
3. Measure the absorbance of the supernatant at 532 nm vs blank of TBA solution and water (500 μl) treated as the sample.

To quantify TBARS, prepare a standard curve with tetramethyl acetate (for *in situ* formation of MDA) in a range of 0.75–7.50 nmol and add to each tube 1 ml TBA solution. Repeat steps 2 and 3 and draw a calibration curve.

LDL-Associated Lipid Hydroperoxides and Hydroxides

Cholesteryl Linoleate Hydroperoxide (CL-OOH) and Cholesteryl Linoleate Hydroxide (CL-OH) Assay

Most of the cholesterol in the LDL particles is present in the ester form, predominantly as cholesteryl linoleate (CL). CL-OOH is formed as an early product during LDL oxidation. It readily decomposes into more stable compounds, but can still be measured directly, or after its reduction to CL-OH. Unlike lipid peroxides, which may be formed from any PUFA, the detection of CL-OOH is a direct measurement of a specific oxidative constituent in Ox-LDL. High-performance liquid chromatography (HPLC) analysis for CL-OOH and for CL-OH assays is as follows.[10]

1. Take 600 μl LDL (100 mg protein/liter), containing 20 μM butylated hydroxyanisole (BHA), to prevent an oxidation during the assay.

---

[9] J. A. Buege and S. D. Aust, *Methods Enzymol.* **52**, 302 (1978).
[10] P. A. Belinky, M. Aviram, B. Fuhrman, M. Rosenblat, and J. Vaya, *Atherosclerosis* **137**, 49 (1998).

2. Extract the LDL sample three times with hexane/2-propanol (3/2, v/v, 1.8 ml each), vortex, and collect the upper phase.

3. Evaporate the organic solvents under nitrogen and dissolve the residue in 200 µl acetone.

4. Inject the sample (loop of 20 µl) to HPLC equipped with a C_8 column (Merck, Germany, 25 cm length, 0.4 cm diameter, 5 µm particle size). The eluent, methanol/water, is used at a flow rate of 1 ml/min with a gradient of 70/30 (v/v) to 94/6 for 10 min. Increase the methanol fraction up to 97% for an additional 15 min. The detector is set at 234 nm for both CL-OOH (Retention time, $Rt = 28.2$ min) and CL-OH ($Rt = 28.8$ min). By using the same chromatographic conditions at 210 nm, cholesteryl esters of arachidonate, palmitate, oleate, linoleate ($Rt = 40$ min), and stearate can be separated and their amounts can be determined from their area under the peak against a calibration curve of appropriate standards.

5. Preparation of standards: CL-OOH is synthesized from CL according to a modification of the method of Browne and Armstrong[11] as follows: CL (15 mg in hexane) is incubated with 5 M tet-butyl hydroperoxide solution in decane (*t*-Bu-OOH, 3.5 ml, Aldrich, Milwaukee, WI) and ferrous sulfate (100 µM) for 5 hr at 37°. Chloroform/methanol is added (2/1, v/v, 10 ml), and the solution is washed with water (3 × 5 ml). The organic layer is separated, dried with Na_2SO_4, and evaporated. The quantity of CL-OOH is calculated from its absorption at 234 nm ($\varepsilon = 29,500 M^{-1}cm^{-1}$).[12] CL-OH is synthesized from CL-OOH (8 mg) in ethanol (1 ml) by reduction with $NaBH_4$ at 4° for 30 min. Ether (5 ml) and water (1 ml) containing 10 mg citric acid are added, and the organic phase is separated and washed twice with water (2 × 1 ml) prior to drying and evaporation.

6. The quantity of CL-OOH and CL-OH in LDL is calculated from the calibration curve of the synthetic standards.

Linoleyl Hydroperoxide (L-OOH) and Linoleyl Hydroxide (L-OH) Assay

During the initial stage of lipid peroxidation, mostly hydroperoxides and hydroxides of esterified fatty acids in LDL are formed, with minor amounts of free oxidized fatty acids. The position of the hydroperoxide and the hydroxide along the fatty acid chain depends on the type of the fatty acid and the nature of the oxidative stress. The major hydroperoxy and monohydroxy derivatives detected in free fatty acids are those of oleic, arachidonic, and especially linoleic acid. Various approaches have been suggested for the quantitative determination of L-OOH, particularly of monohydroxy linoleate, the most direct of which is the HPLC method.

---

[11] R. W. Browne and D. Armstrong, in "Free Radical and Antioxidant Protocols" (D. Armstrong, ed.), p. 139. Humana Press,Totowa, NJ, 1998.

[12] W. Sattler, D. Mohr, and R. Stocker, *Methods Enzymol.* **233,** 469 (1994).

This method[13] is specific and does not require derivatization, as does the more sensitive gas chromatography coupled to mass spectrometry (GC-MS) (esterification, silylation and, in some instances, also a reduction of the double bonds). The HPLC method is as follows.[13]

Linoleic acid (LA), L-OOH, and L-OH are determined by HPLC equipped with a C_{18} column (Merck, Germany, 25 cm length, 0.4 cm diameter, 5-μm particle size), with detection at 210 nm for LA and at 234 for both L-OOH and L-OH. The eluent that separates L-OH and L-OOH is a mixture of water/acetonitrile/tetrahydrofuran/acetic acid (40/40/20/0.25 v/v)[11] at a flow rate of 1 ml/min. Under the conditions described earlier, the retention time, for L-OOH and L-OH are 11.6 and 12.2 min, respectively. L-OOH and L-OH are quantified by calculation from a calibration curve of standards ($\varepsilon = 2.3 \times 10^4$ M^{-1} cm^{-1}, for both L-OOH and L-OH).

Preparation of Standards. Standards for L-OOH and L-OH are synthesized from LA by the same procedure as for the synthesis of CL-OOH and CL-OH.

LDL-Associated Oxysterol Assay

Oxysterols, the products of oxidized cholesterol, are present in human tissues and blood and are involved in the development of atherosclerosis, as well as in several other degenerative diseases.[14] Oxysterols originate from the diet and from enzymatic or nonenzymatic autoxidation of free or esterified cholesterol. They have been shown to affect cholesterol biosynthesis, plasma membrane structure, and cell proliferation. Oxysterols exert cytotoxicity on arterial cells, induce apoptosis, act as chemoattractants, and stimulate the formation of macrophage foam cells and advanced atherosclerotic lesions.[15] Hence, the type and level of oxysterols in the body can be used as markers for oxidative damage to lipoproteins[16,17] and to cell membranes.[18]

Several oxysterols have been detected in human tissues, plasma lipoproteins, and atherosclerotic lesions. Their type and amount vary widely, depending on the origin of the sample and the detection method used. Such methods include thin-layer chromatography (TLC), HPLC, liquid chromatography coupled to mass spectrometry (LC-MS), gas chromatography (GC), and GC-MS.[19]

---

[13] M. Aviram, E. Hardak, J. Vaya, S. Mahmood, S. Milo, A. Hoffman, and M. Rosenblat, *Circulation* **101,** 2510 (2000).

[14] F. Guardiola, R. Codony, P. B. Addis, M. Rafecas, and J. Boatella, *Food Chem. Toxicol.* **34,** 193 (1996).

[15] A. J. Brown and W. Jessup, *Atherosclerosis* **142,** 1 (1999).

[16] A. Sevanian, J. Berliner, and H. Peterson, *J. Lipid Res.* **32,** 147 (1991).

[17] S. K. Peng, G. A. Phillips, G. Z. Xia, and R. J. Morin, *Atherosclerosis* **64,** 1 (1987).

[18] A. Sevanian and L. L. McLeod, *Lipids* **22,** 627 (1987).

[19] J. Vaya, S. Mahmood, T. Hayek, E. Grenadir, S. Milo, A. Hoffman, and M. Aviram, *Free Radic. Res.* **34,** 850 (2001).

LDL Oxysterol Detection by GC-MS

1. Add 50 μl α-cholestane or 19-hydroxycholesterol (internal standard, 50 μg/ml) to 2 ml of native or oxidized LDL (500 mg LDL protein/liter) and extract the lipids with diethyl ether (3 × 4 ml). The diethyl ether should be filtered through aluminum oxide prior to use in order to eliminate possible peroxide contamination. Collect the upper organic layer and dry with Na_2SO_4.

2. Separate the ether solution by filtration and divide it into two parts: one for total oxysterols, as described later (from step 3), and one for free oxysterols (from step 4). Evaporate each part separately.

3. To determine total oxysterol, hydrolyze the first part by dissolving the sample in 2 ml ether and 2 ml 20% KOH in methanol (w/v) at 25° for 3 hr. Add 2 ml 25% citric acid in water (w/v) and collect the organic layer. Wash the aqueous layer (2 ml ether × 2), combine the organic layers, dry, and evaporate to dryness under nitrogen.

4. Silylate the samples to determine both total and free oxysterols with 200 μl N,O-bis(trimethylsilyl)acetamide in 1,4-dioxane (1/1, v/v) at 70° for 30 min. The dioxane should be first filtered through aluminum oxide to eliminate peroxide contamination.

5. Sample analysis is performed by GC fitted with a 30-m HP-5 trace analysis capillary column (0.32 mm i.d., 0.25 μm film thickness, 5% phenyl methyl silicon), coupled to a quadrupole mass spectrometer, and detected in single ion monitor (SIM) mode. The GC is operated in splitless mode for 0.8 min and then with a split ratio of 1:1. Helium is used as a carrier gas, at a flow rate of 0.656 ml/ min, under a pressure of 10.4 psi and a linear velocity of 31 cm/sec. The injector temperature is set to 300°, the detector at 330°, and the column is heated at a gradient, starting at 200°, increasing to 250° at 10°/min and then at 5°/min up to 300° and maintained for an additional 15 min at 300°. The MS transfer line is maintained at 280°.

6. Identification of oxysterols by GC-MS in samples is based on (a) the retention time of the peak in the chromatogram, (b) the most characteristic ions for each oxysterol, and (c) their relative abundance (which should be similar to the corresponding standard). The quantity of each oxysterol is calculated from the calibration curves of the standards (Steraloids Inc., Wilton, NH).

Under the conditions just given, the following oxysterols can be determined (numbers in parentheses indicate most characteristic ions for each oxysterol): α-cholestane (internal standard) (372, 357), 7α-hydroxycholesterol (546, 456), cholesterol (458, 368), 7β-hydroxycholesterol (546, 456), 4β-hydroxycholesterol (546, 474, 456, 384), 5α, 6α and 5β,6β-epoxycholesterol (546, 474, 456, 384), cholestane-3,5,6-triol (triol, after elimination of one silyl group) (546, 456, 403), 25-hydroxycholesterol (546, 472, 456, 382), 7-ketocholesterol (546, 472, 456, 382), and 27-hydroxycholesterol (546, 456, 417). Figure 2 shows a representative GC-MS profile of Ox-LDL oxysterols.

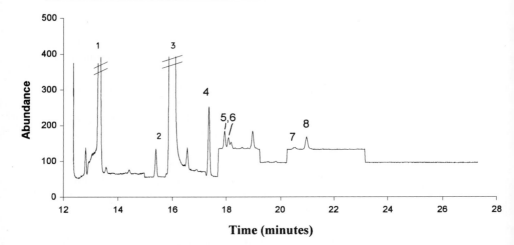

FIG. 2. GC-MS spectrum of oxidized LDL in a single-ion monitor mode. Peak 1, Cholestane (internal standard); peak 2, 7α-hydroxycholesterol; peak 3, cholesterol; peak 4, 7β-hydroxycholesterol; peak 5, 5β,6β-epoxycholesterol; peak 6, 5α,6α-epoxycholesterol; peak 7, 25-hydroxycholesterol; and peak 8, 7-ketocholesterol. In atherosclerotic lesions 27-hydroxycholesterol is also present at time ~22 min.

LDL-Associated Isoprostanes

Isoprostanes (IsoPs) are prostaglandin (PG)-like compounds formed during nonenzymatic oxidation by free radicals of phospholipids containing arachidonyl.[20] Arachidonic acid is a major fatty acid in the LDL phospholipid and cholesteryl ester moieties. Oxidation of LDL *in vitro*, using copper ions, endothelial cells, peroxynitrite, or peroxyl radicals, is followed by a rapid increase in the levels of a major group of IsoPs, which consist of the prostaglandin $F_{2\alpha}$ ($PGF_{2\alpha}$) isomers, referred to as F2-IsoPs. Their formation *in vivo* is not through the cyclooxygenase pathway, but rather via endoperoxide intermediates (Fig. 3), followed by phospholipase hydrolysis of arachidonate esters.[21] They are found in human fluids and tissues and also in atherosclerotic lesions[22] at levels exceeding by an order of magnitude that of PGs. Among the F2-IsoPs, 8-IsoP$_s$-F$_{2\alpha}$ has been shown to be a potent pulmonary and renal vasoconstrictor.

Two major approaches are available for the detection and quantification of IsoPs in human fluids, mainly in urine and plasma. The first method uses GC-MS with

[20] J. A. Lawson, J. Rokach, and G. A. FitzGerald, *J. Biol. Chem.* **274,** 24441 (1999).
[21] J. D. Morrow, K. E. Hill, R. F. Burk, T. M. Nammour, K. F. Badr, and L. J. Roberts, *Proc. Natl. Acad. Sci. U.S.A.* **87,** 9383 (1990).
[22] C. Gniwotta, J. D. Morrow, L. J. Roberts, and H. Kuhn, *Arterioscler. Thromb. Vasc. Biol.* **17,** 3236 (1997).

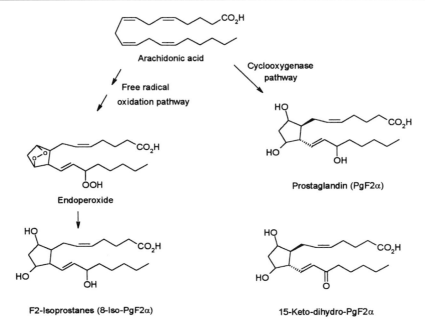

FIG. 3. General scheme for the formation of prostaglandin $PGF_{2\alpha}$ and F2-isoprostanes. The biosynthetic origin of both groups of compounds is arachidonic acid with the first one forming via the cyclooxygenase pathway and the second one via a free radical oxidation process and endoperoxide destruction.

negative ion chemical ionization,[23] and the second is through an immunological approach.[24] The GC-MS method is as follows.

1. Extraction: To prevent *ex vivo* oxidation during sample preparation, fresh samples or samples stored at $-70°$ under nitrogen are used. A biological sample (e.g., LDL) is extracted on ice with a mixture of chloroform/methanol (2/1, v/v) containing 0.005% butylated hydroxy toluene (BHT). The organic layer is separated by centrifugation and is then evaporated.

2. Hydrolysis: The residue in methanol (also containing BHT) is incubated in the presence of 15% aqueous KOH (1/1, v/v) at $37°$ for 30 min, followed by acidification and dilution to pH 3 with 1 M HCl. A deuterated internal standard (200–1000 pg) of $[^2H_4]PGF_{2\alpha}$ is added.

3. Purification: The hydrolyzate is introduced into a C_{18} Sep-Pak, preconditioned with methanol (5 ml), followed by acidic water (5 ml, pH 3.0), and the

---

[23] J. D. Morrow and L. J. Roberts, *Methods Enzymol.* **233**, 163 (1994).
[24] J. D. Morrow and L. J. Roberts, *Prog. Lipid Res.* **36**, 1 (1997).

elution is performed with acidic water (10 ml, pH 3.0), followed by acetonitrile/water (10 ml, 15/85, v/v) and finally by heptane (10 ml). IsoPs are collected with ethyl acetate/heptane (10 ml, 50/50, v/v). The ethyl acetate/heptane fraction is then applied to an additional silica Sep-Pak column, washed with ethyl acetate (5 ml), and the IsoPs are eluted with ethyl acetate/methanol (50/50, v/v), followed by evaporation of the solvent and collection of the residue.

4. Esterification: The just collected residue is esterified, using 10% (w/v) pentafluorobenzylbromide (PFB) (40 μl) in acetonitrile and 10% (w/v) N,N-diisopropylethylamine in acetonitrile (20 μl) with stirring for 30 min at 25°. The reagents are then evaporated under nitrogen.

5. Purification: The ester with PFB obtained during the previous stage is purified on silica TLC, using chloroform/ethanol (93/7, v/v) as a solvent. In parallel, a standard of $PGF_{2\alpha}$ methyl ester is run under identical conditions. The plates are sprayed with phosphomolybdic acid to visualize the TLC area to be scraped ($R_f \sim 0.15$). Ethyl acetate is used as the solvent for the extraction of the IsoPs scraped from the plate and is then evaporated under nitrogen.

6. Silylation of IsoPs–PFB ester: The residue from the previous stage is silylated with 20 μl N,O-bis(trimethylsilyl)trifluoroacetamide and 10 μl dimethylformamide (DMF) by incubation for 20 min at 40°.

7. GC-MS detection: The last sample is dissolved in undecane and injected to GC-negative ion chemical ionization mass spectroscopy (MS) for IsoPs quantification, using SIM mode to monitor the F2-IsoPs carboxylate anion.

Antibodies against Oxidized LDL in Plasma

Heterogeneous groups of antibodies are generated *in vivo* in human and animals in response to lipid peroxidation and can be detected in plasma from atherosclerotic patients. The exact nature of the various antigens formed in the LDL as a result of its modification and their biochemical effects is not fully known. A part of these antigens is formed as a result of the modification of LDL by aldehydes. High titers of autoantibodies in serum or plasma correlate with enhanced LDL oxidation *ex vivo*.

Such antibodies against Ox-LDL can thus serve as a marker for oxidized LDL formation.[25] The procedure for the quantification of Ox-LDL antibodies in plasma is as follows.[26,27]

The Ox-LDL antibodies are detected by an ELISA method, using disposable 96-23II, polystyrene plates (Corning). One milligram of native LDL protein/ml is

[25] J. L. Witztum, *Lancet* **344,** 793 (1994).
[26] J. T. Salonen, S. Yla-Herttuala, R. Yamamoto, S. Butler, H. Korpela, R. Salonen, K. Nyyssonen, W. Palinski, and J. L. Witztum, *Lancet* **339,** 883 (1992).
[27] E. Maggi, R. Chiesa, G. Melissano, R. Castellano, D. Astore, A. Grossi, G. Finardi, and G. Bellomo, *Arterioscler. Thromb.* **14,** 1892 (1994).

kept in PBS (10 mmol/liter) containing 1 mg/ml Na_2EDTA and is used as a control. Ox-LDL is obtained by oxidizing 1 mg of protein/ml LDL with 5 μmol/liter $CuSO_4$ for 18 hr at 30°.

1. Wells are coated with 10 μg Ox-LDL (antigen protein) or native LDL (control) in PBS.
2. Incubate the plates for 4 hr at 37° and then remove the coating solution.
3. Add to each well 250 μl blocking solution on PBS buffer (pH 7.4) containing 3% fetal bovine serum albumin (BSA) (to block any remaining binding sites), incubate for 2 hr at 37°, and remove the solution from the wells.
4. Dilute plasma samples with 220 μl PBS containing 1% BSA (in duplicate) to 1/11–1/21 volume ratio and incubate at 37° for 2 hr. Remove the solution and wash the plate wells four times with PBS buffer containing 21.2 g NaCl/liter and 5 mg/ml Tween 20.
5. Add 100 μl of peroxidase-conjugated antibody specific for IgG or IgM (diluted 1/2000) and incubate for 1 hr at 37°, followed by washing as in step 4.
6. Prepare substrate solution for developing peroxidase activity with phenylenediamine dihydrochloride and H_2O_2 as revealing reagents.
7. Measure the absorption at 492 nm in an automatic microplate reader and calculate antibody titers by means of the ratio between the spectrophotometric reading of antioxidized and antinative LDL wells.

LDL Electrophoretic Mobility

Lipoproteins can be separated according to their density (by ultracentrifugation), their size (by chromatography), or their charge (by electrophoresis). During LDL oxidation, physical, chemical, and biological properties of LDL are modified, and the lipoproteins become more electronegatively charged. It is evident that degradation products of LDL lipid peroxidation (aldehydes, isoprostanes) react with amino groups of lysine, histidine, and cysteine of the LDL apolipoprotein B-100 and with LDL phospholipid to form stable covalently bound adducts.[28,29] The result of these reactions is also an increase in the net negative surface charge of the LDL particle. Table I shows the electrophoretic mobility of oxidized LDL in comparison with native LDL and the effect of several antioxidants on copper ion-induced oxidation.

The electrophoretic mobility of oxidized LDL is compared to that of native LDL. LDL (100 μg of protein/ml) is incubated with 5 μM $CuSO_4$ for 4 hr at 37° in the presence of the compound under test. Electrophoresis of the lipoproteins is performed on 1% agarose by means of a Hydragel-Lipo kit (Sebia, Issy-les-Moulineaux, France), under constant voltage (50 V) and a starting current of

[28] K. Uchida and E. R. Stadtman, *J. Biol. Chem.* **268**, 6388 (1993).
[29] C. J. Brame, R. G. Salomon, J. D. Morrow, and L. J. Robert, *J. Biol. Chem.* **274**, 13139 (1999).

TABLE I
EFFECT OF DIETARY ANTIOXIDANTS ON COPPER ION-INDUCED
LDL OXIDATION AND LDL ELECTROPHORETIC MOBILITY[a]

Antioxidant	LDL oxidation	
	TBARS (nmol MDA/mg protein)	Electrophoretic mobility (cm)
Native LDL	1	0.4
Oxidized LDL	26	1.6
+ Vitamin E	9	0.9
+ Vitamin C	12	1.0
+ β-Carotene	18	1.3
+ Lycopene	16	1.2
+ Catechin	8	0.9
+ Quercetin	6	0.8
+ Glabridin	4	0.9

[a] LDL (100 mg protein/liter) was oxidized using copper ions ($5 \mu M$, $CuSO_4$) by incubation for 3 hr at 37° in the absence or presence of the various dietary antioxidants ($30 \mu M$).

22 mA, in Tris–barbital buffer (7.2 g/liter Tris; 1.84 g/liter barbital; 10.3 g/liter sodium barbital; 0.1 g/liter sodium azide). At the end of the run (90 min), and after drying the gel, the lipids are stained in 6% (w/w) Sudan black for 15 min.

[22] Antioxidant Activity of Hydroxycinnamic Acids on Human Low-Density Lipoprotein Oxidation

By ANNE S. MEYER and EDWIN N. FRANKEL

Hydroxycinnamic Acids and Their Conjugates

Hydroxycinnamic acids (phenylpropanoids) are ubiquitous compounds found in plant foods that may provide important antioxidant components. In fruits, vegetables, and whole grains, hydroxycinnamic acids occur mainly in various bound, conjugated, and esterified forms.[1] In cereals belonging to the family of *Gramineae* (wheat, rye, oat, barley, and corn), *trans*-ferulic acid (4-hydroxy-3-methoxycinnamic acid) and *trans-p*-coumaric acid (4-hydroxycinnamic acid) are

---

[1] K. Herrmann, *Crit. Rev. Food Sci. Nutr.* **28**, 315 (1989).

predominantly esterified to hemicellulosic components by covalent links to the arabinofuranose in the heteroxylans.[2] In fruits and vegetables, ferulic acid is linked to pectins by O-2 bonding to arabinose or by O-6 bonding to galactose,[3,4] and in spinach also by O-2 or O-3 bonding to arabinose–pyranose residues.[3] In both fruits and vegetables, hydroxycinnamates also occur either as free quinic acid esters or as glucose esters.[1,5]

Hydroxycinnamoylquinic acids, notably 5′-caffeoylquinic acid (chlorogenic acid), dominate in most pome and stone fruits, and in many berries.[1,5–7] In plums, 3′-caffeoylquinic acid (neochlorogenic acid) is the major hydroxycinnamic acid species,[1] and in cherries, 3-p-coumaryl quinic acid constitutes a high proportion of the hydroxycinnamates.[1,7] Citrus fruits contain conjugates of glucaric and galactaric acids,[5,7] whereas tartaric acid esters are the dominant hydroxycinnamates in grapes (*Vitis* species).[1,8,9]

Extraction and Purification of Conjugated Hydroxycinnamates

Compositional data for hydroxycinnamates depend very much on the quantitative extraction methods according to the types of phenolic acids and factors affecting recovery. In whole grains, fruits, and vegetables, the bound hydroxycinnamic acids are usually released by either alkali or acid treatment, but they may not be completely solubilized. However, yields may be compromised by excessively harsh sample treatment and extraction conditions. Recovery of phenolic compounds may be influenced in general by the type of solvent, and the solvent contact time. Fruit tissue contains native phenol oxidases (also referred to as "phenolases") that may promote oxidation of hydroxycinnamic acids to form undesirable browning materials. These reactions can be effectively retarded by excluding air, by rapid and cold sample homogenization, and by the use of reducing agents such as high SO_2 and ascorbic acid. Such precautions retain phenolic compounds in their reduced form during purification. However, to evaluate the antioxidant activity of hydroxycinnamates in fruit, vegetable, or grain extracts the use of reducing agents has confounding effects and must be avoided. The rate of oxidation of different hydroxycinnamic acids may vary greatly during sample preparation and extraction procedures. Efforts to obtain high yields of specific compounds may affect the actual composition of hydroxycinnamic acid mixtures in plants. Differences in variety, stage of maturity, and external and environmental factors may also confer

---

[2] I. Mueller-Harvey, R. D. Hartley, P. J. Harris, and E. H. Curzon, *Carbohydr. Res.* **148**, 71 (1986).
[3] S. C. Fry, *Biochem. J.* **203**, 493 (1982).
[4] F. M. Rombouts and J.-F. Thibault, *Carbohydr. Res.* **154**, 177 (1983).
[5] M. N. Clifford, *J. Sci. Food Agric.* **79**, 362 (1999).
[6] B. Schuster and K. Herrmann, *Phytochemistry* **24**, 2761 (1985).
[7] B. Risch and K. Herrmann, *Z. Lebensm. Unters.-Forsch.* **187**, 530 (1988).
[8] V. L. Singleton, C. F. Timberlake, and A. G. H. Lea, *J. Sci. Food Agric.* **29**, 403 (1978).
[9] V. L. Singleton, J. Zaya, E. Trousdale, and M. Salgues, *Vitis* **23**, 113 (1984).

significant quantitative variations in the content of hydroxycinnamic acids among fruits, vegetables, and grains.

No standard methods are available to extract and isolate hydroxycinnamoyl conjugates from plant material and foods. In structural investigations of plant cell walls,[2-4,10] various methods are employed to prepare hydroxycinnamic acid–polysaccharide fragments that contain feruloyl and coumaroyl groups. Although these methods may be used to isolate bound hydroxycinnamoyl groups, they have not yet been applied in antioxidant studies on human low-density lipoprotein (LDL) oxidation. Apart from the hydroxycinnamoyl quinic acids, chlorogenic[11,12] and neochlorogenic acid,[12] the only hydroxycinnamate conjugates tested for antioxidant activity toward LDL oxidation *in vitro* are the three dominant hydroxycinnamoyl tartrates from grapes[12] and two isolated ferulic acid sugar esters from corn bran cell wall fragments.[13,14]

Hydroxycinnamoyl Tartrates from Grapes

According to the original procedures of Singleton *et al.*,[8,15] a portion of freshly harvested grapes is mixed with potassium metabisulfite and ascorbic acid (∼2 g of each reducing agent/kg grapes), crushed, and pressed under a flowing stream of CO_2 (or N_2). The juice obtained is filtered, stirred in the presence of equal amounts of charcoal and diatomaceous earth, and filtered again. The cake is washed with deionized water and the hydroxycinnamates extracted with methanol containing 1% (v/v) acetic acid. The pooled methanolic extracts are then concentrated by rotary evaporation, and the hydroxycinnamoyl tartrates are either separated by chromatography on Sephadex LH-20,[8,15] or preferably purified directly by semipreparative high-performance liquid chromatography (HPLC).[12] The purified fractions are then stored in methanolic aliquots under N_2 in the dark at $-20°$. However, because these materials can deteriorate even under freezing storage, their spectral identity should be confirmed by HPLC-diode array detection before testing for antioxidant activity. The separation by HPLC is performed with a reversed-phase C_{18} column using a ternary acidic–acetonitrile solvent[16] or other solvent gradient systems.[17] The elution order is *trans*-caftaric acid, *trans-p*-coutaric acid, and *trans*-fertaric acid. The spectral properties are *trans*-caftaric acid, $\lambda_{max} = 198, 220, 244, 332$ nm, $\lambda_{min} = 268$ nm; *trans-p*-coutaric acid,

---

[10] M.-C. Ralet, J.-F. Thibault, C. B. Faulds, and G. Williamson, *Carbohydr. Res.* **263**, 227 (1994).

[11] J. A. N. Laranjinha, L. M. Almeida, and V. M. C. Madeira, *Biochem. Pharmacol.* **48**, 487 (1994).

[12] A. S. Meyer, J. L. Donovan, D. A. Pearson, A. L. Waterhouse, and E. N. Frankel, *J. Agric. Food Chem.* **46**, 1783 (1998).

[13] T. Ohta, T. Nakano, Y. Egashira, and H. Sanada, *Biosci. Biotech. Biochem.* **61**, 1942 (1997).

[14] T. Ohta, N. Semboku, A. Kuchii, Y. Egashira, and H. Sanada, *J. Agric. Food Chem.* **45**, 1644 (1997).

[15] V. L. Singleton, J. Zaya, and E. Trousdale, *Vitis* **25**, 107 (1986).

[16] R. M. Lamuela-Raventos and A. L. Waterhouse, *Am. J. Enol. Vit.* **45**, 1 (1994).

[17] A. L. Waterhouse, S. F. Price, and J. D. McCord, *Methods Enzymol.* **299**, 113 (1998).

$\lambda_{max} = 198, 214, 232, 316$ nm, $\lambda_{min} = 252$ nm; and *trans*-fertaric acid, $\lambda_{max} = 198, 220, 240, 330$ nm, $\lambda_{min} = 266$ nm.[12]

Feruloyl Glucosides from Corn Bran

To test the antioxidant activity of cereal cinnamate sugar esters on human LDL, Ohta *et al.*[13,14] isolated the feruloyl glucoside fragments from refined corn bran by hydrolysis with oxalic acid followed by chromatography on a Sephadex LH-20 column eluting with 50% (v/v) methanol and monitoring the absorption of fractions for ferulate conjugates at 320 nm.

Oxidation of Human Low-Density Lipoprotein

Oxidation of low-density lipoproteins is now recognized as a primary, critical event in the atherogenic process in humans.[18] The oxidative susceptibility of LDL is known to be dependent on several factors, including its fatty acid composition and the amount of incorporated lipophilic antioxidants.[19] Epidemiological evidence[20,21] and *in vitro* antioxidant studies[22] have led to the hypothesis that phenolic antioxidants, including flavonoids, hydroxycinnamic, and benzoic acids, present in plant foods and beverages may act *in vivo* to retard the oxidation of LDL into atherogenic particles.[21–24] Several hydroxycinnamic acids and their esterified derivatives occurring in fruits, vegetables, and cereals are very active in reducing LDL oxidation *in vitro*.[11–14,25–31] A wide array of different conditions have been employed to evaluate the antioxidant activity of hydroxycinnamates in inhibiting LDL oxidation *in vitro*. Tables I and II summarize the methodologies used to evaluate LDL oxidation and the order of antioxidant activity of hydroxycinnamates and their conjugates.

[18] D. Steinberg, S. Parthasarathy, T. E. Carew, J. C. Khoo, and J. L. Witzum, *N. Engl. J. Med.* **320,** 915 (1989).
[19] H. Esterbauer, J. Gebicki, H. Puhl, and G. Jürgens, *Free Radic. Biol. Med.* **13,** 341 (1992).
[20] S. Renaud and M. de Lorgeril, *Lancet* **339,** 1523 (1992).
[21] M. H. Criqui and B. L. Ringel, *Lancet* **344,** 1719 (1994).
[22] E. N. Frankel, J. Kanner, J. B. German, E. Parks, and J. E. Kinsella, *Lancet* **341,** 454 (1993).
[23] J. E. Kinsella, E. Frankel, B. German, and J. Kanner, *Food Technol.* **47,** 85 (1993).
[24] J. L. Witzum, *Lancet* **344,** 793 (1994).
[25] C. Castelluccio, G. Paganga, N. Melikian, G. P. Bolwell, J. Pridham, J. Sampson, and C. Rice-Evans, *FEBS Lett.* **368,** 188 (1995).
[26] C. Castelluccio, G. P. Bolwell, C. Gerrish, and C. Rice-Evans, *Biochem. J.* **316,** 691 (1996).
[27] M. Nardini, M. Dáquino, G. Tomassi, V. Gentili, M. D. Felice, and C. Scaccini, *Free Radic. Biol. Med.* **19,** 541 (1995).
[28] J. A. Vinson, Y. A. Dabbagh, M. M. Serry, and J. Jang, *J. Agric. Food Chem.* **43,** 2800 (1995).
[29] R. Abu-Amsha, K. D. Croft, I. B. Puddey, J. M. Proudfoot, and L. J. Beilin, *Clin. Sci.* **91,** 449 (1996).
[30] P. L. Teissedre, E. N. Frankel, A. L. Waterhouse, H. Peleg, and J. B. German, *J. Sci. Food Agric.* **70,** 55 (1996).
[31] F. Natella, M. Nardini, M. Di Felice, and C. Scaccini, *J. Agric. Food Chem.* **47,** 1453 (1999).

TABLE I
ANTIOXIDANT ACTIVITIES OF HYDROXYCINNAMIC ACIDS TOWARD HUMAN LOW-DENSITY LIPOPROTEIN OXIDATION in Vitro[a]

LDL concentration	Prooxidant	HCA solvent	HCA concentration	Oxidation measurement, reaction time, temperature	Trends of antioxidant activity	Ref.[f]
250 μg LDL protein/ml	10 μM Metmyoglobin	Warm water	0.1–100 μM	ApoB surface charge, TBARS (532 nm) 6 hr, 37°	Caffeic > ferulic > p-coumaric	1
200 μg LDL protein/ml	5 μM CuCl$_2$	Hot water	5, 100 μM	Conjugated dienes (234 nm) 4 hr, 37°	100 μM: Caffeic ~ ferulic > p-coumaric 5 μM: Caffeic > ferulic	2
70 μg LDL protein/ml[b]	25 μM Cu(H$_3$C$_2$O$_2$)$_2$	Purified water	1–10 μM[c]	Electrophoresis of apoB TBARS 6 hr, 37°	Caffeic > p-coumaric	3
1 mg LDL protein/ml	80 μM CuSO$_4$	DMSO	5 μM	Hexanal by SH-GC[d] 2 hr, 37°	Caffeic > sinapic	4
0.3 mM cholesterol	8 μM CuCl$_2$	Ethanol/water[e]	0.1–10 μM	Conjugated dienes (234 nm) 4 hr, 37°	Caffeic > p-coumaric	5
1 mg LDL protein/ml	80 μM CuSO$_4$	Warm water	5–20 μM	Hexanal by SH-GC 2 hr, 37°	Caffeic > ferulic > p-coumaric	6
200 μM LDL protein/ml	5 μM CuCl$_2$ 4 mM AAPH		5 μM	Conjugated dienes (234 nm) 4 hr, 37° Trp residue destruction	Caffeic ≥ sinapic > ferulic > p-coumaric	7

[a] In all reports, phosphate-buffered (10 mM, pH 7.4) saline (100 mM) was used. HCA, hydroxycinnamic acids; DMSO, dimethyl sulfoxide.
[b] Fraction contained LDL plus VLDL (20%, w/w).
[c] Phenols diluted according to the Folin method[8] with catechin as standard.
[d] Static headspace gas chromatography.
[e] Ethanol in the final mixture did not exceed 0.1% (v/v).
[f] Key to references: (1) C. Castelluccio, G. Paganga, N. Melikian, G. P. Bolwell, J. Pridham, J. Sampson, and C. Rice-Evans. FEBS Lett. **368**, 188 (1995); (2) M. Nardini, M. Dáquino, G. Tomassi, V. Gentili, M. D. Felice, and C. Scaccini. Free Radic. Biol. Med. **19**, 541 (1995); (3) J. A. Vinson, Y. A. Dabbagh, M. M. Serry, and J. Jang. J. Agric. Food Chem. **43**, 2800 (1995); (4) P. L. Teissedre, E. N. Frankel, A. L. Waterhouse, H. Peleg, and J. B. German, J. Sci. Food Agric. **70**, 55 (1996); (5) R. Abu-Amsha, K. D. Croft, I. B. Puddey, J. M. Proudfoot, and L. J. Beilin, Clin. Sci. **91**, 449 (1996); (6) A. S. Meyer, J. L. Donovan, D. A. Pearson, A. L. Waterhouse, and E. N. Frankel, J. Agric. Food Chem. **46**, 1783 (1998); (7) F. Natella, M. Nardini, M. Di Felice, and C. Scaccini, J. Agric. Food Chem. **47**, 1453 (1999) (8); V. L. Singleton and J. A. Rossi, Am. J. Enol. Vitic. **16**, 144 (1965).

TABLE II
ANTIOXIDANT ACTIVITIES OF HYDROXYCINNAMIC ACID CONJUGATES TOWARD HUMAN LOW-DENSITY LIPOPROTEIN (LDL) OXIDATION in Vitro[a]

LDL concentration	Prooxidant	HCA-C stock solvent	HCA-C concentration	Oxidation measurement, reaction time, temperature	Trends of antioxidant activity	Ref.[e]
30 μg LDL protein/ml	5 mM AAPH	Phosphate buffer	0.5–1.5 μM	O_2 consumption cis-Parinaric acid decay 35 min, 37°	Chlorogenic > caffeic	1
250 μg LDL protein/ml	10 μM Metmyoglobin	Warm water	0.1–100 μM	ApoB surface charge TBARS (532 nm) 6 hr, 37°	Chlorogenic ≈ caffeic	2
70 μg LDL protein/ml[b]	25 μM $Cu(H_3C_2O_2)_2$	Purified water	1–10 μM[c]	TBARS 6 hr, 37°	Caffeic > chlorogenic	3
100 μg LDL protein/ml	20 μM $CuSO_4$		10–40 μM	Conjugated dienes (234 nm) 2 hr, 37°	Feruloylarabinose > Feruloylarabinoxylan > ferulic acid	4
1 mg LDL protein/ml	80 μM $CuSO_4$	Warm water	5–20 μM	Hexanal by SH-GC[d] 2 hr, 37°	Caffeic ≈ caftaric > chlorogenic ≥ neochlorogenic > fertaric > ferulic	5

[a] In all reports, phosphate-buffered (10 mM, pH 7.4) saline (100 mM) was used. HCA-C, hydroxycinnamic acid conjugates.
[b] Fraction contained LDL plus VLDL (20%, w/w).
[c] Phenols diluted according to the Folin method[6] with catechin as standard.
[d] Static headspace gas chromatography.
[e] Key to references (1) J. A. N. Laranjinha, L. M. Almeida, and V. M. C. Madeira, Biochem. Pharmacol. **48**, 487 (1994); (2) C. Castelluccio, G. Paganga, N. Melikian, G. P. Bolwell, J. Pridham, J. Sampson, and C. Rice-Evans, FEBS Lett. **368**, 188 (1995); (3) J. A. Vinson, Y. A. Dabbagh, M. M. Serry, and J. Jang, J. Agric. Food Chem. **43**, 2800 (1995); (4) T. Ohta, N. Semboku, A. Kuchii, Y. Egashira, and H. Sanada, J. Agric. Food Chem. **45**, 1644 (1997); (5) A. S. Meyer, J. L. Donovan, D. A. Pearson, A. L. Waterhouse, and E. N. Frankel, J. Agric. Food Chem. **46**, 1783 (1998); (6) V. L. Singleton and J. A. Rossi, Am. J. Enol. Vitic. **16**, 144 (1965).

Preparation of Human Low-Density Lipoprotein

Human LDL is a lipoprotein particle isolated by ultracentrifugation within a density range of 1.019–1.063 g/ml.[19,32,33] Several procedures have been published to isolate LDL from plasma in the presence of 0.01 % (w/v) EDTA by sequential density ultracentrifugation for 2 to 24 hr or longer.[32,33] After recovery of the LDL fraction, EDTA is removed by dialysis for 18–24 hr in phosphate-buffered (10 mM, pH 7.4) saline (100 mM) purged with N_2.[33] The buffer is changed once or twice during dialysis. More rapid isolation methods have been used to prepare LDL for antioxidant evaluations,[34–36] but these procedures may not yield authentic LDL particles because the nature and composition of lipoprotein fractions may vary with different separation methods. To isolate LDL by affinity column chromatography for antioxidant activity studies is inexpedient because the fraction contains about 20% by weight of very low-density lipoproteins (VLDL) and would be expected to affect the results of antioxidant activity.[36]

Oxidation of Human LDL *in Vitro*

A number of agents have been used to oxidize LDL, including metal ions,[19,37] myoglobin,[38] peroxynitrite,[39] cultured vascular cells,[40] and macrophages.[41] In the studies of antioxidant activity of hydroxycinnamates toward human LDL, Cu^{2+}, metmyoglobin,[25] and AAPH [2,2′-azobis (2-amidinopropane) dihydrochloride][11,29] have been employed to induce oxidation (Tables I and II). AAPH is an artificial hydrophilic azo initiator that is widely used in quantitative studies of oxidation kinetics because it has the advantage of generating localized radicals at a constant rate.[42] However, oxidation initiated by azo compounds is not appropriate to model LDL oxidation in biological systems. Although the use of copper as an initiator for *in vivo* oxidation is debated, copper-induced LDL oxidation is the most intensively studied procedure *in vitro*. This oxidation requires both binding of

---

[32] R. J. Havel, H. A. Eder, and J. H. Bragdon, *J. Clin. Invest.* **34,** 1345 (1955).
[33] J. R. Orr, G. L. Adamson, and F. T. Lindgren, in "Analyses of Fats, Oils, and Lipoproteins" (E. G. Perkins, ed.), p. 524. American Oil Chemists' Society, Champaign, IL (1991).
[34] B. H. Chung, T. Wilkinson, J. C. Geer, and J. P. Segrest, *J. Lipid Res.* **21,** 284 (1980).
[35] K. D. Croft, J. Proudfoot, C. Moulton, and L. J. Beilin, *Eicosanoids* **4,** 75 (1991).
[36] J. A. Vinson and B. A. Hontz, *J. Agric. Food Chem.* **43,** 401 (1995).
[37] E. N. Frankel, J. B. German, and P. A. Davis, *Lipids* **27,** 1047 (1992).
[38] G. Dee, C. Rice-Evans, S. Obeisekera, S. Meraji, M. Jacobs, and K. Bruckdorfer, *FEBS Lett.* **294,** 38 (1991).
[39] V. Darley-Usmar, N. Hogg, N. O' Leary, M. Wilson, and S. Moncada, *Free Radic. Res. Commun.* **17,** 9 (1992).
[40] U. P. Steinbrecher, S. Parthasarathy, D. S. Leake, J. L. Witzum, and D. Steinberg, *Proc. Natl. Acad. Sci. U.S.A.* **81,** 3883 (1984).
[41] S. Parthasarathy, D. J. Printz, D. Boyd, L. Joy, and D. Steinberg, *Arteriosclerosis* **6,** 505 (1986).
[42] E. Niki, T. Saito, A. Kawakami, and Y. Kamiya, *J. Biol. Chem.* **259,** 4177 (1984).

Cu^{2+} ions by apolipoprotein B (apoB) and reduction of copper by LDL.[43] One study suggests that the lipid peroxidation of LDL is initiated by the oxidative degradation of tryptophan residues in apoB.[44] Other studies report binding of copper to histidine residues of apoB, where lipid peroxidation is initiated by the oxidation of histidine to 2-oxohistidine,[45–47] which has a lower affinity for copper than native histidine in apoB.[46,47] *In vitro* LDL oxidation catalyzed by Cu^{2+} is inhibited by ascorbic acid or dehydroascorbic acid and is accompanied by the formation of 2-oxohistidine and release of up to 70% bound copper from the LDL particle.[47]

Antioxidant Mechanisms

Phenolic compounds may act as antioxidants by several different mechanisms, including free radical scavenging, metal chelation, and protein binding.[12,30] In LDL oxidation induced by metmyoglobin, hydroxycinnamates have been suggested to partially work by reacting with ferryl myoglobin produced by oxidation from metmyoglobin.[26] Two other antioxidant mechanisms may be envisaged for the copper-mediated oxidation of LDL *in vitro:* (a) binding of phenols to specific amino acid residues (tryptophans and/or histidines) in apoB to block access to the copper catalyst and (b) limiting the copper binding to LDL by oxidation of histidine to 2-oxohistidine by phenolic radical species. The latter mechanism is based on the observations[47] that both ascorbic acid and dehydroascorbic acid retard copper-induced LDL oxidation *in vitro*.

Measurement of LDL Oxidation *in Vitro*

Several methods have been used to assess the extent of LDL oxidation *in vitro* (Tables I and II). Many widely employed methods, including measurement of thiobarbituric acid reactive substances (TBARS), are not sensitive and specific enough to accurately reflect oxidation of a complex biological particle like human LDL.[48,49] The use of more than one assay is recommended for lipid peroxidation because it is important to distinguish between the activity of antioxidants in inhibiting hydroperoxide formation as well as hydroperoxide decomposition.[49] Monitoring of the conjugated diene absorbance at 234 nm during oxidation to determine the lag phase[50] is useful for many antioxidant tests. Although this method

---

[43] M. Kuzuya, K. Yamada, T. Hayashi, C. Funaki, M. Naito, K. Asai, and F. Kuzuya, *Biochim. Biophys. Acta* **1123,** 334 (1992).
[44] A. Giessauf, E. Steiner, and H. Esterbauer, *Biochim. Biophys. Acta* **1256,** 221 (1995).
[45] P. Wagner and J. W. Heinecke, *Arterioscler. Thromb. Vasc. Biol.* **17,** 3338 (1997).
[46] K. Chen and B. Frei, *Redox Rep.* **3,** 175 (1997).
[47] K. L. Retsky, K. Chen, J. Zeind, and B. Frei, *Free Radic. Biol. Med.* **26,** 90 (1999).
[48] M. C. Gutteridge and B. Halliwell, *TIBS* **15,** 129 (1990).
[49] E. N. Frankel, *Methods Enzymol.* **299,** 190 (1998).
[50] G. Knipping, M. Rothender, G. Striegl, and H. Esterbauer, *J. Lipid Res.* **31,** 1965 (1990).

measures the formation of polyunsaturated hydroperoxides, it is their aldehyde decomposition products that are implicated in the oxidative modifications of apoB. Non-UV-absorbing oxidation products may also be formed that escape spectrophotometric measurement.[50] The determination of hexanal by rapid static headspace chromatography is a useful method for fast, quantitative assessment of LDL oxidation. Hexanal is one of the specific volatile products used as a marker of n-6 polyunsaturated lipids oxidation of LDL. The validity and technical details of this technique have been described previously.[37,49] To evaluate the antioxidant activity of hydroxycinnamates and other antioxidants by this method, we incubate 0.5 ml of freshly dialyzed LDL standardized to a protein concentration of 1.0 mg/ml with 10–80 μM $CuSO_4$ and 2.5–20 μM of the phenolic antioxidant. Hexanal produced during oxidation in special headspace bottles is determined after incubation for 2 hr at 37°. The results are then calculated as percentage inhibition of hexanal formation compared to a control.[12,22,30]

Antioxidant Activity of Hydroxycinnamates and Their Conjugates toward Human LDL

A wide variety of methods were employed to assess the antioxidant activity of hydroxycinnamates using different agents to induce LDL oxidation (Tables I and II). The reported differences in antioxidant activity among individual hydroxycinnamates clearly correlated with the hydroxylation and methylation patterns of the aromatic ring. Thus, the antioxidant efficiency of free hydroxycinnamates on *in vitro* human LDL oxidation decreased in the order of caffeic > sinapic > ferulic > *p*-coumaric acid (Table I). The presence of the *o*-dihydroxy group in the phenolic ring (as in caffeic acid) consistently enhanced the antioxidant activity of hydroxycinnamic acids toward human LDL oxidation *in vitro* (Table I). This structure–reactivity relationship is consistent with the radical-scavenging antioxidant mechanism recognized for flavonoids,[51] involving the ability to donate a hydrogen and the stabilization of the resulting antioxidant radical by electron delocalization. Presence of the *o*-dihydroxy substituents also permits metal chelation.[52] Although this trend in hydroxycinnamate antioxidant activity is also observed with lipid substrates other than LDL,[53,54] it is *not* manifest by the "total antioxidant capacity" TEAC (Trolox equivalent antioxidant activity) assay,[52] which measures the ability of a compound to scavenge the $ABTS^{\bullet+}$ radical cation in an aqueous system.[52] The activity order of hydroxycinnamates is reversed in the TEAC assay. The ranking of antioxidant activity is thus strongly dependent on the test system and on the substrate protected by the antioxidant.

[51] B. J. F. Hudson and J. I. Lewis, *Food Chem.* **10**, 47 (1983).
[52] C. A. Rice-Evans, N. J. Miller, and G. Paganga, *Free Radic. Biol. Med.* **20**, 933 (1996).
[53] M.-E. Cuvelier, H. Richard, and C. Berset, *Biosci. Biotech. Biochem.* **56**, 324 (1992).
[54] E. M. Marinova and N. V. Yanishlieva, *J. Am. Oil Chem. Soc.* **71**, 427 (1994).

Hydroxycinnamic acids may inhibit LDL oxidation by several different and complex mechanisms. In addition to free radical scavenging, metal chelation, and regeneration of endogenous LDL antioxidants, a unique antioxidant mechanism may involve binding to apoB of the LDL particle. Very little is known about the effect of esterification of hydroxycinnamic acids on antioxidant activity and mechanism. Ferulic acid sugar esters from corn bran were demonstrated to be better inhibitors of *in vitro* LDL oxidation than ferulic acid itself[13,14] (Table II). We have also observed that esterification enhances the antioxidant activity of both ferulic and *p*-coumaric acid[12] (Table II). On this basis, we proposed that the esterification of certain hydroxycinnamic acids to acid or sugar moieties may affect antioxidant activity by conferring differences in the ability to bind to apoB in the LDL particle and to block oxidation of tryptophan residues.[12] Esterification and conjugation of hydroxycinnamates can also induce differences in their solubility and phase distribution properties that may affect interfacial oxidation and antioxidation in human LDL.[55]

Conclusions

Hydroxycinnamic acids are universally present in plant foods and occur in relatively high amounts and in a wide range of esterified forms. Further work is needed to understand the effect of conjugation and conjugate type on the antioxidant activity of hydroxycinnamates in various lipoprotein substrates. The relevance of conjugation on antioxidant activity *in vivo* remains to be clarified. The potential interactions with other antioxidants and phytochemicals also need to be explored. The ranking of antioxidant activity *in vitro* is strongly dependent on the test system and on the substrate protected by the antioxidant. Although various metabolites of hydroxycinnamates and flavonoids are formed during absorption in animal and human *in vivo* studies,[56–58] and conjugated quercetin metabolites were potent antioxidants in inhibiting rat lipoprotein oxidation *in vitro*,[57] no information is available on the relative antioxidant activity toward LDL of possible metabolites of hydroxycinnamates. There is an urgent need to better interpret the potential antioxidant and antiatherogenic activities of hydroxycinnamic acids on human LDL oxidation on a molecular basis by using more specific methods to measure LDL oxidation. Standardization of the *in vitro* methodologies is required to relate *in vitro* results to possible *in vivo* effects.[59]

---

[55] E. N. Frankel, S.-W. Huang, J. Kanner, and B. German, *J. Agric. Food Chem.* **42**, 1054 (1994).

[56] A. N. Booth, O. H. Emerson, F. T. Jones, and F. De Eds, *J. Biol. Chem.* **229**, 51 (1957).

[57] C. Morand, V. Crespy, C. Manach, C. Besson, C. Demigne, and C. Remesy, *Am. J. Physiol.* **275**, R212 (1998).

[58] J. L. Donovan, J. R. Bell, S. Kasim-Karakas, J. B. German, R. L. Walzem, R. J. Hansen, and A. L. Waterhouse, *J. Nutr.* **129**, 1662 (1999).

[59] E. N. Frankel and A. S. Meyer, *J. Sci. Food Agric.* **80**, 1925 (2000).

[23] Rapid Screening Method for Relative Antioxidant Activities of Flavonoids and Phenolics

By ANANTH SEKHER PANNALA and CATHERINE RICE-EVANS

Introduction

Considerable interest has emerged over recent years in the potential for screening medicinal plants, herbs, fruit, and vegetable extracts and other phytochemical agents for their ability to function as antioxidants through their H-donating properties. There are several published methodologies for measuring antioxidant activity. Few of these show broad application to the main matrices of investigation, namely pure compounds, food extracts, and biological fluids, such as plasma. The major methods applied have been the Trolox equivalent antioxidant capacity (TEAC) assay[1–3] and the oxygen radical absorbing capacity (ORAC) assay,[4,5] both based on the hydrogen-donating potential of the compounds in question, and the ferric ion reducing antioxidant potential (FRAP).[6] All these assays are based on different chemistries but should, in principle, produce consistent results.

The general consensus is that the chemical designs of the assays are such that the different assays are more appropriate for different applications. The TEAC and ORAC assays are the most appropriate for studying the total antioxidant activities of extracts of plants and foods, and beverages. Although direct comparative assessments have yet to be reported, the findings so far suggest reasonable consistency.[7] The TEAC decolorization assay for the assessment of the antioxidant activity of pure compounds displays broad associations with H-donating properties and reduction potentials.[8] The ORAC and FRAP assays have been applied to the determination of the relative antioxidant potentials of body fluids, although the latter has more limitations than the former.[9]

Measurement of the antioxidant activities, or reducing properties, of flavonoids and phenolic acids over a time scale of minutes, as in the methods alluded to earlier,

---

[1] C. Rice-Evans, N. Miller, and G. Paganga, *Free Radic. Biol. Med.* **20,** 933 (1996).
[2] C. Rice-Evans, N. Miller, G. P. Bolwell, P. M. Bramley, and J. Pridham, *Free Radic. Res.* **22,** 375 (1995).
[3] R. Re, N. Pellegrini, A. Proteggente, A. Pannala, M. Yang, and C. Rice-Evans, *Free Radic. Biol. Med.* **26,** 1231 (1998).
[4] G. Cao, H. M. A. Lessio, and R. G. Cutler, *Free Radic. Biol. Med.* **14,** 303 (1993).
[5] G. Cao, C. P. Verdon, A. Wu, H. Wang, and R. L. Prior, *Clin. Chem.* **41,** 1738 (1995).
[6] I. F. Benzie and S. Strain, *Anal. Biochem.* **239,** 70 (1996).
[7] G. Paganga, N. Miller, and C. Rice-Evans, *Free Radic. Res.* **30,** 153 (1999).
[8] C. Rice-Evans, in "Flavonoids in Health and Disease" (C. Rice-Evans and L. Packer, eds.), p. 199. Dekker, New York (1998).
[9] G. Cao and R. L. Prior, *Clin. Chem.* **44,** 1309 (1998).

would give a result encapsulating both rapid and slower reacting components. The purpose of the rapid TEAC assay designed here is to differentiate between the more effective reducing compounds, defined by the structure of the phenolic compound and the position of the specific hydroxyl group within the structure, and those phenolics with slower reacting reducing components. A quick and convenient method has been devised for measuring the relative antioxidant activities of flavonoids and phenolic acids, rather than absolute values, to gain a marker for the relative potencies of pure compounds, as H-donating antioxidants. The method can also be applied to plant extracts and nutritional mixtures. Thus the fast reactions will be relevant rather than the reactivity of the compounds over longer time scales, during which slower reacting functional hydroxyl groups might also participate in the reaction.

Methodological Principles

The assay is carried out by interacting the phenolics or extracts of interest with a model stable-free radical derived from 2,2′-azinobis(3-ethylbenzothiazoline-6-sulfonic acid) (ABTS). The production of the radical cation is undertaken by preparing a stock solution of 7 mM ABTS in water. To this solution potassium persulfate (2.45 mM final concentration) is added and the solutions are allowed to react for a duration of 12 hr in the dark at ambient temperature. ABTS and potassium persulfate react with a stoichiometry of 1 : 0.5 leading to an incomplete oxidation of ABTS to generate ABTS$^{\bullet+}$. The radical thus generated is stable in the dark at room temperature for 2 days.[3] The final concentration of the ABTS$^{\bullet+}$ radical cation can be calculated using the molar extinction coefficient, $\varepsilon = 16000\ M^{-1}\ cm^{-1}$, at λ_{734}.[3]

The interaction between antioxidants and ABTS$^{\bullet+}$ is carried out by stop-flow kinetics (SFA20, Hi-Tech Scientific, Salisbury, UK). The SFA20 is a two-syringe stop-flow kinetics system capable of mixing two solutions in a spectrophotometer (Hewlett-Packard 8453, Waldbronn, Germany). The two syringes of the SFA20 are each filled with the antioxidant and the ABTS$^{\bullet+}$ solution. Equal volumes of the solutions are withdrawn into the driving syringes followed by rapid mixing of the solutions in the cuvette. Measurement of absorbance change is triggered as soon as the cuvette is filled, which triggers a signal to the spectrophotometer. The time taken for the cuvette to be filled for mixing is 8 msec. Absorbance changes are monitored every 0.1 sec for a duration of 3 sec.

Stock solutions of the antioxidants (1 mM) are prepared in ethanol and diluted subsequently to give initial concentrations of 2, 10, and 20 μM. These are further diluted twofold after mixing in the cuvette for reaction with the ABTS$^{\bullet+}$. Prior to testing with the antioxidants, a baseline is obtained by monitoring the change in absorbance in the reaction between ABTS$^{\bullet+}$ and ethanol for 3 sec. This reading is used as the basal value for calculating the antioxidant activity of the compounds.

Subsequently, three concentrations of the antioxidants (1, 5, and 10 μM, final concentrations) are assessed and applied to estimate the antioxidant activity of the compounds. The absorbance at 734 nm is plotted against time every 0.1 sec for 3 sec. The antioxidant activity is measured in comparison with Trolox, the water-soluble vitamin E analog, as standard. The Trolox equivalent antioxidant activity is defined as the concentration of Trolox required to scavenge the ABTS radical cation to an equivalent extent as a 1 mM concentration of the compound in question, or equivalent concentration of plant extract.

Results and Implications for Structure–Antioxidant Activity Relationships

The structures of the families of phenolic compounds studied, flavonol, flavone, hydroxycinnamate (or phenylpropanoid), anthocyanidin, and flavanol are shown in Fig. 1. The rapid rates of reaction with the ABTS radical cation are assessed by the spectrophotometric monitoring of the change in absorbance at 734 nm at time points up to 3 sec. All data are compared with the ABTS radical cation uninhibited and with the inhibitory effects of the Trolox standard.

The results of the TEAC values at 100 msec and 3 sec are shown in Table I. Under the time scale of these fast reaction conditions, there is negligible contribution to the antioxidant activity from the slower acting A-ring *meta*-hydroxyl groups as demonstrated from the reaction of chrysin, a flavone with an unsubstituted B ring. Table I demonstrates that the reactivity of 1 mM chrysin is equivalent to that of 80 μM Trolox, with the antioxidant activity of 1 mM Trolox being 1, by definition.

The compounds studied are tabulated according to three structural categories of the B ring that predominate in more common flavonoids or in the single ring of the hydroxycinnamates; catechol-rich (including trihydroxy) structures, methoxyl/monohydroxyl structures, and monohydroxyl groups. The catecholic *o*-dihydroxyphenolic structures, in general, are the most potent compounds from the various flavonoid classes tested, and virtually all show a reactivity with the ABTS radical cation greater than or equal to that of Trolox at 3 sec except for cyanidin and taxifolin. Most of the catechol-rich structures demonstrate their reactivity essentially within 0.1 sec: epigallocatechin, epigallocatechin gallate, luteolin, quercetin, caffeic acid, and gallic acid. Other catechol-containing phenolics continue an extended reaction to 3 sec such as taxifolin, epicatechin gallate, catechin, epicatechin, delphinidin, and cyanidin. The catechol group in the B ring of flavonoids is among the major structural considerations underlying antioxidant activity due to the favorable reduction potential.[10–12] Furthermore, phenolics

---

[10] S. V. Jovanovic, S. Steenken, Y. Hara, and M. G. Simic, *J. Chem. Soc. Perkin Trans.* **2**, 2497 (1996).
[11] S. V. Jovanovic, S. Steenken, M. Tosic, B. Marjanovic, and M. G. Simic, *J. Am. Chem. Soc.* **116** (1994).
[12] S. V. Jovanovic, S. Steenken, M. G. Simic, and Y. Hara, in "Flavonoids in Health and Disease" (C. Rice-Evans and L. Packer, eds.) p. 163. Dekker, New York (1998).

a

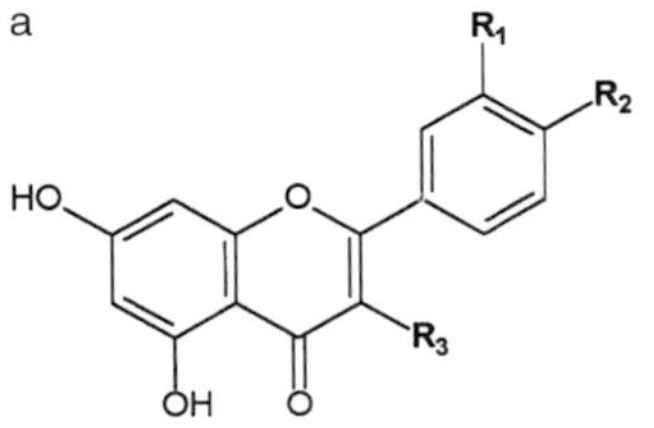

Flavonol and Flavone

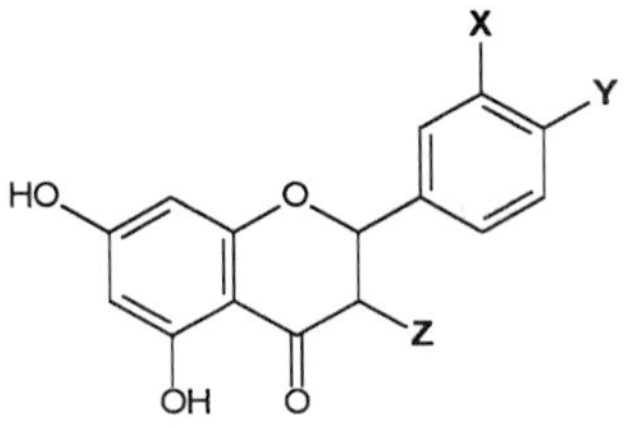

Flavanone

Flavonols	R_1	R_2	R_3
Quercetin	OH	OH	OH
Kaempferol	H	OH	OH
Galangin	H	H	OH
Flavones	R_1	R_2	R_3
Luteolin	OH	OH	H
Apigenin	H	OH	H
Chrysin	H	H	H
Flavanones	X	Y	Z
Taxifolin	OH	OH	OH
Hesperetin	OH	OCH_3	H
Naringenin	H	OH	H

b

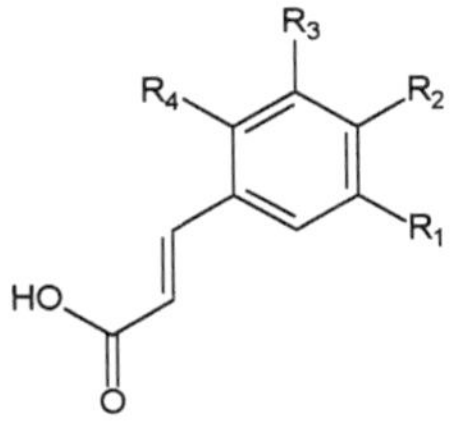

Phenylpropanoids

Phenylpropanoids	R_1	R_2	R_3	R_4
p-Coumaric acid	H	OH	H	H
o-Coumaric acid	H	H	H	OH
m-Coumaric acid	H	H	OH	H
Ferulic acid	H	OH	OCH_3	H
Caffeic acid	H	OH	OH	H
Sinapic acid	OCH_3	OH	OCH_3	H

FIG. 1. Chemical structures and locations of hydroxyl groups of (a) flavonols, flavones, and flavanones; (b) hydroxycinnamates; (c) anthocyanidins; and (d) flavanols and gallic acid.

c

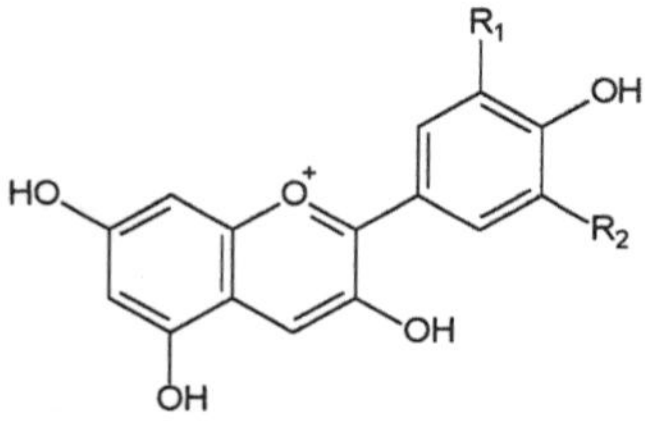

Anthocyanidin

Anthocyanidin	R_1	R_2
Delphinidin	OH	OH
Malvidin	OCH_3	OCH_3
Cyanidin	OH	H
Pelargonidin	H	H

d

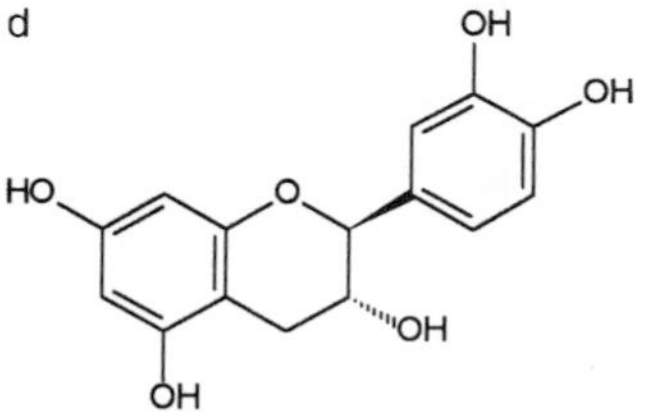

Catechin

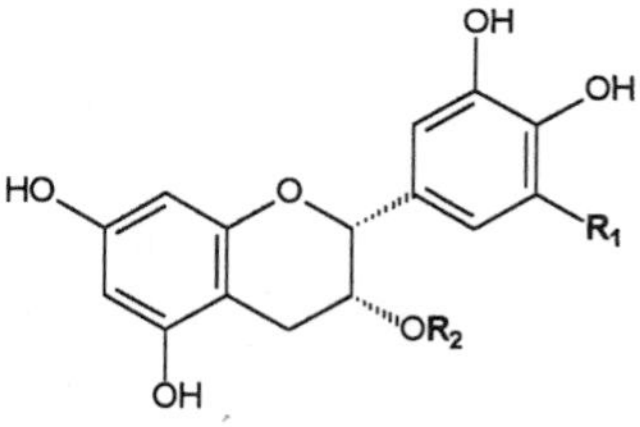

Other Flavanols

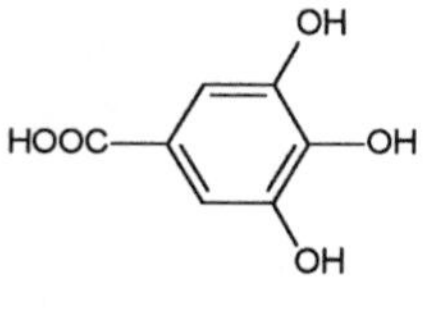

Gallic acid

Flavanols	R_1	R_2
Catechin	-	-
Epicatechin	H	H
EGC	OH	H
ECG	H	Gallate
EGCG	OH	Gallate

FIG. 1. (*continued*)

TABLE I
TEAC VALUES

Compounds	TEAC	
	0.1 sec	3 sec
Catechols (3',4'- or 3,4-dihydroxy structures)		
EGCG	2.070	2.239
EGC	1.770	1.882
ECG	1.640	2.022
Gallic acid	1.208	1.244
Delphinidin	1.157	1.624
Luteolin	1.044	1.008
Quercetin	1.007	1.211
Catechin	0.959	1.403
Epicatechin	0.825	1.439
Caffeic acid	0.817	0.917
Cyanidin	0.685	0.940
Taxifolin	0.650	0.951
3,4'-Dihydroxy structure		
Kaempferol	0.607	0.826
Methoxyl, monohydroxyl structures		
Sinapic acid	0.952	1.117
Ferulic acid	0.615	1.175
Malvidin	0.627	0.797
Hesperetin	0.179	0.448
Monohydroxyl group		
p-Coumaric acid	0.484	0.528
Pelargonidin	0.458	0.509
Apigenin	0.157	0.161
Naringenin	0.112	0.124
m-Coumaric acid	0.044	0.056
Unsubstituted B ring flavone for comparison		
Chrysin	0.080	0.089

containing three adjacent hydroxyl groups such as delphinidin and epigallocatechin are more effective than their dihydroxyl counterparts, cyanidin and epicatechin, respectively, illustrating more ready oxidation of these specific trihydroxy structures.

Compounds containing a 4'-monohydroxyl group on the B ring are less potent antioxidants, the mechanism of action probably being via the formation of a phenoxyl radical.[13] This is especially the case for phenolics in which there is no conjugation with the C ring, i.e., the C ring is either saturated, unsaturated at

---

[13] G. Galati, T. Chan, B. Wu, and P. J. O'Brien, *Chem. Res. Toxicol.* **12**, 521 (1999).

the 2,3 position but lacking a 4-carbonyl group, or for flavone phenolics lacking a hydroxyl group in the 3-position. Apigenin and naringenin have minimal reactivity possibly due to the relatively slow formation of the phenoxyl radical compared to the hindered phenols (see later) *m*-Coumaric acid demonstrates no activity as expected from phenolic hydroxyl groups in the *meta* arrangement. Perlargonidin and *p*-coumaric acid (4-hydroxycinnamic acid) are relatively more reactive, with about 50% of the value of Trolox at 100 msec and 3 sec. Kaempferol, a flavonol with a single 4'-hydroxyl group in the B ring, has a relatively high activity compared with other monohydroxyl compounds studied, possibly due to the potential for conjugation between the 4'-hydroxyl group and the 3-hydroxyl group in the C ring.

The presence of a hindered phenol on the B ring via the presence of a methoxyl group enhances the antioxidant activity or the H-donating properties in the 4- or 4'-position greatly. For example, the hydroxycinnamates, ferulic (3-methoxy-4-hydroxycinnamic) acid and sinapic (3,5-dimethoxy-4-hydroxycinnamic) acid, are approximately twice as effective in scavenging the ABTS radical cation in relation to the monophenolic structures, at 3 sec. The more hindered the phenol, the more rapid the reaction, as shown by sinapic acid and malvidin, in which the phenolic group is hindered by the presence of two methoxyl groups. This possibly increases the rate of the reaction and stabilizes the formation of the phenoxyl radical such that the reaction is almost complete at 100 msec. In contrast, ferulic acid shows a biphasic trend in scavenging the radical cation, a fast initial phase followed by a continued and extensive reaction between 100 msec and 3 sec. The low activity of hesperetin is due to the location of the hydroxyl group in the B ring at the 3'-position akin to a *m*-hydroxyphenolic structure, with little H-donating potential. In contrast, the 4'-position would show more rapid formation of the phenoxyl radical.

Clearly the overall rates of reaction of these types of phenolic compounds depend on the structure of the B ring, which varies between families, and the number and positioning of the hydroxyl groups. This method of analysis is fast and reproducible, and the results described here suggest that the rapid TEAC assay involving the model-free radical, the ABTS radical cation, can be applied usefully to screen compounds and plant extracts for indicators of their relative abilities to act as H-donating antioxidants.

Acknowledgment

Financial support from the Biotechnology and Biological Sciences Research Council for this research is acknowledged.

[24] Nitric Oxide Formation in Macrophages Detected by Spin Trapping with Iron–Dithiocarbamate Complex: Effect of Purified Flavonoids and Plant Extracts

By QIONG GUO, GERALD RIMBACH, and LESTER PACKER

Introduction

Activated macrophages can generate large amounts of nitric oxide (NO) from L-arginine by the action of inducible NO synthase (iNOS). NO is an important intracellular and intercellular regulatory molecule of multiple biological functions, including macrophage-mediated cytotoxicity, neurotransmission, and smooth muscle relaxation.[1,2] Overproduction of NO has been associated with oxidative stress and with the pathophysiology of various diseases such as arthritis, diabetes, stroke, septic shock, autoimmune disease, and chronic inflammation.[3] Cytokines such as interferon-r (IFN-γ), interleukin, (IL-1), or tumor necrosis factor α (TNF-α), and other inflammatory stimuli, including bacterial lipopolysaccharide (LPS), regulate the activity of iNOS in macrophages.[4,5]

Epidemiological reports have indicated that consumption of foods rich in flavonoids is associated with a lower incidence of degenerative diseases. Consistently, experimental data are accumulating regarding phenolic compounds as natural phytochemical antioxidants that possess anti-inflammatory, antiviral, antiproliferative, and anticarcinogenic properties.[6,7]

There is increasing interest in the biological activities of plant extracts such as that obtained from the bark of the French maritime pine (*Pinus maritima*) and *Ginkgo biloba*. Pine bark extract (pycnogenol, PBE) is a unique mixture of phenols and polyphenols, broadly divided into monomers (e.g., catechin, epicatechin and taxifolin), dimers (e.g., procyanidin B1, B2, B3, and B7), trimers (e.g., procyanidin C1, C2), and oligomers up to 5–7 units (Fig. 1). PBE also contains phenolic acids such as caffeic, ferulic, and *p*-coumatic acid as minor constituents.[8] Bioflavonoids derived from PBE participate in the antioxidant network[9] and likely spare endogenous vitamin E[10] and glutathione[11] in cultured cells. The standardized

---

[1] L. J. Ignarro, *Adv. Pharmacol.* **26**, 35 (1994).
[2] J. S. Beckman and W. H. Koppenol, *Am. J. Physiol.* **271**, C1424 (1996).
[3] S. Moncada, R. M. Palmer, and E. A. Higgs, *Pharmacol. Rev.* **43**, 109 (1991).
[4] D. J. Stuehr and M. A. Marletta, *J. Immunol.* **139**, 518 (1987).
[5] S. Narumi, J. H. Finke, and T. A. Hamilton, *J. Biol. Chem.* **265**, 7036 (1990).
[6] C. A. Rice-Evans, N. J. Miller, and G. Paganga, *Free Radic. Biol. Med.* **20**, 933 (1996).
[7] C. A. Rice-Evans and N. J. Miller, *Biochem. Soc. Transact.* **24**, 790 (1996).
[8] L. Packer, G. Rimbach, and F. Virgili, *Free Radic. Biol. Med.* **27**, 704 (1999).
[9] E. Cossins, R. Lee, and L. Packer, *Biochem. Mol. Biol. Int.* **45**, 583 (1998).

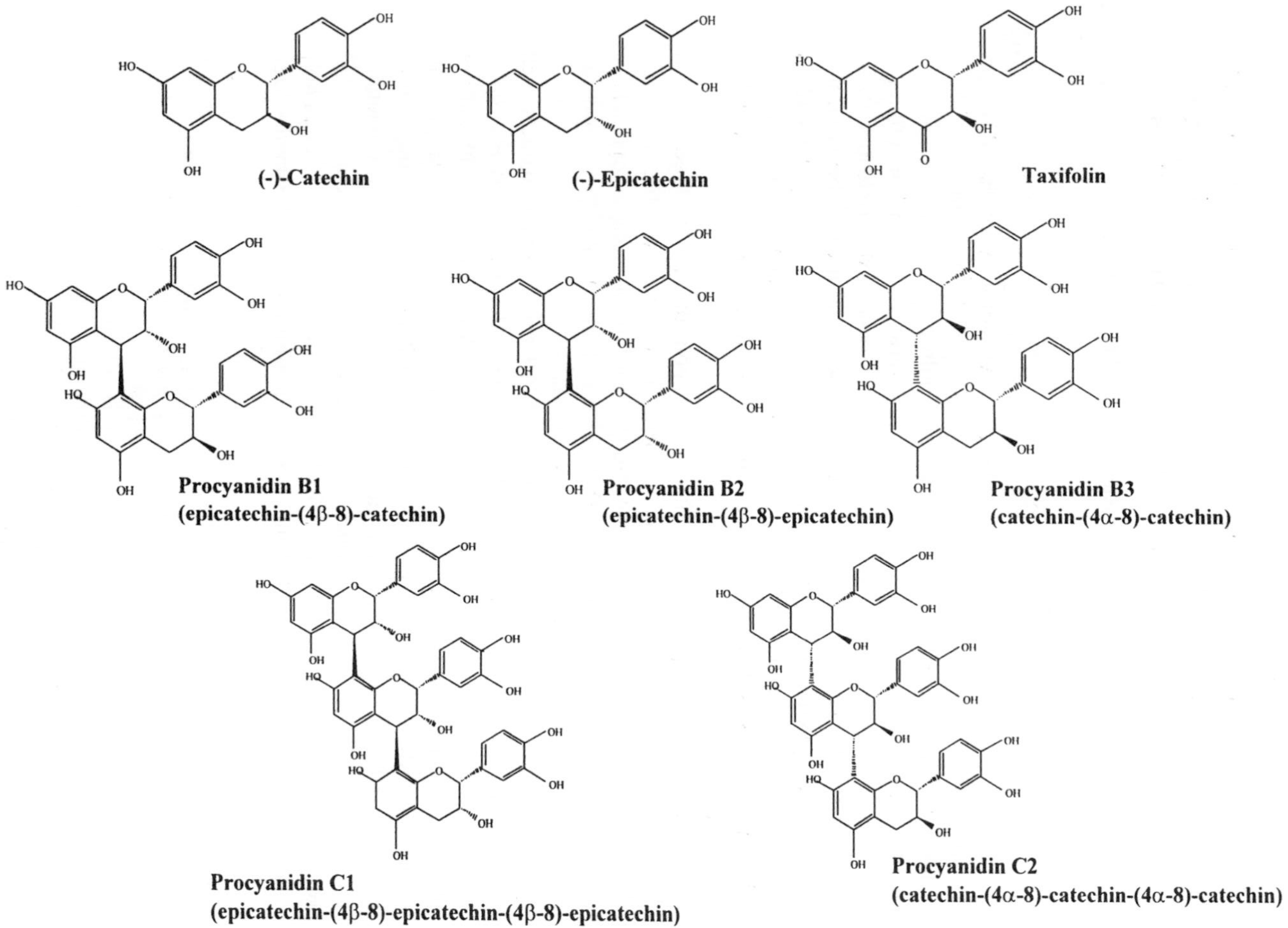

FIG. 1. Molecular structures of monomeric, dimeric, and trimeric flavonoids.

Gi. biloba extract EGb 761 contains 24% flavonoids (ginkgo-flavon glycosides) and 6% terponoids (ginkgolides, bilobalides) as active components. The flavonoid fraction is mainly composed of the favonols quercetin, kaemperol, isorhamnetpferol, and isorhamnetin, which are linked to a sugar molecule.[12] EGb 761 is commonly used in Europe to treat a variety of pathological conditions such as peripheral arterial diseases and organic brain syndromes. Both plant extracts PBE and EGb 761 have been shown to have strong scavenging activity in terms of hydroxyl and superoxide anion radicals and may significantly contribute to low-density lipoprotein (LDL) protection from oxidation.[13,14] Beneficial effects of PBE and EGb 761 may also be mediated by suppressing NO production.

Methods for Nitric Oxide Measurements

Several methods for the quantitation of NO such as chemiluminescence,[15] methemoglobin formation[16] and gas chromatography–mass spectrometry (GC-MS) detection[17] are available. Furthermore, dissolved NO is detectable by potentiometric measurement using a gas-specific electrode.[18,19] Nitrite is monitored most widely with the Griess reaction assay[20] in which NO_2^- is chemically transformed to a colored diazo compound, followed by quantitation with optical spectroscopy. However, the Griess reagent is highly acidic, and addition of the reagent to intact cells decreases their viability rapidly.[21]

In order to investigate interactions between flavonoids and NO in biological systems, sensitive methods for NO detection are necessary. Because NO is paramagnetic, it reacts to form high-affinity nitroso complexes with a variety of metal complexes, such as diethyl dithiocarbamate (DETC) and *N*-methyl-D-glucamine dithiocarbamate (MGD). The distinctive electron paramagnetic resonance (EPR) spectra of these nitroso complexes permit quantitative measurements of NO generation.[22] The disadvantage of DETC is a low water solubility of the

---

[10] F. Virgili, D. Kim, and L. Packer, *FEBS Lett.* **431**, 315 (1998).
[11] G. Rimbach, F. Virgili, Y. C. Park, and L. Packer, *Redox. Rep.* **4**, 171 (1999).
[12] L. Marcocci, J. J. Maguire, M. T. Droylefaix, and L. Packer, *Biochem. Biophys. Res. Commun.* **201**, 748 (1994).
[13] D. F. Fitzpatrick, B. Bing, and P. Rohdewald, *J. Cardiovasc. Pharmacol.* **32**, 509 (1998).
[14] L. J. Yan, M. T. Droy-Lefaix, and L. Packer, *Biochem. Biophys. Res. Commun.* **212**, 360 (1995).
[15] M. J. Downes, M. W. Edwards, T. S. Elsey, and C. L. Walters, *Analyst* **101**, 742 (1976).
[16] M. Kelm and J. Schrader, *Circ. Res.* **66**, 1561 (1990).
[17] R. M. Palmer, A. G. Ferrige, and S. Moncada, *S. Nature* **327**, 524 (1987).
[18] Z. Taha, F. Kiechle, and T. Malinski, *Biochem. Biophys. Res. Commun.* **188**, 734 (1992).
[19] T. Malinski and Z. Taha, *Nature* **358**, 676 (1992).
[20] L. C. Green, D. A. Wagner, J. Glogowski, P. L. Skipper, J. S. Wishnok, and S. R. Tannenbaum, *Anal. Biochem.* **126**, 131 (1982).
[21] Y. Kotake, T. Tanigawa, M. Tanigawa, I. Ueno, D. R. Allen, and C. S. Lai, *Biochim. Biophys. Acta.* **1289**, 362 (1996).
[22] A. M. Komarov and C. S. Lai, *Biochim. Biophys. Acta.* **1272**, 29 (1995).

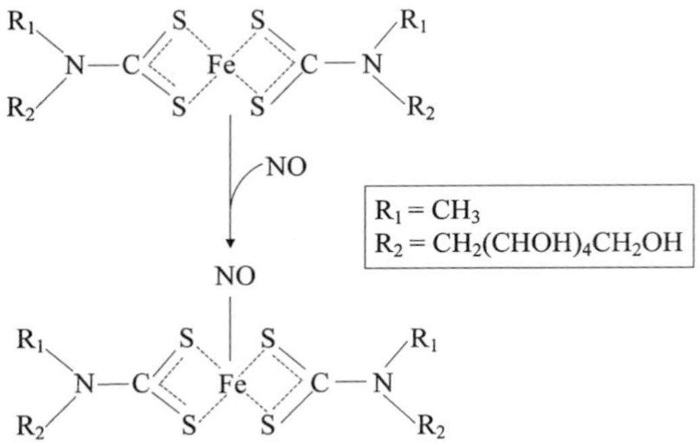

FIG. 2. Chemical structures of *N*-methyl-D-glucamine dithiocarbamate–Fe(II) complex and its NO spin adduct.[26] Modified according to Tsuchiya *et al.*[26]

Fe^{2+}–$(DETC)_2$ complex. Furthermore, excess DETC usually applied to biological systems for the effective trapping of NO inactivates various enzymes involved in NO metabolism.[23,24] The metal–chelator complex consisting of MGD and reduced iron has been successfully established to overcome several drawbacks of DETC.[25] Iron–dithiocarbamate complex spin trap agent [$(MGD)_2$-Fe^{2+}] forms a stable and water-soluble complex with NO^{26} (Fig. 2), which can be detected by EPR in intact cells at ambient temperature. In contrast to DETC, MGD affects neither iNOS enzyme activity and expression nor NF-κB activation in murine macrophages.[27]

In this article, an EPR spin-trapping method,[21,24,25] using [$(MGD)_2$-Fe^{2+}] as a spin trap, is described to continuously detect NO radical formation in stimulated macrophages. The effects of supplementing RAW 264.7 cells with purified flavonoids or plant extracts such as PBE and EGb 761 on real time NO radical formation are also described.

Procedure for Nitric Oxide Detection in Macrophages

Cell Culture

The murine cell line of monocyte macrophages RAW 264.7 (American Type Culture Collection, Rockville, MD) is used to monitor NO radical formation in real time. Cells are grown in 175-cm^2 Corning flasks at 37° and 5% (v/v) CO_2 in RPMI

---

[23] S. V. Paschenko, V. V. Khramtsov, M. P. Skatchkov, V. F. Plyusnin, and E. Bassenge, *Biochem. Biophys. Res. Commun.* **225,** 577 (1996).
[24] S. W. Norby, J. A. Weyhenmeyer, and R. B. Clarkson, *Free Radic. Biol. Med.* **22,** 1 (1997).
[25] A. Komarov, D. Mattson, M. M. Jones, P. K. Singh, and C. S. Lai, *Biochem. Biophys. Res. Commun.* **195,** 1191 (1993).
[26] K. Tsuchiya, M. Yoshizumi, H. Houchi, and R. P. Mason, *J. Biol. Chem.* **275,** 1551 (2000).
[27] Y. Kotake, T. Tanigawa, M. Tanigawa, I. Ueno, D. R. Allen, and C. S. Lai, *Biochim. Biophys. Acta* **1289,** 362 (1996).

1640 medium (GIBCO-BRL Company, Gaithersburg, MD) supplemented with 10% fetal calf serum (University of California, San Francisco Cell Culture Facility, San Francisco, CA), 1% (w/v) penicillin–streptomycin, and 2 mM L-glutamine.

Activation and Treatment of Macrophages

Confluent cells are seeded into 100-mm^2 tissue culture dishes and incubated overnight to allow for adherence. To detect maximal NO production from activated macrophages, RAW 264.7 cells are stimulated with 10 μg/ml LPS (Sigma, St. Louis, MO) and 50 units/ml IFN-γ (Pharmingen, San Diego, CA) for a specific duration from 4 to 10 hr.

In order to measure modulation of NO production by L-NMMA, L-lysine, or SOD (superoxide dismutase), cells are pretreated for 1 hr with indicated concentrations of L-NMMA (Sigma) or L-lysine (Sigma) prior to a 6-hr stimulation with 10 μg/ml LPS and 50 units/ml IFN-γ. Activated cells are then washed twice with Dulbecco's phosphate-buffered saline (DPBS) (GIBCO-BRL Company), harvested, centrifuged, and resuspended in DPBS containing indicated concentrations of L-NMMA, L-lysine, or SOD (Sigma) to a concentration of 4.0×10^7 cells per milliliter.

In order to measure the effect of plant extracts and purified flavonoids on NO production before a 6-hr stimulation with 10 μg/ml LPS and 50 units/ml IFN-γ, cells are pretreated for 1 hr with indicated concentrations of EGb 761 (IPSEN, Paris, France), PBE (Horphag Research Ltd., Guernsey, France), its monomers (taxifolin, catechin, or epicatechin, Sigma), dimers (B1, B2, or B3; Tokyo Research Laboratories, Kyowa Hakko Kogyo, Machida, Japan), or trimers (C1 or C2; Tokyo Research Laboratories). All test compounds are freshly prepared in sodium phosphate buffer (50 mM, pH7.4) containing 0.2% (v/v) dimethyl sulfoxide (DMSO). Activated cells are washed twice with DPBS, harvested, centrifuged, and resuspended in DPBS to a concentration of 4.0×10^7 cells per milliliter.

Preparation of Spin Trap [(MGD)$_2$-Fe^{2+}]

Iron–dithiocarbamate complex spin trap agent [(MGD)$_2$-Fe^{2+}] for trapping NO is prepared by reacting 25 mM MGD (Polyscience Inc., Warrington, PA) with 5 mM FeSO$_4$. The FeSO$_4 \cdot$7H$_2$O solution is freshly prepared in distilled water for each experiment.

EPR Detection of NO

Collected cells are incubated in the presence of the spin trap complex [(MGD)$_2$-Fe^{2+}] with 0.7 mM L-arginine or various concentrations of L-arginine (at 0–3.5 mM to test L-arginine-dependent NO formation). Each sample contains 2.4×10^6 cells in a final volume of 75 μl. The reaction mixtures are incubated in a water bath at 37° for 90 min, and the entire samples are loaded into a quartz capillary and fitted into a quartz glass flat cell for EPR measurement. EPR spectra are recorded using an IBM ER 200D-SRC EPR spectroscopy (Danbury, CT). EPR spectrometer settings: central field 3420 G; modulation frequency 100 kHz; modulation amplitude 3.2 G; microwave power 20 mW; scan width 200 G; gain 6.3×10^5; and temperature 298 K. The relative EPR signal intensity (the height of the first peak in the EPR spectrum of the [(MGD)$_2$-Fe^{2+}-NO] spin adduct) represents the amount of NO production.

Typical Results

Time Kinetics and L-Arginine Dependency of NO Radical Formation in Stimulated RAW 264.7 Macrophages

Macrophages activated with LPS and IFN-γ in the presence of 0.7 mM L-arginine produced a characteristic three-line ESR spectrum of the [(MGD)$_2$-Fe^{2+}-NO] spin adduct (Fig. 3). The spectral parameters of this EPR signal

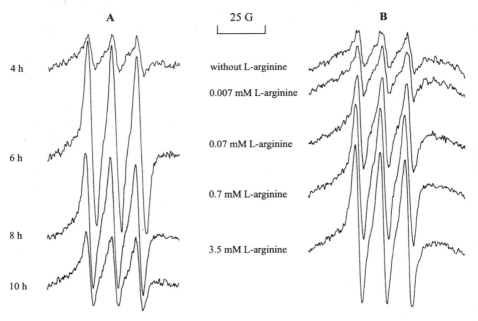

FIG. 3. EPR detection of NO radical formation in stimulated macrophages by the spin trap agent [(MGD)$_2$-Fe^{2+}]. (A) Effect of stimulation time on NO production. RAW 264.7 macrophages were stimulated with 50 U/ml IFN-γ and 10 μg/ml LPS for various time intervals ranging from 4 to 10 hr. The stimulated cells were harvested in DPBS and were then incubated with 0.7 mM L-arginine in the presence of the spin trap agent [(MGD)$_2$-Fe^{2+}] for 90 min. (B) Dose-dependent effect of L-arginine on NO production. Stimulated cells were harvested in DPBS and incubated with increasing concentrations of L-arginine (0–3.5 mM) in the presence of [(MGD)$_2$-Fe^{2+}] for 90 min. The incubation mixtures were then loaded for EPR measurements.

($g = 2.04$, $a^N = 12.5G$) are identical to those reported in the literature when an aqueous solution of authentic NO was added to a solution containing the complex of iron and MGD.[24] Macrophages produced detectable amounts of NO after 4 hr of activation, as indicated by the relatively weak ESR signal (Fig. 3A). The level of NO production peaked at 6 hr and decreased thereafter. In addition, L-arginine, the iNOS substrate, increased NO radical formation in a dose-dependent manner. Concentrations of L-arginine higher than 0.7 mM resulted in only a minor further increase in NO production (Fig. 3B).

Modulation of NO Radical Formation by L-NMMA, L-Lysine, and Superoxide Dismutase

The treatment of macrophages with L-NMMA, the competitive inhibitor of iNOS, dose dependently decreased NO radical formation (Table I). Similarly, L-lysine, the competitive inhibitor of L-arginine transport, reduced the production of NO in RAW 264.7 cells. However, as compared to L-NMMA, at least a 10-fold higher concentration of L-lysine was necessary in order to obtain a similar but significant inhibition of NO production. These results further confirmed that NO production in activated macrophages was via the L-arginine–iNOS pathway.

TABLE I
NITRIC OXIDE PRODUCTION IN ACTIVATED MACROPHAGES VIA
L-ARGININE-INDUCIBLE NO SYNTHASE PATHWAY[a]

Treatment	EPR signal intensity (% of control[b])
Control	100 ± 6.1
+L-NMMA	
0.05 mM	63.3 ± 5.3[c]
0.2 mM	35.5 ± 5.1[c]
+L-lysine	
0.5 mM	76.1 ± 8.2[c]
2.5 mM	39.4 ± 2.2[c]
+SOD	
50 U/ml	124.4 ± 3.5[c]
100 U/ml	144.9 ± 3.5[c]

[a] RAW 264.7 macrophages were pretreated for 1 hr with L-NMMA or L-lysine prior to a 6-hr stimulation with 50 U/ml INF-γ and 10 μg/ml LPS. Activated cells were harvested in DPBS containing indicated concentrations of L-NMMA, L-lysine, or SOD and were then incubated with 0.7 mM L-arginine in the presence of the spin trap agent [(MGD)$_2$-Fe^{2+}] for 90 min. The incubation mixtures were then loaded for EPR measurements.

[b] Values are means ±SD.

[c] $P < 0.05$ compared to control.

In contrast to L-NMMA and L-lysine, treatment of activated macrophages with 100 U/ml SOD increased the NO radical formation up to 40%, indicating that scavenging of superoxide anion radicals effectively prevented the formation of peroxynitrite, thereby leading to a higher NO signal intensity.

Effects of Plant Extracts and Purified Flavonoids on Real Time NO Radical Formation

Flavonoids may affect NO production by different mechanisms such as NO radical scavenging, modulating of iNOS enzyme activity, and gene expression, as well as protein-binding properties. Therefore we sought to investigate whether complex mixtures of flavonoids such as PBE or EGb 761, as well as the purified constituents of PBE, might have any effect on real time NO radical formation in macrophages. Figure 4 shows the effect of PBE on NO radical formation in RAW 264.7 cells stimulated with LPS and IFN-γ. Pretreatment of murine macrophages with PBE was associated with a significant reduction of the generation of NO radicals starting from a concentration of 5 μg/ml in the medium. The inhibition of NO radical formation by PBE was dose dependent, and at a concentration of 100 μg/ml PBE, about 90% of NO production was inhibited as compared to the control. The efficacy of EGb 761 on cellular NO radical formation is given in Fig. 4.

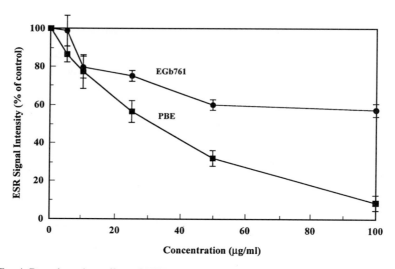

FIG. 4. Dose-dependent effect of PBE and EGb 761 on NO radical formation in stimulated macrophages. RAW 264.7 macrophages were pretreated for 1 hr with various concentrations of PBE or EGb 761 prior to a 6-hr stimulation. Activated cells were harvested in DPBS and were then incubated with 0.7 mM L-arginine in the presence of [(MGD)$_2$-Fe^{2+}] for 90 min. The incubation mixtures were then loaded for EPR measurements.

When macrophages were pretreated with increasing concentrations of EGb 761, a significant suppression in NO radical formation was evident. Up to 50 μg/ml the decrease in NO radical formation due to EGb 761 was dose dependent. However, a higher concentration of EGb 761 did not result in a further suppression of NO production in stimulated macrophages.

Because PBE displayed very strong properties in suppression of NO radical formation, it was of particular interest to investigate which constituent(s) of this complex mixture of flavonoids accounts for its antioxidant effects. Accordingly, the effect of various purified monomeric, dimeric, and trimeric flavonoids on cellular NO radical formation in RAW 264.7 cells was elucidated. The medium was supplemented with the test compound at a concentration of 100 μg/ml, whereas taxifolin, catechin, or epicatechin was added at a concentration of 50 μg/ml. At these concentrations, no cytotoxicity, as measured by the neutral red assay, was evident (data not shown), indicating that the modulation of NO radical formation by the flavonoids was not due to cell death. As shown in Fig. 5, NO radical formation remained largely unaffected by pretreating murine macrophages with the monomeric flavonoids catechin and epicatechin, whereas the addition of 50 μg/ml

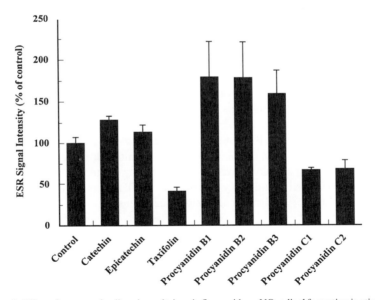

FIG. 5. Effect of monomeric, dimeric, and trimeric flavonoids on NO radical formation in stimulated macrophages. RAW 264.7 macrophages were pretreated for 1 hr with indicated concentrations of monomeric, dimeric, or trimeric flavonoids prior to a 6-hr stimulation. Activated cells were harvested in DPBS and were then incubated with 0.7 mM L-arginine in the presence of [(MGD)$_2$-Fe^{2+}] for 90 min. The incubation mixtures were then loaded for EPR measurements.

taxifolin to the medium significantly decreased NO production in RAW 264.7 cells by about 60%. Interestingly, the dimeric procyanidins B1, B2, and B3 increased NO radical formation up to 80%. However, the pretreatment of macrophages with the trimeric procyanidins C1 and C2 suppressed NO radical signal intensity by about 40 and 50%, respectively.

The EPR spin-trapping technique present herein can monitor real time NO radical formation in murine macrophages stimulated with LPS and IFN-γ using $[(MGD)_2\text{-}Fe^{2+}]$ as a spin-trapping agent. The rate of NO generation peaks 6 hr after stimulation. NO radicals produced by RAW 264.7 macrophages derive from the L-arginine-inducible nitric oxide synthase pathway. Purified flavonoids, as well as plant extracts such as obtained from the bark of *P. maritima* (pycnogenol) and *G. biloba* leaves (EGb 761), act as modulators on NO radical formation in macrophages.

[25] Redox Cycles of Caffeic Acid with α-Tocopherol and Ascorbate

By João Laranjinha

Introduction

Caffeic acid is a phenolic compound from the group of hydroxycinnamates that is derived biosynthetically from phenylalanine in plants.[1] It is found in plant-derived foodstuffs, including fruits, vegetables, flowers, nuts, and seeds, as well as in wine, tea, coffee, and olive oil.[2,3] Data from epidemiological, biochemical, and biological studies suggest the potential beneficial role of caffeic acid as an antioxidant in the prevention of coronary heart disease and cancer. Recent reports also point to the ability of caffeic acid to interfere with intracellular signaling pathways triggered by oxidized low-density lipoproteins (LDL) through mechanisms that are not necessarily dependent on antioxidant activities.[4]

The structure of caffeic acid (Fig. 1) is predictive of its efficient interaction with several types of oxidant radicals. Caffeic acid retains the structural principles established by Bors *et al.*[5] for optimal radical scavenging by flavonoids, namely

---

[1] , "Phytochemical Dictionary: A Handbook of Bioactive Compounds from Plants" (J. B. Harborne and H. Baxter eds.), p. 472. Taylor & Francis, London, 1993.

[2] K. Herrmann, *Crit. Rev. Food Sci. Nutr.* **28,** 315 (1989).

[3] J. J. Macheix, A. Fleuriet, and J. Billot, *in* "Fruit Phenolics." CRC Press, Boca Raton, FL, 1990.

[4] O. Vieira, I. Escargueil-Blanc, O. Meilhac, J. P. Basile, J. Laranjinha, L. Almeida, R. Salvayre, and A. Negre-Salvayre, *Br. J. Pharmacol.* **123,** 565 (1998).

[5] W. Bors, W. Heller, C. Mitchel, and M. Saran, *Methods Enzymol.* **186,** 343 (1990).

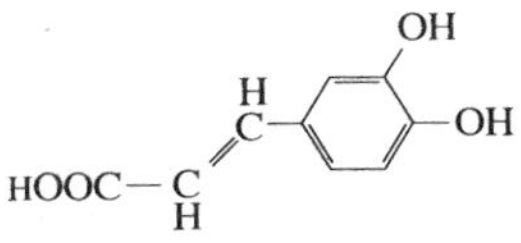

FIG. 1. Structure of caffeic acid.

an *o*-dihydroxy (catechol) structure in conjugation with a lateral double bond and an oxo function. The *o*-dihydroxy structure is typically the radical target site, producing transiently the *o*-semiquinone radical after one-electron oxidation. The lateral double bond conjugated with the catechol ring proportionates an extensive electron delocalization, increasing the stability of the *o*-semiquinone radical and, consequently, antioxidant activity.

The characterization of the antioxidant activity of caffeic acid has been carried out on several systems under diverse oxidant conditions.[6–17] The nature of the interactions of caffeic acid with other antioxidants is fundamental for understanding the potential effects of this phenolic compound *in vivo*.

This article describes the strategy and methods used to study the redox cycles of caffeic acid with α-tocopherol and ascorbate in relation to the inhibition of lipid peroxidation in LDL.

Experimental Procedures

Strategy

The overall strategy to investigate the redox cycles of caffeic with α-tocopherol and ascorbate and the potential concerted antioxidant activity in the protection of LDL from lipid oxidation includes two approaches. First, chemical studies on the interaction of antioxidants in solution and micelles using absorption spectroscopy and separation of products by high-performance liquid chromatography (HPLC)

---

[6] O. P. Sharma, *Biochem. Pharmacol.* **25**, 1811 (1976).
[7] H. Iwahashi, T. Ishii, R. Sugata, and R. Kido, *Arch. Biochem. Biophys.* **276**, 242 (1990).
[8] H. Chimi, J. Cillard, P. Cillard, and M. Rahmani, *J. Am. Oil Chem. Soc.* **68**, 307 (1991).
[9] M. E. Cuvelier, H. Richard, and C. Berset, *Biosc. Biotech. Biochem.* **56**, 324 (1992).
[10] J. Laranjinha, L. Almeida, and V. Madeira, *Arch. Biochem. Biophys.* **297**, 147 (1992).
[11] J. Terao, H. Karasawa, H. Arai, A. Nagao, T. Suzuki, and K. Takama, *Biosc. Biotech. Biochem.* **57**, 1204 (1993).
[12] J. Laranjinha, O. Vieira, L. Almeida, and V. Madeira, *Biochem. Pharmacol.* **48**, 487 (1994).
[13] C. A. Rice-Evans, N. J. Miller, and G. Paganga, *Free Radic. Biol. Med.* **20**, 933 (1996).
[14] J. Laranjinha, O. Vieira, L. Almeida, and V. Madeira, *Biochem. Pharmacol.* **51**, 395 (1996).
[15] M. Nardini, P. Pisu, V. Gentili, F. Natella, M. Di Felici, E. Picconella, and C. Scaccini, *Free Radic. Biol. Med.* **25**, 1098 (1998).
[16] O. Vieira, J. Laranjinha, V. Madeira, and L. Almeida, *Biochem. Pharmacol.* **55**, 333 (1998).
[17] T. Niwa, U. Doi, Y. Kato, and T. Osawa, *FEBS Lett.* **459**, 43 (1999).

in order to follow stable reactants and products. Electron paramagnetic resonance (EPR) spectroscopy, either direct or continuous flow, is used to detect the transient radical intermediates. Second, evaluation of the effect of mixtures of the antioxidants on the inhibition of LDL against lipid peroxidation in terms of potentiation of individual effects. LDL oxidation is followed by measuring the consumption of O_2 and the accumulation of conjugated diene hydroperoxides as well as by the quantitation of cholesteryl linoleate hydroperoxide and 7-ketocholesterol after separation by HPLC. Oxidation reactions are initiated by ferrylmyoglobin, the two-electron oxidation product of metmyoglobin,[18] by peroxyl radicals derived from thermal decomposition of 2,2'-azobis(2-amidinopropane) hydrochloride (AAPH) and by UV irradiation.

Isolation and Treatment of LDL

LDL particles are isolated from fresh human plasma by density gradient ultracentrifugation and simultaneously concentrated and dialyzed at 4° for 45 min by ultrafiltration under N_2 atmosphere as described in a rapid two-step method.[19] This methodology also removes contaminating low molecular weight plasmatic antioxidants, such as uric acid and ascorbate. Ferrylmyoglobin is prepared by mixing metmyoglobin and H_2O_2 according to standard procedures.[18] Horse heart metmyoglobin is dialyzed against phosphate buffer (20 mM phosphate, 110 mM NaCl), pH 7.4, containing 50 μM DTPA (referred to as phosphate buffer) and Chelex 100 (to improve removal of contaminating iron). Stock metmyoglobin and hydrogen peroxide solutions are standardized using $\varepsilon_{632\,nm} = 2.1$ mM^{-1} cm^{-1} and $\varepsilon_{240\,nm} = 43.6\,M^{-1}$ cm^{-1}, respectively. Solutions are prepared in water purified in a Milli-Q apparatus.

The rate of oxygen consumption is followed at 37° with a Clark-type oxygen electrode (YSI Model 5331, Yellow Springs Inst.). The accumulation of conjugated diene hydroperoxides is followed in the UV at 233 nm according to the procedure described by Esterbauer *et al.*[20] Unless otherwise stated, the standard mixture for oxidation measurements consists of 1 ml phosphate buffer containing 180 μg LDL protein, 6 μM metmyoglobin, and 9 μM H_2O_2 to initiate the reactions.

Absorption Spectroscopy

Absorption spectra during the reaction of caffeic and ascorbate with ferrylmyoglobin are acquired with a Perkin-Elmer (Norwalk, CT) lambda 6 spectrophotometer at 37°. The standard reaction mixture consists of 2 ml phosphate buffer

---

[18] C. Giulivi and E. Cadenas, *Methods Enzymol.* **233**, 189 (1994).
[19] O. Vieira, J. Laranjinha, V. Madeira, and L. Almeida, *J. Lipid Res.* **37**, 2715 (1998).
[20] H. Esterbauer, M. Dieber-Rotheneder, G. Striegl, and G. Waeg, *Am. J. Clin. Nutr.* **53**, 314S (1991).

containing 10 μM metmyoglobin and 15 μM H_2O_2 to initiate the reactions. The concentrations of caffeic acid and ascorbate are 20 and 40 μM, respectively. Metmyoglobin and ferrylmyoglobin exhibit similar spectra within the wavelength range used. Therefore, the spectrum of myoglobin is subtracted before the addition of caffeic acid and ascorbate.

Electron Paramagnetic Resonance

Spectra are recorded using a Bruker ECS 106 or a Bruker EMX spectrometer at room temperature. In continuous-flow experiments, an open Pasteur pipette directly connected to a 1-ml mixing cell is used as the cell cavity, and fluxes are typically 15 ml/min. The solutions are partially saturated with He, and the concentrations of α-tocopherol, caffeic acid, and ascorbate are maintained at 1 mM; those of metmyoglobin and H_2O_2 are maintained at 50 and 60 μM, respectively. For direct EPR experiments, the samples are transferred to bottom-sealed Pasteur pipettes and inserted in the EPR cavity for measurements. Typically, the instruments settings are microwave frequency, 9.8 GHz; microwave power, 20 mW; modulation frequency, 100 kHz; modulation amplitude, 2 G; and time constant, 0.65 sec.

HPLC

For the analysis of α-tocopherol, the LDL solution (180 μg LDL protein/ml) in phosphate buffer is incubated with 6 μM of metmyoglobin and 9 μM of H_2O_2 at 37° under gentle stirring in the absence and presence of caffeic acid. Along the time, aliquots of 1 ml are withdrawn chilled in ice, spiked with 100 μg of BHT, and extracted immediately according to the hexane/SDS method.[21] The presence of BHT (a phenol compound) does not affect qualitatively the results and allows reproducible data, avoiding ongoing LDL oxidation and perturbations from the presence of contaminating heme during extraction procedures. The hexane extracts are analyzed using a Beckman, System Gold on a LiChrospher 100 RP-18 (5 μm) column (Merck, Darmstadt, Germany) eluted (1.5 ml/min) with a solvent mixture consisting of 65% methanol and 35% ethanol 2-propanol (95 : 5) and UV detection at 292 nm. Typically, initial levels of α-tocopherol in LDL range from 9 to 14 nmol/mg LDL protein.

When evaluating the effect of ascorbate in the time course of caffeic acid oxidation, 1 ml of medium containing 50 μM DTPA, 20 μM caffeic acid, and 10 μM metmyoglobin in phosphate buffer, pH 7.4, at 37° is supplemented with 15 μM H_2O_2 to initiate the reaction. Then 40 μM ascorbate is added at different times to the reaction mixture followed by immediate treatment with 20 μl $HClO_4$ (85%) and centrifugation at 10,000g. The supernatant is analyzed using the column

---

[21] G. W. Burton, A. Webb, and K. U. Ingold, *Lipids* **20,** 29 (1985).

described earlier eluted with a mixture of 2% acetic acid/25% 2-propanol using a flow of 1 ml/min. Detection is carried out at 320 nm.

Under these conditions, no vestigial contamination of the supernatant with heme is noticed. Cholesteryl linoleate hydroperoxide and 7-ketocholesterol are determined essentially as described by Kritharides et al.[22] One-milliliter aliquots (200 µg of LDL protein) are withdrawn along the incubation time, chilled in ice, are spiked with 20 µM BHT, and lipids are extracted with methanol and n-hexane. The hexane extracts are then analyzed using the column described earlier eluted with acetonitrile/2-propanol/water (44/54/2, v/v/v) at a flow rate of 1 ml/min. Analysis of these compounds is performed by UV detection at 234 nm.

CL-OOH standard is prepared photochemically by irradiation of cholesteryl linoleate in solution in the presence of tetrasulfonated chloraluminum phthalocyanine[23] and quantified by the peroxide-dependent oxidation of iodide to iodine[24] using the molar absorptivity of iodine measured at 365 nm ($\varepsilon = 2.46 \pm 0.25 \times 10^4 \, M^{-1} \, cm^{-1}$).

Reduction of α-Tocopheroxyl Radical by Caffeic Acid

The reduction of α-tocopheroxyl radical (α-TO·) by caffeic acid (CAF-OH) is described by the following reaction:

$$\alpha\text{-TO·} + \text{CAF-OH} \rightarrow \alpha\text{-TOH} + \text{CAF-O·} \tag{1}$$

If reaction (1) occurs, then caffeic acid should prevent the oxidation of α-tocopherol to an extent dependent on the concentration. This has been observed in LDL and also in α-tocopherol containing Triton X-100 micelles.[25] In the presence of phenolic acid, the rate of α-tocopherol depletion in LDL oxidized by ferrylmyoglobin decreases with increasing concentrations of caffeic acid in the medium. In a typical oxidation experiment, α-tocopherol is exhausted at the 15th min of oxidation, but in the presence of 2, 6, and 18 µM of caffeic acid, respectively, 16, 68, and 91% of initial concentration is still present in LDL after 45 min of incubation. Moreover, when added during the course of α-tocopherol oxidation, caffeic acid partially restores α-tocopherol levels in a concentration-dependent way. For example, when α-tocopherol had been depleted to 30% of initial concentration, the addition of 2 and 6 µM caffeic acid resulted in a transient increase in α-tocopherol to 55 and 65% of initial content, respectively.

---

[22] L. Kritharides, W. Jessup, J. Gifford, and T. A. Dean, *Anal. Biochem.* **213,** 79 (1993).
[23] J. Thomas, B. Kalyanaraman, and A. Girotti, *Arch. Biochem. Biophys.* **315,** 244 (1994).
[24] M. El-Saadani, H. Esterbauer, M. El-Sayed, M. Goher, A. Nassar, and G. J. Jurgens, *J. Lipid Res.* **30,** 627 (1989).
[25] J. Laranjinha, O. Vieira, V. Madeira, and L. Almeida, *Arch. Biochem. Biophys.* **323,** 373 (1995).

TABLE I
INHIBITION OF FERRYLMYOGLOBIN-INDUCED OXIDATION OF
LDL BY α-TOCOPHEROL AND CAFFEIC ACID

Sample[a]	Lag phase (min)	Rate of O_2 consumption during lag phase (nmol O_2/min)
nLDL	10	1.65
nLDL + caffeic acid	104	0.52
eLDL	110	0.62
eLDL + caffeic acid	613	0.12

[a] nLDL and eLDL contain 12 and 52 nmol α-tocopherol/mg protein, respectively.

Similar to the redox cycling between ascorbate and α-tocopherol,[26–28] the mixtures of α-tocopherol and caffeic acid induce synergistic protective effects of LDL against oxidation. The experimental design to evaluate this potential effect involved the preparation of two populations of LDL isolated from the same plasma sample and containing different amounts of α-tocopherol; a normal (nLDL) and an α-tocopherol-enriched fraction (eLDL). The enrichment of lipoproteins with α-tocopherol was achieved using the procedure described by Esterbauer et al.[20] that, previous to LDL isolation, involves the incubation of plasma with 250 μM α-tocopherol from a stock solution in dimethyl sulfoxide (DMSO) at 37° under gentle stirring. Table I illustrates quantitatively a typical experiment on the synergism arising from the interaction of caffeic acid and α-tocopherol in the protection of LDL from oxidation induced by ferrylmyoglobin. Notably, caffeic acid was very effective in prolonging the lag phases of LDL oxidation to an extent longer than the sum of individual compounds, particularly in the case of α-tocopherol-enriched particles. Concurrent oxidation of caffeic acid and α-tocopherol by ferrylmyoglobin in the experimental conditions used is likely to occur; the lower rates of O_2 consumption during the lag phases in the presence of caffeic acid, as compared with respective controls, may reflect this phenomena. However, the rate of α-tocopherol depletion in LDL as a function of ferrylmyoglobin concentration[25] also suggests a direct interaction of α-tocopherol with the heme protein with production of the α-tocopheroxyl radical.

The incubation medium contained, in a final volume of 1 ml of phosphate buffer, 90 μg LDL protein, 6 μM metmyoglobin, and, when present, 0.5 μM

[26] J. E. Packer, T. F. Slater, and R. L. Wilson, *Nature* **278,** 737 (1979).
[27] E. Niki, J. Tsuchiya, R. Tanimura, and Y. Kamyia, *Chem. Lett.* 789 (1982).
[28] K. Mukai, M. Nishimur, and S. Kikuchi, *J. Biol. Chem.* **266,** 274 (1991).

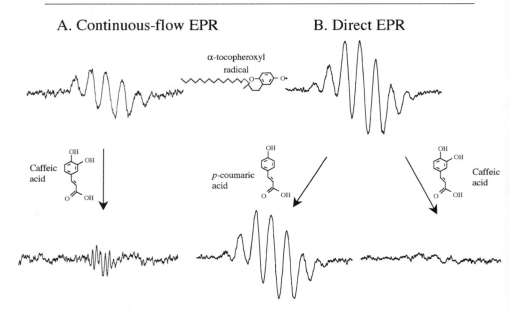

FIG. 2. Continuous-flow and direct EPR measurements of α-tocopherol-containing SDS micelles oxidized by ferrylmyoglobin and UV radiation, respectively. Effect of caffeic and p-coumaric acids.

caffeic acid. The reaction was started at 37° by the addition of 9 μM hydrogen peroxide.

Continuous-flow EPR measurements of α-tocopherol and caffeic acid phenoxyl radicals directly support the reduction of α-tocopheroxyl radical by caffeic acid, suggested by previous data. Tocopherol-containing SDS micelles were used as a convenient model for recycling α-tocopherol, as shown by others.[29] On introduction of ferrylmyoglobin in the mixing cell (located just above the EPR cavity), α-tocopherol gives a seven-line spectrum characteristic of the tocopheroxyl radical[30] (Fig. 2A). Subsequent flowing of caffeic acid in the system resulted in the replacement of tocopherol signal by a signal assigned to the caffeic acid o-semiquinone radical. In separate experiments, the caffeic acid radical was generated by autoxidation at alkaline pH and by incubation with peroxynitrite in order to be clearly identified and simulated.[31]

Studies with direct EPR failed to detect the expected caffeic radical produced in the regeneration of α-tocopherol, which, in agreement with other studies,[29]

---

[29] P. K. Witting, C. Westerlund, and R. Stocker, *J. Lipid Res.* **37,** 853 (1996).
[30] V. E. Kagan, E. A. Serbinova, T. Forte, G. Scita, and L. Packer, *J. Lipid Res.* **33,** 385 (1992).
[31] P. Ashworth, *J. Org. Chem.* **41,** 2920 (1976).

points to a short-lived nature of such species. Figure 2B illustrates this finding and also supports the notion that the recycling of tocopherol cannot be readily extended to the other members of the caffeic acid group, the hydroxycinnamates. As shown, the addition of caffeic acid to tocopheroxyl radical-containing SDS micelles resulted in the suppression of the persistent tocopheroxyl radical, but no other signal was observed. On the contrary, the addition of the biosynthetic precursor of caffeic acid, p-coumaric acid, had no effect on the intensity of the tocopheroxyl radical. In these experiments the radicals were generated by UV irradiation of SDS micelles containing 1 mM α-tocopherol. The effect of phenolic acids was checked in a concentration range between 0.1 and 1 mM.

The redox potential determines the thermodynamic ability of the compounds to reduce the α-tocopheroxyl radical to α-tocopherol. The one-electron reduction potential of the caffeic o-semiquinone radical ($E7 = 0.54$ V) at pH 7 is slightly higher than $E7 = 0.48$ V for the Trolox phenoxyl radical (the water-soluble analog of vitamin E assumed to have identical reduction potential),[32] implying that if the figures can apply, the reduction of the tocopheroxyl radical by caffeic acid is not thermodynamically feasible. However, under such conditions that the concentration of caffeic acid is far more than that of tocopheroxyl radical, these reduction potentials are of limited prognostic value because the reaction may be kinetically pulled, overcoming the thermodynamic constrains. Additionally, it could be speculated that the buried tocopherol molecule in the apolar environment of the membrane, favoring protonation, may actually deviate to higher values of redox potential, as compared with that of Trolox in the water phase. The reduction potential of p-coumaric acid is far higher[25] as compared with that of caffeic acid, which, in agreement with the results depicted on Fig. 2B, suggests that the regeneration of α-tocopherol is unlikely to occur.

The relevance of the recycling mechanism of α-tocopherol by caffeic acid *in vivo* has received experimental support. Nardini *et al*[33] reported that lipoproteins from caffeic acid-fed rats were markedly resistant to oxidative modification and that caffeic acid dietary supplementation resulted in a statistically significant increase of α-tocopherol in both plasma and lipoproteins.

Redox Interaction of Caffeic Acid and Ascorbate

Absorption spectroscopy studies on the decay of ascorbate and caffeic acid oxidized separately with ferrylmyoglobin revealed two convenient spectral details (Fig. 3). First, ascorbate has a very weak absorption at the wavelength that caffeic acid displays an absorption maximum. Second, the pattern of caffeic acid spectral

---

[32] S. Stenken and P. Neta, *J. Phys. Chem.* **86**, 3661 (1982).
[33] M. Nardini, F. Natella, V. Gentili, M. Di Felice, and C. Scaccini, *Arch. Biochem. Biophys.* **342**, 157 (1997).

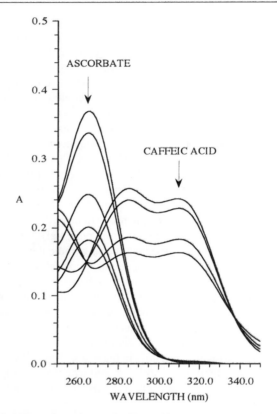

FIG. 3. Spectral changes of ascorbate and caffeic acid under oxidation by ferrylmyoglobin. Data for each compound were obtained in separate experiments.

decay exhibits an isosbestic point at the absorption maximum of ascorbate. These spectral characteristics enable us to follow simultaneously the spectral changes of both compounds in solution under oxidation by ferrylmyoglobin or by AAPH-derived peroxyl radicals. Additionally, the effect of ascorbate on the time course of caffeic acid oxidation by ferrylmyoglobin was analyzed by HPLC. Consistently, under the experimental conditions used, caffeic acid decreases as much as 65% of the initial concentration, and the addition of ascorbate during the ongoing oxidation partially restores caffeic acid levels in a time- and concentration-dependent way.[34]

Direct EPR measurements of the caffeic acid/ascorbate interactions have shown that incubation of phenolic acid in buffer at pH 9.3 or, alternatively, with ferrylmyoglobin at pH 8.3 yielded a seven-line EPR spectrum characteristic of the caffeic

[34] J. Laranjinha and E. Cadenas, *IUBMB Life* **48,** 1 (1999).

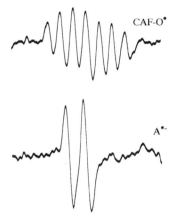

FIG. 4. EPR spectrum of caffeic acid *o*-semiquinone radical generated by incubation of caffeic acid with ferrylmyoglobin at pH 8.3. Effect of ascorbate. Similar results were obtained when caffeic radical was obtained by autoxidation at pH 9.3.

o-semiquinone radical[31] (Fig. 4). The incubation medium contained 0.2 mM caffeic acid and 10 μM metmyoglobin/15 μM H_2O_2. When 0.2 mM ascorbate was introduced in the medium, the EPR spectrum of *o*-semiquinone was replaced by a doublet characteristic of ascorbate radical,[35] suggesting that ascorbate reduces the caffeic radical to the parent phenolic compound. Alternatively, the ascorbate radical can be produced on direct oxidation of ascorbate by ferrylmyoglobin, as reported previously.[36] However, competition reactions for the oxidant are likely to occur because the intensity of the ascorbate radical obtained by incubation with ferrylmyoglobin or by autoxidation at pH 9.3 was attenuated by caffeic acid in a concentration-dependent way.[34] Thus, overall, results support the ability of ascorbate (AH^-) to undergo a redox reaction with caffeic *o*-semiquinone radical regenerating caffeic acid and producing the ascorbate radical, according to reaction (2). Still, other reactions may occur as discussed for this redox pair[34] and, in more detail, for the interaction of flavonoids with ascorbate.[37] Examples incude the reduction of caffeic acid quinone by ascorbate radical and a radical–radical recombination involving ascorbate and caffeic acid radicals.

$$\text{CAF-O}^\bullet + \text{AH}^- \to \text{CAF-OH} + \text{A}^{\bullet-} \tag{2}$$

[35] G. R. Buettner and B. A. Jurkiewicz, *in* "Handbook of Antioxidants" (E. Cadenas and L. Packer eds.), p. 91. Dekker, New York, 1996.
[36] C. Giulivi and E Cadenas, *FEBS Lett.* **332**, 287 (1993).
[37] W. Bors, C. Michel, and S. Schikora, *Free Radic. Biol. Med.* **19**, 45 (1995).

TABLE II
EFFECT OF CAFFEIC ACID ASSOCIATED WITH ASCORBATE ON LAG
PHASES OF CHOLESTERYL LINOLEATE HYDROPEROXIDE (CL-OOH)
AND 7-KETOCHOLESTEROL (7KC) FORMATION IN LDL OXIDIZED
BY FERRYLMYOGLOBIN

Compound	Lag phases^a (min)	
	CLOOH	7KC
None	nd	nd
Ascorbate	2	nd
Caffeic acid	4.5	6.5
Ascorbate + caffeic acid	11	8

^{a} The rate of CL-OOH and 7KC formation during lag phases was null. nd, no measurable lag phase.

Actually, reaction (2) is likely to be involved in the metabolism of caffeic acid in plants. In view of the pseudo-peroxidase activity of ferrylmyoglobin,[18] it is worth mentioning that reaction (2) has been proposed as the basis for a very effective mechanism of detoxification in plants: the facile electron donation to peroxidases by caffeic acid and the subsequent reduction of caffeic radicals by ascorbate can efficiently reduce H_2O_2 without any accumulation of oxidized phenolic products.[38,39]

Further insight into the electron-transfer reactions between caffeic and ascorbate is provided by studies of LDL oxidation promoted by ferrylmyoglobin.[16] A combination of caffeic acid with ascorbate synergistically increased the lag times required for the accumulation of linoleate hydroperoxide and 7-ketocholesterol, major oxidation products in LDL (Table II). Because caffeic acid induced a longer lag phase than ascorbate under the experimental conditions used, the potential recycling of caffeic acid from its phenoxyl radical by ascorbate would provide a constant resupply of caffeic acid and would limit the decay of the caffeic radical by a second-order reaction. Under these conditions, synergistic antioxidant effects can occur as reflected by the length of the lag phase in the presence of a mixture of caffeic acid and ascorbate longer than the sum of individual effects.

Dynamic Interaction of α-Tocopherol, Caffeic Acid, and Ascorbate

The feasibility of coupled redox reactions among caffeic acid, α-tocopherol, and ascorbate was studied further by continuous-flow EPR. Figure 5 shows a

---

[38] U. Takahama and T. Oniki, *Plant Cell Physiol.* **33,** 379 (1992).
[39] H. Yamasaki and S. C. Grace, *FEBS Lett.* **422,** 377 (1998).

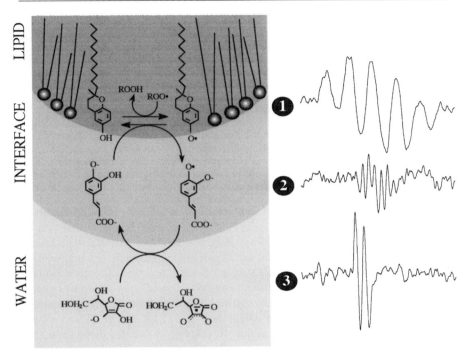

FIG. 5. Continuous-flow EPR measurements of α-tocopherol, caffeic acid, and ascorbate added in sequence to a reaction mixture under oxidation by ferrylmyoglobin. The proposed redox transitions are depicted.

continuous-flow EPR measurement of a mixture of α-tocopherol containing SDS micelles oxidized with ferrylmyoglobin and the effect of sequential introduction of caffeic acid and ascorbate in the medium. The α-tocopheroxyl radical is detected immediately after mixing with ferrylmyoglobin (1). Introducing caffeic acid in the flowing mixture resulted in the quenching of the α-tocopheroxyl radical and appearance of the caffeic o-semiquinone radical (2). Subsequent flowing of ascorbate into the system yielded the doublet spectrum of ascorbate radical (3). A persistent ascorbate radical is observed when mixing ascorbate and ferrylmyoglobin that decreases in intensity as caffeic acid and α-tocopherol-containing SDS micelles are fluxed into the system (not shown). These results are consistent with the role of ascorbate as the terminal reductant[35] and suggest that reduction of the α-tocopheroxyl radical by caffeic acid and ascorbate efficiently competes with their oxidation by ferrylmyoglobin. Thus, a sequence of redox couple reactions can be envisaged whereby the radical character is sequentially transferred from the lipoprotein particle to the aqueous medium through the one-electron reduction of

TABLE III
EFFECT OF CAFFEIC ACID ASSOCIATED WITH ASCORBATE ON LAG PHASES OF
CONJUGATED DIENE ACCUMULATION IN TWO POPULATIONS OF LDL WITH
DIFFERENT CONTENT OF α-TOCOPHEROL OXIDIZED BY FERRYLMYOGLOBIN[a]

LDL	Lag phases (min)			
	No addition	Ascorbate	Caffeic acid	Ascorbate + caffeic acid
nLDL	20	30	80	390
eLDL	150	180	490	1400

[a] nLDL and eLDL contain 10.5 and 36.1 nmol α-tocopherol/mg protein, respectively. The concentration used for caffeic acid and ascorbate was 2 μM.

tocopheroxyl radical by caffeic acid and, in turn, of the caffeic radical by ascorbate (Fig. 5).

It is expected that the sequence of redox cycles amplifies the antioxidant effects of individual compounds. Data on the lag phases of conjugated diene accumulation in LDL depicted in Table III indicate that, under the experimental conditions used, α-tocopherol, caffeic acid, and ascorbate acted synergistically to afford optimal protection of LDL against oxidation. Again, the experimental design to evaluate the synergistic effects involved the preparation of a nLDL and eLDL from same plasma sample. It should be stressed that the rates of conjugated diene formation during the lag phases are very low as compared with those observed during the propagation steps of lipid peroxidation and are similar for all mixtures within each LDL preparation. The standard reaction mixture consisted of 2 ml phosphate buffer containing 50 μM DTPA, 9 μg metmyoglobin/9 μg H_2O_2, and, when present, 2 μM caffeic acid and 2 μM ascorbate.

The evaluation of antioxidant capacities requires information on thermodynamic (redox potentials) and kinetic (rate constants with different types of radicals, stability of the antioxidant-derived radical, stoichiometry) properties of compounds.[5,40] The solubility characteristics of molecules, however, are important in assessing antioxidant behavior in lipid:water systems,[41] particularly when studying synergistic activities supported by recycling mechanisms.

It has been emphasized that the localization of flavonoids at the surface of membranes influences their antioxidant activities.[42–44] Similarly, the amphiphilic properties of caffeic acid are relevant for the analysis of data presented here.

[40] E. Cadenas, in "Free Radicals, Oxidative Stress and Antioxidants" (Ozben ed.), p. 237. Plenum Press, New York, 1998.
[41] E. Niki, *Chem. Phys. Lipids* **44,** 227 (1987).
[42] A. K. Ratty, J. Sunamoto, and N. P. Das, *Biochem. Pharmacol.* **37,** 989 (1988).
[43] A. Saija, M. Scalese, M. Lauza, D. Marzullo, F. Bonina, and F. Castelli, *Free Radic. Biol. Med.* **19,** 481 (1995).
[44] J. Terao and M. K. Piskula, in "Flavonoids in Health and Disease" (C. Rice-Evans and L. Packer, eds.), p. 277. Dekker, New York, 1998.

The mechanistic basis for the well-known synergism between ascorbate and α-tocopherol in the inhibition of lipid peroxidation chains relies on the ability of ascorbate to reduce α-tocopherol radicals rapidly to tocopherol at the water:lipid interface.[26,27] The localization of α-tocopherol at or near the membrane surface[45] and ascorbate in the water phase makes possible the interaction of the two antioxidants at the surface of membranes or lipoproteins. By this way, unpaired electrons are channeled from reactive-free radicals in membranes, such as the lipid peroxyl and α-tocopheroxyl radicals,[46] to ascorbate in the water phase. The ascorbate-free radical is, in turn, eliminated by disproportionation or by reductase activities.[35]

A number of indirect evidence suggest the partition of caffeic acid into membranes.[47,48] The potential localization of caffeic acid at the lipid:water interface in membranes, micelles, and lipoproteins would increase the physical accessibility of the phenolic acid to the lipid compartment, hence improving antioxidant protection by two mechanisms: effective repair of lipophilic α-tocopheroxyl radicals and effective inhibition of attack by free radicals in the aqueous phase.

Concluding Remarks

The physicochemical properties of caffeic acid permit one to consider this compound as an efficient antioxidant and of potential use in the prevention of disorders related to lipid oxidation. The inhibition of LDL oxidation during initial events of atherosclerosis is a prime example.

In addition to the kinetic and thermodynamic characteristics, solubility properties may determine antioxidant effectiveness in recycling mechanisms at lipid:water interfaces. Therefore, research *in vitro* with models such as the one presented, screening molecular mechanisms of antioxidant interplay, may help in the study of compounds that can interact in the "antioxidant network" and in the design of synergistic antioxidant mixtures.

Acknowledgments

These studies were supported by Fundação Ciência e Tecnologia (Grants PRAXIS/PCNA/BIA/160/96 and PRAXIS/P/BIA/1395/1998), Portugal. The helpful comments of Dr. Enrique Cadenas and the technical assistance of João G. Frade are gratefully acknowledged.

---

[45] J. C. Gomez-Fernandez, J. Villalain, and F. J. Aranda, *Ann. N. Y. Acad. Sci.* **570**, 109 (1989).
[46] R. Stocker and V. W. Bowry, *in* "Handbook of Antioxidants" (E. Cadenas and L. Packer eds.), p. 27. Dekker, New York, 1996.
[47] C. A. Tyson, S. E. LeValley, R. Chan, P. D. Hobbs, and M. I. Dawson, *J. Pharmacol. Exp. Ther.* **228**, 676 (1984).
[48] M. Foti, M. Piattelli, M. T. Baratta, and G. Ruberto, *J. Agric. Food Chem.* **44**, 497 (1996).

[26] DNA Damage by Nitrite and Peroxynitrite: Protection by Dietary Phenols

By KAICUN ZHAO, MATTHEW WHITEMAN, JEREMY P. E. SPENCER, and BARRY HALLIWELL

Introduction

Nitric oxide is an essential metabolite in the human body, performing multiple physiological functions.[1] Its final oxidation product is NO_2^-, which is oxidized rapidly to NO_3^- *in vivo*.[2–7] Substantial levels of NO_3^- (e.g., range 0–46 μM in plasma) and measurable levels of NO_2^- (plasma range 0–13 μM) are present in body fluids.[3–7] Saliva is especially rich in NO_2^- (concentration often >100 μM).[3–9] Whereas most attention is paid to the oxidation of L-arginine by NOS enzymes as a source of NO• *in vivo*,[1] there may be substantial formation of oxides of nitrogen by nonenzymatic mechanisms.[9] Thus when NO_2^- in foods and saliva enters the stomach, the low gastric pH will lead to the formation of HNO_2 and hence to oxides of nitrogen.[9,10] It is also possible that ischemic tissues might become sufficiently acidic to form HNO_2 from any NO_2^- present.[9,11,12]

Nitric oxide is poorly reactive with most biomolecules. For example its rate of direct reaction with DNA bases may be at or close to zero. However, NO• reacts extremely fast (rate constants >10^9 M^{-1} sec^{-1})[13–17] with other free radicals,

---

[1] S. Moncada and E. A. Higgs, *FASEB J.* **9,** 1319 (1995).
[2] P. C. Ford, D. A. Wink, and D. M. Stanbury, *FEBS Lett.* **326,** 1 (1993).
[3] L. C. Green, D. A. Wagner, J. Glogowski, P. L. Skipper, J. S. Wishnock, and S. R. Tannenbaum, *Anal. Biochem.* **126,** 131 (1982).
[4] D. Tsikas, I. Fuchs, F. M. Gutzki, and J. C. Frolich, *J. Chromatog. B* **715,** 441 (1998).
[5] A. Wennmalm, G. Benthin, A. Edlund, L. Jungersten, N. Kieler-Jensen, S. Lundin, U. N. Westfelt, A. S. Peterson, and F. Waagstein, *Circ. Res.* **73,** 1121 (1993).
[6] A. Wennmalm, G. Benthin, and A. S. Peterson, *Br. J. Pharmacol.* **106,** 507 (1992).
[7] A. Wennmalm, G. Benthin, L. Jungersten, A. Edlund, and A. S. Peterson, *Biol. Nitric Oxide* **4,** 474 (1994).
[8] A. van der Vliet, J. P. Eiserich, M. K. Shigenaga, and C. E. Cross, *Am. J. Resp. Crit. Care Med.* **159,** 1 (1999).
[9] E. Weitzberg and J. O. N. Lundberg, *Nitric Oxide* **2,** 1 (1998).
[10] C. Oldreive, K. Zhao, G. Paganga, B. Halliwell, and C. Rice-Evans, *Chem. Res. Toxicol.* **11,** 1574 (1998).
[11] J. L. Zweier, P. Wang, A. Samouilov, and P. Kuppusamy, *Nature Med.* **1,** 804 (1995).
[12] A. Samouilov, P. Kuppusamy, and J. L. Zweier, *Arch. Biochem. Biophys.* **357,** 1 (1998).
[13] J. P. Eiserich, J. Butler, A. van der Vliet, C. E. Cross, and B. Halliwell, *Biochem. J.* **310,** 745 (1995).
[14] G. Czapski, J. Holcman, and B. H. J. Bielski, *J. Am. Chem. Soc.* **116,** 11465 (1994).
[15] R. E. Huie and S. Padmaja, *Free Radic. Res. Commun.* **18,** 195 (1993).
[16] S. Padmaja and R. E. Huie, *Biochem. Biophys. Res. Commun.* **195,** 539 (1993).

including Tyr-O˙, ˙OH, and RO$_2$˙. Attention in recent years has tended to focus on the reaction of NO˙ with O_2^- to form ONOO$^-$, a cytotoxic agent,[18] although the overall effects of NO˙ *in vivo* may be antioxidant rather than prooxidant.[19] Both peroxynitrite and higher oxides of nitrogen (especially N_2O_3) attack DNA readily, leading to base deamination.[20–24] Formation of 8-nitroguanine from guanine may be a specific marker of attack by ONOO.[22–24]

Reaction of Nitrite with DNA

Nitrous acid can damage DNA indirectly by nitrosating amines to give mutagenic nitrosamines that can then react with DNA.[25–27] However, this article focuses on direct damage to DNA.

Incubation of DNA with NO_2^-, even at millimolar concentrations, at pH 7.4, does not lead to detectable DNA base modifications.[28,29] Perhaps surprisingly, exposure of a human respiratory tract cell line to NO_2^- caused significant increases in the levels of xanthine and hypoxanthine, presumably resulting from deamination of guanine and adenine, respectively, in DNA subsequently isolated from the cells.[29] This was accompanied by DNA strand breakage. It seems that exposure of cells to NO_2^- may lead to the intracellular generation of reactive nitrogen species capable of deaminating purines, even at physiological pH. Intracellular DNA deamination was detectable in cells incubated at pH 7.4 and proceeds faster at more acidic culture medium pH values.[29] Nitrite can cross membranes, probably as HNO_2.[30]

Deamination presumably results from the nitrosation of $-NH_2$ groups on DNA bases to form diazonium salts, followed by loss of this group and its replacement by $-OH$ from H_2O. Thus cytosine is deaminated to uracil, 5-methylcytosine can form

---

[17] G. V. Buxton, C. L. Greenstock, W. P. Helman, and A. B. Ross, *J. Phys. Chem. Ref. Data* **17**, 513 (1988).
[18] J. S. Beckman and W. H. Koppenol, *Am. J. Physiol.* **271**, C1424 (1996).
[19] B. Halliwell, K. Zhao, and M. Whiteman, *Free Radic. Res.* **31**, 651 (1999).
[20] S. Burney, J. L. Caulfield, J. C. Niles, J. S. Wishnock, and S. R. Tannenbaum, *Mutat. Res.* **424**, 37 (1999).
[21] B. Halliwell, *Mutat. Res.* **443**, 37 (1999).
[22] H. Ohshima, V. Yermilov, Y. Yoshie, and J. Rubio, *in* "Advances in DNA Damage and Repair" (M. Dizdaroglu and A. E. Karakaya, eds.), p. 32. Plenum, New York, 1999.
[23] V. Yermilov, Y. Yoshie, J. Rubio, and H. Ohshima, *FEBS Lett.* **399**, 67 (1996).
[24] J. P. Spencer, J. Wong, A. Jenner, O. I. Aruoma, C. E. Cross, and B. Halliwell, *Chem. Res. Toxicol.* **9**, 1152 (1996).
[25] S. Tamir and S. R. Tannenbaum, *Biochim. Biophys. Acta* **1288**, F31 (1996).
[26] P. B. Farmer and D. E. G. Shuker, *Mutat. Res.* **424**, 275 (1999).
[27] M. Eichholzer and F. Gutzwiller, *Nutr. Rev.* **56**, 95 (1998).
[28] M. Whiteman, J. P. E. Spencer, A. Jenner, and B. Halliwell, *Biochem. Biophys. Res. Commun.* **257**, 572 (1999).
[29] J. P. E. Spencer, M. Whiteman, A. Jenner, and B. Halliwell, *Free Radic. Biol. Med.* **28**, 1039 (2000).
[30] R. Shingles, M. H. Roh, and R. E. McCarty, *J. Bioeng. Biomembr.* **29**, 611 (1997).

thymine, and guanine and adenine can form xanthine and hypoxanthine, respectively. Xanthine is unstable in DNA and can detach to leave an abasic site; in addition, mispairing of xanthine can lead to GC→AT transition mutations. Similarly, uracil and thymine (from 5-methylcytosine) can lead to GC→AT transitions. A mispairing of hypoxanthine with cytosine can produce AT→GC transitions. Cells seem to repair uracil lesions in DNA efficiently.[31,32] They also contain enzymes that can repair xanthine and hypoxanthine lesions,[33,34] although rates of such repair have been reported as slow in a cell line.[29] Deamination of purines seems to proceed faster than that of cytosine in isolated DNA exposed to HNO_2.[35] Exposure of guanine in DNA to HNO_2 additionally leads to the formation of oxanine, another mutagenic lesion, as well as other products.[36]

Measurement of Deamination Products

We use a high-performance liquid chromatography (HPLC)-based method to examine deamination products. DNA solution (0.5 mg/ml) is prepared either in a potassium phosphate buffer (50 mM) for pH 3, 5, and 7 or in sodium carbonate buffer for pH 9 or in hydrochloric acid (0.1 M) for pH 1. After preincubation of DNA (1 ml) at 37° for about 15 min, NO_2^- (50 μl) at various concentrations is added and, following a further incubation at 37° for 2 hr, the mixture is dialyzed exhaustively against deionized water. The concentration of the DNA is determined at 260 nm, and 100 μg of DNA is freeze-dried. The dried DNA is hydrolyzed in 60% formic acid (0.5 ml) by incubation at 145° for 45 min. After removing the acid by freeze-drying, the residue is reconstituted in 1 ml of water and subjected to HPLC analysis of the deaminated products.

HPLC analysis is carried out using a Gynkotek (Gynkotek UK Ltd., Cheshire, England) system consisting of a Model 480 solvent delivery pump, a GINA 50 autosampler, a UVD 340S photodiode array UV detector, and an INTRO electrochemical detector. The separation of DNA bases and their deamination products (illustrated in Fig. 1) is achieved on a reversed-phase Hypersil C_{18} column (250 by 4.6 mm, 5 μm), which is eluted at a flow rate of 1 ml/min with an isocratic mobile phase, a potassium phosphate buffer (50 mM, pH 3) containing 2 mM triethylamine and 0.05 M EDTA. The deaminated products hypoxanthine, uracil, and xanthine are detected by UV spectroscopy at 250 and 265 nm, respectively. Xanthine can also be detected by an electrochemical detector at a potential of 0.9 V. This system

---

[31] R. Savva, K. McAuley-Hecht, T. Brown, and L. Pearl, *Nature* **373**, 487 (1995).
[32] J. D. Domena, R. T. Timmer, S. A. Dicharry, and D. W. Mosbaugh, *Biochemistry* **27**, 6742 (1998).
[33] B. Demple and L. Harrison, *Annu. Rev. Biochem.* **63**, 915 (1994).
[34] P. Fortini, E. Parlanti, O. M. Sidorkina, J. Laval, and E. Dogliotti, *J. Biol. Chem.* **274**, 15230 (1999).
[35] J. L. Caulfield, J. S. Wishnock, and S. R. Tannenbaum, *J. Biol. Chem.* **273**, 12689 (1998).
[36] T. Suzuki, M. Yoshida, M. Yamada, H. Ide, M. Kobayashi, K. Kanaori, K. Tajima, and K. Makino, *Biochemistry* **37**, 11592 (1998).

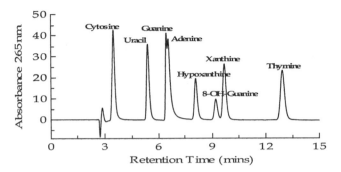

FIG. 1. A representative chromatogram showing separation of normal from deaminated DNA bases and from 8-hydroxyguanine.

also separates 8-hydroxyguanine, which can be detected electrochemically at a potential of 0.65 V or by its UV absorbance at 295 nm.

To study the whole profile of DNA base damage caused by reactive nitrogen species, we use a gas chromatography–mass spectrometry (GC-MS) method that has been described previously.[37] This method enables one to analyze the damage to all four DNA bases simultaneously. DNA bases need to be derivatized before they can be analyzed on GC-MS. After hydrolysis of DNA, the vials containing freeze-dried DNA hydrolyzate are gassed with nitrogen and sealed. The residues are dissolved in 15 μl of acetonitrile and then derivatized by adding 60 μl of bis(trimethylsilyl)trifluoroacetamide (BSTFA) plus 25% ethanethiol.[37] The reaction mixture is left at room temperature for 2 hr before being subjected to GC-MS analysis.

The GC-MS analysis is carried out on a Hewlett-Packard 5890II gas chromatograph interfaced with a Hewlett-Packard 5917A mass selective detector (Hewlett-Packard Ltd., Stockport, Cheshire, UK). Separation of the bases is achieved on a fused silica capillary column (12 m × 0.2 mm) coated with cross-linked 5% phenylmethylsiloxane. Temperatures of the injection port and the GC-MS interface are 250° and 290°, respectively. Column temperature is programmed as follows: Started and held at 125° for 2 min, increased to 175° at 8°/min and from 175° to 220° for 1 min, increased to 290° at 40°C/min, and held at this temperature for 2 min. The temperature in the ion source is kept at 185°, and ion mass is selectively monitored using electron ionization mode at 70 eV. The derivatized DNA bases (1 μl) are injected and eluted using helium as carrier gas at a flow rate of 0.93 ml/min.

Exposure of calf thymus DNA to NO_2^- at pH 3 (as an approximation to gastric pH in healthy human subjects) leads to rapid concentration-dependent formation of hypoxanthine and xanthine. Significant background levels of these products are also seen in the calf thymus DNA used (Fig. 2). Xanthine is formed more

[37] A. Rehman, A. Jenner, and B. Halliwell, *Methods Enzymol.*, **319**, 401 (2000).

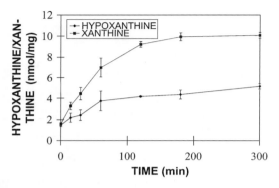

FIG. 2. The time course of deamination of DNA bases in calf thymus DNA (0.5 mg/ml) exposed to 1 mM NaNO$_2$ in potassium phosphate buffer (50 mM, pH 3). The mixture was incubated at 37° for various times as indicated. After exhaustive dialysis of the DNA samples against water, 100 μg DNA was freeze-dried and hydrolyzed in 60% formic acid (0.5 ml). The acid was removed by freeze-drying and the residue was dissolved in water (1 ml) for HPLC analysis.

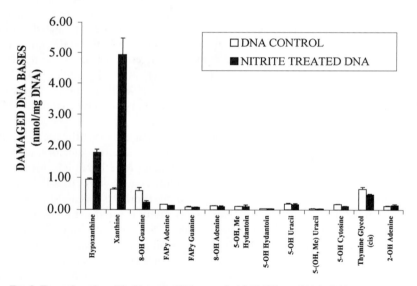

FIG. 3. Formation of modified bases in DNA treated with NaNO$_2$ at pH 3.0. Calf thymus DNA (0.5 mg/ml) was incubated with sodium nitrite (0.5 mM) in a total volume of 1 ml at 37° for 2 hr. Following exhaustive dialysis against water, the DNA (100 μg) was freeze-dried and hydrolyzed in 60% formic acid at 145° for 45 min. After removing the acid by freeze-drying, vials containing the dried hydrolyzate were gassed with nitrogen and sealed. The residues were dissolved in 15 μl of acetonitrile containing 25% ethanethiol and then derivatized by adding 60 μl of BSTFA. The reaction mixture was left at room temperature for 2 hr before being subjected to GC-MS analysis. Note the extensive deamination but essentially no formation of oxidized base products.

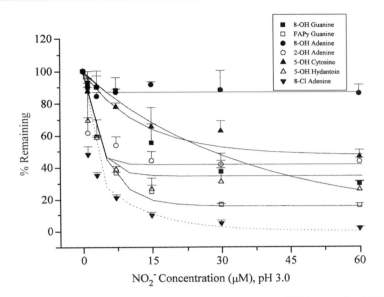

FIG. 4. Loss of 8-hydroxyguanine and other oxidized DNA bases in DNA exposed to HNO$_2$. All solutions were treated with Chelex 100 resin (Bio-Rad) before use. Oxidative DNA base damage was induced by incubating calf thymus DNA (2.0 mg/ml) with H$_2$O$_2$ (28 mM), CuCl$_2$ (100 μM), and ascorbate (1 mM) in 50 mM phosphate buffer (pH 7.4). After a 1-hr incubation the DNA was dialyzed against water (24 hr). Oxidized DNA (0.5 mg/ml) in phosphate buffer (100 mM K$_2$HPO$_4$–KH$_2$PO$_4$, pH 3) was incubated in a water bath at 37° for 15 min and then increasing concentrations of NaNO$_2$ (0–1000 μM) were added to give a final volume of 2.0 ml. The samples were incubated for another hour before further dialysis for 24 hr, acidic hydrolysis, and analysis by GC-MS.[29,37]

readily than hypoxanthine at all concentrations examined. Uracil, the deamination product of cytosine, is also detected in smaller quantities in acidic nitrite-treated DNA. The formation of uracil increases from 0.2 to 1.5 nmol/mg DNA with 0.2–1.0 mM NO$_2^-$, but then falls to a steady level of about 0.3 nmol/mg DNA at higher NO$_2^-$ levels (2–4 mM). Indeed, <0.01% of the starting material is converted to uracil when free cytosine (1.8 mM) is treated with nitrite (up to 7 mM) at pH 3. There is no evidence for the nitration of bases (i.e., no nitroguanine formation) during the exposure of DNA to nitrite. No formation of oxidized DNA bases is observed (Fig. 3). Indeed, the addition of NO$_2^-$ to DNA containing high levels of 8-hydroxyguanine at pH 3 leads to rapid loss of this oxidation product and of several other DNA base oxidation products (Fig. 4). 8-Hydroxyguanine and 8-hydroxy-2'-deoxyguanosine are much more susceptible to oxidative destruction by a range of agents than guanine itself.[28,38,39] Indeed, free 8-hydroxyguanine is lost rapidly on addition of NO$_2^-$ at pH 3.0 (data not shown).

[38] B. Halliwell, *Free Radic. Res.* **29**, 469 (1998).
[39] S. Burney, J. C. Niles, P. C. Dedon, and S. R. Tannenbaum, *Chem. Res. Toxicol.* **12**, 513 (1999).

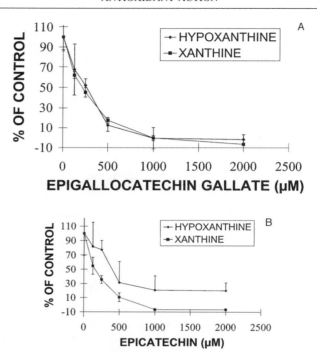

FIG. 5. Inhibition of DNA base deamination by (A) epigallocatechin gallate and (B) epicatechin. Calf thymus DNA (0.5 mg/ml) in potassium phosphate buffer (50 mM, pH 3) was incubated with sodium nitrite (0.5 mM) in the presence of the phenolic compounds at various concentrations as indicated. The final volume of the reaction mixture was 1 ml and the incubation was carried out at 37° for 2 hr. The dialysis and hydrolysis of the DNA and the HPLC analysis of deaminated products were as described in Figure 2 and Table I.

Protection by Phenolic Compounds Found in Fruits and Vegetables

It has been suggested that flavonoids and other phenolic compounds could exert gastroprotective effects by scavenging HNO_2 produced in the stomach from NO_2^-, as NO_2^- is found in saliva[3,9] and in many foods and is still frequently used as a meat preservative.[9,40] Indeed, many phenolic compounds are able to protect individual DNA bases (adenine, guanine) against deamination by HNO_2.[10] They are also able to protect DNA against HNO_2. For example, Fig. 5A shows data for epigallocatechin gallate and Fig. 5B for epicatechin. Inhibition at low levels of these phenolics is very variable as indicated by the size of the error bars on

---

[40] R. Cammack, C. L. Joannou, X. Y. Cui, C. T. Martinez, and M. N. Hughes, *Biochim. Biophys. Acta* **1411**, 475 (1999).

TABLE I
IC$_{50}$ VALUES FOR INHIBITION BY PHENOLIC COMPOUNDS OF
HYPOXANTHINE/XANTHINE FORMATION IN DNA EXPOSED TO NITRITE AT pH 3[a]

Phenol	IC$_{50}$ (μM)	
	Hypoxanthine	Xanthine
Caffeic acid	383 ± 66	154 ± 68
Catechin	107 ± 24	126 ± 15
Epicatechin	351 ± 184	139 ± 37
Epigallocatechin	384 ± 36	105 ± 7
Epigallocatechin gallate	187 ± 48	195 ± 6
Quercetin	390 ± 16	398 ± 94

[a] Data are mean ± SD, $n = 3$. Calf thymus DNA (0.5 mg/ml) was incubated with nitrite (0.5 mM) in the presence of various concentrations of phenolic compounds in a total volume of 1 ml. The incubation was carried out at 37° for 2 hr. The DNA concentration in the reaction mixture was determined following an exhaustive dialysis against water (24 hr with two changes). After freeze-drying, the DNA (100 μg) was hydrolyzed in 60% formic acid (0.5 ml) at 145° for 45 min. The acid was removed by freeze-drying and the residue was dissolved in 1 ml of water for HPLC analysis. The deaminated bases were separated on a Hypersil C$_{18}$ column (250 × 4.6 mm, 5μm) by isocratic elution with potassium phosphate buffer (50 mM, pH 3) containing 2 mM triethylamine and 0.05 mM EDTA.

the figures, but becomes clearer at higher concentrations. Most of the phenolic compounds used in our study inhibited deamination almost completely at about 1 mM. With some, such as caffeic acid and epicatechin, complete inhibition of the formation of hypoxanthine was not achieved up to 2 mM. Previous studies have suggested that the mechanism of the inhibition may proceed via oxidation or nitration/nitrosation of the phenolic compounds as a result of scavenging reactive nitrogen species.[10] Oxidized phenolics may form quinone/hydroxyquinone redox systems, producing reactive species that might themselves cause oxidation or deamination of DNA bases. Nitrosated phenolics could also conceivably react with DNA bases. These events could contribute to the large variations in the inhibition of the nitrite-dependent deamination at low levels of phenolics. By comparison with the free bases, higher IC$_{50}$ values were found for the inhibition of the deamination of DNA. Some evidence for binding of several phenolics to DNA was obtained in our studies (data not shown). Table I summarizes IC$_{50}$ values for inhibition of DNA deamination by phenols. Because many foods are rich in phenols, the concentrations that achieve protection (Figs. 4 and 5, Table I) could easily be achieved in the stomach after a meal rich in plant products.[41]

[41] G. Paganga, N. Miller, and C. A. Rice-Evans, *Free Radic. Res.* **30**, 153 (1999).

Modulation of Hypochlorous Acid-Dependent DNA Damage by NO_2^-

Hypochlorous acid, HOCl, induces formation of a wide range of base lesions in DNA, including thymine glycol, 5-hydroxycytosine, 5-hydroxyuracil, and 5-hydroxyhydantoin.[28,42] The presence of NO_2^- can modulate the formation of these products even at pH 7.4, when NO_2^- itself does not affect DNA.[28,29] Thus with isolated calf thymus DNA the presence of NO_2^- increased the levels of thymine glycol, 5-hydroxyhydantoin, 8-hydroxyadenine, and 5-chlorouracil above those obtained by treatment of the DNA with HOCl alone. Similar effects were observed within cells treated with NO_2^- and HOCl and, in addition, the presence of HOCl decreased the formation of xanthine and hypoxanthine in cellular DNA.[29] Nitrite and hypochlorous acid can react to form nitryl chloride, NO_2Cl, which may account for the change in product distribution.[43–45]

Protection of DNA against Peroxynitrite-Dependent Damage by Phenolic Compounds

Several phenolic compounds are able to inhibit the nitration of tyrosine induced by the addition of HNO_2[10] or of $ONOO^-$.[46–48] It was therefore of interest to examine their ability to protect DNA against damage by $ONOO^-$. Because 8-nitroguanine appears to be a specific product of $ONOO^-$ attack on guanine residues in DNA,[22–24] we measured the effect of phenols on levels of 8-nitroguanine induced in DNA by treatment with $ONOO^-$.

Treatment of DNA with peroxynitrite at pH 7.4 caused nitration of guanine, forming 8-nitroguanine. Previous studies [24] have demonstrated that peroxynitrite can cause multiple damages to DNA bases, including some oxidation as well as nitration. Treatment of free guanine (0.25 mM) with an equimolar concentration of peroxynitrite (0.25 mM) caused formation of about 7 μM 8-nitroguanine. About 56 pmol 8-nitroguanine/mg DNA was formed in calf thymus DNA (1 mg/ml) treated with 250 μM peroxynitrite. Thus free guanine seems to be more susceptible

---

[42] M. Whiteman, A. Jenner, and B. Halliwell, *Chem. Res. Toxicol.* **10,** 1240 (1997).

[43] J. P. Eiserich, C. E. Cross, A. D. Jones, B. Halliwell, and A. van der Vliet, *J. Biol. Chem.* **271,** 19199 (1996).

[44] J. P. Eiserich, M. Hristova, C. E. Cross, A. D. Jones, B. A. Freeman, B. Halliwell, and A. van der Vliet, *Nature* **391,** 393 (1997).

[45] D. W. Johnson and D. W. Margerum, *Inorg. Chem.* **30,** 4845 (1991).

[46] A. S. Pannala, R. Razaq, B. Halliwell, S. Singh, and C. A. Rice-Evans, *Free Radic. Biol. Med.* **24,** 594 (1998).

[47] A. S. Pannala, C. A. Rice-Evans, B. Halliwell, and S. Singh, *Biochem. Biophys. Res. Commun.* **232,** 164 (1997).

[48] G. R. Haenen, J. B. Paguay, R. E. Korthouwer, and A. Bast, *Biochem. Biophys. Res. Commun.* **236,** 591 (1997).

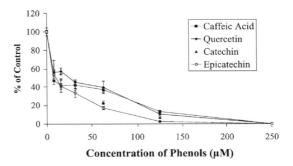

FIG. 6. The inhibitory effect of plant phenols on nitration of guanine induced by ONOO$^-$. Guanine (0.25 mM) in potassium phosphate buffer (100 mM, pH 7.4) was preincubated with various concentrations of the phenolic compounds at room temperature for about 10 min and then peroxynitrite (25 μl) was added to give a final concentration of 0.25 mM. After further incubation at room temperature for 15 min, the reaction mixture was analyzed by the HPLC method as described in Table II. In the control, the level of nitroguanine formed at equimolar concentrations of guanine and peroxynitrite (0.25 mM) was 6.93 ± 0.27 μM. The IC$_{50}$ values were 16.3 ± 1.1 μM for epicatechin, 21.9 ± 2.7 μM for catechin, 22.2 ± 4.6 μM for caffeic acid, and 29.9 ± 1.5 μM for quercetin ($n = 3$ for all the compounds).

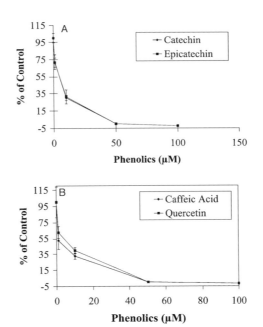

FIG. 7. The inhibitory effect of plant phenols on 8-nitroguanine formation in DNA treated with ONOO$^-$. Methods are as described in Table II. (A) Catechin and epicatechin and (B) caffeic acid and quercetin.

TABLE II
IC$_{50}$ Values of Phenolics for Inhibition of 8-Nitroguanine Formation in Calf Thymus DNA Treated with Peroxynitrite at pH 7.4[a]

Compound	IC$_{50}$ (μM)
Caffeic acid	4.17 ± 1.21
Catechin	4.91 ± 0.32
Epicatechin	4.95 ± 0.42
Quercetin	5.06 ± 0.58

[a] Data are mean ± SD, $n = 3$. The DNA solution (1 mg/ml) was prepared in potassium phosphate buffer (100 mM, pH 7.4). To 1 ml of the DNA solution, the phenolic compound (50 μl) at different concentrations was added. After a 10-min incubation at room temperature, peroxynitrite (10 μl) was added to a final concentration of 450 μM and the mixture was left at room temperature for another 15 min. Then 2 volumes of cold ethanol was added to the reaction mixture to precipitate the DNA. After three washes with 2 ml of 75% ethanol, DNA was freeze-dried. The DNA concentration was determined after reconstitution of the DNA in water (1 ml). Hydrolysis was carried by adding 0.1 ml of HCl (1 M) and then incubating for 30 min at 100°. The hydrolyzate was freeze-dried following neutralization of the acid with 0.2 ml of potassium hydroxide (1 M). 8-Nitroguanine was measured by an HPLC-based method using a Gynkotek system [Gynkotek (UK) Ltd, Cheshire, England], including an M480 mobile-phase delivery pump, a GINA-50 autosampler, a UVS340S photodiode array detector, and an INTRO electrochemical detector. The separation of 8-nitroguanine from other bases was performed on a reverse-phase column Hypercil C$_{18}$ (5 μm, 250 by 46 mm). The column was eluted using an isocratic mobile phase, ammonium formate buffer (50 mM, pH 6.0), at a flow rate of 1 ml/min. 8-Nitroguanine can be detected by UV absorbance at 396 nm or by electrochemical detection at a potential of 0.9 V. The typical retention time of 8-nitroguanine on this system is about 7.81 ± 0.01 min. However, the retention time of 8-nitroguanine is very sensitive to changes in pH of mobile phase and of samples. Lower pH values will cause a longer retention time and so the pH of samples for analysis should be adjusted to pH 6.0–7.4 if needed.

to peroxynitrite-dependent nitration. Plant phenolics were found to be effective inhibitors of the peroxynitrite-dependent nitration of guanine (Fig. 6). They also inhibited its formation in DNA (Table II, Fig. 7).

Preparation of 8-Nitroguanine Standard

Guanine (10 mM) is dissolved by sonication in 20 ml water adjusted to pH 12, and ONOO$^-$ (10 mM) is added followed by mixing. HCl (1 M) is added dropwise to adjust to pH 7.0, and the samples are mixed by vortexing and then filtered to remove insoluble matter. Supernatants are collected and freeze-dried overnight before reconstitution in 2.1 ml water. 8-Nitroguanine is then purified by preparatory HPLC. The sample (2.0 ml) is injected into a Zorbax column (25 cm × 21.2 mm; HPLC Technology Ltd., Macclesfield, Cheshire, England) at a flow rate of 5.0 ml/min in 10 mM ammonium formate, pH 6.0, containing 10% (v/v) methanol. UV absorbance is monitored at 250 nm using a Waters 486 tunable absorbance UV detector. Under these conditions, two closely eluting peaks containing yellow material are detected after 45 min. The top one-third of the second peak is collected and freeze-dried overnight followed by reconstitution in 2.1 ml water. The sample (2.0 ml) is then reinjected onto the Zorbax column and the process is repeated twice more.

After obtaining sufficient freeze-dried material, a small amount is dissolved in methanol, analyzed by HPLC-MS, and found to be >95% pure. On the HPLC-MS analysis, the synthesized product gives a molecular ion of mass 197, consistent with the molecular weight of nitroguanine. When this compound is treated by dithionite, the reduced product gives a molecular ion of mass 167, suggesting that the nitro group is reduced to an amine, giving further evidence of the identity of the compound as nitroguanine.

Conclusion

This article described methods for the analysis of base deamination products, and of 8-nitroguanine, in DNA. Phenolic compounds are powerful inhibitors of DNA damage by HNO_2 and by $ONOO^-$, which may contribute to their beneficial health effects. This may be especially true in the stomach, where phenols from freshly ingested foods could interact with HNO_2 generated when NO_2^- from saliva and foods contacts the acidic gastric contents. This could be one explanation of why the consumption of beverages and foods rich in phenolics is negatively associated with the incidence of gastric carcinoma,[49–51] even though some of these beverages can contain H_2O_2.[52] Flavonoids can also "repair" damage to DNA.[53]

---

[49] S. Kono, M. Ikeda, S. Tokudome, and M. Kuratsume, *Japan J. Cancer Res.* **79,** 1067 (1988).
[50] J. L. Bushman, *Nutr. Cancer* **31,** 151 (1998).
[51] A. A. Botterweck, P. A. van der Brandt, and R. A. Goldbohm, *Am. J. Epidemiol.* **148,** 842 (1998).
[52] L. H. Long, A. N. B. Lan, F. T. Y. Hsuan, and B. Halliwell, *Free Radic. Res.* **31,** 67 (1999).
[53] R. F. Anderson, C. Amarasinghe, L. J. Fisher, W. B. Mak, and J. E. Packer, *Free Radic. Res.* **33,** 91 (2000).

[27] Repair of Oxidized DNA by the Flavonoid Myricetin

By ISABELLE MOREL, VALÉRIE ABALEA, PIERRE CILLARD, and JOSIANE CILLARD

Introduction

Accumulation of DNA lesions can have considerable consequences for the cell in terms of mutagenesis and carcinogenesis. More precisely, oxidative damage to DNA bases generates a wide variety of oxidation products,[1] which have been postulated to be potentially mutagenic.[2-4] Nevertheless, most of oxidized bases can be removed from DNA by base-excision repair (BER) pathways, involving different enzymes, which have been identified recently.[5-7] The removal of modified bases from nuclear DNA plays an important role in the prevention of various degenerative diseases.[8] It is important therefore to find compounds capable of modulating DNA repair pathways and it appears that flavonoids such as myricetin are good candidates for the stimulation of DNA repair mechanisms.[9]

This article describes analytical techniques used to investigate the genoprotective effect of the flavonoid myricetin in rat hepatocyte cultures. For this purpose, we have evaluated the extent of DNA base modifications and the efficiency of BER mechanisms. We have used an experimental culture model that consists of rat hepatocyte cultures treated with iron, as an inducer of oxidative stress, and with the genoprotector myricetin. Chemical characterization of free radical-induced modifications of pyrimidine and purine DNA bases was achieved by capillary gas chromatography/mass spectrometry with selected ion monitoring (GC/MS-SIM). In addition, the efficiency of DNA repair mechanisms was estimated by analyzing the release of DNA base products in culture media of hepatocytes resulting from their removal from cellular DNA. The expression of DNA excision-repair enzymes in cultured cells was evaluated by quantification of mRNA levels using

---

[1] M. Dizdaroglu, *Free Radic. Biol. Med.* **10**, 225 (1991).
[2] Y. Kuchino, F. Mori, H. Kasai, H. Inoue, S. Iwai, K. Miura, E. Ohtsuka, and S. Nishimura, *Nature* **327**, 77 (1987).
[3] T. J. McBride, B. D. Preston, and L. A. Loeb, *Biochemistry* **30**, 207 (1991).
[4] A. A. Purmal, Y. Wah Kow, and S. S. Wallace, *Nucleic Acids Res.* **22**, 72 (1994).
[5] A. R. Collins, M. Ai-guo, and S. J. Duthie, *Mutat. Res.* **336**, 69 (1995).
[6] M. S. Satoh and T. Lindahl, *Cancer Res.* **54**(Suppl),1899 (1994).
[7] R. W. Sobol, J. K. Horton, R. Kuhn, R. K. Singhal, R. Prasad, K. Rajewsky, and S. H. Wilson, *Nature* **379**, 183 (1996).
[8] B. N. Ames, *Free Radic. Biol. Med.* **7**, 121 (1989).
[9] V. A. Abalea, J. Cillard, M. P. Dubos, O. Sergent, P. Cillard, and I. Morel, *Free Radic. Biol. Med.* **26**, 11 (1999).

Northern/dot blot analysis. An example is described herein with the BER enzyme, DNA polymerase β (pol β).[9] Some specific technical precautions are pointed out in the description of the methods used, especially procedures to minimize artifactual DNA oxidation.[10]

Material and Methods

Material

Nitrilotriacetic acid disodium salt (NTA), ribonuclease A (EC 3.1.27.5), ribonuclease T1 (EC 3.1.27.3), proteinase K (EC 3.4.21.64), 6-azathymine (azaT), 8-azaadenine (azaA), 5-hydroxyuracil (5-OH-Ura), 4,6-diamino-5-formamidopyrimidine (FapyAde), guanine (Gua), isoguanine (2-oxo-Ade), xanthine (Xan), anhydrous acetonitrile, and bis(trimethylsilyl)trifluoroacetamide (BSTFA) containing 1% of trimethylchlorosilane, myricetin, and formic acid are from Sigma-Fluka Chemical Co. (Saint Quentin Fallavier, France). 5-Hydroxycytosine (5-OH-Cyt), 5-hydroxymethyluracil (5-OHMe-Ura), and 7,8-dihydro-8-oxoguanine (8-oxo-Gua) are gifts from Dr. J. Cadet, (CEA Grenoble, France). Ferric ammonium citrate is from Merck (Darmstadt, Germany).

Hepatocyte Culture and Treatment

Hepatocytes are isolated by perfusion of the liver of male Sprague–Dawley rats (300–350 g) with a collagenase solution, as described previously.[11] According to the experimental procedure, 20×10^6 hepatocytes are plated in 175-cm^2 Nunclon flasks (Poly Labo, Strasbourg, France). Four hours after cell seeding, a serum-free medium consisting of 75% (v/v) minimum essential medium and 25% (v/v) medium 199 supplemented with 10 μg bovine insulin/ml, 0.1% (v/v) bovine serum albumin, and 100 nM dexamethasone is added to the cultures.

Oxidative stress is induced by iron supplements to the primary rat hepatocyte culture.[12] Ferric nitrilotriacetate (Fe-NTA) is added in culture medium to give a final iron concentration of 100 μM. Control cultures are maintained in the presence of NTA alone (200 μM). The cultures are incubated at 37° under 5% (v/v) CO_2 for 4, 24, or 48 hr. Myricetin (25, 50, 100 μM) is dissolved in dimethyl sulfoxide (DMSO); DMSO is used at 1% (v/v) final concentration in all samples, even in samples that do not contain myricetin.

---

[10] T. Douki, T. Delatour, F. Bianchini, and J. Cadet, *Carcinogenesis* **17**, 347 (1996).

[11] C. Guguen, A. Guillouzo, A. Boisnard, A. Le Cam, and M. Bourel, *Biol. Gastroenterol.* **8**, 223 (1975).

[12] I. Morel, G. Lescoat, J. Cillard, N. Pasdeloup, P. Brissot, and P. Cillard, *Biochem. Pharmacol.* **39**, 1647 (1990).

Measurement of DNA Oxidation

DNA Extraction. In order to prevent artifactual oxidation of DNA resulting from high levels of intracellular iron, DNA isolation is performed in two steps corresponding first to an isolation of nuclei and an elimination of other cell components, and second to an extraction of DNA from isolated nuclei. Moreover, to try to avoid artifactual DNA oxidation during extraction procedure, the iron chelator desferrioxamine [Desferal (Novartis Pharma, Rueil Malmaison, France) 100 μM] is added before and after cell lysis.

For the first step, hepatocytes (20×10^6/point) are collected after washing with 0.01 M Na$_2$HPO$_4$ buffer, pH 7.45, and recovered by centrifugation at 50g for 5 min at 5°. The buffer is discarded and the cells are suspended in 2 ml of buffer Tris–HCl (10 mM), sucrose (320 mM), MgCl$_2$ (5 mM), Desferal (0.1 mM), pH 7.5. Cells are lysed with 40 μl of Triton X-100 (2%). After centrifugation (10 min, 500g), the pellet is resuspended in the same buffer. After another centrifugation, the pellet mostly contains nuclei. In the second step, nuclei are resuspended in 0.6 ml of the following buffer: Tris–HCl (10 mM), EDTA-Na$_2$ (5 mM), Desferal (0.1 mM), pH 8. Nuclei are lysed with 35 μl of SDS (10%). Enzymatic digestion is performed by incubating successively at 50° with a mixture of ribonuclease A (0.5 mg/ml) and ribonuclease T1 (10 IU/ml) for 15 min and then at 37° with proteinase K (1 mg/ml) for 1 hr. DNA is then precipitated with 2 ml of cold 2-propanol. After centrifugation (2500g, 15 min, 5°), DNA is rinsed with 1 ml of cold 70% (v/v) ethanol, allowed to dry under nitrogen, and redissolved in distilled water. DNA content is measured by UV absorbance at 260 nm (absorbance of 1 OD$_{260}$ = 50 μg of dsDNA/ml), a ratio A_{260}/A_{280} higher than 1.9 represents a good purity for DNA extracts. These concentrations are verified by fluorometric analysis using Hoechst dye 33258. DNA samples are then transferred to injection vials before evaporation to dryness in a Speed-Vac evaporator (Bioblock, Illkirch, France). The samples can be kept frozen until further hydrolysis.

DNA Hydrolysis. Hydrolysis of DNA can be performed either by acidic treatment of DNA extracts, leading to free DNA bases, or by enzymatic digestion, leading to nucleosides. Often the nucleosides are not sufficiently volatile to permit analysis by GC/MS. In such cases, they are measured preferentially by high-performance liquid chromatography (HPLC) coupled with electrochemical detection.[13,14]

To perform the acidic hydrolysis of DNA, the internal standards azaT and azaA (4 nmol each) are added to the dried DNA samples, and DNA is hydrolyzed with 400 μl of 60% (v/v) formic acid for 30 min at 130°. Hydrolyzates are dried

---

[13] B. Halliwell and M. Dizdaroglu, *Free Radic. Biol. Med.* **16,** 75 (1992).
[14] I. Morel, C. Hamon-Bouer, V. Abalea, P. Cillard, and J. Cillard, *Cancer Lett.* **119,** 31 (1997).

thoroughly once more and capped tightly. They can be kept frozen in a desiccator until derivatization.

Derivatization. The derivatization procedure must be performed under argon at room temperature to prevent artifactual oxidation of DNA bases. Moreover, care must be taken to avoid any trace of water, which inhibits the derivatization reaction.

DNA bases are derivatized at room temperature for 30 min in reaction vials sealed under argon, after adding 100 μl of a mixture of trifluoroacetic acid, BSTFA containing 1% trimethylchlorosilane, and anhydrous acetonitrile (10 : 80 : 10, v/v/v). The samples are then analyzed directly by GC/MS. Derivatized sample stability does not exceed 2 hr.

Measurement of DNA Repair

Repair of most forms of oxidative DNA damage is mediated via base-excision pathways. DNA glycosylases cleave N-glycosyl bonds linking damaged bases to the deoxyribose–phosphate backbone and release free oxidized bases. These oxidized bases are then excreted from the cells without further modification.

The measurement of DNA repair activity can be performed in three ways: (1) an estimation of the decrease in oxidized bases accumulated in cell DNA; this technique is, however, rarely possible, as a continuous production of oxidized bases often hides elimination of some of them; (2) an estimation of an increase in oxidized base levels in DNA after inhibition of DNA repair enzymes by low temperatures (4°)[15]; and (3) an estimation of oxidized bases released from the cells into culture media after their removal from DNA by repair enzymes.

To perform this last method, culture media (2 ml) are collected, supplemented with internal standards (azaT and azaA), and filtered through a 20,000-Da membrane ultrafilter (Sartorius, France). Aliquots in triplicate (400 μl) of the filtrate are evaporated to dryness and are derivatized by adding 120 μl of a mixture of trifluoroacetic acid, BSTFA containing 1% of trimethylchlorosilane, and anhydrous acetonitrile (20 : 80 : 20, v/v/v). In these experiments, we estimate the release of oxidized bases and not of oxidized nucleosides (no hydrolysis of the culture medium was performed). This prevents the evaluation of other products derived from normal cell turnover of nucleosides, such as those coming from mitochondrial function and purine metabolism. In addition, because some culture media are supplemented with nucleosides and various products of purine and pyrimidine, it is important to prepare controls of pure culture medium that do not incubate cells, in order to estimate a possible oxidation of culture medium components. Furthermore, to perform such experiments, culture media must contain a low level of fetal calf serum, which interferes with ultrafiltration and GC/MS analysis. This technique is not always suited for cell line cultures, which often require high levels of serum to maintain cell functions.

Oxidized DNA Base Estimation by Gas Chromatography/Mass Spectrometry

Oxidized DNA bases are estimated in samples prepared according to the procedures described earlier and corresponding either to DNA extracts (DNA oxidation) or to culture media (DNA repair). The analysis by capillary GC/MS-SIM is performed on an HP 5890 series II gas chromatograph (Hewlett Packard, Les Ulis, France) equipped with a capillary column (0.25 mm, 30 m) coated with 5% phenylmethylsiloxane (HP5-MS; Hewlett Packard). The constant flow rate is 1 ml/min of helium as the carrier gas. Injections (2 µl) are performed by an autosampling injector (6890 series, Hewlett Packard) in the splitless mode with the injection port set at 250°. Temperature of the oven is left at 150° for 3 min, followed by an increase at a rate of 7°/min to 270°. Detection of positive ions is performed by an HP 5972 mass detector using electron impact ionization (70 eV). Chromatograms are collected in the single-ion monitoring mode, recorded, and treated using chromatography software (MS Chemstation, Hewlett Packard). The mass selective detector operates in the SIM mode, which monitors both the molecular ion M^+ and the ion $(M-CH_3)^+$, which is characteristic of trimethylsilylated derivatives. These selected ions are detected at a dwell time of 100 msec in the following elution order: azaT, m/z 256,271; 5-OH-Ura, m/z 329,344; 5-OHMe-Ura, m/z, 343,358; 5-OH-Cyt, m/z, 328,343; azaA, m/z, 265,280; FapyAde, m/z, 354,369; Xan, m/z, 353,368; 2-oxo-Ade, m/z, 352,367; Gua, m/z, 355,369; and 8-oxo-Gua, m/z, 440,455. The concentrations of oxidized base in DNA extracts (in nanomole/mg DNA) are calculated using the ratios of peak areas with azaT and azaA as internal standards, respectively, for pyrimidine- and purine-derived oxidized bases. Calibration curves are constructed for the analyte standards. Sample analytes are identified by the comparison of retention times and relative abundance of confirming ions with the corresponding values for authentic calibrators. Guanine concentration is determined in the same way using a standard curve expressed in nanomoles and then converted to milligrams of DNA in each sample based on the relation 100 µg DNA corresponds to 10.19 µg guanine. This value is correlated with spectrophotochemically and fluorometrically measured DNA concentrations. The number of modified bases per 10^6 bases in DNA is calculated by dividing values expressed in nanomole/mg DNA by 3.14 and then multiplying by 10^3. The concentrations of oxidized bases released in culture media are calculated as described earlier and are usually expressed in picomole/ml of culture medium.

RNA Blot Analysis of DNA Repair Enzyme Expression

Total RNA is extracted from hepatocytes and is purified using the Trireagent kit (Sigma-Fluka Chemical Co, Saint Quentin Fallavier, France). The cDNA probe for rat DNA polymerase β (EC 2.7.7.7) is kindly provided by Dr. R. W. Sobol (Department of Health and Human Services, Laboratory of Structural Biology,

NIH, Research Triangle Park, NC) as pBluescript II SK-plasmids. Prior to RNA blot analysis, specificity of the probe is verified by Northern blot analysis on rat liver RNA extracts. Aliquots of total RNA (10 μg) are electrophoresed on a 1% agarose formaldehyde gel, transferred on a Hybond N+ membrane (Amersham International Ltd., UK), and hybridized overnight at 65° with the DNA probe labeled with [α-^{32}P]dCTP by random priming.

Dot blot analysis is performed as follows: aliquots of total RNA (10 μg) are blotted to a nylon membrane (Amersham International Ltd., UK) and hybridized overnight at 65° with cDNA labeled with [α-^{32}P]dCTP by random priming. Relative mRNA amounts are corrected for differences in RNA loading by comparison with the corresponding 18S rRNA signals.

Results and Conclusion

DNA Oxidation and Repair in Hepatocyte Cultures on Myricetin Treatment

Using the GC/MS-SIM method, the following DNA base products were identified and quantified in cellular DNA: 5-hydroxyuracil (5-OH-Ura), 5-hydroxycytosine (5-OH-Cyt), 5-hydroxymethyluracil (5-OHMe-Ura), 2-oxoadenine (2-oxo-Ade), 4,6-diamino-5-formamidopyrimidine (FapyAde), xanthine (Xan), and 8-oxoguanine (8-oxo-Gua). Among these compounds, uracil derivatives were products of cytosine oxidation, except 5-OHMe-Ura, which derived from thymine modifications. In addition, xanthine and FapyAde corresponded to purine oxidation products.

In general, all experiments performed with this model revealed that oxidized purines were the most abundant oxidation products derived from DNA bases.[15] We have detected a basal level of all oxidized bases in controls, resulting from normal cell metabolism. These levels were increased greatly when the cells were treated with iron, as an inducer of oxidative stress.[15] In addition, we have reported that the high levels of DNA oxidation products in iron-treated samples were reduced greatly in the simultaneous presence of myricetin.[9]

The calculated concentrations of each oxidation product can be summed up to represent global DNA oxidation (Fig. 1). It is expressed in percentage of controls and corresponds to the extent of DNA oxidation on each treatment. As already noted for individual oxidized base, myricetin prevented iron-induced DNA oxidation, as shown by a decrease in the accumulation of oxidation products in genomic DNA (Fig. 1).

Concurrently with the formation of oxidized bases in DNA extracts, their excision from DNA, as a repair mechanism, was estimated based on various treatments.

---

[15] V. A. Abalea, J. Cillard, M. P. Dubos, J. P. Anger, P. Cillard, and I. Morel, *Carcinogenesis* **19,** 1053 (1998).

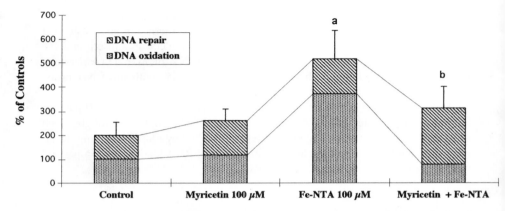

FIG. 1. Cumulative DNA oxidation and repair in rat hepatocyte cultures treated with myricetin and Fe-NTA. Hepatocyte cultures were incubated for 4 hr in the presence of myricetin (100 μM) and/or Fe-NTA (100 μM). Global DNA oxidation of hepatocyte DNA (plain bars) and global DNA repair (hatched bars) were, respectively, expressed as percentage of control values representing 100% and corresponding to basal levels of oxidation and repair ($n = 3$ hepatocyte populations from different animals). (a) DNA oxidation significantly greater than controls, $p < 0.01$; (b) DNA repair significantly greater than controls, $p < 0.01$. [From V. A. Abalea, J. Cillard, M. P. Dubos, O. Sergent, P. Cillard, and I. Morel, *Free Radic. Biol. Med.* **26**, 1457 (1999).]

All oxidized base products were detected in the culture medium of control cultures, showing a basal repair activity that represents 100% of repair (Fig. 1) Global DNA repair was activated in the presence of myricetin added to iron-treated cultures, whereas iron alone did not modify this repair activity (Fig. 1) Moreover, when representing cumulative DNA oxidation and repair (Fig. 1), intense DNA oxidation induced by iron was counterbalanced by the stimulation of DNA repair by myricetin. In the presence of both myricetin and Fe-NTA, cumulative levels of DNA oxidation products (in cells and in culture media) were reduced significantly when compared with Fe-NTA alone, indicating that in addition to an activation of DNA repair by myricetin, another protective mechanism might occur, which quite surely corresponded to the antioxidant properties of myricetin, destroying reactive oxygen species and sequestering iron.[16]

DNA Repair Enzyme Expression on Myricetin Treatment

Accumulation of oxidative modified bases in DNA is governed by a balance between their production via free radical attack and their elimination by specific base excision-repair enzymes.

[16] I. Morel, G. Lescoat, P. Cogrel, O. Sergent, N. Pasdeloup, P. Brissot, P. Cillard, and J. Cillard, *Biochem. Pharmacol.* **45**, 13 (1993).

In order to determine the inducibility of base excision repair enzymes on myricetin treatment, the expression of the specific repair enzyme DNA polymerase β (pol β) was investigated in our experimental model. The technique used for these experiments required a 48-hr delay before notable modifications in expression of the repair enzyme (Fig. 2). We have shown that pol β mRNA levels increased with myricetin in a dose-dependent manner in the presence or absence of iron (Fig. 2). In

DNA Polymerase β mRNA level

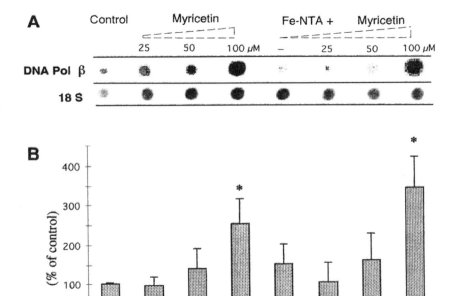

FIG. 2. RNA dot blot analysis of DNA polymerase β in rat hepatocytes treated with Fe-NTA and myricetin (A) Total RNA was prepared from hepatocyte cultures incubated for 48 hr in the presence of Fe-NTA (100 μM) and myricetin (25, 50, and 100 μM). RNA blot analysis was performed using radiolabeled polymerase β cDNA (top) and an oligonucleotide specific for the 18S rRNA (bottom) as probes. (B) mRNA signals corresponding to polymerase β were quantified by densitometry and normalized for differences in RNA loading using the 18S RNA probe. Data were expressed as the percentage of controls representing 100%. Results are representative of three independent experiments. (*) Significantly greater than controls, $p < 0.01$. [From V. A. Abalea, J. Cillard, M. P. Dubos, O. Sergent, P. Cillard, and I. Morel, *Free Radic. Biol. Med.* **26**, 1457 (1999).]

such experiments using primary hepatocyte culture, the increase in mRNA levels following isolation stress often renders analysis difficult up to 24 hr after cell seeding. Moreover, when using cell lines, care must be taken to standardize the delay after cell subculture and cell density at the beginning of the treatments.

Our experiments allowed us to demonstrate that, in iron-treated hepatocyte cultures, the flavonoid myricetin was able to prevent the accumulation of oxidized bases in DNA of the cells. This action resulted both from an antioxidant activity decreasing the generation of DNA oxidation products and from an activation of repair mechanisms, eliminating oxidized products already formed in DNA.

The GC/MS technique described here can be applied to evaluate oxidized base products of DNA. However, these experiments required some precautions to prevent artifactual oxidation of DNA during extraction and derivatization. Other recent techniques do not require such sample preparation and therefore reduce the risk of artifactual DNA oxidation. This is the case for HPLC with tandem mass spectrometry (MS/MS) analysis, which presents advantages in terms of accuracy and sensitivity, although it is also more expensive. In addition, alternative methods can be used to estimate repair of oxidized DNA, such as new technologies in which DNA arrays are developed using cDNA probes of numerous repair enzymes. The choice of technique depends on the function of the materials and the funds available.

Section IV

Biological Activity

[28] Binding of Flavonoids to Plasma Proteins

By Olivier Dangles, Claire Dufour, Claudine Manach, Christine Morand, and Christian Remesy

Introduction

Flavonoids are widely distributed plant polyphenols involved in several biologically important mechanisms such as pigmentation, nitrogen fixation, and chemical defense.[1] Their huge structural diversity essentially rests on the following factors (Scheme I): (1) different types of aglycone that differentiate the flavonoid subfamilies; (2) substitution of the aglycone by hydroxyl and methoxyl groups; and (3) glycosidation of OH groups by a variety of sugars which themselves may be acylated by a variety of aliphatic and aromatic acids

Many flavonoids possess *in vitro* anticancer, antiviral and anti-inflammatory properties, which may be the consequence of their ability to inhibit a broad range of enzymes and to act as potent antioxidants.[2a-2h] In particular, polyhydroxylated flavones and flavonols (3-hydroxyflavones) display electron-rich, highly conjugated nuclei, allowing them to react quickly with toxic reactive oxygen species such as hydroxyl, alkoxyl, and alkylperoxyl radicals and singlet dioxygen[3a-3f]

---

[1] "The Flavonoids, Advances in Research since 1986" (J. B. Harborne, ed.), Chapman and Hall, London, 1994.
[2a] E. Middleton, Jr., and C. Kandaswami, in "The Flavonoids, Advances in Research since 1986" (J. B. Harborne, ed.), p. 619. Chapman and Hall, London, 1994.
[2b] "Flavonoids in Health and Disease" (C. A. Rice-Evans and L. Packer, eds.). Dekker, New York, 1997.
[2c] P. Cos, L. Ying, M. Calomme, J. P. Hu, K. Cimanga, B. Van Poel, L. Pieters, A. J. Vlietinck, and D. Vanden Berghe, *J. Nat. Prod.* **61,** 71 (1998).
[2d] J. R. S. Hoult, M. A. Moroney, and M. Paya, *Methods Enzymol.* **234,** 443 (1994).
[2e] I. Morel, G. Lescoat, P. Cillard, and J. Cillard, *Methods Enzymol.* **234,** 437 (1994).
[2f] W. F. Hodnick, D. L. Duval, and R. S. Pardini, *Biochem. Pharmacol.* **47,** 573 (1994).
[2g] N. Cotelle, J.-L. Bernier, J.-P. Catteau, J. Pommery, J.-C. Wallet, and E. M. Gaydou, *Free Radic. Biol. Med.* **20,** 35 (1996).
[2h] G. Cao, E. Sofic, and R. L. Prior, *Free Radic. Biol. Med.* **22,** 749 (1997).
[3a] S. Steenken and P. Neta, *J. Phys. Chem.* **86,** 3661 (1982).
[3b] W. Bors and M. Saran, *Free Radic. Res. Commun.* **2,** 289 (1987).
[3c] W. Bors, W. Heller, C. Michel, and M. Saran, *Methods Enzymol.* **186,** 343 (1990).
[3d] W. Bors, C. Michel, and M. Saran, *Methods Enzymol* **234,** 420 (1994).
[3e] S. V. Jovanovic, S. Steenken, M. Tosic, B. Marjanovic, and M. G. Simic, *J. Am. Chem. Soc.* **116,** 4846 (1994).
[3f] S. V. Jovanovic, S. Steenken, Y. Hara, and M. G. Simic, *J. Chem. Soc. Perkin Trans.* **2,** 2497 (1996).

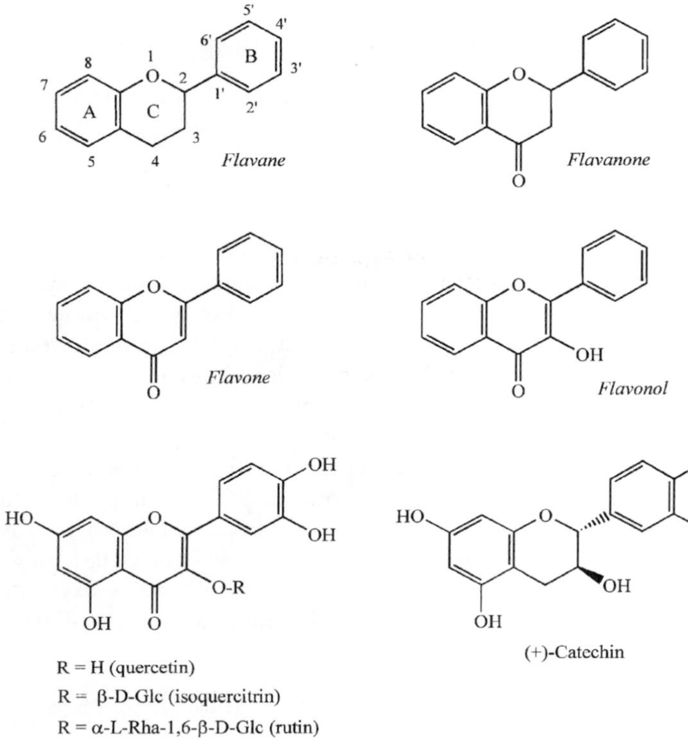

SCHEME I

and efficiently inhibit lipid peroxidation.[4a–4g] Being relatively abundant in the human diet, flavonoids could play an important role in the protective mechanisms against cardiovascular diseases and other pathologies (cancers, neurodegenerescence, aging) in which reactive oxygen species are involved, as suggested by

[4a] J. A. Vinson, Y. A. Dabbagh, M. M. Serry, and J. Jang, *J. Agric. Food Chem.* **43**, 2800 (1995).
[4b] C. V. de Whalley, S. M. Rankin, J. R. S. Hoult, W. Jessup, and D. S. Leake, *Biochem. Pharmacol.* **39**, 1743 (1990).
[4c] S. Miura, J. Watanabe, M. Sano, T. Tomita, T. Osawa, Y. Hara, and I. Tomita, *Biol. Pharm. Bull.* **18**, 1 (1995).
[4d] M. Foti, M. Piattelli, M. T. Baratta, and G. Ruberto, *J. Agric. Food Chem.* **44**, 497 (1996).
[4e] V. A. Roginsky, T. K. Barsukova, A. A. Remorova, and W. Bors, *J. Am. Oil Chem. Soc.* **73**, 777 (1996).
[4f] K. Ioku, T. Tsushida, Y. Takei, N. Nakatani, and J. Terao, *Biochim. Biophys. Acta* **1234**, 99 (1995).
[4g] N. Salah, N. J. Miller, G. Paganga, L. Tijburg, G. P. Bolwell, and C. Rice-Evans, *Arch. Biochem. Biophys.* **322**, 339 (1995).

epidemiological studies.[5a,5b] However, demonstrating the possible relationship between flavonoid-rich diets and health requires a deeper understanding of the way flavonoids are absorbed and metabolized. Quercetin conjugates have been detected in plasma of humans and rats fed a quercetin-rich diet.[6a–6e] Such conjugates (sulfoglucuronides of quercetin and 3′-methylquercetin) were essentially bound to serum albumin[6a,6b] and (at least in their free form) retained potent antioxidant activities.[6d,6e] Hence, flavonoid–albumin complexation may help maintain significant concentrations of flavonoids in plasma, thereby prolonging their antioxidant action.

This article reports on the binding of flavonoids to serum albumin (as a function of the flavonoid structure) and on its possible influence on the properties and bioavailability of flavonoids. The possible complexation of flavonoids by other plasma proteins will also be addressed.

Spectroscopic Methods for Assessing Flavonoid–Albumin-Binding Constants

Fluorescence spectroscopy is a very sensitive tool for investigating flavonoid–serum albumin binding. Two methods have been used. In both, aliquots of concentrated solutions of ligand (flavonoid) in methanol or dimethyl sulfoxide (DMSO) are added repeatedly (without significant dilution) to a pH 7.4 (phosphate buffer) solution of albumin held at constant concentration.

Method 1: Enhancement of Flavonoid Fluorescence

Naturally occurring flavones and flavonols (3-hydroxyflavones) usually bear a OH group at position 5, which is known to completely quench their intrinsic fluorescence.[7] However, albumin binding can promote a saturable fluorescence emission above 500 nm whose plateau intensity is highly dependent on the

---

[5a] M. G. L. Hertog, E. J. M. Feskens, P. C. H. Hollman, M. B. Katan, and D. Kromhout, *Lancet* **342**, 1007 (1993).

[5b] D. S. Leake, in "Phytochemistry of Fruit and Vegetables" (F. A. Tomas-Barberan and R. J. Robins, eds.), p. 287. Clarendon Press, Oxford, 1997.

[6a] C. Manach, C. Morand, O. Texier, M.-L. Favier, G. Agullo, C. Demigné, F. Régérat, and C. Rémésy, *J. Nutr.* **125**, 1911 (1995).

[6b] C. Manach, O. Texier, F. Régérat, G. Agullo, C. Demigné, and C. Rémésy, *J. Nutr. Biochem.* **7**, 375 (1996).

[6c] C. Manach, C. Morand, C. Demigné, O. Texier, F. Régérat, and C. Rémésy, *FEBS Lett.* **409**, 12 (1997).

[6d] C. Morand, V. Crespy, C. Manach, C. Besson, C. Demigné, and C. Rémésy, *Am. J. Physiol.* **275**, R212 (1998).

[6e] C. Manach, C. Morand, V. Crespy, C. Demigné, O. Texier, F. Régérat, and C. Rémésy, *FEBS Lett.* **426**, 331 (1998).

[7] O. S. Wolfbeis, M. Begum, and H. Geiger, *Z. Naturforsch.* **39b**, 231 (1984).

flavonoid substitution.[6a–8] The F vs L_t curves can be fitted against Eqs. (1) and (2) for the determination of optimized values for parameters K_1 and f_1.

$$F = \frac{f_1 K_1 [L] c}{1 + K_1 [L]} \tag{1}$$

$$L_t = [L]\left(1 + \frac{K_1 c}{1 + K_1 [L]}\right) \tag{2}$$

where F is the fluorescence intensity, f_1 is the molar fluorescence intensity of the complex, [L] is the free ligand concentration, L_t is the total ligand concentration, K_1 is the 1 : 1 binding constant, and c is the total albumin concentration.

Method 2: Flavonoid-Induced Changes in Fluorescence of Albumin-Bound Indicators

Quercetin can be used as an indicator for investigating the binding of flavonoids that do not fluoresce in their free or bound forms. Alternatively, competition between flavonoids and ligands known, from X-ray crystallographic studies,[9a,9b] to bind to typical albumin-binding sites helps determine precisely the flavonoid-binding site(s). Assuming pure competitive binding between indicator I and ligand L, Eqs. (3) and (4) can be derived and used in the curve fitting of the F vs L_t curve for the determination of optimized values for parameters K_1 and f_I (after previous determination of K_I in the absence of L).

$$F = f_I \frac{K_I [P] I_t}{1 + K_I [P]} \tag{3}$$

$$c = [P]\left(1 + \frac{K_I I_t}{1 + K_I [P]} + \frac{K_1 L_t}{1 + K_1 [P]}\right) \tag{4}$$

where f_I is the molar fluorescence intensity of the indicator–albumin complex, I_t is the total indicator concentration, K_I is the indicator–albumin-binding constant; [P] is the free protein concentration, c is the total protein concentration, K_1 is the ligand–albumin-binding constant, and L_t is the total ligand concentration.

Alternatively, taking into account the possibility of noncompetitive binding between indicator I and ligand L leads to Eqs. (5)–(8). Their use in the curve-fitting of the F vs L_t curve yields optimized values for parameters f_{IL} and K (f_{IL}: molar fluorescence intensity of the indicator–ligand–albumin ternary complex; K, thermodynamic constant for the formation of the ternary complex from the indicator–albumin complex and free ligand L), K_1, K_I, and f_I being previously determined

---

[8] O. Dangles, C. Dufour, and S. Bret, *J. Chem. Soc. Perkin Trans.* **2**, 737 (1999).
[9a] D. C. Carter and J. X. Ho, *Adv. Protein Chem.* **45**, 153 (1994).
[9b] X. M. He and D. C. Carter, *Nature* **358**, 209 (1992).

from investigations of the ligand–albumin and indicator–albumin systems.

$$F = K_I[\text{I}][\text{P}](f_I + f_{IL} K[\text{L}]) \tag{5}$$

$$I_t = [\text{I}](1 + K_I[\text{P}] + K K_I[\text{L}][\text{P}]) \tag{6}$$

$$L_t = [\text{L}](1 + K_1[\text{P}] + K K_I[\text{I}][\text{P}]) \tag{7}$$

$$c = [\text{P}](1 + K_I[\text{I}] + K_1[\text{L}] + K K_I[\text{L}][\text{I}]) \tag{8}$$

Binding Studies

The affinity of flavonoids for serum albumin is in line with the general ability of this protein to bind small negatively charged ligands.[9a,9b,10] Indeed, most flavonoids display acidic phenolic OH groups, which are dissociated at physiological pH. Because of the generally high homology within the serum albumin family, binding studies typically deal not only with human serum albumin (HSA), but also with the less expensive bovine serum albumin (BSA, 76% homology with HSA).

Binding to serum albumin has been investigated with a series of flavonoids belonging to the main flavonoid subfamilies. Quercetin (3,3′,4′,5,7-pentahydroxyflavone, Scheme I) is a flavonoid aglycone ubiquitous in plants and is relatively abundant in the human diet.[11a–11d] It also displays the structural requirements (electron delocalization on the whole polyphenolic nucleus due the C-2–C-3 double bond, 1,2-dihydroxy substitution on ring B, a OH group at position 3) for a strong antioxidant activity. HSA has been shown to bind quercetin and other structurally related flavonoids strongly *in vitro*.[6a–12] In particular, ultracentrifugation experiments at 20° using [4-^{14}C]quercetin[12] showed that 99.1 ($\pm$0.5)% and 99.4 ($\pm$0.5)% of the quercetin (1.5–15 μM) is bound in plasma and in a pH 7.4 solution of HSA at physiological concentration (40 mg/ml, i.e., about 0.6 mM), respectively. The bound fraction was found independent of the total quercetin concentration in agreement with the nonsaturation of the HSA-binding sites at low quercetin–HSA molar ratios. From these results, the value for the apparent quercetin–HSA-binding constant can be estimated (assuming 1 : 1 binding): $K = 267\,(\pm 33) \times 10^3 M^{-1}$ at 20°.[13] Interestingly, the binding of quercetin to plasma protein is quantitatively

---

[10] T. Peters, *Adv. Protein Chem.* **37**, 161 (1985).
[11a] A. Bilyk and G. M. Sapers, *J. Agric. Food Chem.* **33**, 226 (1985).
[11b] M. G. L. Hertog, P. C. H. Hollman, and M. B. Katan, *J. Agric. Food Chem.* **40**, 2379 (1992).
[11c] M. G. L. Hertog, P. C. H. Hollman, and B. van der Putte, *J. Agric. Food Chem.* **41**, 1242 (1993).
[11d] U. Justesen, P. Knuthsen, and T. Leth, *J. Chromatogr.* **799**, 101 (1998).
[12] D. W. Boulton, U. K. Walle, and T. Walle, *J. Pharm. Pharmacol.* **50**, 243 (1998).
[13] This K value is higher than that deduced from fluorescence measurements (see Table I) and may be marred by inaccuracy due to the quasisaturation conditions used (e.g., changing the fraction of bound quercetin from 99.0 to 99.5% increases the K value by a factor 2). In addition, the reported stronger binding with decreasing temperature bears evidence of an *exothermic* complexation contrary to the authors' claim. Using the K values reported in Ref. 13, we found $\Delta H^0 = -18.4$ kJ mol^{-1}, $\Delta S^0 = 41$ JK^{-1} mol^{-1}.

accounted for by its binding to HSA, thus suggesting that the binding to other plasma proteins is much weaker (see later).

Binding of quercetin to BSA was also investigated quantitatively by fluorescence spectroscopy.[8] At pH 7.4, BSA was found to promote a strong saturable fluorescence emission of quercetin around 530 nm (excitation at 450 nm) as reported for HSA.[6a] Assuming n identical binding sites with a microscopic binding constant K (Scatchard analysis), one gets $n = 0.95\,(\pm 0.04)$, $K = 103\,(\pm 16) \times 10^3 M^{-1}$ at 25°. Hence, the stoichiometry of the quercetin–BSA complex is 1:1. Retreating data using a simple 1:1 binding model (Method 1) gives $K = 135\,(\pm 13) \times 10^3 M^{-1}$. The quercetin–BSA complexation can also be investigated by fluorescence at pH 5.6, despite a much weaker BSA-induced fluorescence enhancement of quercetin. Scatchard analysis of the fluorescence titration curve yields $n = 1.09\,(\pm 0.04)$, $K = 81(\pm 14) \times 10^3 M^{-1}$ at 25°.[8] Hence, neutral quercetin (pH 5.6) and its anions (pH 7.4) roughly display the same affinity for BSA.

The intrinsic fluorescence of albumin (excitation at 295 nm, emission at 340 nm) is due to the tryptophan residue(s) (Trp-214 in HSA, Trp-212 and Trp-134 in BSA). Trp-214 in HSA (Trp-212 in BSA) is conserved in all mammalian albumins and is located strategically in the IIA domain for developing van der Waals interactions with ligands bound to that site.[9] Quercetin actually promotes a strong quenching of the Trp emission at 340 nm (excitation at 295 nm). For quercetin concentrations ensuring complete complexation of albumin (quercetin–albumin molar ratios higher than 2), the intrinsic fluorescence emission is virtually suppressed.[8] Such observations are consistent with a binding occurring in the IIA domain.

Serum albumin is composed of several mostly helical domains, which are structred by disulfide bridges.[9a,9b,10] In particular, domains IIA and IIIA are delimited by a hydrophobic surface on one side and a positively charged surface on the other side that allow them to specifically bind relatively large heterocyclic negatively charged ligands and small aromatic carboxylic acids, respectively. For instance, from X-ray crystallographic studies, warfarin, a coumarin structurally related to flavonoids, has been demonstrated to bind to the IIA domain.[9a,9b] In addition, ultracentrifugation experiments have shown that [4-^{14}C] quercetin (1.5–15 μM) is significantly displaced from HSA (0.6 mM) by (±)-warfarin in large excess (up to 1 mM), whereas (±)-ibuprofen, a IIIA site binder, caused no effect in the same conditions.[12] At first sight, this suggests competition between warfarin and quercetin for binding to the IIA domain. However, the large excess of warfarin required to promote a significant displacement, despite its reported high affinity for HSA[9a] (K in the range 8.9×10^4–$2.5 \times 10^5 M^{-1}$), throws some doubt on the hypothesis of a purely competitive behavior. In addition, 3-hydroxyflavone and its anion have been shown to bind HSA to low-affinity and high-affinity sites, respectively.[14] The efficient energy transfer from Trp-214 to neutral 3-hydroxyflavone was taken as evidence

[14] A. Sytnik and I. Litvinyuk, *Proc. Natl. Acad. Sci. U.S.A.* **93**, 12959 (1996).

that the low-affinity binding site was the IIA domain. Alternatively, the much slower energy transfer from HSA to the 3-hydroxyflavone anion suggested that the high-affinity binding site was the IIIA domain. Hence, additional experiments are needed to precisely locate the flavonols within the albumin structure.

When increasing amounts of warfarin are added to a solution of BSA and quercetin held in constant concentrations (1 : 1 molar ratio), the quenching of the fluorescence of the quercetin–BSA complex cannot be accounted for by assuming a competitive model because of a strong remaining fluorescence at warfarin–quercetin molar ratios higher than 10. Alternatively, when increasing amounts of quercetin are added to a solution of BSA and warfarin held in constant concentrations, the plateau fluorescence intensity of the quercetin–BSA complex is lower than in the absence of warfarin. A more general model taking into account noncompetitive binding allows one to satisfactorily fit both F vs L_t curves. Thus, warfarin was found to bind free BSA and the quercetin–BSA complex with binding constants of 134 (± 8) × 10^3 and 10.3 (± 1.5) × $10^3 M^{-1}$, respectively. The former value is in reasonable agreement with literature values.[9a] Unlike warfarin, ibuprofen only very weakly perturbs the fluorescence of the quercetin–BSA complex, thus confirming that the binding of quercetin occurs more likely in the IIA domain. However, the IIA domain of BSA may simultaneously accommodate quercetin and other IIA domain binders such as warfarin.

The main circulating flavonoids in the plasma of human volunteers fed a quercetin-rich diet are sulfoglucuronides of quercetin and 3'-O-methylquercetin (isorhamnetin).[6e] Hence, the consequences of methylation, sulfation, and glycosidation (as a model for glucuronidation) of quercetin on its affinity for serum albumin deserve special attention. Quercetin 7-O-sulfate retains a high affinity for HSA and BSA, whereas additional sulfation of 4'-OH markedly weakens the binding to both albumins (Table I). Sulfation of 3-OH, methylation of 3'-OH, and deletion of O-3' (kaempferol) all significantly lower the affinity to BSA, with the binding to HSA being much less sensitive to those structural changes. These observations outline the importance of a free OH at position 3' for a strong complexation to BSA. They are consistent with earlier observations that the affinity of BSA for catechol (1,2-dihydroxybenzene) is much higher than for resorcinol (1,3-dihydroxybenzene)[15a,15b] and suggest a bidentate-binding mode between the catechol group and BSA. In the case of quercetin, the most acidic 4'-OH probably forms an intramolecular hydrogen bond with O-3', thereby favoring hydrogen bonding between 3'-OH and a protein acceptor on the one hand and between O-4' and a protein donor on the other hand (cooperative hydrogen

---

[15a] J. P. McManus, K. G. Davis, J. E. Beart, S. H. Gaffney, T. H. Lilley, and E. Haslam, *J. Chem. Soc. Perkin Trans.* **2**, 1429 (1985).

[15b] J. P. McManus, K. G. Davis, T. H. Lilley, and E. Haslam, *J. Chem. Soc. Chem. Commun.* **309**, (1981).

TABLE I
FLAVONOID–ALBUMIN 1 : 1 BINDING CONSTANTS[a]

Flavonoid[b]	$K_1(M^{-1})$ HSA	$K_1(M^{-1})$ BSA
Quercetin	$61 (\pm 16) \times 10^3$ [c]	$135 (\pm 13) \times 10^3$
Quercetin 7-O-sulfate	$71 (\pm 16) \times 10^3$	$143 (\pm 29) \times 10^3$
Quercetin 3-O-sulfate	$47.6 (\pm 4.1) \times 10^3$	$18.1 (\pm 1.4) \times 10^3$
Quercetin 4′,7-O-disulfate	$5.0 (\pm 0.6) \times 10^3$	$21.1 (\pm 1.6) \times 10^3$
3′-O-Methylquercetin	$96 (\pm 34) \times 10^3$ [d]	$17.2 (\pm 2.0) \times 10^3$
Isoquercitrin	$14.6 (\pm 1.1) \times 10^3$	$8.4 (\pm 0.3) \times 10^3$
Kaempferol[e]	$101 (\pm 19) \times 10^3$	$10.8 (\pm 0.7) \times 10^3$
4′-O-Methylquercetin	—	$45.5 (\pm 2.4) \times 10^3$ [f]
Chrysin	—	$155 (\pm 11) \times 10^3$ [f]

[a] Determined by Method 1 unless specified otherwise; pH 7.4 phosphate buffer, 25°.
[b] Available commercially from Aldrich or Extrasynthèse (France) except quercetin 7-O-sulfate and quercetin 4′,7-O-disulfate, which have been synthesized from quercetin according to the literature [D. Barron and R. K. Ibrahim, *Tetrahedron* **43**, 5197 (1987)].
[c] Probably underestimated because of additional 1 : 2 complexation in the range of ligand concentration investigated.
[d] 1 : 1 and 1 : 2 serum albumin–flavonoid bindings. $K_2 = 3.5 (\pm 0.7) \times 10^3 M^{-1}$.
[e] 3,4′,5,7-Tetrahydroxyflavone (3′-deoxyquercetin).
[f] Determined by Method 2 using quercetin as the fluorescence indicator.

bonds, Scheme II). In that view, methylation at O-4′ would cancel cooperativity but still maintain the bidentate-binding mode in agreement with our observation that 4′-O-methylquercetin (tamarixetin) actually binds BSA more strongly than isorhamnetin.

The presence of a glycosyl group at position 3 markedly lowers the binding strength as already pointed out in the case of the BSA complexes of isoquercitrin

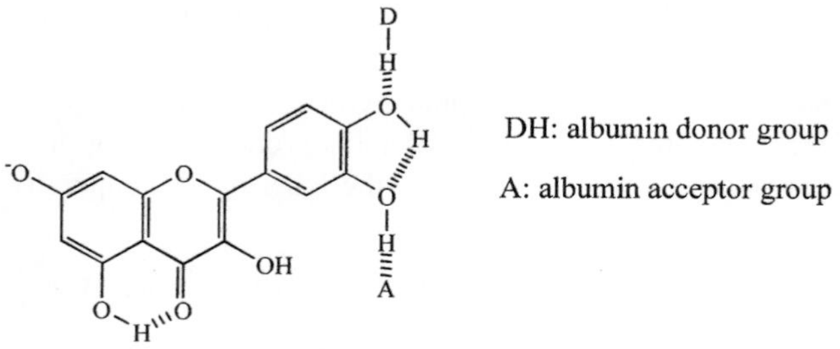

DH: albumin donor group

A: albumin acceptor group

SCHEME II

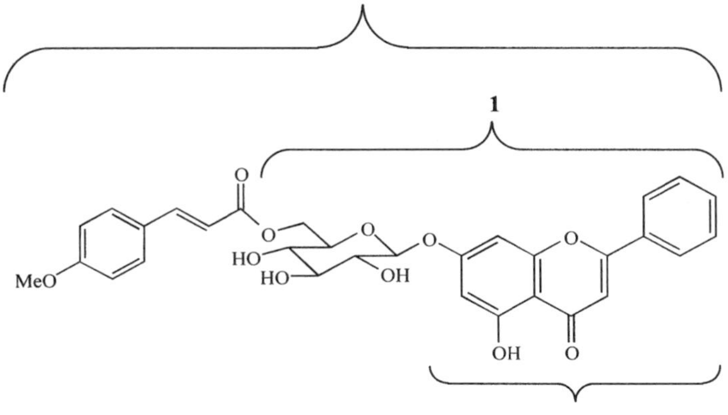

SCHEME III

and rutin (Scheme I).[9a,9b] For instance, the affinity of isoquercitrin for HSA and BSA is, respectively, about 4 and 15 times as low as that of quercetin (Table I).

Chrysin (5,7-dihydroxyflavone) and its chemically synthesized 7-O-β-D-glucopyranoside (**1**) and 7-O-[6-(p-methoxy)cinnamyl]-β-D-glucopyranoside (**2**) (Scheme III) provide an interesting series for investigating the influence of typical flavonoid substituents (sugars, acyl groups) on albumin binding.[16] Because chrysin and compounds **1** and **2** do not fluoresce in their free form or their BSA-bound form, their binding to BSA was studied using quercetin as the fluorescence indicator (Method 2). The gradual quenching of the quercetin fluorescence by chrysin goes on until complete extinction and can be accounted for quantitatively from a simple competitive model, which shows that chrysin roughly displays the same affinity for BSA as quercetin. In contrast, the addition of flavone **1** or methyl-p-methoxycinnamate slightly perturbs the fluorescence of the quercetin–BSA complex, thus suggesting that both ligands only interact weakly with BSA. Surprisingly, on addition of **2**, the fluorescence intensity of the quercetin–BSA complex decreases sharply until it reaches a clearcut plateau (about half the initial value in the absence of **2**) for a 1 : 1 flavone **2**–quercetin molar ratio. Hence, **2** and quercetin seem to bind BSA in a noncompetitive way, which leads to the formation of a weakly fluorescent flavone **2**–quercetin–BSA ternary complex. The flavone nucleus and the aromatic acid moiety of **2** are both required for the formation of the ternary complex. Hence, flavonoid glycosides bearing aromatic acyl groups may act as potent bidentate ligands for serum albumin. Similarly, polyphenols from the

[16] B. Alluis and O. Dangles, *Helv. Chim. Acta.* **82**, 2201 (1999).

gallotannin family bind to BSA with an affinity that increases dramatically with the number of galloyl groups on the D-glucose core.[15a,15b]

Significant bathochromic shifts are observed in the low-energy absorption band of flavonoids when they bind to HSA.[6b] At a HSA–flavonoid molar ratio of 5, shifts in the range of 20 to 30 nm were typically measured for flavonoids having a planar C ring (flavones, flavonols), whereas flavonoids having a nonplanar C ring (flavanones, dihydroflavonols) had their absorption band shifted by less than 10 nm. This does not necessarily mean that planar aglycones are bound more tightly to HSA than nonplanar analogs. Indeed, ultrafiltration experiments with solutions of taxifolin (dihydroquercetin) and flavanone eriodictyol (40 μM) in the presence of HSA (200 μM) showed that more than 90% of both nonplanar flavonoids is actually bound.[6a] This corresponds to an apparent binding constant higher than $55 \times 10^3 M^{-1}$ (pH 7.4, 25°). Apparently, the monomeric tannin catechin (Scheme I) binds albumin very loosely as evidenced by its inefficient quenching of the intrinsic protein fluorescence. In addition, catechin does not affect the fluorescence emission of the quercetin–BSA complex. Hence, if binding occurs to some extent, it does not take place in the IIA domain.

Albumin-Induced Changes in Flavonoid Structure

As a 5-hydroxyflavone, free quercetin fluoresces very weakly in both aqueous and nonaqueous solvents.[7] In contrast, quercetin displays a very strong fluorescence emission band around 520 nm (excitation at 450 nm) when dissolved in a 1 M HBr solution in acetic acid. Similar observations have already been made with flavones and flavonols[7,17a,17b] and interpreted by the protonation of O-4 in strongly acidic conditions and subsequent conversion of the neutral flavonoids into highly fluorescent pyrylium cations. Despite the very different conditions, the emission and excitation spectra of quercetin are very similar in HBr–acetic acid and in a pH 7.4 phosphate buffer containing BSA.[8] However, the excitation spectrum of albumin-bound quercetin is markedly different from its absorption spectrum, thus suggesting that bound quercetin in the ground state is actually a mixture of different forms with very different fluorescence properties. In weakly acidic and neutral aqueous buffers, free quercetin is essentially a mixture of nonfluorescent pyrone (4-oxo) forms. Binding to BSA may favor highly fluorescent pyrylium (4-hydroxy) forms. Although much less stable (in their free form) than normal pyrone tautomers, some highly fluorescent pyrylium tautomers could be specifically stabilized on binding to BSA. Formation of pyrylium tautomers may occur through intramolecular proton transfer from 4′-OH to O-4 (Scheme IV), as no fluorescence enhancement could be observed on binding of tamarixetin, the

---

[17a] R. Schipfer, O. S. Wolfbeis, and A. Knierzinger, *J. Chem. Soc. Perkin Trans.* **2**, 1443 (1981).
[17b] O. S. Wolfbeis, A. Knierzinger, and R. Schipfer, *J. Photochem.* **21**, 67 (1983).

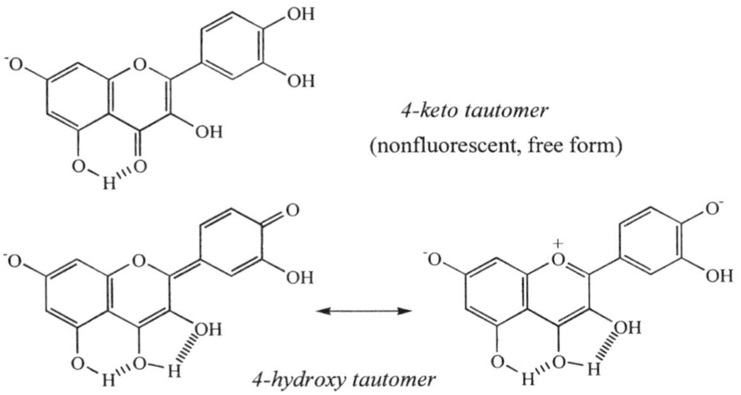

4-keto tautomer
(nonfluorescent, free form)

4-hydroxy tautomer
(fluorescent, stabilized in the albumin complex)

SCHEME IV

4′-methyl analog of quercetin. However, intramolecular proton transfer from 3-OH to O-4 may also contribute because the albumin complexes of luteolin, the 3-deoxy analog of quercetin, and of the 3-O-sulfate and 3-O-glycosides of quercetin are all much less fluorescent than the quercetin complex.

Possible Consequences of Albumin Binding on Properties of Flavonoids

Generally speaking, values for the binding constants rank flavonoids among the ligands displaying an intermediate affinity for albumin.[9a] Taking $10^4 M^{-1}$ as a lower limit of the flavonoid–albumin 1 : 1 binding constant (see Table I) and 0.6 mM as the total albumin concentration in plasma, it may be estimated that even flavonoids having the lowest intrinsic affinity for albumin are 85% bound at 25°. Although the flavonoid–albumin complexation is weaker at 37°,[12] it must remain high enough to retard flavonoid degradation via autoxidation and enzymatic reactions in plasma.

Autoxidation of quercetin might proceed through direct electron transfer from quercetin polyanions (formed at physiological pH and in mildly alkaline aqueous buffers) to the half-occupied π^* orbitals of dioxygen, although catalysis by contaminating trace metal ions is more likely. The aryloxyl radicals thus formed quickly disproportionate into quinones and quinonoid compounds, which add solvent molecules, and are then degraded into benzoic acid derivatives.[18] Binding to

---

[18] O. Dangles, G. Fargeix, and C. Dufour, *J. Chem. Soc. Perkin Trans.* **2**, 1387 (1999).

BSA significantly slows down quercetin autoxidation. For instance, the half-life of quercetin ($5 \times 10^{-5} M$) in a pH 7.4 phosphate buffer at 37° is 5.0 hr in the absence of BSA and 7.7 hr in the presence of a BSA concentration ($2 \times 10^{-4} M$), ensuring the total complexation of quercetin. However, free quercetin was not detected in the plasma of human volunteers after consumption of a quercetin-rich meal.[6e] The only detectable circulating forms of quercetin were sulfoglucuronides of quercetin and its 3'-O-methyl ether (isorhamnetin). Binding experiments reported in this work with quercetin sulfates and quercetin glycosides suggest that the binding of circulating quercetin conjugates to serum albumin may be much weaker than binding of quercetin itself. However, the physiological concentration of serum albumin (about 0.6 mM) is probably large enough to allow the extensive binding of the circulating quercetin conjugates, despite their lower intrinsic affinity for albumin. Hence, although probably less sensitive to autoxidation than quercetin, the circulating quercetin conjugates may have their half-life in plasma prolonged on serum albumin binding.

Flavonoid–albumin complexation may prevent the delivery of flavonoids to specific sites in order to trigger a biological effect. For instance, whole blood incubation experiments showed only a moderate association of quercetin to erythrocytes (more than 80% of quercetin remained in plasma), whereas association was extensive in the absence of plasma proteins (about 4% of free quercetin, only).[12] However, investigations with simple phenols pointed out that binding to BSA was much stronger at pH 6.0 (BSA N form) than at pH 2.2 (BSA F form),[15a,15b] thus suggesting that the binding strength is highly dependent on conformational changes in BSA. Hence, one may speculate that low local pHs at specific sites may help deliver the flavonoids to trigger a biological effect.

Another important question is the influence of albumin binding on the antioxidant properties of flavonoids in plasma. Antioxidants are compounds that, at low concentrations, can protect biomolecules (proteins, nucleic acids, polyunsaturated fatty acids, sugars) from oxidative degradations. Flavonoids may act by a variety of ways, including direct quenching of the reactive oxygen species, inhibition of enzymes involved in the production of the reactive oxygen species, chelation of transition metal ions able to promote radical formation (Fenton reactions), and regeneration of endogenous antioxidants (vitamin E).[2a–4g] Degradation of biomolecules by reactive oxygen species has been implicated in various diseases. In particular, peroxidation of polyunsaturated fatty acids in plasma low-density lipoproteins (LDL) is recognized as an early step in atherosclerosis. Covalent linking between the apoprotein B100 and carbonyl compounds (secondary peroxidation products) is known to trigger an immune response during which the modified LDL are internalized by macrophages. In this process, macrophages become loaded with cholesteryl esters and turn into giant *spumous cells,* which tend to deposit on the artery walls and initiate clots and lesions.[5b] Hence, protection of LDL by antioxidants is a matter of priority.

Several flavonoids, especially quercetin and its glycosides, are known to efficiently inhibit peroxidation (initiated by oxygen-centered radicals, metal ions, or macrophages) in LDL,[4a–4c] and in model solutions of linoleic acid,[4d,4e] as well as in phospholipid bilayers.[4f] Interestingly, peroxidation was shown to be inhibited significantly in LDL isolated from human plasma incubated previously with flavonoid antioxidants (including quercetin and rutin) at 50–200 μM,[19] in LDL isolated from the plasma of rats intragastrically administered with quercetin,[20] and in rat plasma after addition of quercetin at 4–80 μM.[20] These observations suggest that quercetin and its circulating conjugates (sulfoglucuronides) interact to some extent with LDL in plasma. Note, however, that the total concentration of quercetin conjugates measured in the plasma of human volunteers after consumption of a meal rich in plant products[6e] (in the range 0.1–1 μM) is much lower than those used in the incubation of human LDL. However, glucuronides of quercetin and isorhamnetin, as well as quercetin 3-O-sulfate, strongly inhibit lipid peroxidation in LDL (although less efficiently than quercetin) at concentrations lower than 1 μM.[6d,6e] Remarkably, quercetin, isorhamnetin, and their derivatives mentioned previously only retard LDL peroxidation. When peroxidation resumes, its rate is as high as in the absence of antioxidant. Such observations rule out a chain-breaking mechanism but strongly suggest that the flavonols act primarily by regenerating the endogenous LDL-bound amphiphilic antioxidant α-tocopherol. Consistently, α-tocopherol was demonstrated to be spared in LDL of rats administered with quercetin.[20]

From ultracentrifugation experiments using [4-^{14}C]quercetin, quercetin was shown to bind α_1-acid glycoprotein weakly (fraction of bound quercetin about 40%), whereas the binding to LDL was not significant (less than 0.5%).[12] It is reasonable to assume that the binding to LDL of the more polar circulating quercetin sulfoglucuronides is also very weak and that such conjugates are essentially bound to albumin in plasma. Hence, the true flavonoid antioxidants in plasma may well be the albumin–quercetin conjugate complexes themselves. Thus, a mechanism by which the albumin-bound quercetin conjugate transfers its labile H atoms to the hydrophilic phenoxyl head of the α-tocopheryl radical might take place (Scheme V). This mechanism could operate through interactions between the LDL particle and the flavonol-loaded albumin. It requires that the flavonol B ring (catechol group), which is mainly responsible for the antioxidant activity, remains accessible to oxidants and H atom abstracting agents in the albumin complex. Preliminary experiments suggest that it is true. For instance, periodate oxidation of the quercetin–BSA complex in the pH range 5–9 (aimed at investigating the reactivity toward biopolymers of flavonol quinones formed during the antioxidant

[19] J. A. Vinson, J. Jang, Y. A. Dabbagh, M. M. Serry, and S. Cai, *J. Agric. Food Chem.* **43**, 2798 (1995).
[20] E. L. da Silva, M. K. Piskula, N. Yamamoto, J.-H. Moon, and J. Terao, *FEBS Lett.* **430**, 405 (1998).

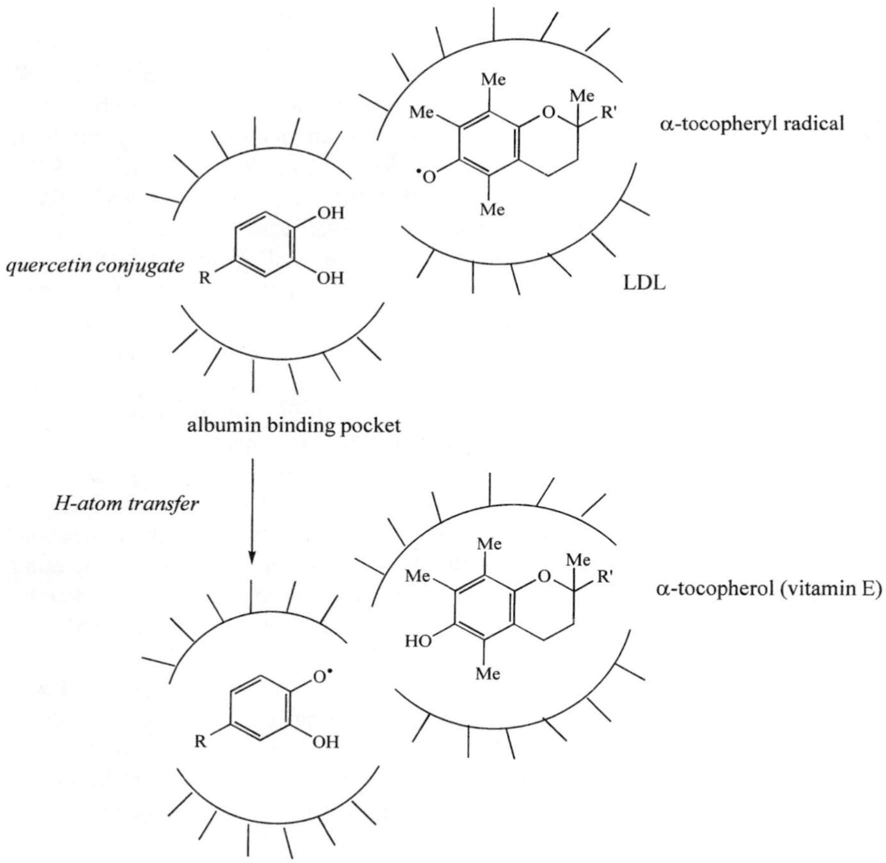

SCHEME V

action) turned out to be as fast as, if not faster than, oxidation of free quercetin.[8] Thus, bound quercetin seems to remain accessible to small oxidizing agents such as the periodate anion.[21] Moreover, H atom transfer from quercetin to the DPPH radical (diphenylpicrylhydrazyl) still takes place when quercetin is bound to BSA,

---

[21] The influence of BSA is only manifested at pH 9 by a strong slowing down of water addition on the bound quinone (in charge transfer interaction with the protein). Size-exclusion chromatography allowed the complete recovery of unmodified BSA (no detectable flavonoid-protein adducts), thus showing that water addition remains the sole significant pathway of quinone reactivity in the complex. This observation suggests that flavonoid quinones are innocuous, rapidly degraded compounds rather than potentially damaging electrophiles or oxidizing agents.

although about 20 times as slowly. In both cases (periodate oxidation, H-transfer to DPPH), flavonoid oxidation is faster than BSA oxidation, thus suggesting that bound flavonoids may also protect serum albumin from oxidative degradation. Finally, flavonoids may also protect albumin-bound polyunsaturated fatty acids from oxidative degradation in the same way as albumin-bound bilirubin, a potent endogenous plasma antioxidant (formed during heme metabolism).[22]

[22] R. Stocker, A. N. Glazer, and B. N. Ames, *Proc. Natl. Acad. Sci. U.S.A.* **84,** 5918 (1987).

[29] Protein Binding of Procyanidins: Studies Using Polyacrylamide Gel Electrophoresis and French Maritime Pine Bark Extract

By HADI MOINI, QIONG GUO, and LESTER PACKER

Introduction

Evidence shows that procyanidins may have beneficial health effects and their antioxidant activity may be responsible for the observed effects.[1–3] However, procyanidins bind with high affinity to extended proteins such as proline-rich proteins and histatins.[4,5] Therefore, the observed biological effects could be, in part, due to procyanidin–protein interactions. Thus, the existence of appropriate methods to determine such interactions is important.

Currently available assays indirectly determine the binding of procyanidins to proteins.[6–9] The competitive binding assay is based on the competition of the protein of interest with radiolabeled bovine serum albumin (BSA) to bind to the procyanidin. Assays based on the ability of procyanidins to inhibit the enzyme activity, such as trypsin or β-glucosidase, measure the reversal of the enzyme inhibition by procyanidins on addition of the protein of interest. However, the competitive binding assay with BSA suffers from the fact that binding of procyanidins to BSA

---

[1] M. Serafini, A. Ghiselli, and A. Ferro-Luzzi, *Lancet* **344,** 626 (1994).
[2] L. Packer, G. Rimbach, and F. Virgilli, *Free Radic. Biol. Med.* **27,** 704 (1999).
[3] L. Bravo, *Nutr. Rev.* **56,** 317 (1998).
[4] J. L. Goldstein and T. Swain, *Phytochemistry* **4,** 85 (1965).
[5] A. E. Hagerman and L. G. Butler, *J. Biol. Chem.* **256,** 4494 (1981).
[6] Q. Yan and A. Bennick, *Biochem. J.* **311,** 341 (1995).
[7] A. E. Hagerman and L. G. Butler, *J. Agric. Food Chem.* **26,** 809 (1978).
[8] H. Makkar, R. K. Dawra, and B. Singh, *J. Agric. Food Chem.* **36,** 523 (1988).
[9] J. Fickel, C. Pitra, B. A. Joest, and R. R. Hofmann, *Comp. Biochem. Physiol. C* **122,** 225 (1999).

is pH dependent, being observed at a pH below 5.0. The drawback of the assays based on the reversal of the enzyme inhibition is that the relationship between enzymatic activity and the binding is not fully understood and the protein of interest could be degraded when proteases are used as the enzyme.

Therefore, methods are needed to directly determining the interaction between procyanidins and proteins. This article describes the application of polyacrylamide gel electrophoresis (PAGE) to determine the binding of procyanidins to proteins. As a source of procyanidins, two highly standardized and defined extracts, namely Pycnogenol and a *Ginkgo biloba* extract (EGb 761), as well as various purified dimeric and trimeric procyanidins, have been used. Pycnogenol, an extract from French maritime pine bark (PBE), consists mainly of procyanidins formed by catechin and epicatechin units having a degree of polymerization of up to a heptamer. PBE also contains catechin, epicatechin, and taxifolin as monomer flavonoids, as well as phenolcarbonic acids and their glycosides as minor components. EGb 761 is composed of flavonoid glycosides, terpenoids, and procyanidins.

Polyacrylamide Gel Electrophoresis

Discontinuous PAGE is carried out under either denaturing or nondenaturing conditions, essentially as described by Laemmli for denaturing conditions.[10] In nondenaturing gel electrophoresis, sodium dodecyl sulfate (SDS), the reducing agent (2-mercaptoethanol), and the stacking gel are omitted. Milk xanthine oxidase (XO) (12.6 μg protein, 12 mU) is subjected to PAGE under nondenaturing conditions in the presence of PBE on the 3–8% NuPAGE precast gel and in the presence of EGb 761 on the 7% gel. Under denaturing condition, XO is subjected to PAGE on a 10% gel. The mixture of the enzyme and test compounds are prepared in 0.1 M phosphate buffer, pH 7.4, containing 0.1 mM EDTA and are loaded into the wells in a final volume of 20 μl. Following electrophoresis, gels are stained for activity.

Xanthine Oxidase Activity Staining in Polyacrylamide Gels

Activity staining on polyacrylamide gels is carried out at room temperature. The reaction mixture contains 50 mM Tris-HCl, pH 7.6, 0.5 mM xanthine, and 0.25 mM nitro blue tetrazolium (NBT).[11] Staining of the gels is continued until the activity band(s) appears on the gels (4–7 min). Then the reaction is stopped by rinsing the gel in a solution containing 15% glacial acetic acid and 15% methanol; the gel is washed five to six times with distilled water and scanned.

[10] U. K. Laemmli, *Nature* **227**, 680 (1970).
[11] N. Ozer, M. Muftuoglu, and H. Ogus, *J. Biochem. Biophys. Methods* **36**, 95 (1998).

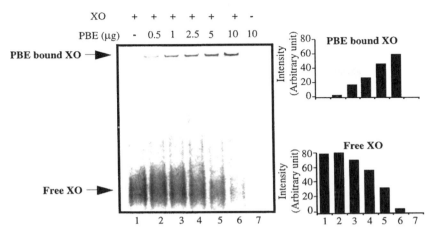

FIG. 1. Effect of PBE on the electrophoretic mobility of XO. (Left) XO and PBE, and a mixture of XO and PBE at indicated concentrations were subjected to PAGE under nondenaturing conditions. Samples were detected by activity staining. The presence of PBE changed the electrophoretic mobility of XO as revealed by a retarded band in the top part of the gel, indicating that PBE binds to XO. (Right) The intensity of PBE-bound XO and free XO bands.

Preparation of PBE, EGb 761, and Pure Flavonoids

A stock solution of 10 mg/ml of PBE and EGb 761 is prepared by dissolving these compounds first in 50 μl of dimethyl sulfoxide (DMSO) and then diluting to 1 ml with phosphate buffer, pH 7.4, containing 0.1 mM EDTA.

Typical Results

Under nondenaturing conditions the presence of PBE with XO caused the appearance in the top part of the gel of a band with a slower electrophoretic mobility, indicating that PBE binds to XO (Fig. 1). Under nondenaturing conditions, the movement of proteins through the polyacrylamide gel matrix depends on a combination of their size, shape, and charge. The change in the electrophoretic mobility of XO in the presence of PBE could possibly be due to the concomitant change in all of the aforementioned parameters rather than molecular weight alone. However, the presence of SDS (2%) and the reducing agent in the sample buffer does not interfere with the binding of PBE to XO under nondenaturing condition.

Due to the diversity of molecular structures among procyanidins and the variety of functional groups present in proteins, the interaction types, such as hydrogen bonding, ionic and hydrophobic interactions can in principle take place.[12] By

---

[12] W. D. Loomis, *Methods Enzymol.* **31,** 528 (1974).

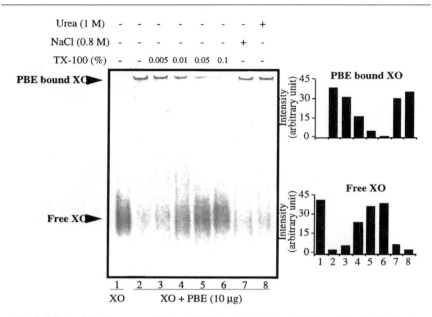

FIG. 2. Effect of NaCl, urea, and Triton X-100 on the binding of PBE to XO. (Left) XO alone or a mixture of XO and PBE were subjected to nondenaturing PAGE in the presence and absence of indicated concentrations of NaCl, urea, or Triton X-100 (TX-100). The concentrations of the agents were calculated according to 20 μl of total loading volume. Samples were detected by activity staining. The retarded band disappears with increasing concentrations of Triton X-100. (Right) The intensity of PBE-bound XO and free XO bands.

adding agents that compete with various types of interactions, to the mixture of procyanidin and protein, the type of interaction could also be investigated. Triton X-100, but not urea or NaCl, dose dependently restored the change in the electrophoretic mobility of XO, suggesting that the dominant mode of interaction between PBE and XO is hydrophobic binding (Fig. 2). Polyethylene glycol (PEG) 400, a nondetergent competitor for hydrophobic interactions, has also restored the change in the electrophoretic mobility of XO, confirming the involvement of hydrophobic interactions in the binding of PBE to XO (data not shown).

Because procyanidins up to trimers do not change the electrophoretic mobility of XO, procyanidins with a degree of polymerization higher than trimer are possibly responsible for the binding of PBE to XO.

PBE does not change the electrophoretic mobility of XO under denaturing conditions, suggesting that the binding is specific for the native structure of the enzyme rather than for an unfolded protein structure.

Under nondenaturing conditions, PBE does not change the electrophoretic mobility of superoxide dismutase. Although the possible structural differences

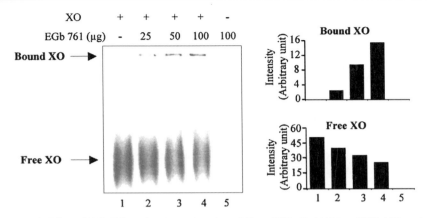

FIG. 3. Effect of EGb 761 on the electrophoretic mobility of XO. (Left) XO and EGb 761, and a mixture of XO and EGb 761 at indicated concentrations were subjected to PAGE under nondenaturing condition. Samples were detected by activity staining. The presence of EGb 761 changed the electrophoretic mobility of XO as revealed by a retarded band in the top part of the gel, indicating that EGb 761 binds to XO. (Right) The intensity of PBE-bound XO and free XO bands.

that cause the binding of PBE to one protein but not to another are not known, overall, selectivity is observed for the binding of PBE to enzymes.

Procyanidins constitute about 75% of PBE, whereas EGb 761 contains 7–9% procyanidins. When compared to PBE, EGb 761 at relatively higher concentrations changed the electrophoretic mobility of XO (Fig. 3).

Synopsis

The application of PAGE to determine the interaction between procyanidins and proteins, as presented here, enables one to directly determine the binding of either a pure or a complex mixture of flavonoids to a particular protein. If the protein of interest is an enzyme, the combination of PAGE with quantitative activity measurements allows identifying whether a change in the enzyme activity is related to the binding. Data presented suggest that PBE and EGb 761 have protein-binding properties, which, in addition to their redox-based effects, could provide a biochemical basis for their action in biological systems.

[30] Characterization of Antioxidant Effect of Procyanidins

By Fulvio Ursini, Ivan Rapuzzi, Rosanna Toniolo, Franco Tubaro, and Gino Bontempelli

Introduction

Polymers of polyhydroxy flavan-3-ol units are defined comprehensively as proanthocyanidins. (+)-Catechin and (−)-epicatechin are the monomeric precursors of procyanidins (Fig. 1), also known as condensed tannins.[1,2] Procyanidins are among the major components of various plants, including grapes. *Vitis vinifera* is a particularly rich source of procyanidins,[3] which are present in grape seeds. As is typical in the case of red wine production, procyanidins are extracted during fermentation in the presence of skins and seeds. Red wine is therefore a major nutritional source of procyanidins. Procyanidin-rich extracts from grape seed[4] and pine bark[5] have been introduced as dietary supplements that might play a protective role on human health.

The rationale for the role of procyanidins in the optimization of nutrition, either as food components or as a supplement, is grounded, mainly on their antioxidant capacity.[6] Prevention or decrease of the risk of atherosclerosis is the most likely biomedical effect of these compounds, which, if fully validated, would contribute to the elucidation of the mechanism of the protective effect of red wines suggested by epidemiological studies.

We used a grape seed extract in a liposome-dispersed form (Leucoselect Phytosome) for treating rabbits in which atherosclerosis was induced by cholesterol feeding.[7] Although the high level of plasma cholesterol remained unchanged, the treatment increased plasma antioxidant capacity and reduced the concentration of plasma lipid hydroperoxides to an undetectable level. This resulted in a dramatic decrease in the area of aorta covered by atherosclerotic plaques. Protection against

---

[1] H. A. Strafford, "Flavonoid Metabolism," CRC Press, Boca Raton, FL, 1990.
[2] L. J. Porter, *in* "The Flavonoids: Advances in Research since 1980" (J. B. Harborne, ed.), p. 21. Chapman & Hill, London, 1988.
[3] R. Maffei Facino, M. Carini, G. Aldini, E. Bombardelli, P. Morazzoni, and R. Morelli, *Arzneim. Forsh. Drug Res.* **44-I,** 592 (1994).
[4] E. Bombardelli and P. Morazzoni, *Fitoterapia* **66,** 291 (1995).
[5] F. Virgili, H. Kobuchi, and L. Packer, *Free Radic. Biol. Med.* **24,** 1120 (1998).
[6] J. M. Ricardo da Silva, M. Darmon, Y. Fernandez, and S. Mitjavita, *J. Agric. Food Chem.* **39,** 1549 (1991).
[7] F. Ursini, F. Tubaro, J. Rong, and A. Sevanian, *Nutr. Rev.* **57,** 241 (1999).

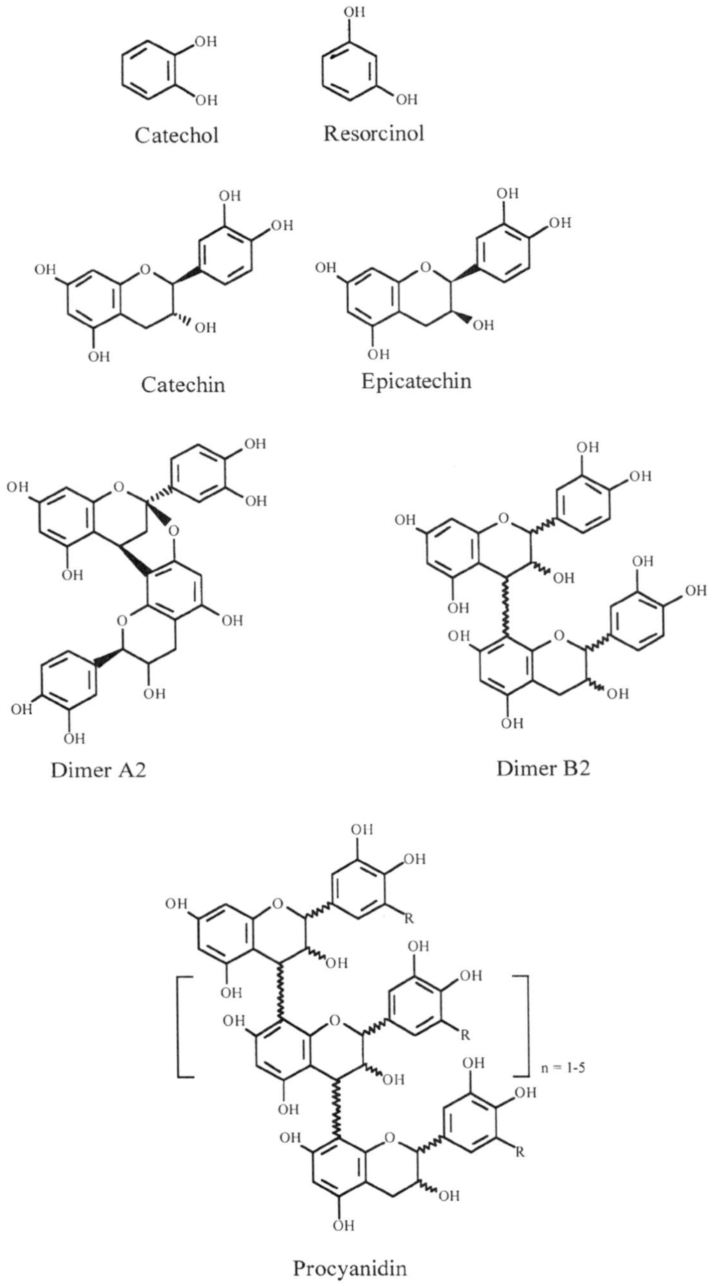

FIG. 1. Compounds used in this study.

experimental atherosclerosis was also observed in another similar study in which it was also reported that catechin was inactive.[8]

This prompted a study on the peculiar features of antioxidant capacity of procyanidins in comparison with catechin using as reference compounds catechol and resorcinol, the expected redox-sensitive moieties of catechin and procyanidins. The possible role of the spatial arrangement of redox groups was investigated by comparing the proanthocyanidin dimers A2 and B2. In these compounds the same redox moieties are arranged differently in space, with the two catechol groups close each other in B2 and far apart in A2 (Fig. 1).

A crude evaluation of antioxidant capacity, measured as a delay in the oxidation of phospholipid dispersions in the presence of peroxy radicals, failed to show any relevant difference among various compounds, with the exception of resorcinol—the antioxidant effect of which was almost undetectable (not shown). Far from being surprising, the similar lag in peroxidation obtained with different compounds was in agreement with the proposed kinetic model[9] in which different antioxidants produce a lag, which is dependent only on the concentration of the antioxidant, irrespective of the rate constant for the reduction of the peroxy radical. This was evaluated, instead, as a relative rate constant by a simple competitive kinetic procedure. Data were corroborated by an electroanalytical approach that measured the thermodynamic constraints of the anodic reaction as well as the elucidation of the oxidation pattern involved.

Kinetic Analysis

The relative rate constant for the reduction of peroxy radicals by antioxidants is calculated by applying a competitive kinetic analysis. The thermal decomposition of 2,2'-azobis(2-amidinopropane)dihydrochloride (ABAP) is used to generate molecular oxygen peroxy radicals.[10,11] These radicals bleach the carotenoid crocin at a constant rate, measured by the specific decrease of absorbance at 443 nm.[12,13] In the presence of an antioxidant that competes with crocin for the reaction with peroxy radicals, the bleaching rate slows down, consistent with the rate constant of the reaction between antioxidant and peroxy radical.

---

[8] J. Yamakoshi, S. Kataoka, T. Koga, and T. Ariga, *Atherosclerosis* **142**, 139 (1999).

[9] F. Tubaro, A. Ghiselli, P. Rapuzzi, M. Maiorino, and F. Ursini, *Free Radic. Biol. Med.* **24**, 1228 (1998).

[10] L. R. C. Barcley, S. J. Locke, J. M. MacNeil, J. VanKessel, G. W. Burton, and K. U. Ingold, *J. Am. Chem. Soc.* **106**, 2479 (1984).

[11] W. A. Pryor, J. A. Cornicelli, L. J. Devall, B. Tait, B. K. Trivedi, D. T. Witiak, and M. Wu, *J. Org. Chem* **58**, 3521 (1993).

[12] W. Bors, C. Michel, and M. Saran, *Biochim. Biophys. Acta* **796**, 312 (1984).

[13] F. Tubaro, E. Micossi, and F. Ursini, *J. Am. Oil Chem. Soc.* **73**, 173 (1996).

Chemicals

Procyanidins (a mixture of three to seven units of catechin), proanthocyanidin A2 and B2, catechin, and epicatechin are generously supplied by Indena Spa (Milan, Italy). ABAP and saffron are from Sigma (St. Louis, MO); Trolox C is from Aldrich (Milwankee, WI); all solvents (HPLC grade) and common reagents are from Merck. Crocin is isolated from saffron by methanol extraction after repeated extraction with ethyl ether to eliminate possible interfering substances. The concentration of crocin is calculated from the molar absorptivity ($\varepsilon = 1.33 \times 10^5 M^{-1}$ cm^{-1} at 443 nm).

Instruments and Procedure

For kinetic measurements, a UV-VIS spectrophotometer (Beckman Model DU-650) is used. Reaction mixtures contain in 2.0 ml of 50 mM phosphate–50 mM Tris at different pH values, 10 μM crocin (from a 0.6 mM solution in methanol), and different amounts of the antioxidant. The reaction is started by the addition of 12.5 mM ABAP (from a fresh 0.5 M solution at 0°) to the complete reaction mixture equilibrated at 40°. The decrease in absorbance is monitored for 10 min, and rates are calculated when the bleaching rate of the carotenoid is linear.

A mathematical transformation of the competition kinetic yields a straight line:

$$\frac{V_o}{V} = \frac{k_c[C] + k_a[A]}{k_c[C]} = 1 + \frac{k_a[A]}{k_c[C]}$$

where V_o is the basal bleaching rate of crocin, V is the bleaching rate of crocin in the presence of antioxidant, k_c is the rate constant of reduction of peroxy radical by crocin, k_a is the rate constant of reduction of peroxy radical by antioxidant, [C] is the concentration of crocin, and [A] is the concentration of antioxidant.

The value of k_a/k_c, calculated from the slope of the linear regression of the plot of [A]/[C] vs V_o/V, indicates the relative capacity of the compound under analysis to interact with peroxy radicals. Results for dimers and procyanidins are calculated using the corresponding molar concentration of catechol moieties.

Results

As expected, the rate constant for the reaction of resorcinol with peroxy radical was very small and was hardly measurable by this approach, thus ruling out any significant antioxidant effect. Catechol and catechin reacted quickly but only at alkaline pH (Fig. 2). Epicatechin has a reactivity identical to that of catechin (not shown). The rate constant was pH dependent for dimer A2 (Fig. 3), whereas a minimal linear pH dependence was observed for the dimer B2 and procyanidins (Figs. 3 and 4). The rate constant for the reaction of Trolox showed a similar linear relationship with pH (Fig. 4).

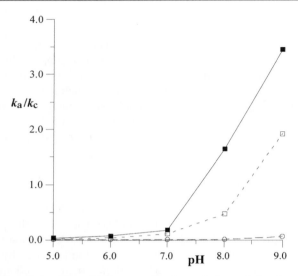

FIG. 2. Kinetic analysis of the antioxidant capacity of resorcinol, catechol, and catechin. The ratio k_a/k_c is calculated from the plot of [A]/[C] vs V_o/V at different pH values for resorcinol (O), catechol (□), and catechin (■). See text for details.

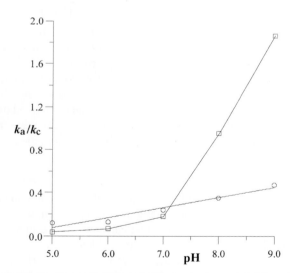

FIG. 3. Kinetic analysis of the antioxidant capacity of proanthocyanidin dimer A2 and dimer B2. The ratio k_a/k_c is calculated from the plot of [A]/[C] vs V_o/V at different pH values for dimer A2 (□) and dimer B2 (O). See text for details.

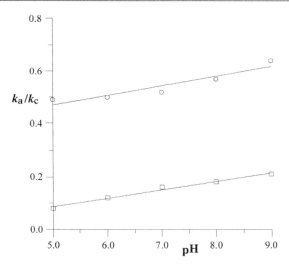

FIG. 4. Kinetic analysis of the antioxidant capacity of procyanidins and Trolox. The ratio k_a/k_c is calculated from the plot of [A]/[C] vs V_o/V at different pH values for procyanidin (■) and Trolox (O). See text for details.

The two patterns of pH dependence are produced because of different mechanisms of reaction. Apparently, in the case of catechol, catechin, and dimer A2, the reduction of peroxy radical takes place through an electron donation, whereas in the case of dimer B2, procyanidins, and Trolox, a hydrogen transfer is most likely involved.

Electroanalysis

Electroanalytical methods are particularly well suited for tackling problems relative to redox processes in that they are able to provide straightforward information not only on the thermodynamic constraints of electron transfers, but also on possible coupled chemical reactions, thus allowing insight into the overall reaction pathway.

Electroanalytical tests on the different polyphenols are carried out in acetonitrile because this polar and scarcely nucleophilic solvent reacts slowly with the unstable cationic species frequently generated in the electrochemical oxidation of many organic compounds. This solvent extends the lifetime of unstable species and facilitates their characterization. At the same time, it increases the specificity of the measurement of oxidation potentials, which is related at best to the sole electron release process because the electrode reaction is not triggered by chemical reactions between the primary electrode product and the solvent. These side reactions account for the observation that electrochemical oxidations occur most frequently at quite less positive potentials (even several hundreds of millivolts) in aqueous media than in acetonitrile.

Chemicals

All the chemicals used are of reagent-grade quality and are employed without further purification. Reagent-grade acetonitrile is purified further by the standard procedure.[14] Nitrogen, used to remove oxygen from the working solutions, is passed through sulfuric acid to remove traces of water and is then equilibrated to the vapor pressure of the solvent.

Instruments and Procedure

Voltammetric, coulometric, and double-pulse amperometric experiments are conducted at 20° in a H-shaped, three-electrode cell with cathodic and anodic compartments separated by a sintered glass disk. In voltammetric and double-pulse amperometric tests, the working electrode is a platinum disk with an apparent geometric area of 0.28 mm^2, mirror polished with graded alumina powder prior to each set of experiments. A platinum gauze is used as the working electrode in coulometric and preparative tests. In all cases, the counter electrode is a platinum sheet, whereas the reference electrode is an aqueous Ag|AgCl,Cl$^-_{(sat)}$ connected to the cell by a salt bridge containing the medium also employed in the test solution.

The voltammetric unit consists of an EG&G PARC Model 273 potentiostat driven by EG&G PARC Model 270 software installed on an IBM 486 computer. In the coulometric and preparative experiments, an AMEL Model 552 potentiostat is used and the associated coulometer is an AMEL integrator Model 731. All tests are carried out at $20 \pm 0.1°$ on 30-ml aliquots of oxygen-free acetonitrile solutions containing the desired species. Typically, voltammetric tests are run at a scan rate of 0.1 V sec^{-1}.

Double-pulse amperometry is accomplished by stepping the potential of the working electrode from the value E_1 (at which the electrooxidation of the analyte occurs) to those required to reduce possible intermediates (E_2) or products (E_3) generated at E_1, and sampling the relevant current signals for 0.01 sec at the end of each potential step. In our experiments, E_1 for different compounds ranges from 980 to 1250 mV, E_2 from 570 to 300 mV, and E_3 from 300 to 0 mV.

Results

Preliminary tests were carried out on the anodic behavior of catechol and resorcinol, the expected electroactive moieties present in catechin and procyanidins.

[14] A. J. Bard (ed.), "Electroanalytical Chemistry," Vol. 4. Dekker, New York, 1970.

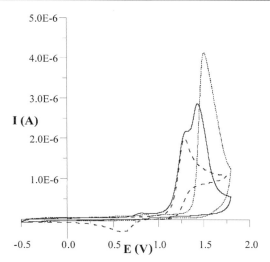

FIG. 5. Cyclic voltammetric curves of resorcinol, catechol, and catechin. Curves were recorded at a scan rate of 0.1 V sec^{-1} with a platinum electrode in a 0.1 M TBAP, CH$_3$CN solution containing of 1 mM resorcinol (dotted line), 1 mM catechol (dashed line), or 1 mM catechin (solid line).

Voltammograms recorded for catechol (Fig. 5, dashed line) showed a single diffusive peak at ca. 1.35 V, which is apparently coupled with a backward cathodic peak located at ca. 0.5 V, whereas resorcinol (Fig. 5, dotted line) displayed an oxidation peak at more anodic potentials (ca. 1.6 V) with a shape typical of a surface-confined process, which led to the formation of a polymeric film that inactivated the working electrode surface[15,16] (Fig. 5, solid line). As expected, the voltammogram for all catechins showed a combination of these two profiles.

Controlled potential coulometric experiments on catechol solutions showed that their oxidation yielded the corresponding o-quinone (the reduction of which occurred at about 0.25 V) after 2 mol of electrons per mole of the starting compound were spent. However, in these electrolyses the anodic current fell to zero when 4 mol of electrons per mole of catechol were transferred, most likely due to the opening of the quinone ring. When these electrolyses were interrupted intentionally after 1 mol of electrons per mole of catechol was released, the resulting voltammetric profile showed that only small amounts of o-quinone (reduction peak at 0.25 V) were formed, whereas a quite high cathodic peak at about 0.5 V was observed (Fig. 6, dotted line). This last peak was unambiguously due to an unstable intermediate

---

[15] G. W. Mengoli and M. M. Musiani, *Progr. Org. Coatings* **24,** 237 (1994).
[16] W. F. Hodnick, E. B. Milosavljevic, J. H. Nelsen, and R. S. Pardini, *Biochem. Pharmacol.* **37,** 2607 (1988).

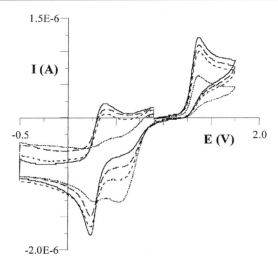

FIG. 6. Disproportionation of catechol radical. Cyclic voltammetric curves were recorded every 10 min at a scan rate of 0.1 V sec^{-1} with a platinum electrode in a 0.1 M TBAP, CH$_3$CN solution after controlled potential oxidation of 1 mM catechol with 1 mol of electrons per mole of catechol. Starting curve: dotted line.

undergoing disproportionation. The height of this peak decreased progressively with time, whereas there was a concomitant increase in both the reduction peak relative to the parent o-quinone (0.25 V) and the oxidation peak of the starting species (1.35 V) (Fig. 6). This finding can be attributed to the phenoxy radical known to be involved as the intermediate in the electrochemical oxidation of o- or p-diphenols.[17–19]

Based on these experiments, the following reaction pathway can be advanced for the oxidative conversion of catechol to the parent o-quinone [C(−2H)]:

$$2C \rightarrow 2C^{\bullet+} + 2e^- \qquad (1)$$

$$2C^{\bullet+} \rightarrow 2H^+ + 2C(-H)^{\bullet} \qquad (2)$$

$$2C(-H)^{\bullet} \rightarrow C + C(-2H) \qquad \text{(slow)} \qquad (3)$$

When voltammograms were run on catechin, dimers A2 and B2, and procyanidins, two anodic peaks were always recorded, the first conceivably due to the catechol moiety and the second attributable to the oxidation of the resorcinol moiety. The half-peak potentials relative to the first oxidation process for all the

[17] B. R. Eggins and J. Q. Chambers, *J. Electrochem. Soc.* **117**, 186 (1970).
[18] S. F. Nelsen, B. M. Trost, and D. H. Evans, *J. Am. Chem. Soc.* **89**, 3034 (1967).
[19] T. G. Edwards and R. Grinter, *Trans. Faraday Soc.* **64**, 1070 (1968).

TABLE I
HALF-PEAK POTENTIALS FOR OXIDATION OF PHENOLIC COMPOUNDS[a]

Compound	$E_{p/2}$(mV)
Catechol	1210
Resorcinol	1510
Catechin	1150
Epicatechin	1150
Proanthocyanidin dimer A2	1160
Proanthocyanidin dimer B2	1020
Procyanidins[b]	1010
Trolox	960

[a] Solutions was 1 mM in CH$_3$CN.
[b] Solution is 1 mM in CH$_3$CN : ethanol, 98 : 2 (v/v), in view of the limited solubility of procyanidins in acetonitrile.

compounds investigated are given in Table I, where they are compared with the half-peak potential for a classical antioxidant such as Trolox. These results pointed out that the oxidation of the catechol moiety present in catechin and in the dimer A2 occurs at potentials fairly close to that for free catechol, whereas this electroactive unit is oxidized more easily when present in the dimer B2 and in procyanidins.

The evidence that catechol is oxidized to the corresponding quinone via a disproportionation of phenoxy radicals obtained as primary oxidation product (Fig. 6) suggested the working hypothesis that the faster rate of its decay in dimer B2 and procyanidins is the driving force that facilitates the overall oxidation process, thus anticipating the relevant potential.

Unfortunately, this hypothesis could not be substantiated by coulometric measurements as carried out on catechol solutions by controlled potential electrolyses. In all these investigations, catechol is accompanied by resorcinol moieties whose oxidation, initiating at potentials partially overlapping those required by the catechol unit, led to the previously mentioned electrode surface inactivation. To overcome this drawback, using catechol as the reference species, other catechol containing compounds were studied by double-pulse amperometry, i.e., by resorting to an electroanalytical approach suitable for short time measurements during which problems related to electrode inactivation are minimized.

Results obtained by double-pulse experiments on catechol, catechin, and dimer B2 are shown in Table II, in which the current ratios E_2/E_1 and E_3/E_1, recorded by this short time approach at the appropriate potentials, are reported.

These data indicate that the lifetimes of the phenoxy radical intermediate for catechol and for catechin are in the same order of magnitude, in agreement with the similar oxidation potentials (Table I). In the case of dimer B2, the absence of any

TABLE II
RATIOS BETWEEN BACKWARD CATHODIC AND FORWARD ANODIC CURRENTS

Compounds	Current ratio	
	E_2/E_1	E_3/E_1
Catechol	0.33	0.36
Epicatechin	0.28	0.39
Dimer B2	0.00	0.33

[a] Sampled at E_2, E_3, and E_1 by double-pulse amperometric experiments. Solution was 1 mM in CH_3CN, 0.1 M tetrabutylammonium perchlorate. See text for details.

signal for this intermediate is strongly indicative of the occurrence of a very fast disproportionation. This quick decay of the primary oxidation product conceivably causes the triggering of the overall anodic process, thus accounting for the less anodic potential observed for this compound. Consistent with this conclusion, the electroanalytical behavior of dimer A2 was identical to that of catechin and that of procyanidins was identical to that of dimer B2.

Conclusions

This study was carried out to elucidate if, and possibly why, polymeric procyanidins are better antioxidants than the corresponding monomers, catechin and epicatechin.

The integration of kinetic and thermodynamic data, obtained using simple and unexpensive procedures, provided evidence for the better antioxidant effect of procyanidins and allowed some general conclusions about the mechanism of the chain-breaking antioxidation.

Results obtained point out that in all polyphenols analyzed, the redox active moiety displaying a real antioxidant effect, measured as the fast reduction of peroxidation driving peroxy radicals, is catechol. Conversely, the resorcinol moiety, also present in the compounds investigated, does not produce an effective contribution to the antioxidant activity because its oxidation occurs at more anodic potentials, producing a cation radical that initiates a chain reaction eventually leading to polymerization, as confirmed by the rapid inactivation of the electrode surface.

No major difference among catechol, catechin, epicatechin, and dimer A2 was observed either by kinetic or by thermodynamic approaches. Moreover, the rate constant for the reduction of peroxy radicals showed a biphasic pH relationship, clearly indicating that an electron transfer mechanism is involved, which requires a dissociated phenol.

The redox behavior of dimers B2, procyanidins, and chromanol (Trolox) is quite different. The rate constant for the reduction of peroxy radicals is not pH dependent, thus suggesting a hydrogen transfer mechanism. The oxidation potential is much lower than that of catechol and catechins due to fast removal from the electrode surface of the decaying primary product of the oxidation process (catechol radical) for procyanidins and dimer B2, as confirmed by double-pulse amperometry. The comparison between dimer A2 and B2, where the catechol moieties are very close to (B2) or far apart from (A2) each other, lends further support to the concept that the oxidation becomes easier when thermodynamically favored by rapid removl of the reaction product by disproportionation.

This could have a general significance for antioxidant reactions. In order to be an active chain-breaking antioxidant, a compound has to fulfill the following thermodynamic and kinetic requirements: (a) they must be characterized by a redox potential as low as possible to make them compatible with oxidation by peroxy radicals with a rate that must be much faster than that of the oxidation of lipids and (b) the product of the antioxidant oxidation has to be a poor oxidizing agent or a good reductant (low redox potential). When the radical formed in the antioxidant reaction is removed rapidly by disproportionation, a twofold advantage is achieved: (1) lowering the concentration of potentially harmful radicals and, concomitantly, (2) decreasing the potential of the first oxidation.

All our electroanalytical data have been obtained in a medium in which the phenolic groups were undissociated. Thus, the correspondence with kinetic data is valid only below the break point. Under these conditions, catechol, catechins, and dimer A2 are rather weak antioxidants, in agreement with their oxidation potentials.

The two patterns of pH dependence obtained for the kinetic results relative to the different phenols suggest a different acid character, together with a different homolytic bond dissociation energy for the O–H of catechol, in the two classes of of compounds. The most reasonable interpretation is that the complex network of hydrogen bonding, in the case of polyphenols with catechol moieties very close to each other (dimer B2 and procyanidins), prevents the dissociation of protons and weakens the O–H bond. Notably, the pH dependence of Trolox for the reduction of peroxy radical is similar to that of procyanidins. Also, in the case of Trolox, the antioxidant effect depends on a hydrogen transfer, and the low homolytic bond dissociation energy of chromanolic O–H is the key element in the final antioxidant effect.

Under biological conditions it is not known whether, in the lipophilic environment where antioxidant reactions take place, the phenolic groups are in a dissociated form, although this appears unlikely. In a hydrogen transfer mechanism, the reduction of lipid peroxy radicals follows the general mechanism of free radical reactions where hydrogen transfer follows the formation of an intermediate between two molecules with a different bond dissociation energy for hydrogen.[20] Instead, the electron transfer mechanism is necessarily less specific. In general terms, we can conclude that an antioxidant active in the dissociated form can be referred to as a general reductant, whereas a chain-breaking antioxidant active by hydrogen transfer is a reductant with specificity for free radicals.

In conclusion, although we do not know whether the different biological effects of procyanidins and catechin are fully accounted for by their different antioxidant capacities, we can state that procyanidins are much better antioxidants than the corresponding monomers.

[20] W. A. Pryor, in "Free Radicals in Biology" (W. A. Pryor, ed.), Vol. 1, p. 1. Academic Press, New York, 1976.

[31] Inhibition of *in Vitro* Low-Density Lipoprotein Oxidation by Oligomeric Procyanidins Present in Chocolate and Cocoas

By DEBRA A. PEARSON, HAROLD H. SCHMITZ, SHERYL A. LAZARUS, and CARL L. KEEN

Introduction

Flavonoids, including the procyanidins, are an integral part of the human diet, being found in a wide variety of plant foods, including vegetables, fruits, and their juices, herbs, wine, tea, and legumes.[1-4] Intense research interest has focused on the potential role of flavonoids as physiologically active antioxidants, particularly with regard to cardiovascular disease. Dietary flavonoid intake has been associated with a lower incidence of coronary heart disease (CHD)[1,5,6] and stroke.[7] Similarly, the consumption of specific flavonoid-rich foods (fruits,

---

[1] M. G. Hertog, E. J. Feskens, P. C. Hollman, M. B. Katan, and D. Kromhout, *Lancet* **342,** 1007 (1993).
[2] J. Kuhnau, *Wld. Rev. Nutr. Diet.* **24,** 117 (1976).
[3] J. F. Hammerstone, S. A. Lazarus, A. E. Mitchell, R. Rucker, and H. H. Schmitz, *J. Agric. Food Chem.* **47,** 490 (1999).
[4] S. A. Lazarus, G. E. Adamson, J. F. Hammerstone, and H. H. Schmitz, *J. Agric. Food Chem.* **47,** 3693 (1999).
[5] M. G. Hertog, D. Kromhout, C. Aravanis, H. Blackurn, R. Buzina, F. Fidanza, S. Giampaoli, A. Jansen, A. Menotti, S. Nedeljkovic, *et al., Arch. Intern. Med.* **155,** 381 (1995).
[6] M. G. Hertog, E. J. Feskens, and D. Kromhout, *Lancet* **349,** 699 (1997).
[7] S. O. Keli, M. G. Hertog, E. J. Feskens, and D. Kromhout, *Arch. Intern. Med.* **156,** 637 (1996).

vegetables, red wine, and tea) has been inversely correlated with CHD mortality[8,9] and risk factors for CHD.[10] Certain cocoas and chocolates have been shown to contain substantial quantities of oligomeric procyanidins,[11] which are members of a subclass of flavonoids increasingly being investigated for their potential health benefits.[12–15]

The oxidative modification of low-density lipoproteins (LDL) is widely recognized to play a pivotal role in the subendothelial formation of foam cells[16] and the subsequent development of atherosclerosis.[17–19] Indeed, the susceptibility of LDL to oxidation may be a biomarker for the presence,[20,21] severity,[22] and progression of the disease.[23] In addition to the direct involvement of oxidized LDL (oxLDL) in atherosclerotic lesion formation, oxLDL may promote atherosclerosis through several biological effects, including cytotoxic effects,[24] enhanced monocyte and smooth muscle chemotaxis,[25,26] increased leukocyte adhesion,[27]

---

[8] A. J. Verlangieri, J. C. Kapeghian, S. El-Dean, and M. Bush, *Med. Hypotheses* **16**, 7 (1985).

[9] S. Renaud and M. De Lorgeril, *Lancet* **339**, 1523 (1992).

[10] I. Stensvold, A. Tverdal, K. Solvoll, and O. P. Foss, *Prev. Med.* **21**, 546 (1992).

[11] G. E. Adamson, S. A. Lazsarus, A. E. Mitchel, R. L. Prior, G. Cao, P. H. Jacobs, B. G. Kremers, J. F. Hammerston, R. B. Rucker, K. A. Ritter, and H. H. Schmitz, *J. Agric. Food Chem.* **47**, 4184 (1999).

[12] N. Haramaki, D. E. Stewart, S. Aggarwal, H. Ikeda, A. Z. Reznick, and L. Packer, *Free Radic. Biol. Med.* **25**, 329 (1998).

[13] T. K. Mao, J. J. Powell, J. Van De Water, C. L. Keen, H. H. Schmitz, and M. E. Gershwin, *Int. J. Immunother.* **XV(1)**, 23 (1999).

[14] M. M. Bearden, D. A. Pearson, D. Rein, K. A. Chevaux, D. R. Carpenter, C. L. Keen, and H. H. Schmitz, ACS Symposium Series, "Caffeinated Beverages: Health Benefits, Physiological Effects and Chemistry." **754**, p. 177. American Chemical Society, Washington, D.C., 2000.

[15] W. Tuckmantel, A. P. Kozikowski, and L. J. Romanczyk, *J. Amer. Chemical Soc.* **121**, 12073 (1999).

[16] R. G. Gerrity, *Am. J. Pathol.* **103**, 181 (1981).

[17] J. L. Goldstein and M. S. Brown, *Annu. Rev. Biochem.* **46**, 897 (1977).

[18] D. Steinberg, S. Parthasarathy, T. E. Carew, J. C. Khoo, and J. L. Witztum, *N. Engl. J. Med.* **320**, 915 (1989).

[19] C. J. Schwartz, A. J. Valente, E. A. Sprague, J. L. Kelley, and R. M. Nerem, *Clin. Cardiol.* **14**, I1 (1991).

[20] L. Cominacini, U. Garbin, A. M. Pastorino, A. Davoli, M. Campagnola, A. De Santis, C. Pasini, G. B. Faccini, M. T. Trevisan, L. Bertozzo, F. Pasini, and V. Lo Cascio, *Atherosclerosis* **99**, 63 (1993).

[21] K. Liu, T. E. Cuddy, and G. N. Pierce, *Am. Heart J.* **123**, 285 (1992).

[22] J. Regnström, J. Nilsson, P. Tornvall, C. Landou, and A. Hamsten, *Lancet* **339**, 1183 (1992).

[23] Y. B. Derijke, C. J. M. Vogelezang, T. J. C. Vanberkel, H. M. G. Princen, H. F. Verwey, A. Van der Laarse, and A. V. G. Bruschke, *Lancet* **340**, 858 (1992).

[24] D. W. Morel, J. R. Hessler, and G. M. Chisolm, *J. Lipid Res.* **24**, 1070 (1983).

[25] M. T. Quinn, S. Parthasarathy, L. G. Fong, and D. Steinberg, *Proc. Natl. Acad. Sci. U.S.A.* **84**, 2995 (1987).

[26] I. Autio, O. Jaakkola, T. Solakivi, and T. Nikkari, *FEBS Lett.* **277**, 274 (1990).

[27] J. A. Berliner, M. C. Territo, A. Sevanian, S. Ramin, J. A. Kim, B. Ramshad, M. Sterson, and A. M. Fogelman, *J. Clin. Invest.* **85**, 1260 (1990).

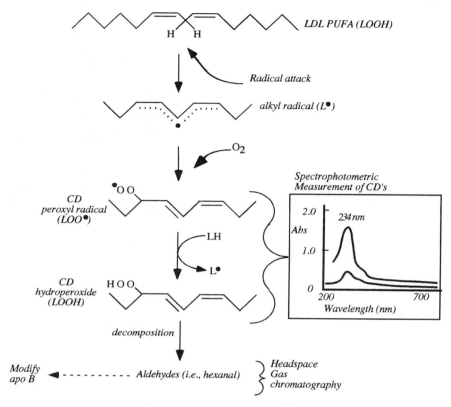

FIG. 1. Lipid peroxidation of LDL and detection of lipid peroxidation products.

and altered eicosanoid and nitric oxide production.[28,29] Thus, researchers have focused on the potential role of synthetic and natural antioxidants to prevent LDL oxidation, thereby preventing or slowing the atherosclerotic process.

Modification of LDL by oxidation of its polyunsaturated lipids is a complex, multistep process that can be inhibited or interrupted at several stages (Fig. 1).[30] Polyunsaturated fatty acids (PUFA) in LDL react with reactive oxygen species or lipid peroxidation products, yielding alkyl radicals, which react rapidly with oxygen to form peroxyl radicals. Peroxyl radicals react with neighboring PUFA to produce lipid hydroperoxides (LOOH), propagating a chain reaction. In the presence of transition metals, LOOH decompose yielding alkoxyl radicals that

[28] D. Daret, P. Blin, B. Dorian, M. Rigaud, and J. Larrue, *J. Lipid Res.* **34,** 1473 (1993).
[29] K. Schmidt, W. F. Graier, G. M. Kostner, B. Mayer, and W. R. Kukovetz, *Biochem. Biophys. Res. Commun.* **172,** 614 (1990).
[30] H. Esterbauer, G. Janusz, H. Puhl, and G. Jurgens, *Free Radic. Biol. Med.* **13,** 341 (1992).

cleave into a variety of aldehydes. It is these aldehyde decomposition products that are often responsible for the damage to biological tissues, including LDL. Aldehyde products react with specific residues of apolipoprotein B_{100} (apo B), resulting in an increase in the net negative charge of apo B and, most importantly, its recognition by the scavenger receptor.[31]

Important to the formation of a proatherogenic form of LDL is the fact that even with relatively large amounts of LOOH produced during oxidation, LDL that is recognizable by the scavenger receptor or appreciably taken up by macrophages is not produced.[32] The LOOH must decompose before LDL becomes recognizable by the scavenger receptor or is taken up by macrophages.[32] The exact mechanism(s) by which LDL is oxidized *in vivo* is unclear, although research evidence suggests several possibilities, including myeloperoxidase, reactive nitrogen species, lipoxygenase, and transition metals, that may mediate the oxidation of LDL at various stages of atherosclerotic lesion formation.[33] Thus thorough evaluation of the role of an antioxidant in protecting LDL should involve more than one method of measurement at different stages of oxidation.

Numerous flavonoids and flavonoids-rich foodstuffs have been studied for their potential to inhibit LDL oxidation and exert cardioprotective effects. A variety of flavonoids protected LDL from oxidation following their *in vitro* addition to oxidizing systems.[34–36] Specific flavonoids or phenolic-rich foodstuffs also inhibited LDL oxidation following *in vivo* supplementation.[37–40] Furthermore, two studies demonstrated significantly decreased LDL oxidizability in association with decreased atherosclerotic lesion formation in apolipoprotein E-deficient mice fed diets supplemented with flavonoid-rich foodstuffs or pure flavonoids compared to control animals.[37,40]

Certain cocoas and chocolate contain significant amounts of the flavan-3-ol monomers (−)-epicatechin and (+)-catechin and several procyanidin oligomers built on these monomeric units (Fig. 2).[3] Several *in vitro* studies and one *ex vivo* study indicate that flavan-3-ols and their associated procyanidins from a

---

[31] U. P. Steinbrecher, J. L. Witztum, S. Parthasarathy, and D. Steinberg, *Arteriosclerosis* **7**, 135 (1987).
[32] W. Jessup, S. M. Rankin, C. V. De Whalley, J. R. Hoult, J. Scott, and D. S. Leake, *Biochem. J.* **265**, 399 (1990).
[33] J. W. Heinecke, *Atherosclerosis* **141**, 1 (1998).
[34] J. A. Vinson, Y. A. Dabbagh, M. M. Serry, and J. Jang, *J. Agric. Food Chem.* **43**, 2800 (1995).
[35] N. Salah, N. J. Miller, G. Paganga, L. Tijburg, G. P. Bolwell, and C. Rice-Evans, *Arch. Biochem. Biophys.* **322**, 339 (1995).
[36] S. Miura, J. Watanabe, T. Tomita, M. Sano, and I. Tomita, *Biol. Pharm. Bull.* **17**, 1567 (1994).
[37] B. Fuhrman, S. Buch, J. Vaya, P. A. Belinky, R. Coleman, T. Hayek, and M. Aviram, *Am. J. Clin. Nutr.* **66**, 267 (1997).
[38] Y. Miyagi, K. Miwa, and H. Inoue, *Am. J. Cardiol.* **80**, 1627 (1997).
[39] T. Ishikawa, M. Suzukawa, T. Ito, H. Yoshida, M. Ayaori, M. Nishiwake, A. Yonemura, Y. Hara, and H. Nakamura, *Am. J. Clin. Nutr.* **66**, 261 (1997).
[40] T. Hayek, B. Fuhrman, J. Yaya, M. Rosenblat, P. Belinky, R. Coleman, A. Elis, and M. Aviram, *Arterioscler. Thromb. Vasc. Biol.* **17**, 2744 (1997).

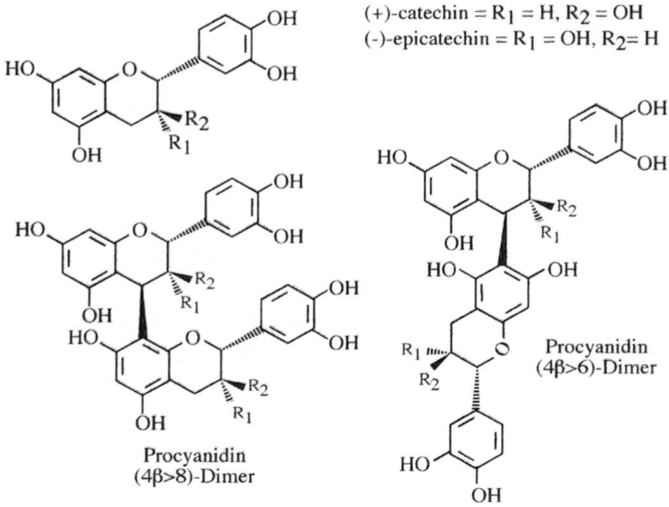

FIG. 2. Flavan-3-ol monomeric and dimeric procyanidins.

(+)-catechin = R_1 = H, R_2 = OH
(−)-epicatechin = R_1 = OH, R_2 = H

Procyanidin (4β>8)-Dimer

Procyanidin (4β>6)-Dimer

variety of foods, including cocoa, possess the following biologically relevant activities: reactive oxygen and nitrogen species scavenging[41–43] and vitamin E sparing activity,[44] inflammation and immune system modulation,[13,45,46] antitumor activity,[47] vasodilatory activity,[48] and platelet adhesion and aggregation modulation.[49] Although this article focuses on the effect of cocoa flavonoids on LDL oxidation, it is worth noting that several of the just-mentioned activities are also important factors that affect the pathogenesis of atherosclerosis.

Experiments have tested the antioxidant activity of cocoa and its polyphenols.[14,50,51] Phenolic extracts from commercial cocoa powder, milk chocolate,

[41] G. E. Arteel and H. Sies, *FEBS Lett.* **462**, 167 (1999).
[42] M. R. Facino, M. Carino, G. Aldini, F. Berti, G. Rossoni, E. Bombardelli, and P. Morazzoni, *Planta Med.* **62**, 495 (1996).
[43] F. Virgili, D. Kim, and L. Packer, *FEBS Lett.* **431**, 315 (1998).
[44] M. R. Facino, M. Carino, G. Aldini, M. T. Calloni, E. Bombardelli, and P. Morazzoni, *Planta Med.* **64**, 343 (1998).
[45] C. Sanbongi, N. Suzuki, and T. Sakane, *Cell. Immunol.* **177**, 129 (1997).
[46] J. Robak, F. Shridi, M. Wolbis, and M. Krolikowska, *Pol. J. Pharmacol. Pharm.* **40**, 451 (1988).
[47] H. U. Gali, E. M. Perchellet, X. M. Gao, J. J. Karchesy, and J. P. Perchellet, *Planta Med.* **60**, 235 (1994).
[48] J. Duarte, F. P. Vizcaino, P. Utrilla, J. Jimenez, J. Tamargo, and A Zarzuelo, *Gen. Pharmacol.* **24**, 857 (1993).
[49] D. Rein, T. G. Paglieroni, T. Wun, D. A. Pearson, H. H. Schmitz, R. Gosselin, and C. L. Keen, *Am. J. Clin. Nutr.* **72**, 30 (2000).
[50] A. L. Waterhouse, J. R. Shirley, and J. L. Donovan, *Lancet* **348**, 834 (1996).
[51] K. Kondo, R. Hirano, A. Matsumoto, O. Igarashi, and H. Itakura, *Lancet* **348**, 1514 (1996).

and baker's chocolate added to LDL *in vitro* inhibited copper-catalyzed oxidation at concentrations similar to other phenol-containing products.[50] Furthermore, LDL isolated from subjects 2 hr after consumption of 35 g defatted cocoa was more resistant to oxidation than LDL isolated prior to cocoa ingestion.[51] Individual oligomeric procyanidin fractions have been isolated from cocoa and shown to inhibit copper-catalyzed LDL oxidation *in vitro*.[14] Interestingly, this research suggested that the individual procyanidin oligomers might have different antioxidant potential.

Little is known about the antioxidant activities of individual procyanidins in cocoa. The use of a LDL oxidation system represents a biologically relevant method for studying relative antioxidant activities. The antioxidant activities of individual procyanidins isolated from cocoa are presented here using a copper-catalyzed and endothelial cell-mediated LDL oxidation system.

Materials and Methods

The cocoa monomer (−)-epicatechin and five cocoa oligomeric procyanidins (dimer, trimer, tetramer, pentamer, and hexamer) and two cocoa procyanidin mixtures (monomer through tetramer and pentamer through decamer) are isolated and purified as described previously[11] and are provided by Mars, Incorporated (Hackettstown, NJ). Human aortic endothelial cells (HAEC) are from Clonetics (San Diego, CA) at passage three. Cells are grown in Clonetics microvascular endothelial cell growth medium containing 5% (v/v) fetal bovine serum, epidermal growth factor (10 ng/ml), hydrocortisone (1 μg/ml), bovine brain extract containing 10 ng/ml heparin, gentamicin (50 μg/ml), and amphotericin B (50 ng/ml). Phenol red-free Ham's F10 media are obtained from GIBCO-BRL (Gaithersburg, MD) and HEPES phosphate buffer from Clonetics (San Diego, CA). The protein Lowry kit and all other reagents are from Sigma (St. Louis, MO).

Preparation of Human LDL

Blood is obtained from healthy, nonsmoking male volunteers by venipunctre into EDTA-containing vacutainer tubes. Plasma is prepared by centrifuging the blood for 30 min at 1500 rpm and 4°. LDL is isolated by sequential density ultracentrifugation.[52] EDTA is removed by dialysis overnight (12 hr) in pH 7.4 phosphate-buffered (10 mM) saline (100 mM) purged with nitrogen. The protein concentration in LDL is determined by the Lowry[53] method using a Sigma protein Lowry kit.

---

[52] J. R. Orr, G. L. Adamson, and F. T. Lindgren, "Analyses of Fats, Oils, and Lipoproteins" (E. G. Perkins, ed.), p. 524. American Oil Chemists' Society, Champaign, IL, 1991.
[53] O. H. Lowry, N. J. Rosebrough, A. L. Farr, and R. J. Randall, *J. Biol. Chem.* **193**, 265 (1951).

Copper-Catalyzed LDL Oxidation

The antioxidant activities of the five purified cocoa procyanidins (mono-, di-, tetra-, penta-, and hexamer) and two cocoa procyanidin mixtures (monomer through tetramer and pentamer through decamer) are assessed by measuring the formation of hexanal using static headspace gas chromatography.[54] Hexanal is the aldehyde decomposition product of ω-6 PUFA lipid hydroperoxides. Omega-6 PUFA, particularly linoleic acid, is the major PUFA in the typical diet in the United States and makes up the majority of PUFA in the LDL used in this system. Thus hexanal is a specific, sensitive marker of the degree of oxidation of LDL in the system. Immediately prior to the oxidation assay, the LDL is dialyzed overnight to remove EDTA and is then diluted with phospate buffer to a final concentration of 1 mg LDL protein/ml, and solutions of the test compounds are made. Ten microliters of a test compound (final concentrations 1 or 5 μM) is added to 250 μl of LDL solution in a 6-ml headspace vial and mixed gently. Five microliters of copper sulfate is added for a final concentration of 10–80 μM. The vial is sealed tightly, vortexed gently, and incubated in a 37° shaking water bath for 2–4 hr. Controls consist of LDL and copper sulfate. All samples and controls are run in triplicate. At the end of the 2- to 4-hr incubation, headspace vials are injected into a Perkin-Elmer (Norwalk, CT) Sigma 3B gas chromatograph equipped with an HS 6 static headspace sampler to measure hexanal production. The gas chromatograph is equipped with a capillary DB-1701 column (30 m × 0.32 mm, 1 μm thickness (J & W, Folsom, CA) that is heated isothermally at 70–80°. The GC conditions are helium carrier gas flow, 20 ml/min (helium pressure gauge of 60 psi); splitless injector temperature, 180°; detector temperature, 200°; an oven temperature, 75°; and headspace sample temperature, 40°. Hexanal is identified by comparison with the retention time of a hexanal standard. Hexanal production in the presence of the test compounds is compared to that of controls. The results are expressed as percentage inhibition of LDL oxidation as determined by the formula: $[(C-S)/C] \times 100$, where C is the hexanal produced by the control and S is the hexanal produced in the presence of the test compounds.

Cell-Mediated LDL Oxidation

HAEC between passages 5 and 7 are seeded onto 35-mm six-well plates and used at confluence. Stock solutions of the test compounds are prepared in dimethyl sulfoxide (DMSO) immediately prior to the assay. Incubation media are also made up immediately prior to use and consist of phenol red-free Ham's F10 containing 200 μg LDL protein/ml and the test compounds (final concentration 1 or 5 μM) with DMSO at a final concentration not exceeding 0.2%. Duplicate wells are washed three times in phenol red-free Ham's F10 (37°) to remove any growth

[54] E. N. Frankel, J. B. German, and P. A. Davis, *Lipids* **27**, 1047 (1992).

medium and serum. Incubation media (1.2 ml) are pipetted into the wells, which are then incubated for 12 hr in a 5% (v/v) CO_2 incubator. Controls consist of cell-free wells containing the incubation medium without test compounds and wells with HAEC containing the incubation medium with DMSO only. The extent of lipid oxidation is determined spectrophotometrically by measuring conjugated diene hydroperoxides at 234 nm. The Ham's F10 medium used in the experiment does not contain fetal bovine serum, linoleic acid, or other polyunsaturated fatty acids. Thus, the measurement of conjugated dienes is a specific marker of LDL oxidation in the system. Aliquots of each incubation medium are diluted 1 : 1 with HEPES phosphate buffer (pH 7.4), and the baseline absorbance reading is measured at 234 nm. After a 12-hr incubation, media are removed and centrifuged to pellet detached cells. The supernatant portion is removed, diluted 1 : 1 with phosphate buffer, and the absorbance read at 234 nm. The level of conjugated dienes is determined by the change in absorbance between the baseline and 12-hr readings. Results are expressed as percentage inhibition of LDL oxidation as determined by the formula: $[(C-S)/C] \times 100$, where C is the conjugated dienes produced by the HAEC–DMSO control and S is the conjugated dienes produced in the presence of the test compounds.

Results and Discussion

Copper-Catalyzed LDL Oxidation

All the cocoa procyanidin compounds tested at 5 μM inhibited copper-catalyzed LDL oxidation (Fig. 3). Each of the pure compounds, monomer through hexamer, inhibited oxidation by approximately 96%. In contrast, the mixtures of monomer through tetramer, and pentamer through decamer, inhibited LDL oxidation by varying degrees in each of the three repeated experiments, suggesting possible instability in the mixture. All the compounds were then tested at a lower dose of 1 μM to reveal potential differences in antioxidant activities between pure compounds. At a concentration of 1 μM, all compounds tested inhibited copper-catalyzed LDL oxidation (Fig. 3). The monomer and dimer inhibited oxidation by approximately 33%. The tetramer and pentamer inhibited oxidation by approximately 47%. The hexamer inhibited oxidation by 63%. These results suggest that the degree of polymerization affects the antioxidant activity of the compound. Thus, at equimolar concentrations, the antioxidant activity, in this copper-catalyzed LDL system, increases with the degree of polymerization.

The exact mechanism by which copper promotes LDL oxidation is poorly understood. Transition metal ions may catalyze the decomposition of lipid hydroperoxides, triggering lipid peroxidation.[55] Some flavonoids act as metal chelators,

---

[55] S. M. Lynch and B. Frei, *J. Lipid Res.* **34**, 1745 (1993).

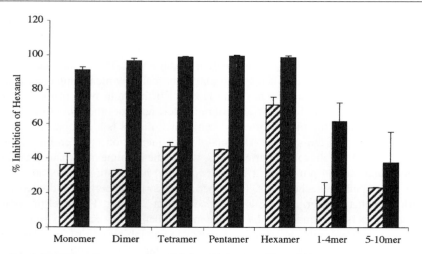

FIG. 3. Inhibition of copper-catalyzed LDL oxidation by 1(▨) or 5(■) μM cocoa procyanidins. Human LDL was incubated with the test compound and copper sulfate for 2 hr at 37°. Hexanal production was measured by static headspace gas chromatography. Results are expressed as mean percentage inhibition of hexanal production ± SD. Percentage inhibition of hexanal production = $[(C-S)/C \times 100]$, where C is hexanal production in the absence of any test compound and S is hexanal production in the presence of the test compounds.

and in this context our findings of increasing antioxidant activity at 1 μM as the number of monomeric units increases in the procyanidin compound are suggestive of a copper-chelating activity. In contrast, the near complete (96%) inhibition of LDL oxidation by all the procyanidin compounds tested at 5 μM, 16-fold less than the concentration of copper sulfate, suggests that their antioxidant activity in this system involves mechanisms other than, or in addition to, metal chelating. Another potential mechanism includes free radical scavenging (i.e., hydroxyl, peroxyl radical scavenging), and the flavan-3-ols of cocoa possess the structural motifs that are critical for this antioxidant activity.[56]

Free radical scavengers are active at concentrations appreciably lower than the oxidizing substrate. The relatively low concentrations of procyanidins that inhibited LDL oxidation are in agreement with a free radical scavenging activity. The cocoa procyanidin antioxidant activities reported here are similar to the reported antioxidant activities of a variety of other well-known flavonoids and phenolic extracts tested in a similar LDL oxidation system. Thus 5 μM catechin inhibited hexanal production approximately 67%,[57] several red wines at 5 μM

---

[56] C. A. Rice-Evans and N. J. Miller, *Free Radic. Biol. Med.* **20,** 933 (1996).

[57] A. S. Meyer, J. L. Donovan, D. A. Pearson, A. L. Waterhouse, and E. N. Frankel, *J. Agric. Food Chem.* **46,** 1783 (1998).

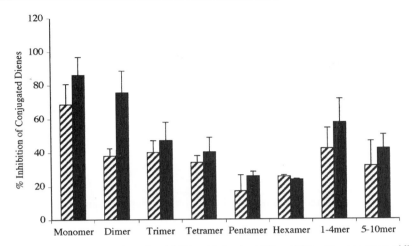

FIG. 4. Inhibition of HAEC-mediated LDL oxidation by 1(◨) or 5(■) μM cocoa procyanidins. Human LDL was incubated with HAEC in Ham's F10 medium in the presence or absence of procyanidins. Absorbance readings at 234 nm were taken initially and at the end of the 12-hr incubation period. Results are expressed as mean percentage inhibition of conjugated dienes ± SD. Percentage inhibition of conjugated dienes = $[(C-S)/C \times 100]$, where C is conjugated diene production in the presence of DMSO and S is conjugated diene production in the presence of procyanidins.

gallic acid equivalents (GAE) inhibited hexanal production by 37–65%,[58] and phenolic extracts from a variety of grapes inhibited hexanal production 22–60% at 10 μM GAE and their relative antioxidant activities correlated with their flavan-3-ol, flavonol, anthocyanin, and hydroxybenzoate content.[59] The results of the copper-catalyzed LDL oxidation experiment demonstrates that purified cocoa procyanidins have antioxidant activity in the physiologically relevant low micromolar range, similar to other flavonoids, although the exact mechanism(s) by which cocoa procyanidins inhibit copper-catalyzed LDL oxidation is not clear.

Cell-Mediated LDL Oxidation

Each of the cocoa procyanidins tested at both concentration, 1 or 5 μM, inhibited HAEC-mediated LDL oxidation as measured by conjugated diene lipid hydroperoxides (Fig. 4). Dose-dependent inhibition was seen for the monomer and dimer only. In contrast to copper-catalyzed LDL oxidation, the monomer and dimer were significantly stronger antioxidants during HAEC-mediated LDL oxidation than were the trimer through hexamer. The monomer through tetramer

---

[58] E. N. Frankel, A. L. Waterhouse, and P. L. Teissedre, *J. Agric. Food Chem.* **43**, 890 (1995).
[59] A. S. Meyer, Y. Ock-Sook, D. A. Pearson, A. L. Waterhouse, and E. N. Frankel, *J. Agric. Food Chem.* **45**, 1638 (1997).

procyanidin mixture inhibited LDL oxidation to a greater extent than the pentamer through decamer mixture (Fig. 4). The exact mechanism by which HAEC mediate LDL oxidation is unknown. The presence of trace amounts of transition metals in the media is required and lipoxygenase activity has been implicated.[60] Flavonoids have been shown to inhibit lipoxygenase.[46] Furthermore, the cocoa procyanidins, monomer through trimer, were significantly stronger inhibitors of lipoxygenase-catalyzed conjugated diene formation from linoleic acid than the procyanidins, tetramer through hexamer.[61] These results are in agreeement with our results and are suggestive of a role of lipoxygenase in the HAEC-mediated LDL oxidation.

Conclusions

We have presented biologically relevant procedures that can be used to assess the antioxidant activity of a wide variety of compounds. In these LDL oxidation systems, purified cocoa procyandins were effective inhibitors of *in vitro* LDL oxidation at concentrations similar to those seen with other well-known phenolic compounds. The results of both copper-catalyzed and HAEC-mediated LDL oxidation experiments demonstrate that purified cocoa procyanidins have antioxidant activities, although their relative activities differ depending on the *in vitro* model used. Their potential activity *in vivo* requires further research, particularly the extent of their absorption and metabolism. A recent study demonstrated that epicatechin from black chocolate was absorbed in human subjects, reaching a peak plasma concentration of 0.7 μM 2 hr after consumption of 80 g.[62] Further research is required to establish the absorption, metabolism, and *in vivo* effects of these compounds.

---

[60] S. Parthasarathy, E. Weiland, and D. Steinberg, *Proc. Natl. Acad. Sci. U.S.A.* **86,** 1046 (1989).
[61] H. H. Schmitz, D. Rein, D. A. Pearson, and C. L. Keen, *FASEB J.* **13,** A546 (1999).
[62] M. Richell, I. Tavazzi, M. Enslen, and E. A. Offord, *Eur. J. Clin. Nutr.* **53,** 22 (1999).

[32] Biological Actions of Oligomeric Procyanidins: Proliferation of Epithelial Cells and Hair Follicle Growth

By TOMOYA TAKAHASHI

Introduction

Hair is composed of epithelial and mesenchymal cells. Interactions between epithelial and mesenchymal cells are considered to be important both in hair morphogenesis in embryonic skin and in the progress of the hair cycle through the anagen, catagen, and telogen phases in adult skin.[1] Inner root sheath cells, outer root sheath cells, and hair matrix cells are all types of epithelial cell. Dermal papilla cells are a type of mesenchymal cell.[2] These mesenchymal–epithelial interactions are assumed to stimulate stem cells to proliferate, to induce the anagen phase in the hair cycle, and to produce hair. Stem cells are hypothesized to exist in the bulge area in the outer root sheath,[3] in the matrix,[4] or between the bulge and the matrix[5]; however, stem cell sites have not been elucidated positively.

Hair growth stimulants are expected to prolong the anagen phase or to promote hair epithelial cell proliferation following anagen induction.[6]

For *in vitro* evaluation of hair growth stimulants, tissue culture methods[7] and organ culture methods using dissected hair follicles[8] or sectioned skin[9] have been reported. For *in vivo* evaluation of hair growth stimulants, methods using murine models such as C3H mouse,[10] C57BL mouse,[11] or nude mouse[12]; rabbit[13]; and stump-tailed macaque[14] have been reported.

[1] U. Gat, R. DasGupta, L. Degenstein, and E. Fuchs, *Cell* **95,** 605 (1998).
[2] R. Paus, S. Müller-Röver, C. van der Veen, M. Maurer, S. Eichmüller, G. Ling, U. Hofmann, K. Foitzik, L. Mecklenburg, and B. Handjiski, *J. Invest. Dermatol.* **113,** 523 (1999).
[3] G. Cotsarelis, T.-T. Sun, and R. M. Lavker, *Cell* **61,** 1329 (1990).
[4] A. J. Reynolds and C. A. Jahoda, *J. Cell Sci.* **99,** 373 (1991).
[5] I. Moll, *J. Invest. Dermatol.* **105,** 14 (1995).
[6] J. Shapiro and V. H. Price, *Dermatol. Clin.* **16,** 341 (1998).
[7] N. Tanigaki-Obana and M. Ito, *Arch. Dermatol. Res.* **284,** 290 (1992).
[8] M. P. Philpott, M. R. Green, and T. Kealey, *J. Cell Sci.* **97,** 463 (1990).
[9] T. Kamiya, A. Shirai, S. Kawashima, S. Sato, and T. Tamaoki, *J. Dermatol. Sci.* **17,** 54 (1998).
[10] M. Hattori and H. Ogawa, *J. Dermatol.* **10,** 45 (1983).
[11] H. Jiang, S. Yamamoto, and R. Kato, *J. Invest. Dermatol.* **104,** 523 (1995).
[12] S. Watanabe, A. Mochizuki, K. Wagatsuma, M. Kobayashi, Y. Kawa, and H. Takahashi, *J. Dermatol.* **18,** 714 (1991).
[13] K. Oba, *Cosmet. Toiletr.* **103,** 69 (1988).
[14] L. Rhodes, J. Harper, H. Uno, G. Gaito, J. Audette-Arruda, S. Kurata, C. Berman, R. Primka, and B. Pikounis, *J. Clin. Endocrinol. Metab.* **79,** 991 (1994).

After an extensive search of natural products for hair growth stimulants, we found procyanidin oligomers to be a potent stimulator of hair epithelial cell growth and to be a potent activator that induces the anagen phase from the telogen phase in the murine hair cycle.[15,16]

This article describes the properties of procyanidin oligomers such as procyanidin B2 and procyanidin C1 by which they promote the proliferation of epithelial cells in murine hair and skin, and also induce the anagen phase in C3H mouse dorsal skin.

Purification Process of Oligomeric Procyanidins

Isolation of Procyanidin B2 [Epicatechin-(4β→8)-epicatechin] and Procyanidin C1 [Epicatechin-(4β→8)-epicatechin)-(4β→8)-epicatechin]

Procyanidin B2 and procyanidin C1 can be isolated with reference to the method reported by Kanda et al.[17] and Yanagida et al.[18]

Unripe apples (5.86 kg, *Malus pumila* Miller var. *domestica* Schneider, Fuji variety, weighing 10 g each on average) are homogenized in 0.1% (w/w) potassium pyrosulfite solution. The homogenate is allowed to stand for 24 hr at 4° and is then centrifuged at 3500g for 30 min at 4°. The supernatant is filtered through a glass filter, and the filtrate (4.4 liter) is passed through a column (5 cm in diameter × 50 cm long) filled with Sepabeads SP-850 resin (Mitsubishi Kasei Co., Tokyo, Japan), equilibrated previously with demineralized water, and then washed with 2 liter of demineralized water. The column is then eluted with 2 liter of 65% (v/v) aqueous ethanol, and the eluted fraction is evaporated to produce 34.28 g of a dry solid. Next, the fraction is dissolved in demineralized water and passed through a column (5 cm in diameter × 50 cm long) filled with Diaion HP-20 resin (Mitsubishi Kasei), equilibrated previously with demineralized water, and then washed with 2 liter of demineralized water and 2 liter of 15% (v/v) aqueous methanol. The column is then eluted with 2 liter of 45% (v/v) aqueous methanol. The eluted fraction is evaporated to produce 20.57 g of a dry solid, which is then dissolved in 200 ml of methyl acetate, stirred for 1 hr at room temperature, and then filtered through filter paper (No. 526, Advantec Toyo, Tokyo, Japan). The residue is reextracted with methyl acetate under the same conditions. Both filtrates are combined and evaporated to produce 12.0 g of a dry solid. This solid is then dissolved in 30 ml of methyl acetate and subjected to preparative high-performance liquid chromatography (HPLC) (5 cm in diameter × 50 cm

---

[15] T. Takahashi, T. Kamiya, and Y. Yokoo, *Acta Derm. Venereol.* **78,** 428 (1998).
[16] T. Takahashi, T. Kamiya, A. Hasegawa, and Y. Yokoo, *J. Invest. Dermatol.* **112,** 310 (1999).
[17] T. Kanda, H. Akiyama, A. Yanagida, M. Tanabe, Y. Goda, M. Toyoda, R. Teshima, and Y. Saito, *Biosci. Biotechnol. Biochem.* **62,** 1284 (1998).
[18] A. Yanagida, T. Kanda, T. Shoji, M. Ohnishi-Kameyama, and T. Nagata, *J. Chromatogr. A* **855,** 181 (1999).

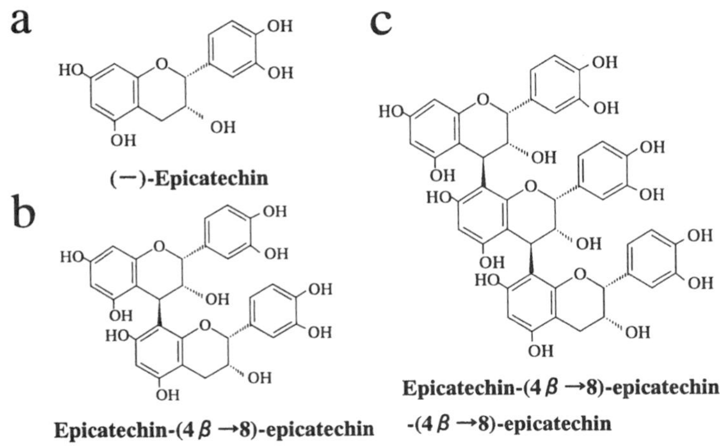

FIG. 1. Structures of procyanidin monomers and oligomers: (a) (−)-Epicatechin, (b) procyanidin B2, and (c) procyanidin C1.

long; Soken Chemical & Engineering Co., Tokyo, Japan; the column is packed with porous spherical silica beads, particle size 15–30 μm, Sokensil s15/30, Soken Chemical & Engineering). The elution conditions are as follows: flow rate, 57 ml/min; room temperature; mobile phase A, hexane : acetone = 35 : 65; mobile phase B, hexane : acetone = 20 : 80; isocratic elution with A for 80 min, and then linear gradient elution from A : B = 100% : 0% to A : B = 0% : 100% in 110 min, followed by isocratic elution with B for 110 min; and monitored using a UV detector at a wavelength of 230 nm. Fractions containing procyanidin B2 are combined to make up fraction A and evaporated to produce 2.17 g of a dry solid. Fractions containing procyanidin C1 are combined to make up fraction B and evaporated to produce 2.06 g of a dry solid. One gram of fraction A is dissolved in 10 ml of demineralized water and subjected to preparative HPLC (30 mm in diameter × 300 mm long; the column is packed with ODS silica gel, 15 μm particle size, μBondasphere, Nihon Waters, Tokyo, Japan). Elution conditions are as follows: flow rate, 20 ml/min; room temperature; mobile phase A, 0.0001% (v/v) aqueous acetic acid; mobile phase B, methanol; isocratic elution with A = 88% (v/v), B = 12% (v/v); monitored by a UV detector at a wavelength of 280 nm. A total of 1475 mg of procyanidin B2 (Fig. 1) of purity greater than 95% (w/w) as dry weight is obtained from 5.86 kg of unripe apples. The product is identified by mass spectrometry, ^{1}H nuclear magnetic resonance (NMR), and ^{13}C NMR.[19,20] One gram of fraction B is dissolved in 10 ml of demineralized water and subjected to preparative HPLC

---

[19] R. S. Thompson, D. Jacques, E. Haslam, and R. J. N. Tanner, *J. Chem. Soc. Perkin Transact. 1 Organ. Bio-organ. Chem.* **1972**, 1387.

[20] S. Morimoto, G.-I. Nonaka, and I. Nishioka, *Chem. Pharm. Bull. (Tokyo)* **34**, 633 (1986).

using the same conditions as described previously for fraction A, except for changing the composition of the eluting solvent (mobile phase A, 0.0001% (v/v) aqueous acetic acid; mobile phase B, methanol; isocratic elution with A = 84% (v/v), B = 16% (v/v)). A total of 412 mg of procyanidin C1 (Fig. 1) of purity greater than 95% (w/w) as dry weight is obtained from 5.86 kg of unripe apples. The product is identified by mass spectrometry, ^{1}H NMR, and ^{13}C NMR.[20,21]

Isolating and Culturing Hair Epithelial Cells

Mouse hair epithelial cells are isolated and cultured according to the method reported by Tanigaki et al.[22] with suitable modifications. The dorsal skin is peeled from 4-day-old C3H/HeNCrj mice (Charles River Japan, Inc., Kanagawa, Japan), cut into about 5-mm widths, washed three times with Eagle's minimum essential medium (MEM) containing 60 mg/liter of kanamycin and 10% (v/v) fetal calf serum (FCS), and dipped into MEM containing 750 IU/ml of dispase (from *Bacillus polymyxa,* Godo Shusei Co., Tokyo, Japan), 60 mg/liter of kanamycin, and 10% (v/v) FCS at 4° for 20 hr. After washing with Dulbecco's phosphate-buffered calcium- and magnesium-free saline containing 50,000 U/liter of penicillin and 50 mg/liter of streptomycin (PBS-PS), the epidermis is peeled off, and the remaining dermis layer is washed three times with PBS-PS and dispersed in Dulbecco's modified Eagle's medium (DMEM) containing 0.25% (w/v) collagenase (from *Streptomyces parvulus,* Nitta Gelatin Co., Osaka, Japan), 50,000 U/liter of penicillin, 50 mg/liter of streptomycin, 0.5% (w/v) bovine serum albumin, and 20% (v/v) FCS at 37° for 1 hr, stirring occasionally. This dermis suspension is filtered through a 212-μm nylon mesh, and the filtrate is centrifuged at 1400 rpm for 7 min. The pellet is resuspended in PBS-PS. The suspension is left to stand for 15 min, allowing the hair follicle tissue to precipitate, after which the supernatant is removed using an aspirator. Hair follicle tissue is resuspended in PBS-PS and then precipitated. This precipitation process is repeated three times. Finally, the hair follicle tissue is incubated in 0.05% (w/v) EDTA–0.25% (w/v) trypsin in Hanks' balanced calcium- and magnesium-free salt solution (HBSS)(Life Technologies, Inc., Rockville, MD) at 37° for 5 min. Hair follicle cells are suspended in DMEM supplemented with 50,000 U/liter of penicillin, 50 mg/liter of streptomycin, and 10% (v/v) FCS at a density of 3×10^5 cells/ml after filtration via a 212-μm nylon mesh. This hair follicle cell suspension is pipetted into a 24-well type I collagen-coated plate (2 cm^2/well, Iwaki Glass Co., Chiba, Japan) at a rate of 1 ml/well and incubated in a humidified atmosphere containing 5% (v/v) CO_2 at 37° for 24 hr. After a 24-hr incubation, the medium is exchanged

---

[21] G.-I. Nonaka, O. Kawahara, and I. Nishioka, *Chem. Pharm. Bull. (Tokyo)* **30,** 4277 (1982).
[22] N. Tanigaki, H. Ando, M. Ito, A. Hashimoto, and Y. Kitano, *Arch. Dermatol. Res.* **282,** 402 (1990).

with MCDB153[23] (Sigma, St. Louis, MO) containing 5 mg/liter of bovine insulin, 5 μg/liter of mouse EGF, 40 mg/liter of bovine pituitary extract, 10 mg/liter of human transferrin, 0.4 mg/liter of hydrocortisone, 0.63 μg/liter of progesterone, 14 mg/liter of O-phosphorylethanolamine, 6.1 mg/liter of ethanolamine, 50,000 U/liter of penicillin, and 50 mg/liter of streptomycin after washing with Dulbecco's phosphate-buffered calcium- and magnesium-free saline (PBS). It is then incubated further in a humidified atmosphere containing 5% (v/v) CO_2 at 37° for 5 days. During incubation, the medium is removed and replaced with fresh medium every other day.

Isolating and Culturing Mouse Epidermal Keratinocytes

The dorsal skin is peeled from 4-day-old C3H/HeNCrj mice (Charles River Japan, Inc.), cut into about 5-mm widths, washed three times with MEM containing 60 mg/liter of kanamycin and 10% (v/v) FCS, and dipped into MEM containing 750 IU/ml of dispase (from *B. polymyxa,* Godo Shusei Co.), 60 mg/liter of kanamycin, and 10% (v/v) FCS at 4° for 20 hr. After washing with PBS-PS, the epidermis is peeled off, washed five times with PBS-PS, and immersed in 0.05% (w/v) EDTA–0.25% (w/v) trypsin in HBSS at 37° for 10 min, stirring occasionally. After 10 min, DMEM supplemented with 50,000 U/liter of penicillin, 50 mg/liter of streptomycin, and 10% (v/v) FCS is added, and the suspension is filtered through a 212-μm nylon mesh, followed by centrifugation at 1500 rpm for 5 min. The keratinocytes thus obtained are suspended in DMEM supplemented with 50,000 U/liter of penicillin, 50 mg/liter of streptomycin, and 10% (v/v) FCS, pipetted into a 24-well type I collagen-coated plate (2 cm^2/well, Iwaki Glass Co., Chiba, Japan) at an initial cell density of 5×10^4 $cells/cm^2$, and incubated in a humidified atmosphere containing 5% (v/v) CO_2 at 37° for 24 hr. After a 24-hr incubation, the medium is exchanged with MCDB153 containing the same additives as those for the hair epithelial cells after washing with PBS. It is then incubated further in a humidified atmosphere containing 5% (v/v) CO_2 at 37° for 6 days. During incubation, the medium is removed and replaced with fresh medium every other day.

Colorimetric Assay for Cell Proliferation by MTT

The degree of cell growth is determined by means of an MTT [3-(4,5-dimethylthiazol-2-yl)-2,5-diphenyltetrazolium bromide] assay.[24] MTT reagent is dissolved in PBS at a concentration of 5 mg/ml, filtered through a 0.45-μm membrane filter (cellulose acetate, DISMIC-13cp, Advantec, Tokyo, Japan), and added

---

[23] S. T. Boyce and R. G. Ham, *J. Invest. Dermatol.* **81,** 33S (1983).
[24] J. Carmichael, W. G. DeGraff, A. F. Gazdar, J. D. Minna, and J. B. Mitchell, *Cancer Res.* **47,** 936 (1987).

10% (v/v) to the culture medium. The culture plate is incubated further in a humidified atmosphere containing 5% (v/v) CO_2 at 37° for 4 hr. After removing the medium, the formed dye is extracted with acidic 2-propanol containing 0.04 N HCl (adding 1.0 ml per 2 cm^2 well), and the absorbance is measured at 570 nm relative to 640 nm.

Preparation of Topically Applied Agents for *in Vivo* Evaluation

Seventy percent (w/w) ethanol, 1% (w/w) procyanidin oligomers, 10% (w/w) 1,3-butylene glycol, 0.5% (w/w) N-acetylglutamine isostearyl ester (Kyowa Hakko Kogyo Co., Tokyo, Japan), 0.25% (w/w) polyoxyethylene (25) glyceryl monopyroglutamate monoisostearate (Nihon Emulsion Co., Tokyo, Japan), and 18.25% (w/w) pure water are mixed together, and the solids are dissolved to prepare a sample solution for the *in vivo* mice test. Vehicle without procyanidins is used as the control.

Test for Hair-Growing Activity to Induce Anagen Phase in Mice

C3H mouse dorsal hair is known to have a time-synchronized hair cycle. From about 18 to 21 days of age and 47 to 95 days of age, the dorsal hairs are in the telogen phase.[10] The test compound is applied topically from the 8th to the 10th week (19-day application) during the second telogen phase, and the hair-covered area at the 10th week is evaluated.

The degree of hair-growing activity inducing the anagen phase in mice is measured with reference to the method introduced by Hattori and Ogawa.[10] In this test, 8-week-old male C3H/HeSlc mice (Japan SLC, Inc., Shizuoka, Japan) whose hair cycle is in the telogen phase are used. The mice are placed into several groups, each containing four or five mice. The hair on the back of each mouse is shaved carefully, first with an electric clipper (Thrive 6000AD, Natsumeseisakusho, Tokyo, Japan) using blade size A (0.5 mm) and then using blade size B (0.1 mm), followed by an electric shaver (National ES467, Matsushita Electric Works, Shiga, Japan) so as not to injure or stimulate the skin and influence the results. Two hundred microliters daily of test sample is applied to the shaved area. On the 19th day of the test, the mouse back skin is observed and photographed; the skin is then peeled from the back of each mouse and photographed.

Mouse Hair Epithelial Cell Growth-Promoting Activity

Procyanidin B2 showed growth-promoting activity of about 300% (30 μM) relative to controls (=100%) in a 5-day culture of hair epithelial cells. Procyanidin C1 also showed growth-promoting activity of about 220% (3 μM) relative to controls (=100%). The maximum growth-promoting activity of procyanidin C1

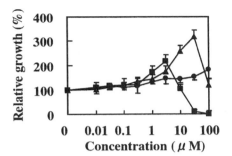

FIG. 2. Growth-promoting activities on murine hair epithelial cells. The activities are shown as relative to controls (=100%). Procyanidins were added to the culture medium during the last 5 days. Medium without procyanidins was used as the control. ●, (−)-epicatechin; ▲, procyanidin B2; ■, procyanidin C1.

relative to controls (=100%) was lower than that of procyanidin B2; however, at concentrations of 0.3 to 3 μM, the growth-promoting activity of procyanidin C1 exceeded that of procyanidin B2. The activity of (−)-epicatechin was lower than that of procyanidin B2 and procyanidin C1 (Fig. 2).

Mouse Keratinocyte Growth-Promoting Activity

Procyanidin B2 showed growth-promoting activity of about 250% relative to controls (=100%) at concentrations of 0.3 to 100 μM. The growth-promoting activity of procyanidin C1 was about 210% relative to controls (=100%) at concentrations of 0.1 to 10 μM. The activity of (−)-epicatechin was somewhat lower than the activities of procyanidins B2 and C1 (Fig. 3).

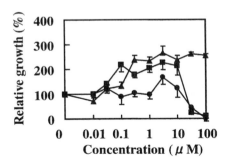

FIG. 3. Growth-promoting activities on murine keratinocytes. The activities are shown as relative to controls (=100%). Procyanidins were added to the culture medium during the last 6 days. Medium without procyanidins was used as the control. ●, (−)-epicatechin; ▲, procyanidin B2; ■, procyanidin C1.

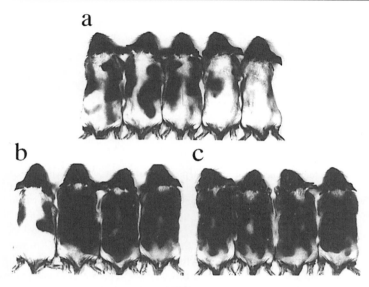

FIG. 4. *In vivo* hair-growing activity evaluated using the C3H mouse model. Photographs were taken after the topical application of test agents for 19 days. Test agents were applied to male 8-week-old C3H telogen mice at 200 µl/day/mouse. (a) Vehicle, (b) 1% (w/w) procyanidin B2, and (c) 1% (w/w) procyanidin C1.

In Vivo Hair-Growing Activity Evaluated Using C3H Mouse Model

The group to which 1% (w/w) procyanidin B2 had been applied showed a hair growth area of 69.6% ± 21.8% (average ± SD), and the group to which 1% (w/w) procyanidin C1 had been applied showed a hair growth area of 78.3% ± 7.6% (average ± SD) on day 19. The control group to which vehicle was applied, however, showed a hair growth area of 41.7% ± 16.3% on day 19 (Fig. 4).

Acknowledgments

I thank Dr. T. Kanda (Nikka Whiskey Distilling Co., Ltd.) and Dr. A. Yanagida (Tokyo University of Pharmacy and Life Science) for their instruction and support, especially in the establishment of a purification process for procyanidins. I also thank my scientific colleagues at Tsukuba Research Laboratories and the Tokyo Research Laboratories of Kyowa Hakko Kogyo Co., in particular Dr. T. Kamiya, Mr. A. Shirai, Ms. A. Kamimura, Ms. T. Hamazono, and Dr. Y. Yokoo for their helpful discussions. I am grateful to Dr. H.-F. Leu for her support and to Ms. Y. Ohishi for technical assistance.

[33] Effect of Polyphenolic Flavonoid Compounds on Platelets

By Dhanansayan Shanmuganayagam and John Folts

The protective effect of fruit, vegetable, and red wine consumption against coronary artery disease is partly attributed to the polyphenolic compounds, such as flavonoids, present in these foods.[1–4] It is believed that one of the ways by which flavonoids offer this protection is through the inhibition of the activity of platelets, which play a significant role in the development and progression of coronary artery disease.[5]

Platelets not only contribute to the initiation and progression of the atherosclerotic process by several mechanisms,[6] but also initiate the clot that kills.[7] Thus, there is considerable interest in studying the effect of various flavonoid compounds on platelet activity. As there are many structural and functional variations among the various subclasses of these compounds, studies must be designed carefully such that the relevance of the biological effects of the flavonoids can be evaluated accurately. This may be done in several steps, beginning with studies measuring their direct effect on platelets (*in vivo* incubation studies), then proceeding to examine their effect on the interaction of platelets with one another and the arterial wall (*in vivo* and *ex vivo* studies), and then finally their effect when taken orally either acutely or daily for a specific length of time (*ex vivo* studies). This article presents two techniques—whole blood impedance platelet aggregometry and the Folts cyclic flow reduction model—for accomplishing the experimental approach outlined earlier.

Basic Principle of Platelet Aggregometry

Whole blood impedance platelet aggregometry provides a relatively accurate and effective method for studying the *ex vivo* and *in vitro* effect of a flavonoid

---

[1] S. C. Renaud, A. D. Beswick, A. M. Fehily, D. S. Sharp, and P. C. Elwood, *Am. J. Clin. Nutr.* **55**, 1012 (1992).
[2] M. G. Hertog, D. Kromhout, C. Aravanis, H. Blackburn, R. Buzina, F. Fidanza, S. Giampaoli, A. Jansen, A. Menotti, and S. Nedeljkovic, *Arch. Intern. Med.* **155**, 381 (1995).
[3] M. G. Hertog, E. J. Feskens, and D. Kromhout, *Lancet* **349**, 699 (1997).
[4] P. Knekt, R. Jarvinen, A. Reunanen, and J. Maatela, *BMJ* **312**, 478 (1996).
[5] M. J. Davies and A. Thomas, *N. Engl. J. Med.* **310**, 1137 (1984).
[6] L. E. Rabbani and J. Loscalzo, in "Thrombosis and Hemorrhage" (J. Loscalzo and A. I. Schafer, eds.), p. 771. Blackwell, Publications, Boston, 1994.
[7] R. Ross, *N. Engl. J. Med.* **314**, 488 (1986).

source on the activity of platelets.[8] The whole blood aggregometer (Chrono-Log Corp., Havertown, PA) measures the impedance between two electrodes that are placed in a blood sample. Because of the presence of electrolytes, blood is very conductive and thus has low electrical impedance. When the platelets in the sample are stimulated by the addition of a known platelet agonist (e.g., collagen, ADP), they adhere and aggregate onto the electrodes, increasing the electrical impedance measured between them. The change in impedance is proportional to the aggregation of the platelets on the electrodes. The extent and rate of aggregation are reflective of the level of activity of the platelets. The platelet response is taken to be the increase in impedance measured 7 min after the addition of the agonist (Fig. 1). The change in the platelet response after exposure of the blood to a flavonoid source would reveal the effect of the compound on the platelets. Platelet aggregation studies can also be conducted using platelet-rich plasma[8,9] and washed resuspended platelets[10]; however whole blood aggregometry allows for the study of platelets in a more natural environment.

Subject Selection: Blood Donors

All subjects used in platelet studies should refrain from taking any known platelet inhibitors. Primary exclusions include aspirin and other nonsteroidal antiinflammatory drugs, Ticlid, Plavix, and flavonoid sources such as wine, beer, fruit juices, jams, jelly, and tea that are thought to inhibit platelet activity. If they had been taking such a drug/compound, they must stop taking it for a period long enough for its effect to be eliminated (usually 5–10 days). As the effect of flavonoids on platelets in the primary focus of this experimental design, vegetarians are excluded, as they would be expected to have a relatively high intake of flavonoids from their diet that may mask any effect of the flavonoid source being tested. One would also wish to avoid using subjects taking vitamin supplements, especially those with antioxidant properties (i.e., vitamin $E \geq 100$ IU/day or vitamin $C \geq 300$ mg/day) as they may inhibit platelet activity.[10]

Blood Sampling

Blood is drawn into a syringe containing 3.8% sodium citrate (1 : 9; citrate : blood). It is imperative that the blood is drawn slowly and at a consistent rate through a needle with a relatively large bore (19- to 21-gauge). It is also important that the subject is not stressed excessively during the draw as stress elevates plasma

---

[8] R. Abbate, S. Favilla, M. Boddi, G. Costanzo, and D. Prisco, *AJCP* **86,** 91 (1986).
[9] D. C. Cardinal and R. J. Flower, *J. Pharmacol. Methods* **3,** 135 (1980).
[10] J. E. Freedman, J. H. Farhat, J. Loscalzo, and J. F. Keaney, Jr., *Circulation* **94,** 2434 (1996).

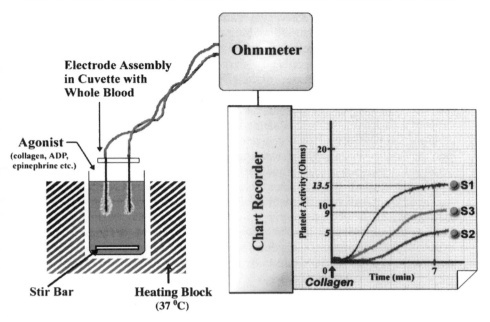

FIG. 1. A schematic representation of a whole blood impedance aggregometer. A cuvette containing a blood sample is placed in the aggregometer where it is warmed and stirred. An electrode assembly is then placed in the blood sample, and the electrical impedance between the two electrodes in the assembly is measured continuously by an ohmmeter and recorded onto a chart. Once a baseline impedance recording is achieved, a specific amount of a platelet agonist is added to the sample. The agonist stimulates the platelets to aggregate between the two electrodes. Because the aggregating platelets act like an insulator between the electrodes, the electrical impedance measured increases. The extent and rate of change of measured impedance are proportional to the aggregation of the platelets between the electrodes. This aggregation is dependent on the resting activity of the platelets before the addition of the agonist. If the platelets had been treated with a platelet inhibitor, the addition of the same amount of agonist would result in a lower change in impedance (S2 or S3) than caused by untreated platelets (S1). The lower the change in impedance, the more potent the inhibitor.

catecholamines, especially epinephrine. An elevated epinephrine level activates platelets to varying degrees and thus adds an uncontrollable variable to the study.[11]

A small portion of the blood sample is used to determine the subject's hematocrit (HCT) and whole blood platelet count (WBPltCt). For any aggregation study, it is important that the HCT and WBPltCt are within the physiological range[12] [HCT 42–50% (men) or 36–45% (women); WBPltCt 150–450 × 10^9/liter]; for any study where multiple samples are drawn from the same subject for comparison of

[11] J. D. Folts and G. G. Rowe, *Thromb. Res.* **50**, 507 (1988).
[12] M. M. Wintrobe, "Wintrobe's Clinical Hematology." Williams & Wilkins, Baltimore, 1999.

platelet activity, these parameters should be kept relatively constant between the blood draws. The blood sample is then diluted with an equal volume of preservative-free saline and kept at room temperature. The sample is mixed very gently before each use.

Aggregometry

One milliliter of the blood sample is placed into a siliconized cuvette containing a siliconized stir bar. The cuvette is then placed in the heated well of the aggregometer and warmed to 37° for 5 min. After 5 min, the electrode assembly is placed into the cuvette and the blood is kept at 37° and stirred (10,000–12,000 rpm) for another 5 min. This is sufficient time for obtaining a baseline impedance measurement. Once the baseline is obtained, the agonist can be injected into the cuvette to trigger aggregation of platelets.

For this description of platelet aggregometry, we will use collagen as the agonist of choice. A working collagen concentration is prepared such that injecting 4 μl of the solution into the cuvette will give a final collagen concentration of 1 μg/ml in the blood. A large enough dose of collagen (usually 50 μl) is added to the sample to obtain a maximal platelet response to the agonist (Fig. 1). A response is considered maximal if using higher doses of collagen does not elicit higher responses. Once this maximal response is achieved, using a smaller dose of collagen (usually 8 μl, but it could be 4–8 μl), a response that is approximately 60–70% of the maximal, is obtained. We will refer to this response as the submaximal response. The effect of a compound on platelet activity will be judged by its effect on collagen-induced platelet aggregation induced by the same dose of collagen (8 μl) that elicited the submaximal response. This procedure is imperative to obtain a submaximal response to confirm that the dose of agonist being used for the study is not eliciting a maximal response. The change in the platelet response to this submaximal dose of agonist (8 μl) after treatment of the subject or the blood with a flavonoid source is the indicator of the effect of the flavonoid source on the platelets. It is assumed that if a flavonoid is able to significantly inhibit this magnitude of platelet response, it may be considered a platelet inhibitor (Fig. 1). The platelet response to any given dose of agonist must be duplicated and the duplicate values averaged. All aggregation studies must be completed within 2–3 hr of blood drawing.

Other agonists such as ADP, thrombin, and phorbol 12-myristate 13-acetate (PMA) can also be used to induce platelet aggregation. As different agonists induce different pathways of platelet activation, using more than one agonist can provide information about the mechanism of platelet inhibition that may be produced by a flavonoid source. For example, as PMA activates platelets through a protein kinase C-dependent pathway, an inhibition of PMA-induced platelet aggregation by a flavonoid source indicates that the flavonoid source is acting through the protein kinase C-dependent pathway.[10]

In Vitro (Incubation) Studies Using Whole Blood Platelet Aggregometry

In order to study the *in vitro* effect of a flavonoid on whole blood platelet activity, it is important to obtain a proper solution of the flavonoid. This presents some difficulty, as many of the flavonoids are insoluble or poorly soluble in water or saline. Thus, other solvents such as ethanol or dimethyl sulfoxide (DMSO) must be used. It is advisable that the amount of these solvents used to obtain the final solution be kept minimal as they alone can inhibit platelets at a higher dose. Often it is possible to first dissolve the flavonoid in a very small volume of these solvents and then add water or saline to obtain the desired final concentration of the flavonoid. It is also important that during each experiment the effect of the solvent by itself (at the amount used in the solution) on platelet activity is determined.

The blood drawing and preparation are done as described previously. Once the dose of collagen needed to produce a submaximal response is determined, the effect of preincubating the blood with the flavonoid on the platelet response stimulated by this dose of collagen can be measured. Often the flavonoid is added to the blood 5 min before the collagen is added. However, if a longer preincubation time is required to ensure that the flavonoid has enough time to act on the platelets, it is crucial that the blood sample used for determining the maximal and the submaximal responses are also warmed and stirred for the additional length of the preincubation time.

In Vivo Studies of Platelet-Inhibiting Properties of Flavonoids

Many polyphenolic compounds undergo structural modifications or degradation in the gastrointestinal (GI) tract.[13] The degree and rate of gastrointestinal absorption of polyphenolic compounds and their modified forms are not well understood. This is mainly due to the lack of techniques for accurately determining the specific polyphenolic compounds and their levels in a sample. If a compound is absorbed, it may be further modified during its first pass through the liver or cleared at a rate much faster than the rate of absorption such that the plasma levels of the compounds may not reach high enough concentration to exert an effect on the platelets. The Folts cyclic flow model of experimental arterial thrombosis facilitates the study of the effect of polyphenolic compounds when administered either intravenously (iv) or intragastrically (ig), thus revealing more information about their bioavailability.

The methodology for preparing an animal for the Folts cyclic flow model has been described previously in detail.[14,15] This model mimics the conditions under which thrombus formation occurs in the damaged and narrowed (stenosed) arteries

---

[13] P. C. Hollman and M. B. Katan, *Food Chem. Toxicol.* **37,** 937 (1999).

[14] J. D. Folts, *Cardiovasc. Rev. Rep.* **11,** 10 (1990).

[15] J. D. Folts, *Circulation* Suppl. IV, IV3 (1991).

of patients experiencing unstable angina.[16] It has been shown previously that the rate of thrombus formation is proportional to the rate of platelet accumulation in the stenosed artery.[17] If an effective platelet inhibitor, like aspirin, is given, the thrombus formation is abolished[14,15,18] While acute platelet-mediated thrombus formation can be studied in arteries of rabbits, pigs and monkeys,[14,15] in this article, the general methodology for studying thrombus formation in the coronary artery of dogs is outlined.

Dogs are premedicated with 3 mg/kg of morphine sulfate given intramuscularly and anesthetized with 25 mg/kg of iv pentobarbital sodium. Blood pressure is measured by placing a catheter in the femoral artery and attached to a pressure transducer.[14,15] After establishing positive pressure ventilation, a thoracotomy is performed and the chest is entered at the fourth intercostal space. The heart is then exposed and placed in a pericardial cradle. The left circumflex coronary artery is dissected out and either an electromagnetic (Spectramed Inc., Oxnard, CA) or Doppler ultrasound flow probe (Transonic Systems Inc., Ithaca, NY), of appropriate size, is placed on the artery to measure blood flow continuously (Fig. 2).[14,15] Then distal to the flow probe, the circumflex coronary artery is dissected out and clamped three to four times to produce damage to the intimal and medial layers of the vessel. Finally, a plastic cylinder is placed around the outside of the damaged artery to produce a 60–70% reduction in the arterial diameter. The construction and placement of the cylinders on the coronary artery have been described previously.[14,19]

Once the artery is damaged and stenosed, acute platelet-mediated thrombus formation occurs. The developing thrombus causes the measured coronary blood flow to decline to near zero flow levels. The occluding thrombus can then be made to embolize distally by shaking the plastic cylinder gently.[15] The thrombus formation followed by distal embolization produces cyclical flow reduction (CFRs) in the measured coronary blood flow, as shown in Fig. 3. These CFRs have been described as an on-line, *in vivo* bioassay for platelet activity.[20] They will continue for 4–6 hr if no intervention is done.[15,16,21] Thus, once the CFRs are established, a drug or a polyphenolic compound can be administered and the CFRs monitored continuously for several hours. If the CFRs are diminished significantly in frequency or abolished, the drug/compound can be assumed to have significant antiplatelet property.[18,19,22]

[16] J. T. Willerson, P. Golino, J. Eidt, W. B. Campbell, and L. M. Buja, *Circulation* **80**, 198 (1989).
[17] N. Maalej, J. E. Holden, and J. D. Folts, *J. Thromb. Thrombol.* **5**, 231 (1998).
[18] J. D. Folts, *Cardiovasc. Drugs Ther.* **9**, 31 (1995).
[19] J. D. Folts, in "Animal Models of Vascular Injury" (D. Simon and C. Rogers, eds.), p. 127. Humana, Totowa, NJ, 2000.
[20] L. R. Bush and R. J. Shebuski, *FASEB J.* **4**, 3087 (1990).
[21] S. P. Roux, K. S. Sakariassen, V. T. Turitto, and H. R. Baumgartner, *Arterioscl, Thromb.* **11**, 1182 (1991).
[22] H. S. Demrow, P. R. Slane, and J. D. Folts, *Circulation* **91**, 1182 (1995).

[33] EFFECT OF POLYPHENOLIC FLAVONOID COMPOUNDS ON PLATELETS 375

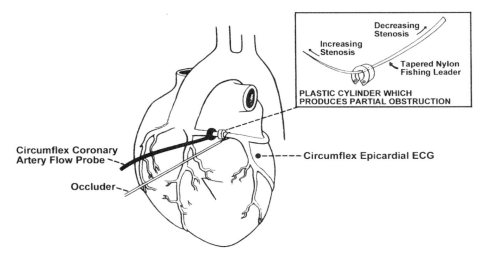

FIG. 2. Animal model for studying conditions similar to humans with coronary artery disease: Technique for producing fixed partial obstruction of the left circumflex coronary artery. The artery is clamped with a surgical clamp three to four times to produce moderate intimal damage. A plastic cylinder 4 mm in length is then placed so as to encircle and constrict the coronary artery, producing a 60–70% reduction in diameter. Smooth, tapered nylon fishline is placed between the inside wall of the plastic cylinder and the outside wall of the coronary artery. The fishline is then pulled in either direction to make slight increases or decreases in the amount of stenosis. Platelet thrombi collect periodically in the stenosed lumen followed by embolization, producing cyclical changes in flow measured by the flow probe. ECG, electrocardiogram.

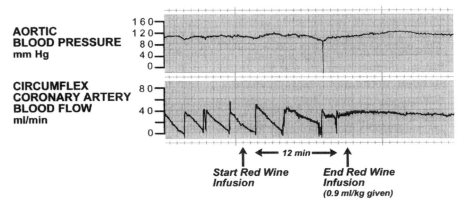

FIG. 3. Representative tracing of the hemodynamic effect of intravenous (iv) administration of red wine. (Top) Mean aortic blood pressure. (Bottom) Blood flow in the stenosed left circumflex artery. Cyclic flow reductions were eliminated 12 min after the red wine–saline infusion was started. The amount of wine administered to this animal was 0.9 ml/kg, and the peak blood alcohol concentration was 0.03 g/dl.

The minimum effective dose required to achieve plasma levels of a flavonoid substance, such as red wine, that would inhibit platelets can be determined by slowly administering incremental doses of the substance intravenously until the CFRs are abolished as in Fig. 3.[22] This method is more efficient for determining the minimum effective dose than the usual pharmacological approach that is used to obtain a dose–response curve. The pharmacological approach would involve giving a specified dose and then measuring the change in *ex vivo* whole blood aggregation and repeating the process with new animals and higher doses until an inhibition of *ex vivo* platelet aggregation is observed or the dose is too high to be clinically relevant. Such a method would be expensive and labor-intensive.

The effective iv dose of a flavonoid can be compared to that estimated from *in vitro* platelet aggregation studies. If the iv dose is found to be much greater, it may be possible that the flavonoid is being modified or excreted rapidly by the liver. Once the iv dose is determined, information about the bioavailability of the flavonoid source can be determined by administering the compound via a tube placed surgically within the proximal duodenum. Often there is considerable difference between the effective iv and ig dose. For example, a red wine when given iv abolishes CFRs at 1.62 ± 1.12 ml/kg (Fig. 4).[22] However, 4 ml/kg of red wine was required to abolish CFRs 113 ± 32 min after an ig administration.[22] This amount of wine is approximately equivalent to 10 ounces (300 ml; approximately two glasses) for a human weighing 80 kg, and falls well within the range that would be considered as moderate alcoholic beverage consumption. The average blood alcohol content in these dogs was 0.03 ± 0.01 g/dl, which also reflects a level achieved with moderate alcoholic beverage consumption.[23]

In order to determine whether pure alcohol could cause an antithrombotic effect, it was studied previously in the same model. A blood alcohol content (BAC) greater than 0.25 g/dl was required to abolish CFRs.[24] This suggests that alcohol alone is not a good platelet inhibitor except at very high doses.[25] It is believed that the flavonoids possess most of the platelet inhibitory property in red wine. An equal amount of a white wine, which contains 7–10 times lower amounts of flavonoids than a red wine,[26] was also tested. While the white wine produced a peak BAC identical to that produced by red wine of 0.03 g/dl, it failed to abolish CFRs.[22] In addition, purple grape juice (10 ml/kg), which contains many of the same flavonoids found in red wine, but without any alcohol, also abolished CFRs 2 hr after ig administration (Fig. 5).

---

[23] J. D. Folts, H. Demrow, and P. Slane, *J. Myocard. Ischemia* **6,** 33 (1994).
[24] J. W. Keller and J. D. Folts, *Cardiovasc. Res.* **22,** 73 (1988).
[25] S. C. Renaud and M. DeLorgeril, *Lancet* **339,** 1523 (1992).
[26] A. L. Waterhouse and P. Teissedre, in "Wine: Nutritional and Therapeutic Benefits" (T. R. Watkins, ed.) 661 Ed., p. 12. American Chemical Society, Chicago, IL, 1997.

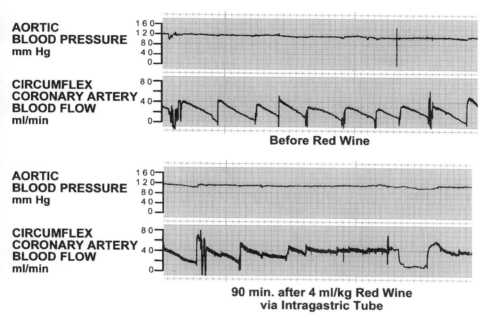

FIG. 4. Representative tracing of the hemodynamic effect of intragastric (ig) administration of red wine. (Top) Mean aortic blood pressure and blood flow in the stenosed left circumflex artery. Cyclic flow reductions (CFRs) in blood flow were observed before the red wine was administered. (Bottom) CFRs are eliminated 90 min after the administration of 4 ml/kg of red wine.

One can also combine *ex vivo* aggregation studies (described later) with the study of CFRs in the anesthetized dog. A control blood sample is obtained for *ex vivo* aggregation studies, when CFRs are occurring. Then the flavonoid source is administered. If CFRs become diminished significantly in frequency or are abolished, then a second blood sample is drawn and the *ex vivo* whole blood aggregation studies are repeated. One has a greater assurance that the flavonoid source is indeed a platelet inhibitor if the inhibition of *ex vivo* aggregation is observed along with the abolishment of CFRs. For example, in the study of purple grape juice, *ex vivo* platelet aggregation studies confirmed that platelet aggregation was reduced approximately 2 hr after the ig administration.[22]

Ex Vivo (Acute or Feeding) Studies

These studies examine the effect of acute administration (oral or intravenous) or daily feeding (7 days–3 months) of flavonoids on platelet activity in animal or human subjects. The blood drawing and preparation are done as described

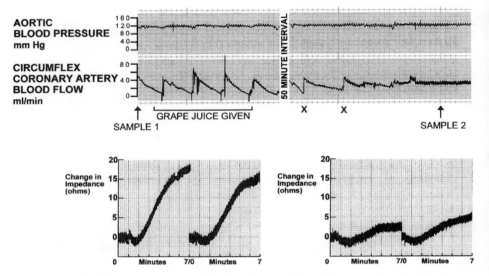

FIG. 5. (Top) Representative tracing of the hemodynamic effects of intragastric (ig) administration of grape juice. "X" denotes a spontaneous platelet thrombus embolization that did not require the manual shaking of the occluding cylinder. (Bottom) Whole blood aggregation tracings from the animal in the top tracing. Sample 1 was drawn before the animal was given the grape juice, and sample 2 was drawn when the cyclic flow reductions were eliminated. The decrease in the collagen induced-aggregation corresponds to the decrease in the frequency of cyclic flow reductions. The whole blood aggregation for each sample was duplicated. A diminished response to collagen (shown on the right) indicates significant inhibition of *ex vivo* aggregation by the administered grape juice.

previously. In a representative study, platelet aggregation studies were conducted on the blood samples drawn from human volunteers before and 2 hr after the consumption of either 300 ml (approximately two glasses) of red or white wine or 600 ml (approximately four glasses) of purple grape juice. While there was a significant decrease in platelet activity after consuming the red wine and the purple grape juice, there was no change in platelet activity after the consumption of the white wine.[27] In another study, the antiplatelet effect of consumption of purple grape juice (5–7 ml/kg) was compared to that of orange juice and grapefruit juice in human subjects.[28] The grape juice inhibited platelet activity while an equal amount of the orange and grapefruit juices did not.[28] These studies allow for determining the effect of the flavonoid on platelet activity induced by several different agonists (e. g., collagen, ADP).

[27] J. D. Folts, *Pharm, Biol.* **36,** (1998).
[28] J. Keevil, H. E. Osman, J. Reed, and J. D. Folts, *J. Clin. Nutr.* **130,** 53 (2000).

Conclusion

When studying the antiplatelet effects of various polyphenolic compounds, a systematic approach to experimentation is necessary to address their bioavailability, metabolism, and function. For years, claims on the potential health benefits of flavonoid sources, such as tea, have been extrapolated from *in vitro* studies of a specific flavonoid found in the flavonoid source. For example, *in vitro* studies of a single compound found in a flavonoid source may show antiplatelet or antioxidant properties, but this is not sufficient for making health claims about the flavonoid source, as these studies would not reveal the function and bioavailability of the compound when taken orally. A good example of a compound that differs greatly in its effectiveness when administered through different routes is quercetin.

Quercetin is a flavonoid commonly found in red wine and other grape products. It has been touted as a platelet inhibitor based on a number of *in vitro* studies.[29-32] Several epidemiological studies that observed an inverse relationship between the flavonoid content of diet and the occurrence of cardiovascular disease attributed quercetin to be one of the flavonoids responsible for this effect based on previous *in vitro* studies.[2,4] The approximate intake of quercetin is thought to be 30 mg/day.[33]

Quercetin inhibited platelet activity significantly and abolished CFRs at a dose of 12 mg/kg given iv in the Folts CFR model. However, when given ig, it required at least 26 mg/kg, which would equate to a daily intake of over 2000 mg in an 80-kg human.[34] This suggests that a flavonoid source such as quercetin may inhibit platelet activity *in vitro*, or *in vivo* when given iv, but it may not be bioavailable when given orally. This notion is supported by two studies that measured the effect of quercetin taken orally for 2–4 weeks on *ex vivo* platelet aggregation.[35,36] In the study by Conquer *et al.*,[35] quercetin (1000 mg/day) was given to human subjects. *Ex vivo* platelet aggregation studies using collagen as an agonist were performed at baseline and after 28 days of quercetin consumption. However, there was no change in platelet aggregation after 28 days of consuming 1000 mg/day of quercetin. Plasma levels of quercetin increased from 0.10 ± 0.09 to 1.5 μmol/

---

[29] S. H. Tzeng, W. C. Ko, F. N. Ko, and C. M. Teng, *Thromb. Res.* **64**, 91 (1991).

[30] N. C. Cook and S. Samman, *J. Nutr. Biochem.* **7**, 66 (1996).

[31] E. Corvazier and J. Maclouf, *Biochim. Biophys. Acta* **835**, 315 (1985).

[32] R. J. Gryglewski, J. Robak, and J. Swies, in "Drugs Affecting Leukotrienes and Other Eicosanoid Pathways" (B. Samuelsson, F. Berti, G. C. Folco, and G. P. Velo, eds.), 149. Plenum, New York, 1985.

[33] M. G. Hertog, E. J. Feskens, P. C. Hollman, M. B. Katan, and D. Kromhout, *Lancet* **342**, 1007 (1993).

[34] P. R. Slane and J. D. Folts, *Clin. Res* **42**, 162A (1994).

[35] J. A. Conquer, G. Maiani, E. Azzini, A. Raguzzini, and B. J. Holub, *J. Nutr.* **128**, 593 (1998).

[36] K. Janssen, R. P. Mensink, F. J. Cox, J. L. Harryvan, R. Hovenier, P. Hollman, and M. B. Katan, *Am. J. Clin. Nutr.* **67**, 255 (1998).

liter.[35] The achieved plasma level is much lower than what was required (mmol/liter range) to inhibit platelets *in vitro*.[35]

In the study by Janssen *et al.*, 220 g of onions containing 114 mg of quercetin was given to human subjects for 14 days. The plasma level of quercetin measured was also approximately 1 μmol/liter. However, there was no measurable inhibition of platelet activity by the onions with their content of quercetin.[36] In the same study, *in vitro* tests showed significant inhibition of platelet activity (80–90%) with a quercetin concentration of 2500 μmol/liter. The authors suggested that a very high concentration of quercetin is required to inhibit platelets and that this level cannot be achieved readily by daily consumption of a quercetin source.[36] This demonstrates the importance of (1) determining the bioavailability of a given flavonoid source and (2) doing *in vitro, in vivo,* or *ex vivo* studies with a given flavonoid source.

The flavonoid source must be absorbed from the GI tract and not be degraded by the first pass through the liver. It also must achieve a sufficient level in the plasma to inhibit platelet activity. Many interesting studies have been done with polyphenolic compounds such as quercetin, showing *in vitro* antiplatelet, antioxidant, and anticancer effects.[36] However, they must be confirmed with *in vivo* or *ex vivo* tests where a reasonable dose of the compound is consumed orally.

[34] Assessing Bioflavonoids as Regulators of NF-κB Activity and Inflammatory Gene Expression in Mammalian Cells

By Claude Saliou, Giuseppe Valacchi, and Gerald Rimbach

Introduction

Inflammation is a multifactorial response that is of fundamental importance in the maintenance of cellular homeostasis when the organism is challenged by noxious agents (e.g., bacteria, viruses, parasites) or by tissue mechanical injury. Inflammatory processes are associated with a dramatic rise in the number of polymorphonuclear leukocytes and monocytes in the affected tissue and the subsequent release of various molecules such as prostaglandins and cytokines. Under ideal conditions, inflammation results in the complete recovery of the integrity of the affected tissue. However, if the response of the triggering stimulus is not subjected to a tight regulation, a condition known as chronic inflammation can be established. Indeed, a chronic inflammatory-like environment characterizes the pathogenesis of various diseases, such as atheriosclerosis, rheumatoid arthritis, and Crohn's disease, and is thought to be among the causative factor of more than 30% of human cancer.[1]

The transcription factor NF-κB, in cooperation with others, has been suggested to coordinate the expression of genes encoding proteins that are involved in inflammatory processes. In particular, NF-κB contributes to the production of interleukin (IL)-1, IL-6, tumor necrosis factor (TNF)-α, lymphotoxin, GM-CSF, and interferon (IFN)-γ. Furthermore, some of these cytokines, e.g., IL-1 and TNF-α, activate NF-κB themselves, thus initiating an autoregulatory feedback loop.[2] NF-κB activation by various stimuli occurs on its dissociation from the inhibitory protein I-κB and its subsequent nuclear translocation. Several lines of evidence, including the inhibition by various antioxidants, suggest that NF-κB activity is subject to redox regulation.[3] Because of the pivotal role in inflammatory response a significant effort has focused on developing therapeutic agents that regulate NF-κB activity.

Epidemiological reports have indicated that consumption of foods rich in flavonoids is associated with a lower incidence of degenerative diseases.[4] Consistently, experimental data are accumulating regarding polyphenolic compounds as natural phytochemical antioxidants that possess anti-inflammatory properties by downregulating NF-κB (Table I). To better understand the effectiveness of bioflavonoids in modulating NF-κB, both accurate and sensitive methods are warranted. The activity of NF-κB can be monitored by different approaches summarized in Table II. The different methodological approaches lead to different information about NF-κB activity, such as DNA binding, nuclear translocation, transactivation capacity, I-κB degradation, and phosphorylation, as well as NF-κB-dependent gene expression. This article focuses on methods to monitor NF-κB–DNA binding and its ability to activate transcription.

Materials and Methods

Materials

Trypsin, as well as culture media, is obtained from Life Technologies (Gaithersburg, MD). Fetal calf serum (FCS), L-glutamine, sodium pyruvate, penicillin, and streptomycin are from the University of California San Francisco (UCSF) cell culture facilities (San Francisco, CA). IFN-γ is from Genzyme, (Cambridge, MA). All other chemicals are obtained from Sigma (St. Louis, MO) unless otherwise specified.

---

[1] L. Packer, G. Rimbach, and F. Virgili, *Free Radic. Biol. Med.* **27,** 704 (1999).
[2] S. Ghosh, M. J. May, and E. B. Kopp, *Annu. Rev. Immunol.* **16,** 225 (1998).
[3] L. Flohé, R. Brigelius-Flohé, C. Saliou, M. Traber, and L. Packer, *Free Radic. Biol. Med.* **22,** 1115 (1997).
[4] A. S. St. Leger, A. L. Cochrane, and F. Moore, *Lancet* **1,** 1017 (1979).

TABLE I
FLAVONOIDS AND FLAVONOID-RELATED COMPOUNDS SUPPRESSING NF-κB ACTIVITY IN CELL CULTURE STUDIES

Name	Concentration (duration)	Inducers	Cell line	Ref.
Apigenin	25 μM (4 hr)	TNF-α, TNF-α+IFN-γ	HUVEC	8
Caffeic acid phenetyl ester (*propolis*)	25 μg/ml (2 hr)	TNF-α, PMA, ceramide-C_8, okadaic acid, H_2O_2	U937	9
Epigallocatechin 3-gallate (green tea)	15μM (coincubation with inducer)	LPS	Mouse peritoneal macrophages	10
	100 μM (2 hr)	LPS	RAW 264.7	11
Genistein (soy, clover)	148 μM (1–2hr)	TNF-α	U937, Jurkat, HeLa	12
		Okadaic acid	U937	12
Ginkgo biloba extract	100–400μg/ml (18 hr)	H_2O_2	PAEC	13
Quercetin (wine, onion)	265 μM (1 hr)	TNF-α	U937	12
	10μM (coincubation with inducer)	H_2O_2	HepG2	14
Silymarin (*Silybum marianum*)	12.5μg/ml (24 h)	Ultraviolet	HaCaT	15
		Okadaic acid, LPS	HepG2	16
		TNF-α	Würzburg	16
	24 μM (2 h)	TNF-α	U937, HeLa, Jurkat	17
Taxifolin (pine bark)	303 μM (24 hr)	IFN-γ	RAW 264.7	18
Theaflavin 3, 3′-digallate (black tea)	10 μM (coincubation with inducer for 1 hr)	LPS	RAW 264.7	19

Cell Lines

The effect of various flavonoids on NF-κB activity was studied in different cell lines. HaCaT[5] (spontaneously transformed human keratinocytes) are from Dr. N. E. Fusenig (Heidelberg, Germany) HepG2 (human hepatoblastoma-derived cell line) and RAW 264.7 (murine monocyte-macrophage cell line) are obtained from the American Type Culture Collection (Rockville, MD).

[5] P. Boukamp, R. T. Petrussevska, D. Breitkreutz, J. Hornung, A. Markham, and N. E. Fusenig, *J. Cell Biol.* **106,** 761 (1988).

TABLE II
ASSESSMENT OF NF-κB ACTIVITY

Assay	Parameter investigated
Electrophoretic mobility shift assay	DNA binding
	Nuclear translocation (if using nuclear protein extract)
Supershift assay	NF-κB subunits bound to cognate DNA sequences
Reporter assay	Transactivation capacity of NF-κB
Cytokine release	Expression of cytokines under the transcriptional control of NF-κB
p65 Western blot, immunocytochemistry	Nuclear translocation
IκB Western blot	IκB degradation
IκB-phosphoserine Western blot	IκB phosphorylation

HaCaT keratinocytes are grown in Dulbecco's modified Eagle medium (DMEM) supplemented with 10% fetal bovine serum (FBS) and antibiotics (100 U/ml of penicillin, 100 μg/ml of streptomycin). HaCaT cells are stimulated with 150 mJ/cm^2 provided by a solar ultraviolet simulator (Oriel, Stratford, CT).

HepG2 are grown in Eagle's minimum essential medium (MEM) supplemented with Earle's salts, 2 mM L-glutamine, 1 mM sodium pyruvate, 1.5 g/liter sodium bicarbonate, 0.1 mM nonessential amino acids, 100 μg/ml steptomycin, 100 U/ml of penicillin, and 10% (v/v) FBS. HepG2 cells are stimulated with various NF-κB inducers, including phosphatase inhibitor okadaic acid (25–100 nM), bacterial lipopolysaccharides (LPS; 100 ng/ml), and proinflammatory cytokine TNF-α-(1–25 ng/ml).

RAW 264.7 cells are cultured in RPMI 1640 containing 10% FBS, 2 mM glutamine, 1 mM sodium pyruvate, and antibiotics. Macrophages are detached by vigorous pipetting and, after centrifugation, are seeded and stimulated with 10 U/ml IFN-γ.

Nuclear Extraction and EMSA

Nuclear extracts are prepared according to Olnes and Kurl[6] with slight modifications. In the case of adherent cells (HaCaT and HepG2), the cell monolayer is washed twice with ice-cold phosphate-buffered saline (PBS). Cells are subsequently scraped with a cell lifter to bring them in suspension. In the case of suspension cells (RAW 264.7), an adequate volume of culture is harvested, spun down, and washed in ice-cold PBS. The cells are pelleted again and finally resuspended in ice-cold PBS and transferred in a microtube if not done previously.

The cell suspension is spun down at 14,000 g, and the cell pellet is resuspended in 400 μl of freshly prepared buffer A [10 mM HEPES (pH 7.9), 10 mM KCl, 1 mM

[6] M. I. Olnes and R. N. Kurl, *Biotechniques* **17,** 828 (1994).

MgCl$_2$, 0.5 mM EDTA (pH 8.0), 0.1 mM EGTA (pH 8.0), 5% (v/v) glycerol, 1 mM dithiothreitol (DTT), 0.5 mM phenylmethylsulfonyl fluoride (PMSF) 5 μg/ml leupeptin, 1 mM benzamidine, and 1% (v/v) aprotinin]. This suspension can be frozen in liquid nitrogen and then transfered at −80° for later processing.

In the case of frozen samples, thaw in a water bath (37°) for 1 min and then continue as follows. The cell suspension is incubated on ice for 15 min before the addition of 25 μl of 10% nonidet P-40 (NP-40) and mixed by vortexing for 5 sec. Incubation is continued on ice for an additional 30 sec to 5 min (according to the cell type: e. g., 30 sec for Jurkat and 5 min for HaCaT), followed by centrifugation at 15,000 g at 4° for 30 sec (longer centrifugation would break the nuclei). The pellet (nuclei) is suspended in 40–150 μl of buffer C [20 mM HEPES (pH 7.9), 400 mM NaCl, 1 mM EDTA (pH 8.0), 1 mM EGTA (pH 8.0), 1 mM DTT, 1 mM PMSF, 10 μg/ml leupeptin, 1% aprotinin, 1 mM benzamidine, and 20% (v/v) glycerol)]. Note that the volume of buffer C should be kept to a minimum to have a high protein concentration, making it easier to handle in subsequent steps and preventing protein degradation as well. Next, the suspension is incubated at 4° for 15 min before mixing on a vortex for 15 min and finally centrifuging for 20 min as indicated earlier. The supernatant (nuclear extract) is collected, split into two aliquots (one of 13 μl for protein determination and the other with the rest), and frozen at −80°. Protein concentration is measured using Bio-Rad protein assay I (Bio-Rad, Richmond, CA).

Electrophoretic mobility shift assays are performed as described previously. For EMSAs, the vertical gel apparatus from GIBCO-BRL is recommended. It is easy to use, fast to mount, and its gasket system prevents leaks. Prepare the polyacrylamide solution on ice [use a 30 or 40% concentrated mix (acrylamide : bisacrylamide, 29 : 1) to which 0.2% (v/v) of TEMED and 0.1% (w/v) of ammonium persulfate are added immediately prior to pouring]. The gel can be poured using a large syringe or a 25-ml pipette. Place the comb when the gel is poured completely, avoiding the formation of bubbles at the base of the teeth. Once polymerized, the gel is submitted to a prerun (200 V for 60 min) during which the samples are prepared.

Equal amounts of the nuclear protein extracts (2.5–15 μg) (caution: the more protein, the more nonspecific bands) are incubated with NF-κB-specific ^{32}P-labeled double-stranded oligonucleotide (Promega, Madison, WI). Binding reactions are typically performed in a 20- to 30-μl volume containing the nuclear extract, 4–6 μl of 5× binding buffer [50 mM Tris–HCl (pH 7.6), 5 mM EDTA (pH 8.0), 250 mM NaCl, 5 mM DTT, and 50% glycerol], 0.1–0.25 μg/μl poly(dI-dC)·poly(dI-dC) (Pharmacia, Piscataway, NJ) as a nonspecific DNA competitor, and 50,000 cpm of the labeled oligonucleotide. The 5× binding buffer can be prepared in bulk and stored at −20°. In some experiments, a cold mutant oligonucleotide (100-fold in excess) is added to the reaction to determine the binding specificity [mutant oligonucleotides are purchased from Santa Cruz

Biotechnology (Santa Cruz, CA)]. Alternatively, polyclonal antibodies against specific NF-κB subunits can be added to the binding reaction mixture. The oligonucleotides is labeled with [γ-^{32}P]ATP using T4 polynucleotide kinase [the authors recommend the enzyme from New England Biolabs (Beverly, MA), at any rate, it has to be used fresh because it loses its activity rapidly with repeated freeze–thaw cycles] and then purified on Chroma-spin-10-TE (Clontech, Palo Alto, CA). The binding reaction is carried out at room temperature for 30 min. Then the samples are loaded onto the 6% nondenaturing polyacrylamide gel and run with a 0.5 × TBE buffer, pH 8.0. After completion of the run, gels are dried on a Hoeffer gel drier and autoradiographed overnight (the length of exposure will depend on the probe activity and the amount of NF-κB in the sample). The autoradiography can be scanned for further densitometry analysis using NIH Image 1.61 (or later version).

Reporter Gene Assays

Using Superfect Reagent. Cells are plated at 30–40,000 cells (HepG2 and HaCaT) per cm^2 in 12-well plates. Twenty four hours later cells are cotransfected transiently with the plasmids pGL3-4κB-Luc and pRL-TK (plasmid reference containing a *Renilla* luciferase gene driven by a thymidine kinase promoter) using the Superfect reagent (Qiagen, Valencia, CA) according to the manufacturer's protocol. The plasmid pGL3-4κB-Luc (kindly provided by Dr. Takashi Okamoto, Nagoya City University, Medical School, Japan) is constructed by inserting four copies of the κB sequences (GGGACTTTCC) from the HIV-1-LTR into the pGL3 basic vector (Promega, Madison, WI) as described previously.[7] A similar plasmid can be obtained from a variety of vendors, including Stratagene (La Jolla, CA) and Clontech.

---

[7] T. Sato, K. Asamitsu, J.-P. Yang, N. Takahashi, T. Tetsuka, A. Yoneyama, A. Kanagawa, and T. Okamoto, *AIDS Res. Hum. Retrovir.* **14**, 293 (1998).
[8] M. E. Gerritsen, W. W. Carley, G. E. Ranges, C. P. Shen, S. A. Phan, G. F. Ligon, and C. A. Perry, *Am. J. Pathol.* **147**, 278 (1995).
[9] K. Natarajan, S. Singh, T. R. Burke, Jr., D. Grunberger, and B. B. Aggarwal, *Proc. Natl. Acad. Sci. U.S.A.* **93**, 9090 (1996).
[10] Y. L. Lin and J. K. Lin, *Mol. Pharmacol.* **52**, 465 (1997).
[11] F. Yang, W. J. de Villiers, C. J. McClain, and G. W. Varilek, *J. Nutr* **128**, 2334 (1998).
[12] K. Natarajan, S. K. Manna, M. M. Chaturvedi, and B. B. Aggarwal, *Arch. Biochem. Biophys.* **352**, 59 (1998).
[13] Z. Wei, Q. Peng, B. H. Lau, and V. Shah, *Gen. Pharmacol.* **33**, 369 (1999).
[14] C. A. Musonda and J. K. Chipman, *Carcinogenesis* **19**, 1583 (1998).
[15] C. Saliou, M. Kitazawa, L. McLaughlin, J.-P. Yang, J. K. Lodge, T. Tetsuka, K. Iwasaki, J. Cillard, T. Okamoto, and L. Packer, *Free Radic. Biol. Med.* **26**, 174 (1999).
[16] C. Saliou, B. Rihn, J. Cillard, T. Okamoto, and L. Packer, *FEBS Lett.* **440**, 8 (1998).
[17] S. K. Manna, A. Mukhopadhyay, N. T. Van, and B. B. Aggarwal, *J. Immunol.* **163**, 6800 (1999).
[18] Y. C. Park, G. Rimbach, C. Saliou, G. Valacchi, and L. Packer, *FEBS Lett.* **465**, 93 (2000).
[19] Y. L. Lin, S. H. Tsai, S. Y. Lin-Shiau, C. T. Ho, and J. K. Lin, *Eur. J. Pharmacol.* **367**, 379 (1999).

TABLE III
TRANSFECTION CONDITIONS

Cell Line	Cells number (cell/cm^2)	Transfection reagent (volume/μg)	PGL3-4kB-Luc (ng)	pRL-TK (ng)
HaCaT	30–40,000	Superfect (15 μl)	125	125
		FuGENE6	440	190
HepG2	30–40,000	Superfect (10 μl)	300	300
RAW 264.7	200,000	FuGENE6	350	150

Briefly, the transfection mixture (100 μl), containing the amounts of plasmids specified in Table III, is mixed with either Superfect or FuGENE 6 and subsequently added to the cell monolayer. In the case of Superfect, the medium has to be changed 2 hr after transfection in order to avoid any cytotoxic effects.

Using Fugene 6 Reagent. RAW 264.7 are plated at 200,000 cells per cm^2 in 12-well plates and 24 hr later are cotransfected transiently with the plasmids pGL3-4κB-Luc and pRL-TK using the FuGENE 6 reagent (Roche Molecular Biochemicals, Indianapolis, IN) (see Table III). Briefly, the transfection mixture containing 350 ng of pGL3-4κB-Luc and 150 ng of pRL-TK (see Table III) is mixed with the FuGENE 6 reagent and added to the cells. There is no need to remove the reagent/DNA complex from the cells prior to assay, particularly when testing the gene expression 24–48 hr after transfection.

Cells are treated with various flavonoids 18–24 hr after transfection with the Superfect and FuGENE 6 reagent, respectively.

Cell Lysis. For cell lysis, discard the culture medium. Wash cells twice with ice-cold PBS. Dilute passive lysis buffer (PLB) 5× to 1× with distilled water. Add 200 μl of 1× PLB to each well. Transfer the plate on a tube shaker for 15 min at 4°. If the dual luciferase assay is performed immediately after, spin samples down at 14,000 g at 4° for 5 min using a cup centrifuge. Transfer the supernatant to a new tube and discard the tube with the pellet.

If the dual luciferase assay has to be done on another day, freeze the cell lysate in liquid nitrogen and transfer samples to a −80° freezer until further analysis.

Measuring Lucifease Activities. After cell lysis, luciferase activities are measured using the dual luciferase reporter assay system (Promega, Madison, WI) with a LKB/Wallac luminometer 1250 (Wallac Inc., Gaithersburg, MD). The dual luciferase system is based on the subsequent measurement of firefly (from pGL3-4κB-Luc) and *Renilla* (from pRL-TK) luciferase activities in the same tube with the same extract. The firefly luciferase activity is normalized in that system with the *Renilla* luciferase activity evaluated simultaneously to correct for differences in transfection efficiency.

Synopsis

This article focused on two methods to measure the activity of NF-κB. Both methods evaluate "post-IκB phosphorylation" stages in the NF-κB activation cascade. In fact, EMSA performed with nuclear extracts provides an information only on NF-κB nuclear translocation and its ability to bind κB-DNA sequences. Likewise, the reporter gene assay is limited to assessing NF-κB-dependent gene expression no matter the mechanism that originally activated NF-κB. Nevertheless, the latter assay represents a more physiological and more reproducible way of measuring NF-κB activity in mammalian cells than the EMSA does.

In order to obtain further insights into NF-κB signal transduction pathways, investigating IκB degradation and phosphorylation are recommended. The cloning and characterization of IκB kinases provided new testing possibilities based on the measure of their activity.

[35] Interaction between Cultured Endothelial Cells and Macrophages: *In Vitro* Model for Studying Flavonoids in Redox-Dependent Gene Expression

By GERALD RIMBACH, CLAUDE SALIOU, RAFFAELLA CANALI, and FABIO VIRGILI

Introduction

A coculture model composed of human endothelial cells and murine macrophages mimicking the proinflammatory environment characterizing atherosclerotic plaque formation is described. Interferon-γ (IFN-γ)-activated macrophages generate reactive oxygen and nitrogen species (ROS and RNS). In turn, ROS and RNS induce the activation of the redox-sensitive transcription factor NF-κB, its transactivation activity, and the expression of monocyte chemotactic protein-1 (MCP-1), a gene containing NF-κB in its promoter side.

We utilized this coculture model to study the role of nutritional antioxidants, such as polyphenols, in molecular events associated with the early stages of atheroma development. This article describes the effect of the flavonoid-rich extract from *Ginkgo biloba* leaves, EGb761, on redox-sensitive gene expression. Results described herein suggest that the model is suitable to assess: (i) the protective effect of flavonoids on oxidative stress induced by reactive oxygen and nitrogen species and (ii) the effect of flavonoids on the activation of signaling pathways leading to the expression of specific response genes, such as MCP-1.

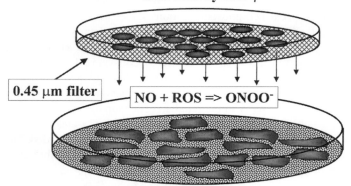

FIG. 1. Scheme of the coculture model: RAW 264.7 cells are grown on a 0.45-μm filter. Following activation with IFN-γ, RAW 264.7 cells are placed on top of a ECV/HUVEC dish in direct contact with the culture medium (1.5 ml). After an appropriate coincubation time, during which underlying cells are challenged by the reactive oxygen and nitrogen species generated by activated macrophages, the filter is removed and specific parameters (transcription factors activation, mRNA expression, cellular antioxidant levels) are assessed.

A Coculture System to Study Role of ROS and RNS in Macrophage–Endothelial Cell Interaction

The coculture model, illustrated in Fig. 1, is composed of primary human endothelial cells (HUVEC) and the monocyte–macrophage murine cell line RAW 264.7. It is characterized by a long-term, sustained generation of high levels of ROS and RNS, thus mimicking the occurrence of a complex though mild "physiological" oxidative stress. This is confirmed by the significant accumulation of nitrite and nitrate (NO_2^- and NO_3^-) in the medium after cell activation, which is an index of both nitric oxide (NO·) generation and NO· reaction with superoxide (see Fig. 2). This mode of generation of RNS appears somehow different from that induced by the addition of NO· donors such as 3-morpholinosydnonimine-N-ethylcarbamide (SIN-1), sodium nitroprusside (SNP), and compounds either containing or generating reactive oxygen (H_2O_2, organic peroxides, transition metals).[1] In fact, macrophages activated by the bacterial wall component lipopolysaccharide (LPS) or by cytokines such as IFN-γ express the enzymatic machinery that generates both ROS and RNS. In such a condition, a direct interaction of NO·

[1] G. Rimbach, F. Virgili, Y. C. Park, and L. Packer, *Redox Rep.* **4**, 171 (1999).

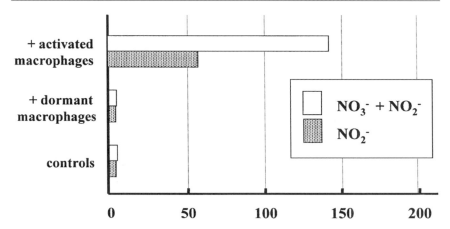

FIG. 2. NO_2^- and NO_3^- accumulation in ECV/HUVEC medium (μM) after a 24-hr exposure to a filter containing RAW 267.4 monocyte–macrophages activated by IFN-γ. NO_2^- and NO_3^- levels in the medium are quantified using the automated NOx analyzer based on Griess reagent. Nitrate is determined after reduction to nitrite using an A7200 copperized cadmium reduction column and is then quantified by Griess reagent.

with $O_2^{\bullet-}$ is likely to occur, leading to $ONOO^-$ formation.[2–6] This reaction can be observed in different cell lines and tissues such as macrophages, neutrophils, and cultured human endothelial cells.[3,7] $ONOO^-$ is a powerful oxidant molecule, which, due to its weak O–O bond strength, decomposes spontaneously to form the hydroxyl radical and nitrogen dioxide.[8] $ONOO^-$ may react directly with many biological targets, such as lipids and sulfhydryl groups. It may also nitrate tyrosine in proteins, thereby forming 3-nitrotyrosine residues. Nitration of tyrosine can result in an inhibition of protein tyrosine phosphorylation and thus inhibit signal transduction.[2,9]

We have already utilized this coculture model to demonstrate that the exposure to activated macrophages is associated with a significant decrease of α-tocopherol and glutathione levels in endothelial cells. In the same studies, we demonstrated

---

[2] G. E. Arteel, K. Briviba, and H. Sies, *FEBS Lett.* **445**, 226 (1999).
[3] M. C. Carreras, G. A. Pargament, S. D. Catz, J. J. Poderoso, and A. Boveris, *FEBS Lett.* **341**, 65 (1994).
[4] G. L. Squadrito and W. A. Pryor, *Free Radic. Biol. Med.* **25**, 392 (1998).
[5] S. Moncada, R. M. Palmer, and E. A. Higgs, *Pharmacol. Rev.* **43**, 109 (1991).
[6] J. S. Beckman and W. H. Koppenol, *Am. J. Physiol.* **271**, C1424 (1996).
[7] N. W. Kooy and J. A. Royall, *Arch. Bioche. Biophys.* **310**, 352 (1994).
[8] W. H. Koppenol, J. J. Moreno, W. A. Pryor, H. Ischiropoulos, and J. S. Beckman, *Chem. Res. Toxicol.* **5**, 834 (1992).
[9] H. Ischiropoulos, *Arch. Biochem. Biophys.* **356**, 1 (1998).

that procyanidins extracted from the bark of *Pinus maritima* protect the cellular antioxidant network from macrophage-induced oxidative challenge.[1,10]

Some Methodological Remarks Regarding Cell Lines

In the coculture model reported herein, a combination of two nonhomo-specific cell lines in the experimental design (RAW 264.7, a murine-derived cell line, with either ECV304 or primary endothelial cells of human origin) was chosen in order to avoid the activation of endothelial cells by the cytokine utilized for macrophage activation. In fact, no changes were observed either on antioxidant levels or other functional cell parameters when ECV304 was exposed to the medium utilized for RAW 264.7 culture after activation with IFN-γ,[1,10] confirming that recombinant murine IFN-γ utilized in our studies displays a species-specific activity.

However, reports alerting the utilization of ECV304 as an appropriate model for the endothelium have appeared. In fact, several distinct differences between ECV304 and human umbilical vein-derived endothelial cells (HUVEC), as well as a strong genetic similarity between ECV304 and T24/83, a human bladder cancer cell line, have been reported.[11] The authors suggested that even though ECV304 displays some endothelial characteristics and may be a useful tool for the study of receptor pharmacology, it should not be considered equivalent to HUVEC and is therefore an inappropriate cell line to study endothelial cell biology. These observations led us to reconsider our model; we now use primary endothelial cells obtained from collagenase treatment of the human umbilical vein known as HUVEC instead. Therefore, data presented here only refer to the coculture of RAW 264.7 cells together with HUVEC. However it is worth mentioning that some of the features and responses displayed by "authentic" HUVEC are quite similar to those we have already reported for ECV304. This observation allows us to suggest that even taking into account the limitations mentioned earlier, some of the studies performed utilizing ECV304 as a model for endothelial cells should still be considered valid.

Cell Culture

Human Umbilical Vein Endothelial Cells

Primary human endothelial cells are obtained from the umbilical cord vein. According to the method of Jaffe and co-workers,[12] the main vein of umbilical cords (kindly provided by the nursery of the University of Roma, "La Sapienza") is

---

[10] F. Virgili, D. Kim, and L. Packer, *FEBS Lett.* **431,** 315 (1998).
[11] J. Brown, S. J. Reading, S. Jones, C. J. Fitchett, J. Howl, A. Martin, C. L. Longland, F. Michelangeli, Y. E. Dubrova, and C. A. Brown, *Lab. Invest.* **80,** 37 (2000).
[12] E. Jaffe, R. L. Nachmann, C. G. Becker, and C. R. Minick, *J. Clin. Invest.* **153,** 5008 (1973).

washed thoroughly with phosphate-buffered saline (PBS) and then treated with 10 ml collagenase (0.2%, dissolved in Medium 199) from *Clostridium histiolyticum* (type XI, Sigma St. Louis, MO) for 20 min at 37°. The endothelial cell layer is removed from the vein using 50 ml of Hanks' buffer (Sigma). Cells are then pelleted by centrifugation and cultured in six-well culture dishes (3.5 cm diameter, Falcon, Franklin Lakes, NJ) pretreated with 1.5% gelatine (type B from bovine skin, Sigma). Cells are grown utilizing 199 Medium (GIBCO-BRL, Rockville, MD) supplemented with 20% fetal calf serum (FCS, University of California cell culture facilities, San Francisco, CA), 2 mM glutamine, 1 mM sodium pyruvate, 10 mM HEPES, 100 μg/ml heparin, streptomycin–penicillin, and 50 μg/ml endothelial cell growth supplement (ECGS, Sigma). HUVEC cells are utilized for experiments at 90–100% apparent confluence and within passage 3 through 6. Passages are performed according to standardized protocols and by diluting the cell population 1 to 3.

RAW 264.7 Murine Monocyte–Macrophage

RAW 264.7, a murine cell line of monocyte–macrophages, are obtained from the American Type Culture Collection (ATCC, Rockville, MD). Cells are seeded on 0.45-μm pore-size, polyethylene terephthalate cell culture inserts (Falcon) and cultured in Dulbecco's modified Eagle's medium (DMEM, GIBCO-BRL) containing 10% FCS, 2 mM glutamine, 1 mM sodium pyruvate, and streptomycin–penicillin. RAW 264.7 are stimulated with 10 U/ml IFN-γ (Genzyme, Cambridge, MA). Three hours after stimulation, the filters containing macrophages are placed on top of HUVEC cells in direct contact with the culture medium. Cells are then cocultured at 37° and, after an appropriate coincubation time, the filter is removed and specific parameters (transcription factors activation, mRNA expression) are assessed on HUVEC cells.

Ginkgo biloba Extract EGb 761

EGb 761 is a standardized extract of *G. biloba* leaves. It is used commonly in Asia, Europe, and the United States to treat a variety of pathological conditions, such as peripheral arterial diseases and organic brain syndromes. EGb 761 contains 24% flavonoids (ginkgo-flavon glycosides) and 6% terpenois (ginkgolides, bilobalides) as active components. The flavonoid fraction is composed mainly of the flavonols quercetin, kaempherol, isorhamnetpferol, and isorhamnetin, which are linked to a sugar moiety.[13] This plant extract has been shown to have strong scavenging activity toward hydroxyl radicals, superoxide anion radicals, and NO˙. EGb 761 may also contribute significantly to LDL protection from oxidation *in*

---

[13] O. Sticher, *Planta Med.* **59**, 2 (1993).

vitro and to a lesser extent *in vivo*.[14] Furthermore, it has been demonstrated in our laboratory and by others that E*Gb* 761 inhibits the activation of NF-κB in endothelial cells and macrophages challenged by different stimuli.[15] All these features render E*Gb* 761 potentially able to contribute to the protection from the development of artheriosclerotic lesion both by acting as a bona fide antioxidant and by modulating the signaling pathway that leads to an endothelial cell response to prooxidation and dysfunction.

This article reports on some observation obtained utilizing the macrophage–endothelial cell coculture model to study the effect of E*Gb* 761 in the modulation of the endothelial cell response to the exposure to activated macrophages. In the experiments described herein, E*Gb* 761 is dissolved in dimethyl sulfoxide (DMSO) and added to HUVEC culture medium at the indicated concentrations. Cells are supplemented with the extract for 16 hr prior to any further treatment. The final DMSO concentration in the medium is less than 0.1%. Control cells, i.e., those not supplemented with E*Gb* 761, are treated with identical amounts of DMSO.

Coculturing Macrophages and Endothelial Cells as Suitable Model for Study of Flavonoids in Activation of NF-κB Pathway Induced by Oxidative–Nitrosative Stress

Coincubation with Activated RAW 264.7 Monocytes/Macrophages to Induce Transfer of NF-κB into Nucleus and DNA Binding in HUVEC

Figure 3 shows the electromobility shift assay (EMSA) performed according to Suzuki and co-workers [16] on the nuclear extract of HUVEC cells after exposure to RAW 264.7 monocytes/macrophages either quiescent or activated by treatment with 10 U/ml murine IFN-γ. In the basal condition (lane 1, Fig. 3), HUVEC cells do not show significant DNA-binding activity, which is induced strongly by TNF-α (lane 2, Fig. 3). Similarly, following the exposure to the medium utilized to culture RAW 264.7 macrophages, transferred to the top of HUVEC cells 3 hr after the addition of murine IFN-γ, no significant changes are observed in the level of NF-κB binding to the specific DNA consensus sequence (lane 3, Fig. 3). After the exposure to nontreated RAW 264.7 cells, the level of NF-κB-binding activity to DNA increases slightly (lane 5, Fig. 3). The increase is more dramatic after the coincubation with macrophages activated by IFN-γ (lane 4, Fig. 3). This result indicates that activated macrophages significantly induce the activation of the signaling pathway leading to NF-κB transfer to the nucleus and possibly the expression of downstream genes in HUVEC cells. It is noteworthy that this observation

---

[14] L. J. Yan, M. T. Droy-Lefaix, and L. Packer, *Biochem. Biophys. Res. Commun.* **212**, 360 (1995).
[15] Z. Wei, Q. Peng, B. H. Lau, and V. Shah, *Gen. Pharmacol.* **33**, 369 (1999).
[16] Y. J. Suzuki, M. Mizuno, and L. Packer, *J. Immunol.* **153**, 5008 (1994).

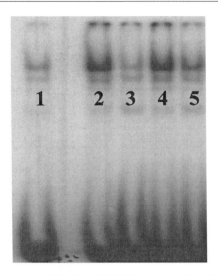

FIG. 3. DNA-binding activity of NF-κB in HUVEC cocultured for 24 hr with nonstimulated or IFN-γ-stimulated RAW 264.7 macrophages. One representative experiment, out of three, is presented. Lane 1, nonstimulated HUVEC cells; lane 2, HUVEC cells stimulated by 10 ng/ml TNF-α; lane 3, HUVEC exposed to the medium used to culture RAW 264.7 cells and containing 10 units/ml murine IFN-γ; lane 4, HUVEC cells after exposure to RAW 264.7 macrophages stimulated by 10 units/ml IFN-γ; and lane 5, HUVEC cells after exposure to nonactivated RAW 264.7 macrophages.

overlaps with unpublished experiments conducted on the endothelial-like cell line ECV 304, suggesting that the immortalized cell line ECV 304 displays a similar pattern of activation of NF-κB than authentic primary endothelial cells, HUVEC.

Coincubation with Macrophages to Induce NF-κB Transactivation Activity in HUVEC: Effect of Preincubation with Flavonoid-Rich Plant Extract (EGb 761)

EMSA provides information about activation of the NF-κB pathway. However, this assay is not suitable to understand whether transactivation activity occurs subsequently to the binding of the transcription factor to the consensus sequence of target genes. An expedient tool to study NF-κB transactivation is a reporter gene assay based on the simultaneous transfection of HUVEC with two plasmids containing either firely or *Renilla* luciferase hooked up to a multiple NF-κB consensus sequence and to a simple thymidine kinase promoter, respectively. The details of the method are given elsewhere.[17]

[17] C. Saliou, M. Kitazawa, L. McLaughin, J. P. Yang, J. K. Lodge, T. Tetsuka, K. Iwasaki, J. Cillard, T. Okamoto, and L. Packer, *Free Radic. Biol. Med.* **26**, 174 (1999).

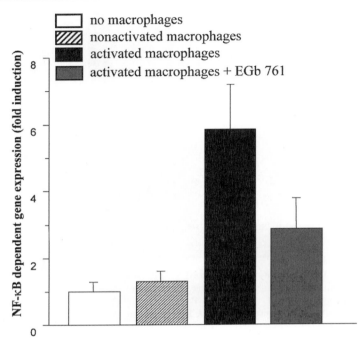

FIG. 4. NF-κB transactivation activity in HUVEC coincubated with RAW 264.7 macrophages treated with 10 U/ml IFN-γ. Twelve hours before the exposure to macrophages, HUVEC cells are transfected transiently with two plasmids containing either firefly or *Renilla* luciferases hooked up to a multiple NF-κB consensus sequence and to a simple thymidine kinase promoter, respectively. Twelve hours after coincubation, HUVEC cells are collected and both *Renilla* and firefly luciferase activities are measured in cell lysate. The effect of the preincubation with *Ginkgo biloba* extract EGb 761 is also shown. Results are expressed as the ratio of *Renilla* to firefly luciferase activity in the cell lysate.

The coincubation of HUVEC with nonstimulated macrophages is associated with a slight but significant increase in the ratio between firefly to *Renilla* luciferase activity in comparison with control endothelial cells (Fig. 4). In agreement with results obtained by EMSA, the observed increase suggests that some transactivation and NF-κB-driven gene expression occur in the presence of nonstimulated macrophages. As suggested earlier, this effect can be due to the presence of a small but significant amount of active macrophages in the population of RAW 264.7, which generate both ROS and RNS and possibly other molecules triggering the cellular response mediated by NF-κB signaling.

After stimulation with IFN-γ, RAW 264.7 are able to induce a much stronger expression of NF-κB transactivation, as measured by luciferase activity, with respect to the basal value. Thus, the exposure of HUVEC to stimulated macrophages is associated with an increase of NF-κB transactivating activity, which eventually leads to the expression of specific response genes. The preincubation with E*Gb* 761 significantly affects the expression of NF-κB-dependent luciferase gene in HUVEC exposed to activated macrophages, whereas no effect is observed both in control cells and in cells exposed to quiescent macrophages. Thus, our observations indicate that the interaction of activated macrophages with endothelial cells is associated with both the transfer of NF-κB to the nucleus and to the activation of its transactivating activity.

Coincubation with Macrophages to Induce Expression of MCP-1 in HUVEC Cells: Effect of Preincubation with Flavonoid-Rich Plant Extract (EGb 761)

Different genes have been reported to be expressed in response to proinflammatory stimuli partially depending on the activation of the NF-κB transcription factor. Among these genes, MCP-1 plays a central role in encoding for a secretory protein, which further recruits monocytes in the inflammatory environment.[18] This recruitment is followed by a complex process of recognition, adhesion, diapedesis, and activation, which eventually leads to the infiltration of macrophages in the subluminal area.[19] If this process is not perfectly controlled and regulated, a vicious circle may be triggered and a chronic inflammation may be established. MCP-1 has different NF-κB sites in the promoter, along with other transcription factors; NF-κB is one of the major regulatory units for the expression of MCP-1.[20] We have therefore assayed the coincubation of macrophages with endothelial cells to study its impact on the expression of the mRNA encoding for the MCP-1 protein. In order to investigate this issue, HUVEC cells are coincubated with either quiescent macrophages or macrophages activated by IFN-γ. After 6 hr of coincubation, total RNA is isolated to study the expression of MCP-1-mRNA by a reverse transcription polymerase chain reaction (RT-PCR). Some of the methodological details of this approach are described elsewhere.[21]

Figure 5 shows that HUVEC express a moderate although detectable level of the MCP-1 gene under basal conditions (lane 1). Preincubation with 100 and 200 μg E*Gb* 761 (lanes 2 and 3, Fig. 5) slightly affected the expression of the MCP-1 gene. Following 6 hr of exposure to RAW 264.7 macrophages activated by IFN-γ, MCP-1 expression increase significantly (lane 4, Fig. 5). Interestingly, the preincubation with E*Gb* 761 reduces the expression of the MCP-1 gene at both concentrations tested, but not in a dose-dependent fashion (lanes 5 and 6, Fig. 5).

[18] J. M. Wang, W. P. Shen, and S. B. Su, *Trends Cardiovas. Med.* **8**, 169 (1998).
[19] J. P. Cai, S. Hudson, M. W. Ye, and Y. H. Chin, *Cell Immunol.* **167**, 269 (1996).
[20] U. Widmer, K. R. Manogue, A. Cerami, and B. Sherry, *J. Immunol.* **150**, 4996 (1993).
[21] F. Virgili, H. Kobuchi, and L. Packer, *Free Radic. Biol. Med.* **24**, 1120 (1998).

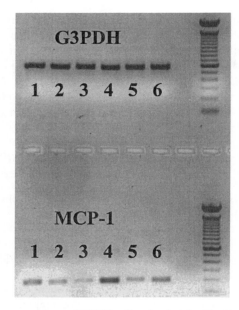

FIG. 5. MCP-1 gene expression in HUVEC cells coincubated with either dormant RAW 264.7 macrophages or those treated with 10 U/ml IFN-γ. After 6 hr of coincubation, HUVEC are collected and total RNA is isolated to study the expression of MCP-1-mRNA using a RT-PCR approach. The effect of preincubation with EGb 761 is also shown. Lane 1, control HUVEC; lanes 2 and 3, preincubation with 100 and 200 μg EGb 761, respectively; lane 4, following 6 hr of exposure to RAW 264.7 macrophages activated by IFN-γ; and lanes 5 and 6, same as lane 4 but in the presence of 100 and 200 μg/ml EGb 761, respectively. Results are shown as the gel photograph of PCR-amplified cDNAs from MCP-1 and G3PDH mRNAs.

Conclusions

Artheriosclerosis is a complex multifactorial disease in the course of which a series of critical events occurs. This includes endothelial dysfunction, infiltration of inflammatory cells into the vessel wall, alteration of the vascular cell phenotype, and vascular remodeling. Before the manifestation of typical dramatic outcomes, such as stenosis or occlusive thrombosis of the arterial lumen, a number of subtle dysfunctions are displayed at cellular and molecular levels at the beginning of disease progression. These events lead to the loss of homeostatic functions of the endothelium and include the modification of the pattern of gene expression, cell proliferation, and apoptosis.[22]

Macrophages are the principal inflammatory cell type in the artherioscle-rotic microenvironment. In the early stages of atheroma formation, macrophages

---

[22] G. H. Gibbons and V. J. Dzau, *Science* **272,** 689 (1996).

interplay with endothelial cells. A shift from normal homeostasis possibly triggers a vicious circle, which finally leads to endothelial dysfunction.[23,24] Macrophage activity displays the features of a proinflammatory condition, including the generation of reactive oxygen and nitrogen species. These events are likely to induce a condition of oxidative–nitrosative stress. Indeed, oxidative stress induced injury has been observed in the environment of artheriosclerotic plaque, suggesting that free radicals may be important in the etiology of disease.[22] Also, the oxidation of low-density lipoprotein has been described to play a central role in the development of the atheriosclerotic lesion.[25] Epidemiological surveys strongly suggest a negative association between the dietary consumption of antioxidants such as polyphenols and the incidence of cardiovascular disease, including atherosclerosis.[26] Experimental studies have given ground to this evidence.[27,28] In fact, polyphenols are attracting growing interest due to their strong antioxident capacity. Phenolic acids and flavonoids have been demonstrated to participate in the cellular antioxidant network and to protect cells from injury induced by both oxygen and nitrogen reactive species.[29–31]

The coculture system described herein provides an expedient experimental tool to study the complex interaction between macrophages and endothelial cells and to better understand the molecular basis of the beneficial role of antioxidant flavonoids in the protection from the risk of artheriosclerotic disease.

---

[23] J. L. Witztum and D. Steinberg, *J. Clin. Invest.* **88,** 1785 (1991).
[24] J. A. Berliner, M. Navab, A. M. Fogelman, J. S. Frank, L. L. Demer, P. A. Edwards, A. D. Watson, and A. J. Lusis, *Circulation* **91,** 2488 (1995).
[25] I. Jialal and S. Devaraj, *J. Nutr.* **126,** S1053 (1996).
[26] A. S. St Leger, A. L. Cochrane, and F. Moore, *Lancet* **8124,** 1017 (1979).
[27] C. H. Hennekens, *Circulation* **97,** 1095 (1998).
[28] S. Hercberg, P. Preziosi, P. Galan, H. Faure, J. Arnaud, N. Duport, A. Malvy, A. M. Roussel, S. Briancon, and A. Favier, *Food Chem. Toxicol.* **37,** 925 (1999).
[29] C. A. Rice-Evans, N. J. Miller, P. G. Bolwell, P. M. Bramley, and J. B. Pridham, *Free Radic. Res.* **22,** 375 (1995).
[30] J. A. Vinson and Y. A. Dabbagh, *FEBS Lett.* **433,** 44 (1998).
[31] W. Bors and C. Michel, *Free Radic. Biol. Med.* **27,** 1413 (1999).

[36] Determination of Cholesterol-Lowering Potential of Minor Dietary Components by Measuring Apolipoprotein B Responses in HepG2 Cells

By ELZBIETA M. KUROWSKA

Introduction

High blood concentrations of low density lipoprotein (LDL) cholesterol have been implicated as one of the major risk factors associated with coronary heart disease (CHD). The association is largely due to the importance of LDL cholesterol in the formation and development of atherosclerotic plaque, the underlying pathological condition of CHD. Current dietary strategies aimed to lower LDL cholesterol include reduced intake of saturated fat and cholesterol, and increased intake of fiber and soy protein.[1] During recent years, a number of studies have suggested that the improvement in the blood lipid profile could also be achieved by increasing the dietary intake of certain minor food components or their mixtures. Natural compounds reported as cholesterol-lowering include tocotrienols (a form of vitamin E found in palm oil and cereal grains)[2] and citrus flavonoids, hesperetin from oranges and naringenin from grapefruit.[3]

The ability of food products and their minor constituents to modulate total and LDL cholesterol is usually assessed in small animals prior to testing in clinical trials. The commonly used animal models include the hamster and rabbit. Both develop hypercholesterolemia associated with elevation of LDL cholesterol, similar to that found in humans after feeding semipurified, casein-based diets.[4,5] Apolipoprotein E-knockout mice with genetically induced high VLDL and LDL cholesterol are also useful, but generally more often in studies focused on prevention and treatment of atherosclerosis.[6] In addition, some research is still being done in cholesterol-fed rats, although hypercholesterolemia in this model is associated with relatively high blood levels of HDL cholesterol.[7] The *in vivo* assessment of cholesterol-lowering potential becomes very expensive when a large number of

---

[1] S. L. Connor and W. E. Connor, *in* "Current Perspectives on Nutrition and Health" (K. K. Carroll, ed.), p. 59. McGill-Queen's University Press, Montreal, PQ, 1998.
[2] A. Theriault, J.-T. Chao, Q. Wang, A. Gapor, and K. Adeli, *Clin. Biochem.* **32,** 309 (1999).
[3] S.-H. Bok, S.-H. Lee, Y.-B. Park, K.-H. Bae, K.-H. Son, T.-S. Jeong, and M.-S. Choi, *J. Nutr.* **129,** 1182 (1999).
[4] S. M. Potter, J. Pertile, and M. D. Berber-Jimenez, *J. Nutr.* **126,** 2007 (1996).
[5] E. M. Kurowska, J. M. Hrabek-Smith, and K. K. Carroll, *Atherosclerosis* **78,** 159 (1989).
[6] J. Reckless, J. C. Metcalfe, and D. J. Grainger, *Circulation* **95,** 1542 (1997).
[7] S. Rajendran, P. D. Deepalakshmi, K. Parasakthy, H. Devaray, and S. N. Devaray, *Atherosclerosis* **123,** 235 (1996).

minor components or their combinations need to be screened. Because at least some minor components are likely to affect LDL metabolism directly in the liver, their mechanisms of action have been studied *in vitro,* in primary hepatocytes or more often in human hepatoma HepG2 cells.[8] Changes in cholesterol metabolism induced by compounds in cultured hepatic cells appear to be qualitatively consistent with cholesterolemic responses produced by the same compounds *in vivo.* This article describes a method of screening compounds for cholesterol-lowering potential by measuring their ability to reduce the medium content of apolipoprotein B (apoB), the structural protein of LDL, in human hepatoma HepG2 cells.

Model

HepG2 cells (American Type Culture Collection, Rockville, MD) resemble normal human liver cells with regard to their ability to secrete and catabolize lipoproteins. The apoB-containing lipoproteins produced by HepG2 cells are smaller than LDL and triglyceride poor. This is due to a defect in lipoprotein assembly, which prevents the addition of substantial amounts of triglycerides to the primordial lipoprotein.[9] Despite this defect, HepG2 cells are generally considered an acceptable model for measuring qualitative changes in the secretion of apoB-containing lipoproteins in response to a variety of factors.[10] The rates of production and catabolism of apoB-containing lipoproteins are regulated at many levels. The rate of production can be altered by the availability of cholesterol, cholesteryl esters, triglycerides, and phospholipids required for lipoprotein assembly and secretion, by activity of the microsomal triglyceride transfer protein, and by co- and posttranslational proteolytic degradation of apoB.[8] The rate of catabolism is dependent on the LDL receptor-specific and nonspecific uptake and degradation of lipoprotein particles.[11] Candidate compounds capable of reducing medium concentrations of apoB-containing lipoproteins may therefore exert their effects via different mechanisms.

Cell Culture

All tissue culture reagents and plasticware are purchased from Life Technologies (Burlington, Canada). Other reagents are from Sigma (St. Louis, MO) unless otherwise stated. For the experiment, HepG2 cells are seeded in 24-well plates (6×10^5 cells/well) and grown in minimum essential medium (MEM) containing 10% fetal bovine serum (FBS). After reaching confluence (usually 7 days after

---

[8] Z. Yao, K. Tran, and R. S. McLeod, *J. Lipid Res.* **38,** 1937 (1997).
[9] X. Wu, A. Shang, H. Jiang, and H. N. Ginsberg, *J. Lipid Res.* **37,** 1198 (1996).
[10] A. Graham, J. L. Wood, and L. J. Russell, *Biochim. Biophys. Acta* **1302,** 46 (1996).
[11] L. Izem, E. Rassart, L. Kamate, L. Falstrault, D. Rhainds, and L. Brissette, *Biochem. J.* **329,** 81 (1998).

plating), cells are preincubated for 24 hr with MEM containing 1% bovine serum albumin (BSA) instead of FBS. This lipoprotein-free medium inhibits cell proliferation and stimulates the biosynthesis of apoB-containing lipoproteins. Cells are subsequently exposed to a range of nontoxic concentrations of tested compounds (as determined by the viability assay described later) for another 24 hr (250 μl medium/well, 4 wells per concentration). Controls (4 wells on each side of the plate) contain 250 μl/well of MEM with BSA only. At the end of the incubation, media are collected and aliquots are frozen at $-70°$ for the determination of apoB by an enzyme-linked immunosorbent assay (ELISA) (described later). For soluble cellular protein determination, cells are washed three times with ice-cold Ca^{2+}- and Mg^{2+}-free phosphate-buffered saline (PBS), and protein is extracted with 0.1 N NaOH and measured using the Coomassie plus protein assay (Pierce, Rockford, IL), according to the manufacturer's directions. ApoB concentrations in the media are calculated per milligram of total cellular protein and are converted to percentage of control.

Evaluation of ApoB-Lowering Potential by Enzyme-Linked Immunosorbent Assay

The competitive ELISA assay kit for the determination of human apoB (Ortho Diagnostics, LaJolla, CA) measures medium apoB concentrations within the range of 0.4–0.031 μg/ml. In this assay, microtiter 96-well plates are coated with monoclonal antibody against human apoB, which binds the apoB epitope from human LDL (standard curve, control sera with high and low content of human apoB) or from LDL-like lipoproteins present in the medium of HepG2 cells (tested samples, usually diluted twice with PBS). After the exposure, a fraction of the antibody is blocked, and the remaining amount is captured by a subsequent incubation with conjugate containing a secondary antibody (against antihuman apoB) coupled with horseradish peroxidase. The following incubation with horseradish peroxidase substrate, o-phenylenediamine, produces a yellow color change proportional to the amount of apoB present, and the optical density of samples is measured at 490–492 nm against the blank PBS/conjugate wells. To determine whether evaluated compounds cross-react with the ELISA assay, in a separate study, various doses of the compounds should be tested in the presence of cell-free culture media.

Preparation of Stock Solutions

Flavonoids and tocotrienols are added to HepG2 cell culture media after solubilization in dimethyl sulfoxide (DMSO). Some compounds may require several hours of sonication and heating to solubilize. Stock solutions in DMSO should be stored frozen at $-20°$, usually up to 3 months. The final concentration of DMSO in culture media should not exceed 0.3% by volume to prevent any effects of DMSO on apoB metabolism. For each experiment, equal concentrations of DMSO should

TABLE I
HIGHEST 100% VIABILITY-SUSTAINING CONCENTRATIONS FOR SELECTED COMPOUNDS[a]

Compound	Highest 100% viability-sustaining concentration
α-Tocotrienol	50 μg/ml (120 μM)
γ-Tocotrienol	20 μg/ml (50 μM)
δ-Tocotrienol	20 μg/ml (50 μM)
Hesperetin	60 μg/ml (200 μM)
Naringenin	60 μg/ml (200 μM)

[a] Incubated for 24 hr with HepG2 cells.

be added to all treatment media. For compounds soluble in ethanol, stock solutions could also be prepared by conjugation to BSA before addition to media to enhance uptake by the cells.[12] This is done by dissolving a compound in a small volume of ethanol in a screw-top glass vial. Subsequently, sterile 10% BSA is added and samples are incubated at 37° until they reach a translucent appearance.

Viability Assay

For each compound, the highest concentration sustaining $100 \pm 10\%$ cell viability is predetermined by the 3-(4,5-dimethylthiazol-2-yl)-2,5-diphenyltetrazolium bromide (MTT) viability assay.[13] In this assay, MTT is converted into a blue formazan product by the mitochondrial succinate dehydrogenase of viable cells. Cells are seeded in 96-well plates (8×10^4 cells/well) and grown in MEM containing 10% FBS until optimal density is reached (usually 90–95% confluence). They are subsequently incubated in the same medium, with or without test compound (in quadruplicate), at wide range of concentrations for 24 hr (100 μl medium/well). Twenty-five microliters of sterile MTT reagent (5 mg MTT/ml PBS) is added to each well. The plate is wrapped with aluminum foil and returned to the incubator for 3 hr. After this time, 100 μl of extraction buffer is added (20% sodium dodecyl sulfate in 50% dimethyl formamide–50% double deionized water at pH 4.7) and the incubation continues overnight. Absorbance at 570 nm is measured using the microplate reader. For $100 \pm 10\%$ viable cells at optimum density, the absorbance should be in the range of 0.7–1.0. Cells treated with the highest nontoxic concentrations of test compound should have average absorbance not lower than controls. To determine whether test compounds cross-react with the MTT reagent, various doses of the compounds should be tested in the presence of cell-free media. Examples of the highest nontoxic concentrations of tocotrienol isomers and citrus flavonoids are presented in Table I.

[12] P. W. Sylvester, H. P. Brikenfeld, H. L. Hosick, and K. P. Briski, *Exp. Cell Res.* **214**, 145 (1994).
[13] M. B. Hansen, S. E. Nielsen, and K. Berg, *J. Immunol. Method.* **119**, 203 (1989).

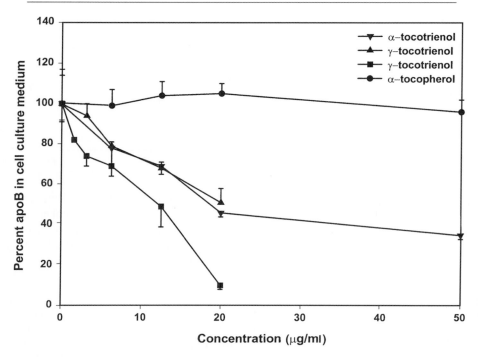

FIG. 1. Effect of increasing doses of α-tocopherol and α-, γ-, and δ-tocotrienols on apoB accumulation in the media of HepG2 cell cultures.

Determination of ApoB IC_{50} Concentrations

For each compound, a dose–response curve is constructed by plotting the percentage change of medium apoB against concentrations used in the assay. Figure 1 shows examples of such curves obtained for tocotrienol isomers and, for comparison, the response produced by α-tocopherol. Figure 2 demonstrates dose–response effects of principal citrus flavonoids: hesperetin and naringenin. The apoB IC_{50} concentration for each compound is determined as treatment dose of the compound that inhibits the net apoB production by 50%. The apoB IC_{50} concentrations for selected minor components are presented in Table II.

Discussion

Evaluation of the apoB-lowering potential in HepG2 cells selects compounds that might produce cholesterol-lowering responses *in vivo,* providing that they are bioavailable and capable of reaching the liver intact. The lack of the effect, if observed, does not exclude the possibility of action at different levels. The

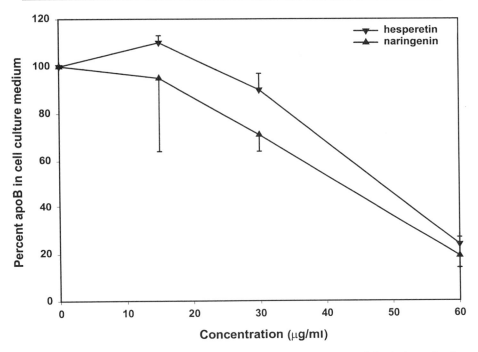

FIG. 2. Dose–response of hesperetin and naringenin effects on apoB accumulation in the media of HepG2 cell cultures.

alternative mechanisms include influencing cholesterol absorption in the intestine or modulating hepatic lipoprotein metabolism indirectly via breakdown products or derivatives released during transport across the intestinal wall.

A number of studies suggest that the ability of minor components to reduce net apoB production from HepG2 cells coincides with cholesterol-lowering potential *in vivo*. For instance, tocotrienols have been shown to induce apoB-lowering

TABLE II
APOB IC_{50} CONCENTRATIONS OF SELECTED COMPOUNDS

Compound	apoB IC_{50} concentration (μg/ml)
α-Tocotrienol	18.0
γ-Tocotrienol	21.0
δ-Tocotrienol	8.8
Hesperetin	43.0
Naringenin	48.5

responses in cells and also reduced serum total and LDL cholesterol in animals and in humans whereas α-tocopherol had no effect on apoB levels *in vitro* and no cholesterol-lowering activity *in vivo*.[2] Similarly, principal citrus flavonoids, hesperetin and naringenin, reduced medium apoB in HepG2 cells,[14] and in agreement, naturally occurring glucosides of these compounds, hesperidin and naringin, lowered blood cholesterol *in vivo*.[3] However, some synthetic inhibitors of hydroxymethylglutaryl (HMG)-CoA reductase, known to reduce hypercholesterolemia, decreased apoB production in HepG2 cells only after the stimulation of cellular triglyceride synthesis by the addition of oleate.[15]

The relationship between apoB IC_{50} concentrations of compounds determined in HepG2 cells and their biopotency *in vivo* needs to be further investigated; however, some data suggest that apoB IC_{50} concentrations may reflect differences in the cholesterol-lowering potential, at least for compounds that are closely related. In support, γ- and δ-tocotrienol, which have lower apoB IC_{50} values than α-tocotrienol, have also been shown to produce more pronounced cholesterol-lowering responses than α-tocotrienol in some previous animal studies.[16] A similar association between apoB IC_{50} concentrations and cholesterol-lowering potential may exist for other minor components.

---

[14] N. M. Borradaile, K. K. Carroll, and E. M. Kurowska, *Lipids* **34,** 591 (1999).
[15] M. W. Huff and J. R. Burnett, *Curr. Opin. Lipidol.* **8,** 138 (1997).
[16] N. Quereshi and A. A. Quereshi, *in* "Vitamin E in Health and Disease" (L. Packer and J. Fuchs, eds.), p. 247. Dekker, New York, 1993.

Author Index

A

Abalea, V. A., 308, 309(9), 310, 313, 313(9), 315
Abanukov, G. A., 195
Abbate, R., 370
Abian, J., 157(24), 158
Abu-Amsha, R., 259, 262(29)
Acar, D., 157
Adams, G. E., 170
Adamson, G. E., 46, 48, 49, 50(7), 51(7),
    57(7; 19), 351, 355(11)
Adamson, G. L., 261, 355
Addis, P. B., 250
Adeli, K., 398, 404(2)
Aeschbach, R., 203
Aeschbacher, H. U., 188
Aggarwal, B. B., 382(9; 12; 17), 385
Aggarwal, R., 217, 219(1), 220(1), 221(1),
    222(1), 223(1)
Aggarwal, S., 174, 351
Agullo, G., 102, 321, 322(6a; 6b), 323(6a; 6b),
    324(6a), 328(6a; 6b)
Ai-guo, M., 308
Aikens, M. L., 131
Aisen, P., 196(22), 197
Akiyama, H., 157, 232, 362
Akman, S. A., 232
Alaranta, S., 27
Albasini, A., 174, 175
Alcaraz, M. J., 157, 157(22), 158
Aldercreutz, H., 112
Aldini, G., 338, 354
Alejandre-Duran, E., 188
Alfassi, Z. B., 176
Alfthan, G., 97, 102(7)
Algee, S. E., 197
Allaway, S., 132, 203
Allen, D. R., 275, 276, 276(21)
Allina, S. M., 76
Alluis, B., 327
al Makdessi, S., 157
Almeida, L. M., 258, 259(11), 262(11), 282,
    283, 284, 286, 287(25), 292(16)

Alsonso-Moraga, A., 188
Amarasinghe, C., 307
Ames, B. A., 233
Ames, B. N., 308, 333
Amic, D., 174
Amiot, M. J., 58
Amrhein, N., 72
Andersen, O. M., 5
Anderson, D., 144
Anderson, R. F., 307
Ando, H., 364
Andrare, P. B., 75
Angelini, M., 142, 145
Anger, J. P., 313
Appel, H. M., 172
Appert, C., 72
Arai, H., 283
Aranda, F. J., 295
Aravanis, C., 103, 350, 369, 379(2)
Ariga, T., 340
Ariniemi, K., 97, 102(7)
Ariza, R. R., 188
Armhein, N., 72
Armstrong, M. A., 130, 249
Arnaud, J., 397
Aro, A., 97, 102(7)
Arora, A., 186
Arteel, G. E., 354, 389
Aruoma, O. I., 242, 297, 304(24)
Arveiler, D., 122
Asai, K., 263
Asamitsu, K., 385
Ascherio, A., 131
Ashworth, P., 288
Assen, N. A., 132
Astore, D., 254
Athar, M., 232
Aubert, S., 58
Audette-Arruda, J., 361
Augustin, M., 83, 92(8)
Aust, S. D., 248
Autio, I., 351
Avdeef, A. E, 202

Aviram, M., 112, 132, 203, 244, 245, 248, 250, 353
Ayaori, M., 353
Azzini, E., 125, 379, 380(35)

B

Badr, K. F., 252
Bae, K.-H., 398, 404(3)
Bae, Y. S., 84, 94(24)
Baert, T., 48, 49(6)
Baker, J., 144
Balas, L., 82
Balentine, D. A., 107, 181
Baranac, J. M., 172
Baratta, M. T., 295, 320, 330(4d), 331(4d)
Barcley, L. R. C., 340
Bard, A. J., 344
Barker, T. P., 57
Barnes, S., 117
Barron, D., 326
Barros, S., 203
Barsukova, T. K., 320, 330(4e), 331(4e)
Barz, W., 70, 71(3)
Basile, J. P., 282
Basolo, F., 195
Bassenge, E., 276
Bast, A., 157(25), 158, 175, 183, 197, 304
Bate-Smith, E. C., 70, 83
Baumgartner, H. R., 374
Bäumker, P. A., 77
Baxter, D., 181
Bearden, M. M., 351, 354(14), 355(14)
Beart, J. E., 325, 328(15a), 330(15a)
Becana, M., 157(24), 158
Becher, A., 80
Becker, C. G., 390
Beckman, J. S., 230, 273, 297, 389
Begley, M. J., 84
Begum, M., 321, 322(7), 323(7), 328(7)
Beilin, L. J., 132, 259, 261, 262(29)
Bekker, R., 172
Belinky, P., 112, 248, 353
Bell, J. R., 107, 115, 116, 118(1), 121(1), 265
Bellomo, G., 254
Bennick, A., 333
Benthin, G., 296
Benza, R. L., 131
Benzie, I. F., 266

Berber-Jimenez, M. D., 398
Berdnikof, V. M., 235
Berg, K., 401
Berger, J.-L., 133
Bergeron, R. J., 197
Berliner, J. A., 244, 250, 351, 397
Berman, C., 361
Bernier, J. L., 157(29), 158, 164, 319, 330(2g)
Berset, C., 264, 283
Bertelli, A. A. E., 144, 146, 151(12), 153(12; 13)
Berti, R., 354
Bertozzo, L., 351
Bertrand, A., 49
Beslo, D., 174
Besson, C., 118, 119(7), 121(7), 153, 265, 321, 322(6d), 323(6d), 331(6d)
Beswick, A. D., 369
Beutler, E., 134, 138(36), 141(36)
Bianchini, F., 309
Bickers, D. R., 232
Bielski, B. H. J., 296
Bijsman, M. N., 154
Bilio, A. J. D., 157(32), 158
Billot, J., 282
Bilyk, A., 323
Bing, B., 275
Bingham, A., 122
Birkhofer, L., 80(61), 81
Blach, A. L., 194
Blache, A., 140, 144(41)
Blackburn, H., 103, 350, 369, 379(2)
Bladt, S., 5
Blin, P., 351
Bloor, S. J., 3, 6
Boatella, J., 250
Boddi, M., 370
Boisnard, A., 309
Boiteux, S., 232
Bok, S.-H., 398, 404(3)
Bokern, M., 81
Bolwell, G. P., 73, 170, 171(24), 186, 203, 259, 260, 261, 262(25), 263(26), 266, 320, 330(4g), 353, 397
Bolwell, P. G., 157(31), 158, 164(31), 181
Bombardelli, E., 27, 338, 354
Bonet, B., 188
Bonina, F., 294
Bonjar, G. H. S., 169
Bonomo, R. P., 157(32), 158, 191(5), 192
Bontempelli, G., 172, 338

AUTHOR INDEX

Booth, A. N., 265
Booyse, F. M., 131
Borel, J. P., 157
Borradaile, N. M., 404
Bors, W., 27, 97, 158, 164(34), 166, 167, 169(2), 170, 170(2), 171, 171(19; 20), 172, 172(19), 173(30), 174, 175, 175(19; 21), 176, 177, 177(30), 178(22), 179(19; 36), 180(36), 232, 282, 291, 294(5), 319, 320, 330(3b; 3c; 3d; 4e), 331(4e), 340, 397
Bose, P., 103
Botterweck, A. A., 307
Boucher, J. L., 230
Boukamp, P., 382
Boulton, D. W., 323, 329(12), 330(12), 331(12)
Bouma, B. N., 132
Bourel, M., 309
Bourne, L. C., 123, 181, 190
Bourzeix, M., 18
Boveris, A., 166, 389
Bowry, V. W., 295
Boyce, S. T., 365
Boyd, D., 261
Bozida, S. S., 103
Bragdon, J. H., 261
Brall, S., 80
Brame, C. J., 255
Bramley, P. M., 157(31), 158, 164(31), 181, 182, 266, 397
Brandt, E. V., 172
Braquet, P., 157
Bravo, L., 333
Breitkreutz, D., 382
Breslow, J. L., 131
Bret, S., 322, 323(8), 324(8), 328(8), 332(9)
Briancon, S., 397
Brikenfeld, H. P., 401
Briski, K. P., 401
Brissette, L., 399
Brissot, P., 309, 314
Briviba, K., 389
Broadhurst, R. B., 83
Brödenfeldt, R., 72
Brotherton, J. E., 83, 86(6), 91(6)
Brouillard, R., 172
Brown, A. J., 250
Brown, B. R., 64
Brown, C. A., 390
Brown, D. G., 194

Brown, J. E., 27, 157, 157(23), 158, 191(9), 192, 390
Brown, J. P. A., 203
Brown, M. S., 351
Brown, T., 298
Browne, R. W., 249
Bruckdorfer, K., 261
Bruno, A., 14, 37
Bruschke, A. V. G., 351
Buch, S., 353
Buchanan, R. M., 194
Buege, J. A., 248
Buettner, G. R., 291, 293(35)
Buijsman, M. N. C. P., 98, 103(9)
Buja, L. M., 374
Buring, J. E., 130, 131
Burk, R. F., 252
Burke, T. R., Jr., 382(9), 385
Burnett, J. R., 404
Burney, S., 297, 301
Burroughs, L. F., 57
Burton, G. W., 285, 340
Bush, K. M., 230
Bush, L. R., 374
Bush, M., 350(8), 351
Bushman, J. L., 307
Bustos, G., 157
Butler, J., 296
Butler, L. G., 83, 86(6), 91(6), 333
Butler, S., 244, 254
Butt, V. S., 73
Buxton, G. V., 296(17), 297
Buysman, M. N. C. P., 98, 101(8), 102(8), 103(9; 11), 107, 117
Buzina, R., 103, 350, 369, 379(2)

C

Cabanis, J.-C., 49, 143
Caccetta, R. A., 132
Cadenas, E., 284, 290, 291, 292(18), 294
Cadet, J., 309
Cahill, P. A., 131
Cai, J. P., 395
Cai, S., 114, 331
Cai, Y., 84
Calderazzo, F., 194
Calloni, M. T., 354
Calomme, M., 319, 330(2c)

Camargo, C. A., Jr., 130
Cambou, J. P., 122
Cammack, R., 302
Campagnola, M., 351
Campbell, W. B., 374
Canada, A. T., 188
Canali, R., 387
Candeias, L. P., 182
Cao, E., 232, 233, 234(18), 236(15; 18), 243(18)
Cao, G., 25, 49, 57(19), 104, 106, 107(6), 169, 181, 266, 319, 330(2h), 351, 355(11)
Cao, R., 217
Cao, Y., 217
Carando, S., 49, 143
Cardinal, D. C., 370
Carew, T. E., 108, 132, 244, 259, 351
Carey, F., 157(22), 158
Carini, M., 338
Carino, M., 354
Carley, W. W., 382(8), 385
Carmichael, J., 365
Carpenter, D. R., 351, 354(14), 355(14)
Carrano, C. J., 202
Carraway, D. T., 73
Carreras, M. C., 389
Carroll, K. K., 398, 404
Carter, D. C., 322, 323(9a), 324(9a), 325(9a), 327(9a), 329(9a)
Casper, R. F., 153
Castellano, R., 254
Castelli, F., 294
Castelluccio, C., 259, 260, 261, 262(25), 263(26)
Cattear, J. P., 157(29), 158, 164
Catteau, J.-P., 319, 330(2g)
Catz, S. D., 389
Caulfield, J. L., 297, 298
Cenas, N., 169, 172(16), 173(16), 188
Cerami, A., 395
Cerutti, P. A., 232
Chacra, A. P., 157
Chambers, J. Q., 346
Chan, B. G., 83, 91(12)
Chan, P. T., 165
Chan, R., 295
Chan, T., 188, 271
Chang, C. W., 174
Chang Sing, P. D. G., 174
Chao, J.-T., 398, 404(2)
Chapple, C. C. S., 73
Chari, V. M., 14

Chaturvedi, M. M., 382(12), 385
Cheminat, A., 86
Chen, C. P., 169, 170(7), 171(7)
Chen, J., 190
Chen, K., 263
Chen, L., 107
Chen, Z. Y., 165, 175
Cheng, K.-L., 145
Cheng, S. J., 217, 232
Chen Liu, K. C. S., 174
Cheon, S. H., 157
Cheong, H., 157
Cherksov, V. K., 195
Chevaux, K. A., 351, 354(14), 355(14)
Cheynier, V., 49, 51(20), 63, 82, 84, 85, 86, 89(17), 93, 170, 171(25), 172
Chiang, V. L., 73
Chiavari, G., 48
Chien, P. Y., 146
Chiesa, R., 254
Chimi, H., 283
Chin, Y. H., 395
Chipman, J. K., 382(14), 385
Chisolm, G. M., 351
Choi, M.-S., 398, 404(3)
Chrisey, L. A., 169
Christova, A. G., 169, 175(14)
Chun, T. Y., 146
Chung, B. H., 261
Chung, H. Y., 175
Chung, K. T., 169
Chung, M.-I., 145
Ciccia, G., 203
Cillard, J., 157(26), 158, 283, 308, 309, 309(9), 310, 313, 313(9), 314, 315, 319, 330(2e), 382(15; 16), 385, 393
Cillard, P., 157(26), 158, 283, 308, 309, 309(9), 310, 313, 313(9), 314, 315, 319, 330(2e)
Cimanga, K., 319, 330(2c)
Clapperton, J., 49, 53(14)
Clarkson, R. B., 276, 279(24)
Clifford, M. N., 257
Cnossen, P. J., 98, 103(9), 154
Cochran, W. G., 136, 137(38), 138(38)
Cochrane, A. L., 131, 381, 397
Codony, R., 250
Coetzee, J., 172
Cogrel, P., 314
Colditz, G. A., 131
Coleman, R., 112, 353

Collins, A. R., 308
Combien, F., 122
Cominacini, L., 351
Conkerton, E. J., 15
Conn, E. E., 71
Connor, S. L., 398
Connor, W. E., 398
Conquer, J. A., 379, 380(35)
Constant, J., 181
Cook, N. C., 122, 379
Cooper, S. R., 202
Cordes, A., 157
Cornicelli, J. A., 340
Corsaro, C., 157(32), 158, 191(5), 192
Corvazier, E., 379
Cos, P., 319, 330(2c)
Cossins, E., 273
Costantino, L., 174, 175
Costanzo, G., 370
Cotelle, N., 157(29), 158, 164, 319, 330(2g)
Cotsarelis, G., 361
Coussio, J. F., 166, 203
Covas, M.-I., 112
Coward, L., 117
Cox, F. J., 379, 380(36)
Crastes de Paulet, A., 174
Crespy, V., 118, 119(7), 121, 121(7), 144, 153, 265, 321, 322(6d; 6e), 323(6d; 6e), 330(6e), 331(6d; 6e)
Criqui, M. H., 259
Croft, K. D., 132, 259, 261, 262(29)
Cross, C. E., 296, 297, 304, 304(24)
Cruickshank, A., 132
Cuddy, T. E., 351
Cui, X. Y., 302
Cuo, C., 25
Curzon, E. H., 257, 258(2)
Cusumano, J. C., 73
Cutler, R. G., 266
Cuvelier, M.-E., 264, 283
Czapski, G., 296
Czichi, U., 72
Czochanska, Z., 65, 82, 84, 89(19)

D

Dabbagh, Y. A., 109, 110, 112, 112(22; 23), 114, 259, 260, 261, 320, 330(4a), 331, 331(4a), 353, 397

Daigle, D. J., 15
da Luz, P. L., 157
Damon, M., 174
Dangles, O., 172, 319, 322, 323(8), 324(8), 327, 328(8), 329, 332(9)
Dáquino, M., 259, 260
Daret, D., 351
Darley-Usmar, V., 261
Darmon, M., 338
Das, N. P., 143, 153, 165, 166, 181, 191(4), 192, 294
DasGupta, R., 361
da Silva, E. L., 117, 331
D'Aurora, V. D., 232
Davidovic-Amic, D., 174
Davies, M. J., 369
Davin, L. B., 70(7), 71
Davis, G. C., 131
Davis, K. G., 325, 328(15a; 15b), 330(15a; 15b)
Davis, P. A., 261, 264(37), 356
Davoli, A., 351
Dawra, R. K., 333
Dawson, M. I., 295
Day, B. W., 177
Dean, T. A., 286
De Carolis, E., 75, 76(38)
De Cooman, L., 6
Dedon, P. C., 301
Dee, G., 261
De Eds, F., 265
Deepalakshmi, P. D., 398
De Feudis, F. V., 27
Degenstein, L., 361
DeGraff, W. G., 365
de Groot, M. J., 175
Delalande, L., 58
de la Torre-Boronat, M. C., 112, 144, 146, 153(14)
Delatour, T., 309
De Leeuw, I., 109
de Lorgeril, M., 122, 131, 203, 259, 350(9), 351, 376
Demacker, P. N., 132
Demer, L. L., 244, 397
Demigné, C., 102, 118, 119(7), 121, 121(7), 144, 153, 265, 321, 322(6a; 6b; 6c; 6d; 6e), 323(6a; 6b; 6c; 6d; 6e), 324(6a; 328(6a; 6b), 330(6e), 331(6d; 6e)
Demple, B., 298
Demrow, H. S., 374, 376, 376(22), 377(22)

Deng, X. S., 190
Dengler, H. J., 143
de Pascual-Teresa, S., 170, 171(25)
Derijke, Y. B., 351
de Rijke, Y. B., 132
De Santis, A., 351
Deschner, E. E., 157
Desmarchelier, C., 203
deTakats, P. G., 144
Devall, L. J., 340
Devaraj, S., 397
Devaray, H., 398
Devaray, S. N., 398
de Villiers, W. J., 157, 382(11), 385
de Vries, J. H. M., 98, 103(9; 10; 12), 107, 117, 143, 144, 144(50), 154
de Whalley, C. V., 320, 330(4b), 331(4b), 353
Diamandis, E. P., 122, 123, 132, 133, 140, 142, 145, 146, 154
Diaz, M. N., 122
Di Bilio, A. J., 191(5), 192
Dicharry, S. A., 298
Dickancaité, E., 169, 172(16), 173(16), 188
Dieber-Rotheneder, M., 284
Dietz, K., 157
Di Felice, M., 259, 283, 289
Dimayuga, M. L., 174
Dimitric-Markovic, J. M., 172
Dixon, R. A., 76
Dizdaroglu, M., 242, 308, 310
Dobbin, P. S., 190
Doco, T., 49, 85, 93
Dogliotti, E., 298
Doi, U., 283
Domena, J. D., 298
Dong, E., 169, 170(7), 171(7)
Dong, Z. G., 232
Donike, M., 80(61), 81
Donnelly, D. M. X., 89
Donn-Op den Kelder, G., 175
Donovan, J. L., 107, 115, 116, 118(1), 120, 121(1), 258, 259, 259(12), 260, 263(12), 264(12), 265, 265(12), 354, 355(50), 358
Doré, J. C., 174
Dorevski, K., 174
Dorfman, L. M., 170
Dorian, B., 351
Doroshow, J. H., 232
Douglas, C. J., 76
Douki, T., 309

Downes, M. J., 275
Dreosti, I. E., 217
Drilleau, J.-F., 49, 57, 58, 58(7), 59, 61(2; 7), 62(2), 63, 64(14), 65(14), 66(7; 14), 67(2; 7), 84, 93(21)
Dripier, J.-C., 230
Drouis, R., 232
Droy-Lefaix, M. T., 27, 275, 392
Duarte, J., 354
Dubos, M. P., 308, 309(9), 313, 313(9), 315
Dubrova, Y. E., 390
Ducrocq, C., 230
Dufour, C., 319, 322, 323(8), 324(8), 328(8), 329, 332(9)
Duke, S. O., 74, 75(35)
Duport, N., 397
Durand, P., 140, 144(41)
Durst, F., 73, 74(32)
Duthie, G. G., 204
Duthie, S. J., 308
Duval, D. L., 319, 330(2f)
Dwyer, J., 48
Dzau, V. J., 396, 397(22)

E

Eaton, D. R., 194
Ecker, D. J., 202
Eder, H. A., 261
Edlund, A., 296
Edwards, M. W., 275
Edwards, P. A., 244, 397
Edwards, R., 76
Edwards, T. G., 346
Egashira, Y., 258, 259(13; 14), 265(13; 14)
Eggins, B. R., 346
Eichholzer, M., 297
Eichmuller, S., 361
Eidt, J., 374
Eiserich, J. P., 296, 304
Elangovan, V., 157
El-Dean, S., 350(8), 351
Elis, A., 112, 353
Elkin, Y., 76
Ellis, B. E., 76
Ellison, R. C., 130
El-Saadani, M., 247, 286
El-Sayed, M., 247, 286

Elsey, T. S., 275
Elwood, P. C., 369
Emerson, O. H., 265
Enslen, M., 107, 360
Erben-Russ, M., 176
Erez, A., 72
Eriksson, L. A., 174
Erlund, I., 97, 102(7)
Erslev, A. J., 134, 138(36), 141(36)
Escargueil-Blanc, I., 282
Escobar, J., 183
Escribano-Bailon, M. T., 49, 51(20), 172
Es-Safi, N. E., 172
Esterbauer, H., 112, 122, 123(2), 246, 247, 247(7), 259, 262(19), 263, 264(50), 284, 286, 352
Estrella, I., 58
Etemad-Moghadam, A., 187
Evans, A., 122
Evans, D. H., 346
Evans, F. J., 84
Evans, G. L., 146
Evans, P. J., 157(20), 158, 188
Everaert, E., 6

F

Faccini, G. B., 351
Fache, P., 6
Facino, M. R., 354
Fadrus, H., 203
Falshaw, C. P., 84
Falstrault, L., 399
Faresjö, Å. O., 232
Fargeix, G., 329
Farhat, J. H., 370, 372(10)
Farmer, P. B., 297
Farr, A. L., 355
Faulds, C. B., 258
Faure, H., 397
Favier, A., 397
Favier, M.-L., 102, 321, 322(6a), 323(6a), 324(6a), 328(6a)
Favilla, S., 370
Fay, L. B., 188
Fedorova, O. S., 235
Fehily, A. M., 369
Feldman, K. S., 172
Felice, M. D., 259, 260

Feresjö, Å. O., 232
Fernandez, Y., 338
Fernandez de Simon, B., 58
Ferrandiz, M. L., 157
Ferraro, G. E., 166
Ferreira, D, 172
Ferreira, M. A., 75
Ferreira, S., 157
Ferrell, J. E., 174
Ferrige, A. G., 275
Ferro-Luzzi, A., 125, 333
Ferry, D. R., 144
Feskens, E. J., 103, 203, 321, 350, 369, 379
Fessenden, R. W., 176
Fickel, J., 333
Fidanza, F., 103, 350, 369, 379(2)
Finardi, G., 254
Finke, J. H., 273
Fisher, L. J., 307
Fitchett, C. J., 390
Fitós, M., 112
FitzGerald, G. A., 252
Fitzpatrick, D. F., 275
Flahery, J. T., 229
Flannery, B. P., 198
Fleuriet, A., 282
Flohé, R., 381
Floriani, C., 194
Flower, R. J., 370
Floyd, R. A., 233
Fogelman, A. M., 244, 351, 397
Foitzik, K., 361
Folsom, A. R., 97
Folts, J. D., 157, 369, 371, 374, 374(14; 15), 376, 376(22), 377(22), 378, 379
Fong, L. G., 351
Foo, L. Y., 65, 82, 84, 89(19), 94(24)
Ford, P. C., 296
Forder, R. A., 157(22), 158
Forkmann, G., 167
Forte, T., 288
Fortini, P., 298
Foss, O. P., 351
Foti, M., 295, 320, 330(4d), 331(4d)
Francia-Aricha, E. M., 172
Francis, F. J., 13
Francis, G. W., 5
Francois-Collange, M., 187
Frank, J. S., 244, 397

Frankel, E. N., 108, 122, 132, 143, 165, 181, 203, 256, 258, 259, 259(12), 260, 261, 263, 263(12; 30), 264(12; 22; 30; 37; 49), 265, 265(12), 356, 358, 359
Freedman, J. E., 370, 372(10)
Freeman, B. A., 230, 304
Frei, B., 111, 112(26), 122, 181, 245, 263, 357
Fricke, U., 157
Friedman, G. D., 130
Frolich, J. C., 296
Fry, S. C., 257, 258(3)
Fuchs, E., 361
Fuchs, I., 296
Fuhrman, B., 112, 132, 203, 248, 353
Fujiki, H., 181
Fujino, T., 169
Fujita, Y., 188, 204
Fujita, T., 180
Fukuhara, K., 203
Fukui, S., 27, 169, 172(15), 173(15)
Fukuoka, M., 169
Fulcrand, H., 85, 172
Funaki, C., 263
Fund, K. P., 165
Furtado, M., 157
Fusenig, N. E., 382
Fyfe, D. W., 144

G

Gaag, M. S., 117
Gaag, M. V. D., 144
Gabriac, B., 73, 74(32)
Gaffney, S. H., 325, 328(15a), 330(15a)
Gaito, G., 361
Galan, P., 397
Galati, G., 188, 271
Gali, H. U., 354
Galletti, G. C. J., 48
Gang, Y. J., 188
Gao, X. M., 354
Gapor, A., 398, 404(2)
Garbin, U., 351
Gardana, C., 14, 34, 46(14), 122, 126
Gardner, P. T., 204
Garovic-Kocic, V., 133, 145
Gaspar, J., 169
Gat, U., 361
Gaydou, E. M., 157(29), 158, 164, 319, 330(2g)

Gazdar, A. F., 365
Gaziano, J. M., 130, 131
Gebicki, J., 112, 259, 262(19)
Geer, J. C., 261
Gehm, B. D., 146
Geiger, H., 6, 13(13), 14, 16(13), 321, 322(7), 323(7), 328(7)
Gentili, V., 259, 260, 283, 289
German, B., 132, 165, 203, 265
German, J. B., 107, 115, 116, 118(1), 121(1), 143, 259, 261, 263(30), 264(22; 30; 37), 265, 356
Gerrish, C., 259, 263(26)
Gerritsen, M. E., 382(8), 385
Gerrity, R. G., 351
Gershwin, M. E., 351, 354(13)
Ghiselli, A., 125, 333, 340
Ghosh, S., 381
Giampaoli, S., 103, 350, 369, 379(2)
Giannella, E., 188
Gibbons, G. H., 396, 397(22)
Giessauf, A., 263
Gifford, J., 286
Gil, B., 157
Ginsberg, H. N., 399
Giovannini, L., 144, 146, 151(12), 153(12; 13)
Giovannucci, E. L., 131
Girotti, A., 286
Giulivi, C., 284, 291, 292(18)
Glazer, A. N., 333
Glebas, Y. Y., 84
Glogowski, J., 275, 296, 302(3)
Glories, Y., 83, 92(8)
Glynn, R. J., 130
Gniwotta, C., 252
Goda, Y., 157, 362
Goher, M., 247, 286
Gold, L. S., 233
Goldberg, D. M., 18, 122, 123, 130, 132, 133, 140, 142, 145, 146, 146(1), 151(1), 152(1), 153(1), 154, 154(1)
Goldbohm, R. A., 307
Goldhaber, S. Z., 131
Goldman, R., 177
Goldstein, J. L., 83, 333, 351
Golino, P., 374
Gollob, L., 62
Gomez-Cardoves, C., 58
Gomez-Fernandez, J. C., 295
Goodman, D. W., 144

Goodman, M. T., 113
Gordon, M. H., 165
Gordon, M. S., 190
Gosselin, R., 354
Goto, K., 204
Gotoh, N., 166
Govindasamy, S., 157
Grace, S. C., 292
Graham, A., 399
Graham, D. R., 232
Graier, W. F., 351
Grainger, D. J., 398
Grand, C., 73
Granit, R., 123, 132, 143
Grass, L., 142, 145
Gräwe, W., 80
Gray, W. G., 146
Grayer, R. J., 157(24), 158
Green, L. C., 275, 296, 302(3)
Green, M. R., 361
Greenstock, C. L., 296(17), 297
Grenadir, E., 250
Grenett, H. E., 131
Griffin, B. A., 123, 132
Griffioen, D. H., 157(25), 158, 197
Griffiths, A., 120
Griffiths, L. A., 143
Grimm, R., 79
Grinter, R., 346
Grisebach, H., 70(6), 71
Grobbee, D., 131
Grodstein, F., 130
Gross, G. G., 77, 78
Grossi, A., 254
Grunberger, D., 382(9), 385
Gryglewski, R. J., 102, 379
Guardiola, F., 250
Guerra, M. T., 172
Gugler, R., 143
Guguen, C., 309
Guillouzo, A., 309
Guissani, A., 230
Gump, B. H., 187
Gunasegaran, R., 157
Guo, Q., 157(27), 158, 204, 217, 219, 219(12; 13), 220, 221(11), 223(11), 224(11), 225(11), 226(13), 227(13), 228(13), 231(13), 232, 273, 333
Guo, Q. N., 180
Gutierrez, P., 157

Gutteridge, J. M. C., 242, 263
Gutzki, F. M., 296
Gutzwiller, F., 297
Guyot, S., 49, 57, 58, 58(7), 59, 61(2; 7), 62(2; 7), 63, 64(14), 65(14), 66(7; 14), 67, 67(2), 84, 85, 93(21)

H

Hackett, A. M., 143
Haenen, G. R., 175, 183, 304
Hagerman, A. E., 333
Hahlbrock, K., 70(4), 71, 72, 76
Hahn, S. E., 130, 132, 133, 145
Hale, A., 132, 203
Hall, A. D., 190
Halliwell, B., 157(20), 158, 181, 183, 184, 185, 187(25), 188, 188(18; 24), 189, 189(24; 25; 31), 190, 190(31), 232, 242, 263, 296, 297, 298(29), 299, 301, 301(28; 29; 37), 302(10), 303(10), 304, 304(10; 24; 28; 29), 307, 310
Ham, R. G., 365
Hamilton, T. A., 273
Hammerstone, J. F., 46, 48, 49, 50(7; 12), 51(7), 53(14), 57(7; 19), 94, 350, 351, 353(3), 355(11)
Hamon-Bouer, C., 310
Hamsten, A., 351
Han, C., 190
Han, D., 174
Hanasaki, Y., 27
Handjiski, B., 361
Hanh, S. E., 145
Hansen, M. B., 401
Hansen, R. J., 107, 115, 116, 118(1), 121(1), 143, 265
Hanson, K. R., 71, 72
Hao, Y., 106, 107(10), 109(10), 112(10)
Hara, Y., 173, 203, 204, 268, 268(12), 271, 319, 320, 330(3f; 4c), 331(4c)
Haraldsdottir, J., 190
Haramaki, N., 174
Harborne, J. B., 3, 12, 17, 70, 70(5), 71, 166
Hardak, E., 250
Harel, S., 123, 143
Harper, J., 361
Harris, D. C., 196(22), 197
Harris, P. J., 257, 258(2)

Harris, W. R., 202
Harrison, L., 298
Harryvan, J. L., 379, 380(36)
Hart, H. C., 132
Hartley, R. D., 257, 258(2)
Hasegawa, A., 362
Hashimoto, F., 172, 364
Haslam, E., 46, 48, 59, 82, 83, 325, 328(15a; 15b), 330(15a; 15b), 363
Haslan, E., 84
Hatano, T., 204
Hatta, A., 245
Hattori, M., 361, 366(10)
Häusler, E., 73
Havel, R. J., 261
Havir, E. A., 72
Hayashi, T., 263
Hayek, T., 112, 250, 353
Haza, Y., 123
He, X. M., 322, 323(9b), 324(9b), 327(9b)
Hecht, S. M., 169
Hedrick, S. A., 76
Heilemann, J., 78
Heinecke, J. W., 263, 353
Heinonen, 203
Heiser, C., 190
Heller, W., 27, 73, 76, 97, 158, 164(34), 167, 169(2), 170, 170(2), 171, 171(19; 20), 172(19), 174, 175(19; 21), 178(22), 179(19), 232, 282, 294(5), 319, 330(3c)
Helman, W. P., 296(17), 297
Hemingway, R. W., 64, 84, 87(22), 94
Hemm, M. R., 73
Hendrickson, R. J., 131
Henichart, J. P., 164
Hennekens, C. H., 130, 131, 397
Henry, Y., 230
Henzi, R., 194
Hercberg, S., 397
Hernandez, T., 58
Herrera, E., 188
Herrmann, K., 256, 257, 257(1), 282
Hertog, M. G. L., 17, 97, 101(6), 102(6), 103, 117, 203, 321, 323, 350, 369, 379, 379(2)
Hervé du Penhoat, C. L. M., 89
Hessler, J. R., 351
Hider, R. C., 157(23), 158, 190, 191, 191(9), 192, 195, 197, 199, 199(26; 27), 202(27)
Higgs, E. A., 273, 296, 389
Higuchi, T., 75

Hill, K. E., 252
Hirano, R., 354, 355(51)
Hirayama, T., 169, 172(15), 173(15)
Ho, C. T., 382(19), 385
Ho, J. X., 322, 323(9a; 9b), 324(9a; 9b), 325(9a), 327(9a; 9b), 329(9a)
Ho, K. Y., 165
Ho, Y., 143
Hobbs, G. A., 110
Hobbs, P. D., 295
Hodnick, W. F., 191(6), 192, 319, 330(2f), 345
Hoffman, A., 250
Hofmann, R. R., 333
Hofmann, U., 361
Hogg, N., 261
Hohlfeld, H., 78, 79
Høj, P. B., 79
Holcman, J., 296
Holden, J. E., 374
Hollingworth, T. A., 187
Hollman, P. C. H., 17, 97, 98, 101(6; 8), 102(6; 8), 103, 103(9–12), 107, 117, 143, 144, 144(50), 154, 203, 321, 323, 350, 373, 379, 380(36)
Holm, R. H., 194
Holmguist, G. P., 232
Holub, B. J., 379, 380(35)
Hontz, B. A., 106, 107(12), 109(12), 261
Hornung, J., 382
Horton, J. K., 308
Hosick, H. L., 401
Hou, J.-W., 219(13), 220, 226(13), 227(13), 228(13), 230, 231(13; 23), 232
Houchi, H., 276
Hoult, J. R. S., 157(20–22), 158, 188, 319, 320, 330(2d; 4b), 331(4b), 353
Hovenier, R., 379, 380(36)
Howard, A. N., 132
Howl, J., 390
Howlin, B., 195
Hrabek-Smith, J. M., 398
Hramaki, N., 351
Hristova, M., 304
Hrstich, L. N., 83, 91(12)
Hsu, F. L., 174
Hsu, K. Y., 143
Hsuan, F. T. Y., 184, 188(24), 189(24), 307
Hu, H., 232
Hu, J. P., 319, 330(2c)
Huang, C. S., 232

Huang, N.-N., 229
Huang, S.-W., 265
Huang, Y. W., 169
Hudson, B. J. F., 264
Hudson, S., 395
Huff, M. W., 404
Hughes, M. N., 302
Hughes, P., 181
Huie, R. E., 296
Humphreys, J. M., 73
Hunter, E. P. L., 176

I

Ibrahim, R. K., 75, 76(38), 326
Ide, H., 298
Iemoli, L., 34, 46(14)
Igarashi, O., 354, 355(51)
Ignarro, L. J., 273
Ikeda, H., 174, 351
Ikeda, M., 307
Imam, M. P., 175
Ingold, K. U., 285, 340
Inoue, H., 132, 308, 353
Ioku, K., 206, 320, 330(4f), 331(4f)
Iraga, C. G., 166
Ischiropoulos, H., 389
Ishii, T., 283
Ishikawa, T., 353
Itakura, H., 354, 355(51)
Ito, M., 361
Ito, T., 353
Iwahashi, H., 283
Iwai, S., 308
Iwasaki, K., 382(15), 385, 393
Izem, L., 399

J

Jaakkola, O., 351
Jacob, R., 157
Jacobs, M., 261
Jacobs, P. H., 49, 57(19), 351, 355(11)
Jacques, D., 48, 59, 83, 363
Jaffe, E., 390
Jahoda, C. A., 361
Jaiswal, A. K., 169, 172(16), 173(16), 188
James, K. A., 131, 133(16)

Jameson, J. L., 146
Jang, J., 109, 112(22), 114, 259, 260, 261, 320, 330(4a), 331, 331(4a), 353
Jankun, J., 217
Jansen, A., 103, 350, 369, 379(2)
Janssen, K., 379, 380(36)
Janusz, G., 352
Järvinen, R., 103, 369, 379(4)
Javanovic, S. V., 203
Jenner, A., 297, 298(29), 299, 301(28; 29; 37), 304, 304(24; 28; 29)
Jensen, O. N., 180
Jeong, T.-S., 398, 404(3)
Jessup, W., 250, 286, 320, 330(4b), 331(4b), 353
Jialal, I., 397
Jiang, H., 361, 399
Jiang, W., 230, 231(23)
Jimenez, J., 354
Joannou, C. L., 302
Joest, B. A., 333
Johnson, D. W., 304
Jolivet, A., 153
Jones, A. D., 304
Jones, F. T., 265
Jones, M. M., 276
Jones, P. R., 79
Jones, S., 390
Jones, W. T., 83
Joshi, C. P., 73
Jovanovic, S. V., 123, 173, 191(8), 192, 268, 268(12), 271, 319, 330(3e; 3f)
Joy, L., 261
Juan, M. E., 144, 146, 153(14)
Jungersten, L., 296
Jurd, L., 13
Jurgens, G., 112, 247, 259, 262(19), 286, 352
Jurkiewicz, B. A., 291, 293(35)
Justesen, U., 323
Jütte, M., 77
Juurlink, B. H., 157

K

Kagan, V. E., 177, 288
Kahie, Y. D., 6
Kaiser, C., 80(61), 81
Kalyanaraman, B., 286
Kamate, L., 399
Kamendulin, L. M., 190

Kamiya, T., 361, 362
Kamiya, Y., 160, 180, 261
Kamsteeg, J., 73
Kamyia, Y., 287
Kanagawa, A., 385
Kanaori, K., 298
Kanda, T., 84, 94(25), 157, 362
Kandaswami, C., 26, 97, 103, 157, 319, 330(2a)
Kang, S. A., 188
Kannel, W. B., 130
Kanner, J., 123, 132, 143, 259, 264(22), 265
Kapeghian, J. C., 350(8), 351
Kaplan, I. V., 110
Karasawa, H., 283
Karchesy, J. J., 84, 94(24), 354
Karumanchiri, A., 140, 145, 154
Karwatzki, B., 73
Kasai, H., 308
Kasim-Karakas, S., 107, 115, 116, 118(1), 121(1), 265
Katan, M. B., 97, 98, 103, 103(9; 10; 12), 117, 132, 143, 144, 144(50), 154, 203, 321, 323, 350, 373, 379, 380(36)
Kataoka, S., 340
Katiyar, S. K., 217, 219(1), 220(1), 221(1), 222(1), 223(1), 232
Kato, M., 203
Kato, R., 361
Kato, Y., 283
Kaul, R., 167, 172(5)
Kaur, H., 232
Kawa, Y., 361
Kawahara, O., 364
Kawahara, R., 180
Kawakami, A., 261
Kawakishi, S., 157(30), 158
Kawashima, S., 361
Kealey, T., 361
Keaney, J. F., Jr., 122, 370, 372(10)
Keen, C. L., 350, 351, 354, 354(13; 14), 355(14), 360
Keevil, J., 378
Keli, S. O., 350
Keller, H., 78, 80
Keller, J. W., 376
Kelley, J. L., 351
Kelm, M., 275
Kennedy, J. A., 195, 197
Kerr, D. J., 144
Khan, W. A., 232

Khodr, H. H., 27, 157(23), 158, 190, 191(9), 192, 197, 199(27), 202(27)
Khoo, J. C., 108, 132, 244, 259, 351
Khramtsov, V. V., 276
Kiatgrajai, P., 62
Kido, R., 283
Kiechle, F., 275
Kieler-Jensen, N., 296
Kikuchi, S., 287
Kim, D., 273(10), 275, 354, 390
Kim, J. A., 351
Kim, K. M., 157
Kindl, H., 72
King, R., 174
Kinsella, J. E., 122, 132
Kirk, M., 117
Kitagaki, H., 181
Kitano, Y., 364
Kitazawa, M., 382(15), 385, 393
Klatsky, A. L., 130, 131
Klaunig, J. E., 190
Klaus, W., 157
Klironomos, G., 144
Klopman, G., 174
Klucas, R. V., 157(24), 158
Klurfeld, D. M., 131
Knekt, P., 103, 369, 379(4)
Kneusel, R. E., 73, 75(22), 76
Knierzinger, A., 328
Knipping, G., 263, 264(50)
Knobloch, K.-H., 76
Knogge, W., 77
Knuthsen, P., 323
Ko, C. M., 379
Ko, F.-N., 145, 157
Ko, W. C., 157, 379
Kobayashi, M., 361
Kobuchi, H., 27, 338, 395
Koch, U., 73, 76
Koga, T., 340
Kohdr, H. H., 199
Kojima, M., 73, 74, 75(34), 78(62), 80(62), 81
Komagoe, K., 204
Komarov, A. M., 275, 276
Kondasmami, C., 203
Kondo, K., 174, 203, 206, 207, 208, 209, 210, 211, 211(20), 212, 354, 355(51)
Kono, S., 307
Koop, D. R., 157
Kooy, N. W., 389

Kopp, E. B., 381
Koppenol, W. H., 273, 297, 389
Korpela, H., 254
Korthouwer, R. E., 304
Kosmol, H., 80(61), 81
Köster, J., 70, 71(3)
Kostner, G. M., 351
Kotake, Y., 275, 276, 276(21)
Koukol, J., 71
Krbavcic, A., 174
Kremers, B. G., 49, 57(19), 351, 355(11)
Kringle, R. O., 136
Kritchevsky, D., 131
Kritharides, L., 286
Krolikowska, M., 354, 360(46)
Kromhout, D., 103, 203, 321, 350, 369, 379, 379(2)
Kruse, A., 176
Kuchii, A., 258, 259(14), 265(14)
Kuchino, Y., 308
Kudo, M., 143
Kuhn, H., 252
Kuhn, R., 308
Kuhnau, J., 26, 350
Kühnl, T., 73, 76
Kukovetz, W. R., 351
Kuo, M. C., 107
Kuppusamy, P., 296
Kurata, S., 361
Kuratsume, M., 307
Kurihara, M., 174, 203
Kurl, R. N., 383
Kurowska, E. M., 398, 404
Kuroyanagi, M., 169
Kuruhara, M., 206, 211(20), 212
Kushi, L. H., 97
Kuyl, J. M., 157
Kuzuya, F., 263
Kuzuya, M., 263
Kwee, D. C. T., 185, 187(25), 189(25)

L

Labarbe, B., 82
Laemmli, U. K., 334
Lagune, L., 83, 92(8)
Lai, C. S., 275, 276, 276(21)
Laitao, R., 75
Lamb, C. J., 76
Lamuela-Raventós, R. M., 104, 106(8), 112, 144, 146, 153(14), 258
Lan, A. N. B., 184, 188(24), 189(24), 307
Landou, C., 351
Lapidot, T., 123, 143
Lapierre, B. C., 59, 64(11), 65(11), 89
Laraba, D., 58, 61(2), 62(2), 67(2)
Laranjinha, J. A. N., 258, 259(11), 262(11), 282, 283, 284, 286, 287(25), 290, 292(16)
Larrue, J., 351
Latorre, L. R., 203
Lau, B. H., 382(13), 385, 392
Laughton, M. J., 157(20), 158, 188
Lauza, M., 294
Laval, J., 232, 298
Lavy, A., 132, 203
Lawson, J. A., 252
Lazarus, S. A., 46, 48, 49, 50(7; 12), 51(7), 57(7; 19), 94, 350, 351, 353(3), 355(11)
Lea, A. G. H., 58(8), 59, 257, 258(8)
Leake, D. S., 169, 261, 320, 321, 330(4b; 5b), 331(4b), 353
Leanderson, P., 232
Lebwohl, M., 232
Le Cam, A., 309
le Doucen, C., 174
Lee, C. Y., 49
Lee, M. J., 107
Lee, R., 273
Lee, S.-H., 398, 404(3)
Lee, S. S., 174
Lee, Y. L., 143
Le Guernevé, C., 63, 64(14), 65(14), 66(14), 84, 93(21)
Leinweber, F. J., 184
Leong-Morgenthaler, P. M., 188
Lepoivre, M., 230
Le Roux, E., 85, 93
Leschik, M., 143
Lescoat, G., 157(26), 158, 309, 314, 319, 330(2e)
Lesgards, G., 140, 144(41)
Lessio, H. M. A., 266
Leth, T., 323
LeValley, S. E., 295
Levesque, M., 132
Levinson, S. S., 110
Levker, R. M., 361
Lewis, J. I., 264
Lewis, N. G., 70(7), 71

Ley, R. D., 217, 220(2), 221(2), 222(2), 223(2)
Li, L., 73, 235
Li, M., 157(27), 158, 204, 229, 232
Li, M.-F., 180, 219, 221(11), 223(11), 224(11), 225(11), 229
Li, X.-J., 217
Liang, X., 114
Liang, Y. C., 157
Liardon, R., 188
Lien, E. J., 174
Ligon, G. F., 382(8), 385
Lilley, T. H., 82, 325, 328(15a; 15b), 330(15a; 15b)
Lim, D., 49, 53(14)
Limasset, B., 174
Lin, C.-N., 145
Lin, J. H., 180
Lin, J. K., 157, 180, 382(10; 19), 385
Lin, J. Y., 174
Lin, M. T., 174
Lin, Y. L., 169, 382(10; 19), 385
Lindahl, T., 308
Lindgren, F. T., 261, 355
Ling, G., 361
Linscheid, M., 81
Lin-Shiau, S. Y., 157, 180, 382(19), 385
Lissi, E. A., 183
Litvinyuk, I., 324
Liu, C. W., 174
Liu, D. L., 143
Liu, D. Y., 199
Liu, K., 351
Liu, S., 202
Liu, Z. D., 190, 197, 199, 199(27), 202(27)
Lo Cascio, V., 351
Locher, R., 131
Locke, S. J., 340
Lockwood, R., 49, 53(14)
Lodge, J. K., 382(15), 385, 393
Loeb, L. A., 308
Loew, G., 174
Loft, S., 190
Logemann, E., 72
Loliger, J., 188
Long, L. H., 181, 184, 185, 187(25), 188(24), 189(24; 25), 307
Longland, C. L., 390
Loomis, W. D., 335
Lopez-Torres, M., 174
Loreau, N., 140, 144(41)

Loscalzo, J., 369, 370, 372(10)
Lowry, O. H., 355
Lozano, Y., 93
Lu, M. L., 143
Lu, S. L., 199
Luc, G., 122
Lucic, B., 174
Lundberg, J., 296, 302(9)
Lundin, S., 296
Lusis, A. J., 244, 397
Luthria, D. L., 120
Lynch, S. M., 357

M

Ma, W. Y., 232, 233, 234(18), 236(18), 243(18)
Maalej, N., 157, 374
Maatela, J., 103, 369, 379(4)
Mabry, T. J., 3, 13(5)
MacGregor, J. T., 169
Macheix, J. J., 282
MacIouf, J., 379
MacNeil, J. M., 340
Madeira, V. M. C., 258, 259(11), 262(11), 283, 284, 286, 287(25), 292(16)
Maffei Facino, R., 338
Maggi, E., 254
Maguire, J. J., 275
Mahmood, S., 250
Maiani, G., 125, 379, 380(35)
Maiorino, M., 188, 340
Mak, W. B., 307
Makino, K., 298
Makino, M., 180
Makkar, H., 333
Malan, E., 172
Malinski, T., 275
Malkhandi, J., 144
Malvy, D., 397
Maly, J., 203
Manach, C., 102, 115, 118, 119(7), 121, 121(7), 144, 153, 265, 319, 321, 322(6a; 6b; 6c; 6d; 6e), 323(6a), 324(6a), 328(6a; 6b), 330(6e), 331(6d; 6e)
Mangas, J. J., 58
Manktelow, B. W., 131, 133(16)
Manna, S. K., 382(12; 17), 385
Manogue, K. R., 395
Manso, C. F., 175

Manson, J. E., 130
Mansour, J. M., 174
Mansour, T. E., 174
Mantaka, A., 176
Mao, T. K., 351, 354(13)
Maran, A., 146
Marcocci, L., 275
Mareschi, J. P., 187
Margerum, D. W., 304
Marinova, E. M., 264
Marjanovic, B., 191(8), 192, 268, 319, 330(3e)
Mark, J. L., 223
Marketos, D. G., 176
Markham, A., 382
Markham, K. R., 3, 5, 6, 7, 8, 12(6), 13(5; 6), 14, 102
Markovic, J., 188
Marletta, M. A., 273
Marnet, N., 57, 58, 58(7), 59, 61(2; 7), 62(2; 7), 63, 64(14), 65(14), 66(7; 14), 67(2), 84, 93(21)
Marques-Vidal, P., 122
Marrugat, J., 112
Marsh, V. H., Jr., 72
Martin, A., 390
Martinez, C. T., 302
Martino, V. S., 166
Marzullo, D., 294
Mason, R. P., 188, 276
Massiot, P., 58
Matern, U., 71, 73, 75(22), 76
Matinier, H., 58
Matoba, T., 181
Matsumoto, A., 354, 355(51)
Matthews, S., 59, 64(11), 65(11), 89
Mattson, D., 276
Matzanke, B. F., 199
Maurer, M., 361
Mauri, P. L., 14, 26, 31, 34, 37, 46(14)
Mavandad, M., 76
Maxwell, S. R., 132, 181, 188(11)
May, M. J., 381
Mayer, B., 351
McAndrews, J. M., 146
McArdle, J. V., 202
McAuley-Hecht, K., 298
McBride, T. J., 308
McCarty, R. E., 297
McClain, C. J., 157, 382(11), 385

McCord, J. D., 15, 258
McCusker, J. K., 174
McGhie, T. K., 14
McGraw, G. W., 64, 94
McLaughlin, L., 382(15), 385, 393
McLeod, L. L., 250
McLeod, R. S., 399
McManis, J. S., 197
McManus, J. P., 325, 328(15a; 15b), 330(15a; 15b)
McMurrough, I., 48, 49(6)
McPhail, D. B., 204
Mecklenburg, L., 361
Mehta, C., 165
Meilhac, O., 282
Meinhard, J., 73
Melikian, N., 259, 260, 261, 262(25)
Melissano, G., 254
Mengelers, M. J. B., 98, 103(11; 12), 107, 117, 143, 144, 144(50)
Mengoli, G. W., 345
Menotti, A., 103, 350, 369, 379(2)
Mensink, R. P., 379, 380(36)
Mentasti, E., 195
Meo, V. A., 187
Meraji, S., 261
Metcalfe, J. C., 398
Metodiewa, D., 169, 172(16), 173(16), 188
Meyboom, S., 98, 103(10)
Meyer, A. S., 203, 256, 258, 259, 259(12), 260, 263(12), 264(12), 265(12), 358, 359
Meyer, K., 73, 97
Meyer, M., 79
Michael, B. D., 176
Michaelis, G., 80(61), 81
Michel, C., 27, 97, 158, 164(34), 166, 167, 169(2), 170, 170(2), 171, 171(19; 20), 172, 172(19), 173(30), 174, 175, 175(19), 176, 177, 177(30), 179(19; 36), 180(36), 232, 291, 319, 330(3c; 3d), 340, 397
Michelangeli, F., 390
Micossi, E., 340
Middleton, E., Jr., 26, 97, 103, 157, 203, 319, 330(2a)
Mila, I., 59, 64(11), 65(11), 89
Milbury, P. E., 15
Milgrom, E., 153
Miller, E. E., 165
Miller, J. R., 195

Miller, N. J., 16, 24(3), 104, 110, 157(31; 33), 158, 164(31; 33), 170, 171(24), 175, 181, 182, 183, 186, 187(6), 203, 264, 266, 273, 283, 303, 320, 330(4g), 353, 358, 397
Milo, S., 250
Milosavljevic, E. B., 191(6), 192, 345
Minick, C. R., 390
Minna, J. D., 365
Mira, L., 175
Mitchel, C., 282, 294(5)
Mitchell, A. E., 49, 57(19), 94, 350, 351, 353(3), 355(11)
Mitchell, J. B., 365
Mitjavita, S., 338
Miura, K., 308
Miura, S., 203, 320, 330(4c), 331(4c), 353
Miura, Y. H., 169, 172(15), 173(15)
Miwa, K., 132, 353
Miyagi, Y., 132, 353
Miyase, T., 172
Miyata, N., 174, 206, 211(20), 212
Miyazawa, T., 143
Mizuno, M., 392
Mochizuki, A., 361
Modak, B., 183
Mohan, R. R., 217, 219(1), 220(1), 221(1), 222(1), 223(1)
Mohd-Nor, A. R., 195
Mohler, E. R., 190
Mohr, D., 249
Mohr, H., 72
Mohr-Nor, A. R., 195
Moini, H., 333
Moll, I., 361
Molle, D., 58(7), 59, 61(7), 62(7), 66(7)
Møller, B. L., 79
Monboisse, J. C., 157
Moncada, S., 261, 273, 275, 296
Monteiro, H. P., 157
Monty, K. J., 184
Moon, J.-H., 331
Moore, F., 131, 381, 397
Moran, J. F., 157(24), 158
Morand, C., 102, 115, 118, 119(7), 121, 121(7), 144, 153, 265, 319, 321, 322(6a; 6c; 6d; 6e), 323(6a; 6c; 6d; 6e), 324(6a), 328(6a), 330(e), 331(6d; 6e)
Morazzoni, P., 27, 338, 354

Morel, D. W., 351
Morel, I., 157(26), 158, 308, 309, 309(9), 310, 313(9), 314, 315, 319, 330(2e)
Morelli, R., 338
Moreno, J. J., 58, 389
Mori, F., 308
Mori, K., 204
Morimoto, S., 46, 363, 364(20)
Morin, R. J., 250
Morita, N., 169
Moroney, M. A., 157(20–22), 158, 319, 330(2d)
Morrison, I. E. G., 195
Morrow, J. D., 252, 253, 255
Mosbaugh, D. W., 298
Motian, A. R., 186
Moulton, C., 261
Moutounet, M., 49, 82, 84, 85, 86, 172
Mueller-Harvey, I., 257, 258(2)
Muftuoglu, M., 334
Mukai, K., 287
Mukhopadhyay, A., 382(17), 385
Mukhtar, H., 217, 219(1), 220(1), 221(1), 222(1), 223(1), 232
Muller, G., 199
Muller, O., 188
Muller-Rover, S., 361
Munday, J. S., 131, 133(16)
Munro, M. H. G., 197
Muraishi, K., 143
Murakami, H., 169
Murphy, B. J., 72
Musiani, M. M., 345
Musonda, C. A., 382(14), 385

N

Nachmann, R. L., 390
Nagao, A., 283
Nagao, M., 169, 188
Nagao, S., 188
Nagata, T., 84, 94(25), 362
Nagel, C. W., 83
Nair, M. G., 186
Naito, M., 263
Nakagawa, K., 143
Nakamura, H., 353
Nakano, T., 258, 259(13), 265(13)

Nakatani, N., 206, 320, 330(4f), 331(4f)
Nakayama, T., 157(30), 158
Nammour, T. M., 252
Nanjo, F., 204
Nardini, M., 259, 260, 283, 289
Narumi, S., 273
Nassar, A. Y., 247, 286
Natarajan, K., 382(9; 12), 385
Natella, F., 259, 283, 289
Natori, S., 169
Navab, M., 244, 397
Nedeljkovic, S., 103, 350, 369, 379(2)
Negre-Salvayre, A., 282
Neirinck, L., 144
Nelsen, S. F., 346
Nelson, J. H., 191(6), 192, 345
Nerem, R. M., 351
Neta, P., 176, 289, 319, 330(3a)
Newman, R. H., 65, 82, 84, 89(19)
Newmark, H. L., 157
Ng, E., 154
Nguyen, T. D., 188
Nicaud, V., 122
Nicolas, J., 58
Nicolay, K., 73, 75(22)
Nie, G., 232
Nielsen, S. E., 190, 401
Nigdikar, S. V., 123, 132
Niki, E., 157, 158, 160, 160(35), 166, 181(15), 182, 261, 287, 294
Nikkari, T., 351
Nikolovska-Coleska, Z., 174
Niles, J. C., 297, 301
Nilsson, J., 351
Nishimur, M., 287
Nishimura, S., 308
Nishioka, I., 46, 169, 170(7), 171(7), 172, 363, 364, 364(20)
Nishiwake, M., 353
Nishiyama, K., 169
Niwa, T., 283
Noguchi, N., 157, 166
Nonaka, G.-I., 46, 169, 170(7), 171(7), 172, 363, 364, 364(20)
Norby, S. W., 276, 279(24)
Norwarth, R. E., 83
Nourooz-Zadeh, J., 184, 188(23)
Noward, A. N., 123

Nurmann, G., 77(65), 81
Nyyssonen, K., 254

O

Oak, M. H., 157
Oba, K., 361
Obeisekera, S., 261
O'Brien, P. J., 188, 271
Ock-Sook, Y., 359
O'Connor, T. R., 232
Oda, O., 188
Offord, E. A., 107, 360
Ogawa, H., 361, 366(10)
Ogawa, S., 27
Ogus, H., 334
Ohashi, H., 75
Ohm, S., 78
Ohnishi-Kameyama, M., 84, 94(25), 362
Ohshima, H., 297, 304(22; 23)
Ohta, T., 258, 259(13; 14), 265(13; 14)
Ohtsuka, E., 308
Ojala, S., 112
Ojasoo, T., 174
Okabe, S., 181
Okamoto, T., 382(15; 16), 385, 393
Okuda, S. S., 131
Okuda, T., 204
Okumura, T., 204
Oldreive, C., 296, 302(10), 303(10), 304(10)
O'Leary, N., 261
Oliveira, M. B., 75
Olkin, S. E., 235
Olnes, M. I., 383
O'Malley, P. J., 174
Omura, H., 169
Omura, T., 74
Oniki, T., 292
Orr, J. R., 261, 355
Orthofer, R., 104, 106(8)
Osakabe, K., 73
Osawa, T., 283, 320, 330(4c), 331(4c)
Osawsa, T., 157(30), 158
Osman, H. E., 157, 378
Oszmianski, J., 18, 49
Otero, P., 188
Ozawa, T., 204
Ozer, N., 334

P

Pace-Asciak, C. R., 132, 133, 145
Packer, J. E., 287, 307
Packer, L., 6, 7, 8, 27, 48, 157(28), 158, 174, 273, 273(10; 11), 275, 288, 319, 330(2b), 333, 338, 351, 354, 381, 382(15; 16; 18), 385, 388, 390, 390(1), 392, 393
Padmaja, S., 296
Paganga, G., 16, 24(3), 110, 143, 153, 157(33), 158, 164(33), 170, 171(24), 181, 186, 187(6), 203, 259, 260, 261, 262(25), 264, 266, 273, 283, 296, 302(10), 303, 303(10), 304(10), 320, 330(4g), 353
Paglieroni, T. G., 354
Paguay, J. B., 304
Pakusch, A.-E., 76
Palinski, W., 244, 254
Palmer, R. M., 273, 275
Palmer, W. A., 389
Pan, H. Y., 143
Pan, M. H., 180
Pannala, A., 104, 175, 183, 185(20), 266, 267(3), 304
Parasakhty, K., 398
Parathasarathy, S., 108
Pardini, R. S., 191(6), 192, 319, 330(2f), 345
Pargament, G. A., 389
Park, M., 188
Park, Y.-B., 398, 404(3)
Park, Y. C., 273(11), 275, 382(18), 385, 388, 390(1)
Parkes, J. G., 130, 145
Parlanti, E., 298
Parthasarathy, S., 132, 244, 259, 261, 351, 353, 360
Paschenko, S. V., 276
Pascual-Martinez, L., 49
Pasdeloup, N., 309, 314
Pasini, C., 351
Pasini, F., 351
Pastorino, A. M., 351
Paterson, P. G., 157
Patterson, L. K., 176
Paus, R., 361
Paya, M., 157, 157(21), 158, 319, 330(2d)
Pearl, L., 298

Pearson, D. A., 203, 258, 259, 259(12), 260, 263(12), 264(12), 265(12), 350, 351, 354, 354(14), 355(14), 358, 359, 360
Pearson, R. G., 195
Pecker, L., 395
Pedersen, J. A., 180
Pekkarinen, M., 103
Pele, H., 165
Peleg, H., 259, 263(30), 264(30)
Pelizzetti, E., 195
Pellegrini, N., 175, 183, 185(20)
Pellerini, N., 104
Peng, Q., 382(13), 385, 392
Peng, S. K., 250
Perchellet, E. M., 354
Perchellet, J. P., 354
Pereira, T. A., 165
Perez-Ilzarbe, J., 58, 63, 84, 89(17)
Perry, C. A., 382(8), 385
Pertile, J., 398
Peterlongo, F., 27
Peters, T., 323, 324(10)
Petersen, M., 71, 73
Peterson, A. S., 296
Peterson, H., 250
Peterson, J., 48
Petranovic, N. A., 172
Petrussevska, R. T., 382
Phan, S. A., 382(8), 385
Phillips, G. A., 250
Phillips, J. G., 187
Phillipson, J. D., 84
Philpott, M. P., 361
Piattelli, M., 295, 320, 330(4d), 331(4d)
Picconella, E., 283
Pierce, G. N., 351
Pierpont, C. G., 194
Pieters, L., 319, 330(2c)
Pietrogrande, M. C., 6
Pietta, P. G., 14, 26, 27, 28, 30, 31, 34, 34(11), 37, 46(14), 122, 126
Pikounis, B., 361
Pileggi, F., 157
Pinicelli, A., 58
Piskula, M. K., 117, 169, 294, 331
Pisu, P., 283
Pitra, C., 333
Planas, J. M., 144, 146, 153(14)

Plumb, G. W., 170, 171(25)
Plyusnin, V. F., 276
Poderoso, J. J., 389
Pollet, B., 59, 64(11), 65(11), 89
Pommery, J., 157(29), 158, 319, 330(2g)
Popko, J. L., 73
Porrini, M., 34, 46(14)
Porter, L. J., 4, 46, 65, 82, 83, 84, 87(22), 89(19), 91(12), 197, 338
Potter, S. M., 398
Poulsen, H. E., 190
Powell, H. K. J., 195, 197
Powell, J. J., 351, 354(13)
Prasad, R., 308
Pratt, D. E., 165
Press, W. H., 198
Preston, B. D., 308
Preziosi, P., 397
Price, M. L., 83, 86(6), 91(6)
Price, S. F., 15, 258
Price, V. H., 361
Pridham, J. B., 157(31), 158, 164(31), 181, 259, 260, 261, 262(25), 266, 397
Prieur, C., 49, 51(20), 84
Pri-Hadash, A., 76
Primka, R., 361
Princen, H. M. G., 351
Printz, D. J., 261
Prior, R. L., 25, 49, 57(19), 104, 106, 107(6), 169, 181, 266, 319, 330(2h), 351, 355(11)
Prisco, D., 370
Proch, J., 103, 106, 107(9), 110(9), 112(9), 114
Proteggente, A., 104, 175, 183, 185(20), 266, 267(3)
Proudfoot, J. M., 259, 261, 262(29)
Pryor, W. A., 340, 349(20), 350, 389
Puddey, I. B., 132, 259, 262(29)
Pueyo, C., 188
Puhl, H., 112, 122, 123(2), 246, 247(7), 259, 262(19), 352
Purmal, A. A., 308

Q

Qin, J., 232, 233, 234(18), 236(15), 236(18), 243(18)
Quereshi, A. A., 404

Quereshi, N., 404
Quesne, M., 153
Quideau, S., 172
Quinn, M. T., 351

R

Rabbani, L. E., 369
Radi, R., 230
Rafecas, M., 250
Raguzzini, A., 379, 380(35)
Rahmani, M., 283
Rajendran, S., 398
Rajewsky, K., 308
Ralet, M.-C., 258
Ramin, S., 351
Ramos, T., 18
Ramshad, B., 351
Rand, M. L., 131
Randall, R. J., 355
Randoux, A., 157
Ranges, G. E., 382(8), 385
Rankin, S. M., 320, 330(4b), 331(4b), 353
Rapuzzi, I., 172, 338, 340
Rassart, E., 399
Rastelli, G., 174, 175
Ratke, R., 157
Ratliff-Thompson, K., 197
Ratty, A. K., 166, 181, 191(4), 192, 294
Rava, A., 37
Raymond, K. N., 199, 202
Razaq, R., 304
Razuvaev, G. A., 195
Re, R., 175, 183, 185(20), 266, 267(3)
Reading, S. J., 390
Reckless, J., 398
Recknagel, R. O., 175
Redmond, E. M., 131
Reed, J., 378
Rees, L. V. C., 195
Reeve, V. E., 217, 220(2), 221(2), 222(2), 223(2)
Regenbrecht, J., 79
Régérat, F., 102, 121, 144, 153, 321, 322(6a; 6b; 6c; 6e), 323(6a; 6b; 6c; 6e), 324(6a), 328(6a; 6b), 330(6e), 331(6e)
Regnström, J., 351
Rehman, A., 190, 299, 301(37)
Rein, D., 351, 354, 354(14), 355(14), 360

Rémésy, C., 102, 115, 118, 119(7), 121, 121(7), 144, 153, 265, 319, 321, 322(6a; 6b; 6c; 6d; 6e), 323(6a; 6b; 6c; 6d; 6e), 324(6a), 328(6a; 6b), 330(6e), 331(6d; 6e)
Remorova, A. A., 320, 330(4e), 331(4e)
Ren, S., 174
Renaud, S. C., 122, 131, 133, 203, 259, 350(9), 351, 369, 376
Repetto, M., 203
Retsky, K. L., 263
Reunanen, A., 103, 369, 379(4)
Reynolds, A. J., 361
Reznick, A. Z., 174, 351
Rhainds, D., 399
Rhodes, L., 361
Ribak, O., 76
Riberéau-Gayon, P., 83
Ricardo da Silva, J. M., 63, 84, 86, 89(17), 338
Rice-Evans, C. A., 6, 7, 8, 16, 24(3), 27, 48, 104, 110, 123, 143, 153, 157(23; 31; 33), 158, 164(31; 33), 170, 171(24), 175, 181, 182, 182(9), 183, 185(20), 186, 187(6), 190, 191(9), 192, 203, 259, 260, 261, 262(25), 263(26), 264, 266, 267(3), 273, 283, 296, 302(10), 303, 303(10), 304, 304(10), 319, 320, 330(2b; 4g), 353, 358, 397
Richard, H., 264, 283
Richell, M., 360
Richelle, M., 107
Richoz, J., 188
Rider, R. C., 27
Rigaud, J., 49, 51(20), 63, 84, 85, 86, 89(17)
Rigaud, M., 351
Rihn, B., 382(16), 385
Rimbach, G., 48, 157(28), 158, 273, 273(11), 275, 333, 380, 381, 382(18), 385, 387, 388, 390(1)
Rimm, E. B., 131
Ringel, B. L., 259
Rinprecht, J. T., 194
Risch, B., 257
Riso, P., 34, 46(14)
Ritter, K. A., 49, 57(19), 351, 355(11)
Rivas-Gonzalo, J. C., 172
Robak, J., 354, 360(46), 379
Roberta, R., 104
Roberts, L. J., 252, 253, 255
Roberts, M. F., 84
Robinson, D., 132, 203

Robinson, E. E., 181, 188(11)
Rocha, R., 175
Rodriguez, J. H., 174, 232
Rodriguez, R., 232
Roedig-Penman, A., 165
Roffi, G., 176
Rogers, I. M., 153
Roginsky, V. A., 320, 330(4e), 331(4e)
Roh, M. H., 297
Rohdewald, P., 275
Rohrscheid, F., 194
Rokach, J., 252
Rolando, C., 89
Romanczyk, L. J., 49, 53(14)
Rombouts, F. M., 257, 258(4)
Rong, J., 338
Rose, I. A., 71
Rosebrough, N. J., 355
Rosen, R., 157
Rosenblat, M., 112, 248, 250, 353
Rosenfeld, M. E., 244
Rosner, B., 131
Ross, A. B., 296(17), 297
Ross, R., 369
Rossi, J. A., 259, 260
Rossoni, G., 354
Rothender, M., 263, 264(50)
Rotheneder, M., 246, 247(7)
Rounova, O., 133, 145
Roussel, A. M., 397
Roux, S. P., 374
Roveri, A., 188
Rowe, G. G., 371
Roy, S., 27
Royall, J. A., 389
Ru, B., 235
Ruberto, G., 295, 320, 330(4d), 331(4d)
Rubin, R., 131
Rubio, J., 297, 304(22; 23)
Rucker, R. B., 49, 57(19), 94, 350, 351, 353(3), 355(11)
Ruf, J. C., 133
Ruiz-Laguna, J., 188
Ruiz-Larrea, M. B., 186
Rump, A. F., 157
Ruperto, J., 157
Russell, L. J., 399
Rustan, I., 140, 144(41)
Ryu, S. Y., 157

S

Sabbarini, G., 31
Saija, A., 294
Saito, N., 204, 261
Saito, Y., 157, 362
Sakai, M., 204
Sakane, T., 354
Sakariassen, K. S., 374
Sakata, K., 175
Salagoïty-Auguste, M.-H., 49, 53(14)
Salah, N., 170, 171(24), 203, 320, 330(4g), 353
Salgues, M., 257
Saliou, C., 380, 381, 382(15; 16; 18), 385, 387
Saliuo, C., 393
Salomon, R. G., 255
Salonen, J. T., 254
Salonen, R., 254
Salvayre, R., 282
Samman, S., 122, 379
Samouilov, A., 296
Sampson, J., 182, 259, 260, 261, 262(25)
Sanada, H., 258, 259(13; 14), 265(13; 14)
Sanbongi, C., 354
Sandermann, H., Jr., 171
Sano, M., 172, 203, 320, 330(4c), 331(4c), 353
Sanoner, P., 57, 58, 58(7), 59, 61(2; 7), 62(2; 7), 66(7), 67(2)
Santos-Buelga, C., 170, 171(25), 172
Sanz, M. J., 157
Sapers, G. M., 323
Saran, M., 97, 158, 164(34), 170, 171(19; 20), 172(19), 175, 175(19; 21), 176, 179(19), 232, 282, 294(5), 319, 330(3b; 3c; 3d), 340
Sarkanen, S., 70(7), 71
Sarkar, S. K., 83
Sarni Manchado, P., 93
Sasaki, K., 172
Sato, R., 74
Sato, S., 361
Sato, T., 385
Satoh, M. S., 308
Sattler, W., 249
Saucier, C., 83, 92(8)
Sauer, M. C., 176
Savouret, J. F., 153
Savva, R., 298
Sawai, Y., 175
Scaccini, C., 259, 260, 283, 289
Scalbert, A., 59, 64(11), 65(11), 89, 91(28)

Scalese, M., 294
Scalia, M., 191(5), 192
Scarrow, R. C., 202
Schantz, S. P., 107
Scheek, L. M., 109
Scheel, D., 78, 79
Schikora, S., 172, 173(30), 177(30), 291
Schipfer, R., 328
Schmid, J., 72
Schmidt, K., 351
Schmitz, H. H., 46, 48, 49, 50(7), 51(7), 57(7; 19), 350, 351, 353(3), 354, 354(13; 14), 355(11; 14), 360
Schrader, J., 275
Schuchmann, H. P., 176
Schulenburg, D. H., 157
Schuler, R. H., 176
Schürmann, W., 79
Schussler, M., 157
Schuster, B., 257
Schwartz, C. L., 351
Scita, G., 288
Sclabert, A., 89
Scott, J., 353
Seabra, R. M., 75
Segrest, J. P., 261
Segura-Aguilar, J., 169, 172(16), 173(16), 188
Sekar, N., 157
Selman, S. H., 217
Semboku, N., 258, 259(14), 265(14)
Sen, C. K., 27
Serafini, M., 125, 333
Serbinova, E. A., 288
Sergent, O., 308, 309(9), 313(9), 315
Serrano Junior, C. V., 157
Serry, M. M., 109, 112(22), 114, 259, 260, 261, 320, 330(4a), 331, 331(4a), 353
Seshadri, T. R., 165
Seto, R., 204
Sevanian, A., 250, 338, 351
Sfakianos, J., 117
Shah, V., 382(13), 385, 392
Shang, A., 399
Shanmuganayagam, D., 157, 369
Shapiro, J., 361
Sharma, O. P., 283
Sharp, D. S., 369
Shaw, I., 120
Shaw, M. R., 64
Shebuski, R. J., 374

Shen, C. P., 382(8), 385
Shen, J.-G., 229
Shen, S.-R., 157(27), 158, 180, 204, 217, 219, 221(11), 223(11), 224(11), 225(11), 232
Shen, W. P., 395
Shen, Z., 84
Sherry, B., 395
Shi, H., 157, 158, 160(35), 181(15), 182
Shigenaga, M. K., 296
Shimada, M., 75
Shimazu, Y., 204
Shimizu, M., 169
Shimizu, T., 78(62), 80(62), 81
Shindo, Y., 174
Shingles, R., 297
Shirahata, S., 169
Shirai, A., 361
Shirley, J. R., 354, 355(50)
Shirota, T., 153
Shlafer, M. J., 229
Shoji, T., 362
Shridi, F., 354, 360(46)
Shuker, D. E. G., 297
Sibonga, J. D., 146
Sichel, G., 157(32), 158, 191(5), 192
Sidorkina, O. M., 298
Sies, H., 354, 389
Sigman, D. S., 232
Silva, M., 175
Silver, J., 195
Simic, M. G., 123, 173, 176, 191(8), 192, 203, 268, 268(12), 271, 319, 330(3e; 3f)
Simonetti, P., 122, 126
Singh, B., 304, 333
Singh, P. K., 276
Singh, S., 382(9), 385
Singhal, R. K., 308
Singleton, V. A., 104, 106(8), 259, 260
Singleton, V. L., 257, 258, 258(8)
Siren, H., 97, 102(7)
Siripitayananon, J., 190
Sitzmann, J. V., 131
Skatchkov, M. P., 276
Skipper, P. L., 275, 296, 302(3)
Skrzyperak-Jankun, E., 217
Slane, P., 376
Slane, P. R., 374, 376(22), 377(22), 379
Slater, T. F., 287
Sloots, L. M., 132
Smeltzer, R. H., 73

Smith, A., 144
Smith, H. H., 94
Snedecor, G. G., 136, 137(38), 138(38)
Sobol, R. W., 308
Sofen, S. R., 202
Sofic, E., 25, 106, 169, 319, 330(2h)
Solakivi, T., 351
Soleas, G. I., 122, 123, 132, 142, 145, 146, 146(1), 151(1), 152(1), 153(1), 154, 154(1)
Soleas, G. J., 18, 130, 132, 133, 140
Solmajer, T., 174
Solvoll, K., 351
Somerville, C., 73
Someya, K., 143
Son, K.-H., 398, 404(3)
Sothy, S. P., 153
Souquet, J.-M., 49, 51(20), 85
Spadone, J. C., 188
Spague, E. A., 351
Spanos, G. A., 49
Spencer, J. P. E., 296, 297, 298(29), 301(28; 29), 304(24; 28; 29)
Squadrito, G. L., 389
St. Leger, A. S., 131, 381, 397
Stadler, R. H., 188
Stadtman, E. R., 255
Stafford, H. A., 167, 172(4)
Stahl, S., 146
Stalenhoef, A. F., 132
Stampfer, M. J., 130, 131
Stanbury, D. M., 296
Steenken, E., 203, 268, 268(12), 271, 319, 330(3a; 3e; 3f)
Steenken, S., 123, 173, 191(8), 192
Stein, G., 176
Steinberg, D., 97, 108, 132, 244, 259, 261, 351, 353, 360, 397
Steinbrecher, U. P., 261, 353
Steiner, E., 263
Stenken, S., 289
Stensvold, I., 351
Stern, A. M., 232
Sterson, M., 351
Stettmaier, K., 27, 166, 167, 169(2), 170(2), 172, 175, 179(36), 180(36)
Stewart, D. B., 174
Stewart, D. E., 351
Steynberg, J. P., 64, 94
Sticher, O., 391
Stocker, R., 249, 288, 295, 333

Stöckigt, J., 80
Stonestreet, E., 83
Strack, D., 70, 71, 71(3), 77, 77(65), 78, 79, 80, 80(63), 81, 84(64)
Stradi, R., 144, 146, 151(12), 153(12; 13)
Strafford, H. A., 338
Strain, S., 266
Strasburg, G. M., 186
Stremple, P., 120
Striegl, G., 246, 247(7), 263, 264(50), 284
Strube, M., 175, 183
Stuehr, D. J., 273
Stutte, C. A., 72
Su, S. B., 395
Su, X., 106, 107(10), 109(10), 112(10)
Suaez, B., 58
Suganuma, M., 181
Sugata, R., 283
Sugimura, T., 169, 188
Sugiura, K., 180
Sullivan, J. J., 187
Sun, T.-T., 361
Sunamoto, J., 294
Sunatomo, J., 181
Suschetel, M., 187
Suter, P. M., 131
Suturkova, L., 174
Suzukawa, M., 353
Suzuki, M., 172, 174, 204
Suzuki, N., 191(7), 192, 354
Suzuki, T., 206, 211(20), 212, 283, 298
Suzuki, Y. J., 392
Svendsen, C. N., 121
Swain, T., 83, 333
Sweidan, H., 157
Swierez, R., 217
Swies, J., 379
Sylvester, P. W., 401
Syms, J., 132, 203
Sytnik, A., 324

T

Tabengwa, E. M., 131
Tacchini, M., 58
Tagesson, C., 232
Taha, Z., 275
Tait, B., 340
Tait, D., 176

Tajaddini-Sarmadi, J., 184, 188(23)
Tajima, K., 298
Takahama, U., 292
Takahashi, N., 385
Takahashi, T., 361, 362
Takama, K., 283
Takamura, H., 181
Takei, Y., 206, 320, 330(4f), 331(4f)
Takeuchi, W., 74, 75(34)
Tamaoki, T., 361
Tamargo, J., 354
Tamasaki, H., 292
Tamir, S., 297
Tanabe, M., 157, 362
Tanaka, M., 73
Tanaka, T., 169, 170(7), 171(7)
Tanigaki, N., 364
Tanigaki-Obana, N., 361
Tanigawa, M., 275, 276, 276(21)
Tanigawa, T., 275, 276, 276(21)
Taniguchi, H., 176
Tanimura, R., 287
Tannenbaum, S. R., 275, 296, 297, 298, 301, 302(3)
Tanner, R. J. N., 48, 59, 83, 363
Tavazzi, I., 107, 360
Taya, M. S., 165
Taylor, M. C., 195
Teissedre, A. L., 165, 259, 359
Teissedre, P.-L., 49, 143, 181, 259, 263(30), 264(30), 376
Teng, C. M., 145, 157, 379
Terao, J., 117, 169, 181, 206, 283, 294, 320, 330(4f), 331, 331(4f)
Terencio, M. C., 157
Territo, M. C., 351
Teshima, R., 157, 362
Testolin, G., 122
Tetsuka, T., 382(15), 385, 393
Teukolsky, S. A., 198
Teutsch, H., 73, 74(32)
Texier, O., 102, 121, 144, 153, 321, 322(6a; 6b; 6c; 6e), 323(6a; 6b; 6c; 6e), 324(6a), 328(6a; 6b), 330(6e), 331(6e)
Tham, L., 140
Theile, J. J., 174
Theilmann, D. A., 76
Theisohn, M., 157
Theriault, A., 398, 404(2)
Thibault, J.-F., 257, 258, 258(4)

Thomas, A., 369
Thomas, J., 286
Thomas, M. B., 3, 13(5)
Thompson, J., 83, 363
Thompson, K. G., 131, 133(16)
Thompson, R. S., 48, 59
Thorpe, G. H., 132, 181, 188(11)
Tie, J., 235
Tijburg, L., 109, 170, 171(24), 203, 320, 330(4g), 353
Tikkanen, M. J., 112
Tilbrook, G. S., 197, 199(26; 27), 202(27)
Tillement, J.-P., 144, 146, 151(12), 153(12; 13)
Timberlake, C. F., 257, 258(8)
Timmer, R. T., 298
Tokudome, S., 307
Tomas-Barberan, F. A., 14
Tomassi, G., 259, 260
Tomita, I., 169, 172(15), 173(15), 203, 320, 330(4c), 331(4c), 353
Tomita, T., 203, 320, 330(4c), 331(4c), 353
Toniolo, R., 172, 338
Tornquist, K., 27
Tornvall, P., 351
Torres, R., 183
Toshima, H., 103
Tosic, M., 191(8), 192, 268
Toyoda, M., 157, 174, 206, 211(20), 212, 362
Traber, M., 381
Tran, K., 399
Trevisan, M. T., 351
Trinajstic, N., 174
Trivedi, B. K., 340
Tromp, M. N., 157(25), 158, 175, 197
Trost, B. M., 346
Trousdale, E., 257, 258
Tsai, S. H., 157, 382(19), 385
Tsao, C. C., 73
Tsikas, D., 296
Tsuchihashi, H., 166
Tsuchiya, J., 160, 287
Tsuchiya, K., 276
Tsugawa, M., 181
Tsushida, T., 206, 320, 330(4f), 331(4f)
Tsuyumu, S., 180
Tubaro, F., 172, 338, 340
Tuckman, 351
Tuo, J., 190
Turesky, R. J., 188
Turitto, V. T., 374

Turner, R. T., 146
Tverdal, A., 351
Tyson, C. A., 295
Tzeng, S. H., 157, 379

U

Uchida, K., 255
Ueda, J., 204
Uehara, I., 204
Ueno, I., 275, 276, 276(21)
Umezawa, T., 73
Uno, H., 361
Urien, S., 144, 146, 153(13)
Ursini, F., 172, 188, 338, 340
Urzua, A., 183
Utrilla, P., 354

V

Vachereau, A., 144
Valacchi, G., 380, 382(18), 385
Valente, A. J., 351
Van, N. T., 382(17), 385
van Acker, S. A. B. E., 157(25), 158, 175, 197
van Beek, T. A., 27
van Bennekom, W. P., 157(25), 158, 197
Vanberkel, T. J. C., 351
van Bladeren, P. J., 190
Van Brederode, J., 73
Vande Casteele, K., 6, 13(13), 16(13)
van den Berg, D. J., 157(25), 158, 175, 197
van den Berg, H., 175, 183
van den Berg, R., 175
Vanden Berghe, D., 319, 330(2c)
VanDenburgh, M., 131
van der Brandt, P. A., 307
van der Gaag, M. S., 98, 103(11), 107
Vanderhelm, D., 190
Van der Laarse, A., 351
van der Putte, B., 323
van der Veen, C., 361
van der Vijgh, W. J. F., 157(25), 158, 175, 197
van der Vliet, A., 296, 304
Van De Water, J., 351, 354(13)
van de Wiel, A., 132
Van Gaal, L., 109
van Gameren, Y., 98, 103(9), 154

Van Golde, P. H., 132
van Jaarsveld, H., 157
VanKessel, J., 340
van Leeuwen, S. D., 98, 103(12), 143, 144(50)
Van Nigtevecht, Z., 73
Van Poel, B., 319, 330(2c)
van Poppel, G., 190
van Staveren, W. A., 98, 103(10)
Van Sumere, C., 6, 13(13), 16(13)
van Tol, A., 109
van Trijp, J. M. P., 98, 101(8), 102(8), 103(11), 107, 117, 144
Varilek, G. W., 157, 382(11), 385
Varner, J. E., 76
Vaughan, P. F. T., 73
Vaughn, K. C., 74, 75(35)
Vaya, J., 112, 244, 248, 250, 353
Venema, D. P., 17, 97, 101(6), 102(6), 117
Venkatramani, L., 190
Vercauteren, J., 82
Verdon, C. P., 266
Verhagen, H., 190
Verlangieri, A. J., 350(8), 351
Vermeulen, W. P., 132
Verschuren, P. M., 73
Vertommen, J., 109
Verwey, H. F., 351
Vetter, W., 131, 176
Vetterling, W. T., 198
Viana, M., 188
Vieira, O., 282, 283, 284, 286, 287(25), 292(16)
Vihma, V., 112
Villalain, J., 295
Vinson, J. A., 103, 106, 107(9–12), 109, 109(10; 11), 110, 110(9; 10), 112, 112(9; 10; 22; 23), 114, 259, 260, 261, 320, 330(4a), 331, 331(4a), 353, 397
Virgili, F., 48, 157(28), 158, 273, 273(10; 11), 275, 333, 338, 354, 381, 387, 388, 390, 390(1), 395
Vita, J. A., 122
Vitali, P., 48
Vivas, N., 83, 92(8)
Vizcaino, F. P., 354
Vlietinck, A. J., 319, 330(2c)
Vogel, G. C., 194
Vogelezang, C. J. M., 351
Vogt, T., 79

von Sonntag, C., 176
Voruela, H., 27

W

Waagstein, F., 296
Wadsworth, T. L., 157
Waeg, G., 284
Wag, G., 122, 123(2)
Wagatsuma, K., 361
Wagner, D. A., 275, 296, 302(3)
Wagner, H., 5
Wagner, P., 263
Wahala, K., 112
Wah Kow, Y., 308
Wakabayashi, K., 188
Wakabayashi, Y., 131
Walker, R., 187
Wallace, S. S., 308
Walle, T., 323, 329(12), 330(12), 331(12)
Walle, U. K., 323, 329(12), 330(12), 331(12)
Wallet, J. C., 157(29), 158, 164, 319, 330(2g)
Walters, C. L., 275
Walzem, R. L., 107, 115, 116, 118(1), 121(1), 265
Walzem, R. M., 143
Wan, Q., 229
Wang, H., 181, 266
Wang, I. K., 180
Wang, J., 165
Wang, J.-C., 217
Wang, J. M., 395
Wang, P., 296
Wang, Q., 398, 404(2)
Wang, W., 113
Wang, Z. Y., 232
Wardman, P., 177
Wargovich, M. J., 217
Watanabe, J., 353
Watanabe, S., 361
Watanabe, T., 169, 172(15), 173(15), 203, 320, 330(4c), 331(4c)
Waterhouse, A. L., 15, 107, 115, 116, 118(1), 120, 121(1), 122, 143, 165, 181, 258, 259, 259(12), 260, 263(12; 30), 264(12; 30), 265, 265(12), 354, 355(50), 358, 359, 376
Watson, A. D., 244, 397
Webb, A., 285
Wei, C. I., 169

Wei, H., 232
Wei, T., 232
Wei, Z., 382(13), 385, 392
Weiland, E., 360
Weisfeldt, M. L., 229
Weissenböck, G., 77, 78, 80
Weitzberg, E., 296, 302(9)
Wekell, M. M., 187
Wellmann, E., 73, 76
Wellons, J. D., 62
Weltring, K.-M., 70, 71(3)
Wendisch, D., 81
Wennmalm, A., 296
Werck-Reichhart, D., 73, 74(32)
Westerlund, C., 288
Westfelt, U. N., 296
Wetti, D. H., 188
Weyhenmeyer, J. A., 276, 279(24)
Wheeler, D. E., 174
White, J. D., 62
Whitehead, T. P., 132, 203
Whiteman, M., 296, 297, 298(29), 301(28; 29), 304, 304(28; 29)
Widmer, U., 395
Wiegand, J., 197
Wielders, J. P., 132
Wiermann, R., 77
Wiid, N. M., 157
Wilkinson, T., 261
Willems, M. I., 190
Willerson, J. T., 374
Willett, W. C., 131
Williams, C. A., 17, 70(5), 71
Williams, N. R., 123, 132
Williams, V. M., 84, 87(22)
Williams, W. J., 134, 138(36), 141(36)
Williamson, G., 170, 171(25), 258
Wilson, E. L., 48
Wilson, M., 261
Wilson, R. D., 5
Wilson, R. L., 287
Wilson, S. H., 308
Wink, D. A., 296
Winter, R., 176
Wintrobe, M. M., 371
Wiseman, H., 232
Wiseman, S. A., 109, 181
Wishnock, J. S., 275, 296, 297, 298, 302(3)
Witiak, D. T., 340
Witte, L., 80

Witting, P. K., 288
Witztum, J. L., 108, 132, 244, 254, 259, 261, 351, 353, 397
Wolbis, M., 354, 360(46)
Wolfbeis, O. S., 321, 322(7), 323(7), 328, 328(7)
Wolleb, U., 188
Wollenweber, M., 197
Wong, G., 157
Wong, J., 297, 304(24)
Wong, T. Y., 169
Wong, Y. F., 175
Wood, J. L., 399
Woolf, S. P., 184, 188(23)
Wray, V., 81
Wrolstad, R. E., 49
Wu, A., 266
Wu, B., 188, 271
Wu, M., 340
Wu, X., 399
Wun, T., 354

X

Xia, G. Z., 250
Xin, W.-J., 157(27), 158, 180, 204, 217, 219, 219(12; 13), 220, 221(11), 223(11), 224(11), 225(11), 226(13), 227(13), 228(13), 229, 230, 231(13; 23), 232
Xu, P. P., 143
Xu, Y., 190

Y

Yahagi, T., 169
Yamada, K., 169, 263
Yamada, M., 157(30), 158, 160, 298
Yamaguchi, T., 181
Yamakoshi, J., 340
Yamamoto, N., 254, 331
Yamamoto, S., 361
Yamanaka, N., 188
Yan, J., 130, 145, 146(1), 151(1), 152(1), 153(1), 154(1)
Yan, L. J., 275, 392
Yan, Q., 333
Yanagida, A., 84, 94(25), 157, 362
Yang, C. S., 107, 217, 232
Yang, F., 157, 382(11), 385

Yang, F.-J., 217
Yang, J., 114
Yang, J.-P., 382(15), 385, 393
Yang, M., 104, 175, 183, 185(20), 266, 267(3)
Yang, W.-D., 229
Yanishlieva, N. V., 264
Yankah, E., 131
Yao, Z., 399
Yaya, J., 353
Ye, M. W., 395
Ye, Z. H., 76
Yeap, F. L., 197
Yermilov, V., 297, 304(22; 23)
Ying, L., 319, 330(2c)
Yla-Herttuala, S., 244, 254
Yochum, L., 97
Yokoo, Y., 362
Yokozawa, T., 169, 170(7), 171(7)
Yonemura, A., 353
Yoneyama, A., 385
Yoo, G. S., 157
Yordanov, N. D., 169, 175(14)
Yoshida, M., 298
Yoshida, T., 204, 353
Yoshida, V. M., 157
Yoshie, Y., 297, 304(22; 23)
Yoshihira, K., 169
Yoshino, K., 172
Yoshioka, H., 180
Yoshizumi, M., 276
Young, J. F., 190
Yow, S., 49, 53(14)

Z

Zamburlini, A., 188
Zarzuela, A., 354
Zaya, J., 257, 258
Zeind, J., 263
Zenk, M. H., 72, 77, 80, 84
Zgainski, E. M., 5
Zhang, A., 109
Zhang, H. Y., 174
Zhang, J.-Z., 229, 232, 233, 234(18), 236(15; 18), 243(18)
Zhang, M., 146
Zhang, X., 232
Zhang, Y., 232, 236(15)
Zhang, Z. S., 175, 232, 236(15)
Zhao, B.-L., 157(27), 158, 180, 204, 217, 219, 219(12; 13), 220, 221(11), 223(11), 224(11), 225(11), 226(13), 227(13), 228(13), 229, 230, 231(13; 23), 232
Zhao, J., 232
Zhao, K., 296, 297, 302(10), 303(10), 304(10)
Zhao, Y., 230, 231(23)
Zhou, Z. C., 232
Zhu, H.-L., 229
Zhu, Q. Y., 175
Zhu, Z., 144
Zimmermann, E., 79
Zock, P. L., 98, 103(10)
Zou, X.-L., 229
Zubik, L., 106, 107(9; 10), 109(10), 110(9), 112(9; 10)
Zweier, J. L., 229, 296

Subject Index

A

AAPH, see 2,2′-Azinobis(2-amidinopropane) hydrochloride
ABAP, see 2,2′-Azinobis(2-amidinopropane) hydrochloride
ABTS, see 2,2′-Azinobis(3-ethylbenzothiazoline 6-sulfonate)
Acid hydrolysis, flavonoid analysis, 12–13, 17, 117
1-O-Acylglucose acyltransferase, hydroxycinnamoyltransferase
  assay, 80
  extraction, 79
  purification, 80
Aggregometry, see Platelet function, flavonoid effects
Albumin–flavonoid binding
  antioxidant activity effects, 330–331
  chrysin, 327
  fluorescence titration for binding constant determination
    flavonoid fluorescence enhancement, 321–322, 328
    fluorescent albumin ligand changes, 322–323
    intrinsic tryptophan fluorescence, 324
  3-hydroxyflavone, 324–325
  prevalence of complexes, 329
  quercetin
    autoxidation prevention, 329–330
    binding affinity, 321, 323–324, 330
    fluorescence assay, 324
    isoquercitrin, 326–327
    low-density lipoprotein protection consequences, 331–333
    stoichiometry, 324
    structural changes on albumin binding, 328–329
    sulfoglucuronides, 325–326
    warfarin competition studies, 324–325
  radioassay, 333–334
  tannin catechin, 328

Alkaline hydrolysis, flavonoid analysis, 12–13
Aminochelin–iron(III) interaction, spectrophotometric analysis, 199–200, 202
ApoB, see Apolipoprotein B
Apolipoprotein B
  HepG2 cell system for antioxidant response studies
    cell culture, 399–400
    enzyme-linked immunosorbent assay of apolipoprotein B, 400
    factors affecting lipoprotein turnover, 399
    hesperetin response, 402–404
    naringenin response, 402–404
    proliferation assay with 3-(4,5-dimethylthiazol-2-yl)-2,5-diphenyltetrazolium bromide, 401
    stock solution preparation, 400–401
    tocopherol and tocotrienol response, 402–404
  low-density lipoprotein association, 399
Apple, procyanidin analysis with reversed-phase high-performance liquid chromatography chromatography
  conditions, 61
  peak identification, 61–62
  direct thiolysis of samples
    comparison withh other techniques, 66–68
    direct solvent extraction before thiolysis, 61
    freeze-dried samples, 59–61
    rationale, 58, 69–70
    yield, 64–66
  juices versus freeze-dried musts, 68–69
  polyphenol classes and cider quality, 57–58
  quantitative analysis
    native samples, 66
    thiolyzed samples, 64–66
  sampling, 58–59
  structural characterization, 62–64
Ascorbic acid, caffeic acid redox cycles and low-density lipoprotein oxidation inhibition
  absorption spectroscopy, 284–285

dynamic interaction with α-tocopherol and ascorbic acid, 292–295
electron paramagnetic resonance, 285, 288, 290–292
experimental approach, 283–284
low-density lipoprotein preparation, 284
redox interactions, 289–292
reversed-phase high-performance liquid chromatography, 285–286
Atherosclerosis, see Coronary heart disease; Endothelial cell–macrophage interaction; Lipoprotein oxidation
2,2′-Azinobis(2-amidinopropane) hydrochloride, catechin studies of radical scavenging, 225–226
2,2′-Azinobis(3-ethylbenzothiazoline 6-sulfonate)
 antioxidant assays, overview, 182–183
 food and beverage antioxidant testing
  persulfate system assay, 185–186
  preservative effects, 187–188
  Trolox as standard, 183–184
 rapid Trolox equivalent antioxidant capacity assay, 267

B

Blood flavonoid analysis
 bioavailability using reversed-phase high-performance liquid chromatography with postcolumn fluorescence derivatization
  advantages, 103
  chromatography conditions, 99–100
  extraction, 99, 102
  overview, 97–98
  postcolumn reaction, 101
  precision, 100–101
  recovery, 101
  sensitivity, 101
  structural requirements for complexation, 102
 conjugate analysis in rat
  acid hydrolysis, 117
  antioxidant activities of conjugates, 121
  dietary effects, 116–117
  enzymatic hydrolysis, 117–118
  extraction, 116
  gas chromatography/mass spectrometry, 121
  reversed-phase high-performance liquid chromatography
   electrochemical detection, 120–121
   ultraviolet detection, 120
  sampling, 115–116
  standards, preparation
   catechol O-methyltransferase modification, 119–120
   glucuronidation with microsomal preparation, 118–119
   glucuronidation with pure enzyme, 119
   O-methylation with chemicals, 120
   sulfation with cytosolic liver extract, 119
 gas chromatography/mass spectrometry of red wine polyphenols
  advantages, 144–145
  calibration curve, 134–136
  distribution among blood components, 140–141
  linearity, 137–138, 143
  precision, 138–140
  rationale, 133
  recovery, 136–137, 143
  trans-resveratrol absorption study, 141–142
  running conditions, 134
  sample preparation and derivatization, 133–134
  sensitivity, 136
 plasma protein binding of flavonoids, see Albumin–flavonoid binding
 platelet function, see Platelet function, flavonoid effects

C

Caffeic acid, see also Hydroxycinnamic acids
 food distribution, 282
 plasma detection
  absorption, 127–128
  dose–response relationship, 128
  reagents, 123–124
  recovery, 125
  reversed-phase high-performance liquid chromatography, 124–127
  sample preparation, 124
  standards, 124

total radical-trapping antioxidant parameter
determination, 125–126, 128–129
volunteers, 123
red wine consumption marker, 123, 130
redox cycles and low-density lipoprotein
oxidation inhibition
absorption spectroscopy, 284–285
ascorbic acid redox interactions, 289–292
dynamic interaction with α-tocopherol and
ascorbic acid, 292–295
electron paramagnetic resonance, 285, 288,
290–292
experimental approach, 283–284
low-density lipoprotein preparation, 284
reversed-phase high-performance liquid
chromatography, 285–286
α-tocopheroxyl radical reduction by caffeic
acid, 286–289
structure–radical scavenging activity
relationship, 282–283
Caffeic acid/5-hydroxyferulate
O-methyltransferase
assay, 76
extraction, 75
Caffeoyl-CoA/5-hydroxyferuloyl-CoA
O-methyltransferase, assay and
purification, 76
Camellia sinensis, see Green tea
Capillary electrophoresis, medicinal plant
flavonoid analysis, 30–31
Catechins, *see also* Procyanidins
antioxidant actions of epicatechin, epicatechin
gallate, epigallocatechin, and
epigallocatechin gallate
comparison of scavenging activities, 214,
217
cyclic voltammetry, 205, 213
DNA damage prevention, *see* DNA
oxidative damage
electron spin resonance studies of radical
scavenging
2,2′-azinobis(2-amidinopropane)
hydrochloride radicals, 225–226
1,1-diphenyl-2-picrylhydrazyl radical,
226–227
hydroxyl radicals from Fenton reaction,
219–220
hydroxyl radicals from hydrogen
peroxide photolysis, 220–221
iron chelation effects, 231

lipid radicals from iron-induced
peroxidation of synaptosome,
334–225
lipid radicals from lipoxidase-catalyzed
peroxidation of lecithin, 223–224
methyl radical from peroxynitrite
oxidation of dimethyl sulfate,
230–231
oxygen radicals from
ischemic–reperfusion myocardium,
229–230
oxygen radicals from phorbol myristate
acetate-stimulated
polymorphonuclear leukocytes,
222–223
singlet oxygen, 227–228
superoxide radicals from
riboflavin/EDTA irradiation,
221–222
liposomal phospholipid peroxidation, 206
mass spectrometry, 204, 207, 209, 211
molecular orbital calculations of bond
dissociation enthalpy, 205, 212–213
superoxide formation assay, 205, 212
ultraviolet–visible spectroscopy, 204–205,
209, 211
(+)-catechin
excretion, 153
gas chromatography/mass spectrometry of
human blood or urine
advantages, 144–145
calibration curve, 134–136
distribution among blood components,
140–141
linearity, 137–138, 143
precision, 138–140
rationale, 133
recovery, 136–137, 143
running conditions, 134
sample preparation and derivatization,
133–134
sensitivity, 136
DNA damage by nitrogen compounds,
protection
nitrite-induced deamination, 302–303
peroxynitrite damage, 304, 306
structural–antioxidant relationship, 24, 214,
217, 219, 231, 243
Chemiluminescence, DNA oxidative damage
assay of green tea polyphenol protection

advantages and limitations, 243
calculations, 237–238
comparison of polyphenols, 239
DNA oxidation, 235
factors affecting chemiluminescence, 236–237
incubation conditions, 235
principle, 234–235
specificity, 243
Chocolate, see Cocoa
Cholesterol
  flavonoid effects on levels, see Apolipoprotein B
  modulation of levels by apolipoproteins, 398–399
Chrysin, albumin binding, 327
Cider, see Apple
β-D-Cinnamate glucosyltransferase, purification, 78
Cinnamate 4-hydroxylase
  extraction and purification, 73–74
  function, 72–73
Cocoa
  lipoprotein oxidation protection assay with procyanidin oligomers
    cell-mediated oxidized lipoprotein effects, 359–360
    copper-oxidized lipoprotein effects, 357–360
    low-density lipoprotein
      cell-mediated oxidation, 356–357
      copper-catalyzed oxidation, 356
      preparation, 355
    materials, 355
    overview, 354–355
  polyphenol content and health benefits, 353–354
  procyanidin analysis with reversed-phase high-performance liquid chromatography/mass spectrometry
    extraction, 51, 53
    gel-permeation chromatography of extract, 53
    oligomer analysis, 50–51, 54–55
    preparative normal phase high-performance liquid chromatography, 53–54
Coenzyme A thioester acyltransferase, hydroxycinnamoyltransferase assay, 79
  extraction, 78
  purification, 78–79
Computer-automated structure evaluation, structure–antioxidant activity relationships of polyphenols, 174
Conjugated dienes assay, oxidized low-density lipoprotein, 246
Coronary heart disease, see also Lipoprotein oxidation
  flavonoid prevention, 97, 103, 112, 122, 319–321, 350–353, 369, 397
  French paradox, 122, 130–131
  low-density lipoprotein oxidation and pathogenesis, 244–245, 351–352, 398
4-Coumarate 3-hydroxylase
  assay, 75
  extraction, 74
  function, 74
4-Coumarate-CoA ligase
  assay, 76–77
  function, 76
4-Coumaroyl-CoA 3-hydroxylase, extraction and assay, 75
Cyclic voltammetry, catechin antioxidant action studies, 205, 213

D

Dimethylaminocinnamaldehyde, proanthocyanidin polymerization assay, 83, 86, 92–94
1,1-Diphenyl-2-picrylhydrazyl radical
  detection of quenching, 175
  electron spin resonance, catechin studies of radical scavenging, 226–227
  polyphenol scavenging assay, 169–170, 181
DNA damage, nitrogen oxides
  hypochlorous acid-dependent damage by nitrite, 304
  nitric oxide oxidation products, 296–297
  nitrite-induced deamination
    gas chromatography/mass spectrometry, 299
    kinetics of product formation, 299, 301
    mechanisms, 297–298
    protection by polyphenols, 302–303
    reversed-phase high-performance liquid chromatography of products, 298–299
  peroxynitrite damage

SUBJECT INDEX

8-nitroguanine standard preparation, 306–307
   protection with polyphenols, 304, 306
DNA oxidative damage
   biomarkers and detection, 233
   green tea polyphenol protection
      chemiluminescence assay
         advantages and limitations, 243
         calculations, 237–238
         comparison of polyphenols, 239
         DNA oxidation, 235
         factors affecting chemiluminescence, 236–237
         incubation conditions, 235
         principle, 234–235
         specificity, 243
      electron spin resonance assay
         calculations, 239
         comparison of polyphenols, 239, 242
         incubation conditions, 238
         mechanisms, 243
         overview, 232–233
         reagents for assay, 234
   mutagenesis, 308
   myricetin stimulation of repair
      DNA oxidation assay
         derivatization, 311
         extraction, 310
         gas chromatography/mass spectrometry, 312–313, 316
         hydrolysis, 310–311
         myricetin response, 313–314
      hepatocyte culture and treatment, 309
      materials, 309
      overview, 308–309
      repair assay, 311
      repair enzyme expression, RNA blot analysis and myricetin response, 312–316
   1,10-phenanthroline–copper induction, 232–233, 235, 242
DPPH, *see* 1,1-Diphenyl-2-picrylhydrazyl radical

E

EGb 761 extract, *see* Ginkgo biloba
Electrochemical detection, flavonoids in high-performance liquid chromatography
   advantages, 16
   amperometric electrochemical detection principle, 19–20
   aromatic substitution relationship with voltammetric properties, 23–24
   array configuration, 20–21
   channel ratios, 21
   fingerprinting, 25
   instrumentation and chromatography, 18–19
   oxidation potentials, 21, 23–24
   resolution factors, 22–23
   sensitivity, 26, 97
   structural–antioxidant relationship, 24
   total redox signal versus total oxidative radical absorbency capacity, 25–26
Electron paramagnetic resonance
   caffeic acid redox cycles and low-density lipoprotein oxidation inhibition
      absorption spectroscopy, 284–285
      ascorbic acid redox interactions, 289–292
      dynamic interaction with α-tocopherol and ascorbic acid, 292–295
      electron paramagnetic resonance, 285, 288, 290–292
      experimental approach, 283–284
      low-density lipoprotein preparation, 284
      reversed-phase high-performance liquid chromatography, 285–286
      α-tocopheroxyl radical reduction by caffeic acid, 286–289
   DNA oxidative damage assay of green tea polyphenol protection
      calculations, 239
      comparison of polyphenols, 239, 242
      incubation conditions, 238
   green tea catechin studies of radical scavenging
      2,2′-azinobis(2-amidinopropane) hydrochloride radicals, 225–226
      1,1-diphenyl-2-picrylhydrazyl radical, 226–227
      hydroxyl radicals from Fenton reaction, 219–220
      hydroxyl radicals from hydrogen peroxide photolysis, 220–221
      lipid radicals from iron-induced peroxidation of synaptosome, 334–225
      lipid radicals from lipoxidase-catalyzed peroxidation of lecithin, 223–224
      methyl radical from peroxynitrite oxidation of dimethyl sulfate, 230–231

oxygen radicals from ischemic–reperfusion myocardium, 229–230
oxygen radicals from phorbol myristate acetate-stimulated polymorphonuclear leukocytes, 222–223
singlet oxygen, 227–228
superoxide radicals from riboflavin/EDTA irradiation, 221–222
nitric oxide assay in macrophages
activation of macrophages, 277–279
arginine-dependence of signal, 279
cell culture, 276–277
data acquisition, 277–278
flavonoid effects
advantages of assay, 282
Ginkgo biloba EGb 761 extract, 280–281
incubation conditions, 277
pine bark extract, 280–282
iron–dithiocarbamate complex spin trap agent preparation, 277
lysine effects, 279
nitric oxide synthase inhibitor effects, 279
spin traps, 275–276
superoxide dismutase treatment effects, 280
polyphenol antioxidant assays, 179–180
Electrophoretic mobility shift assay, flavonoid suppression of nuclear factor-κB
binding reaction, 384–385
electrophoresis, 384–385
nuclear extract preparation, 383–384
ELISA, *see* Enzyme-linked immunosorbent assay
EMSA, *see* Electrophoretic mobility shift assay
Endothelial cell–macrophage interaction, coculture model
antioxidant levels, 389
applications, 387
atheroma modeling, 387, 396–397
cell lines, 390
generation of reactive oxygen and nitrogen species, 388–389
human umbilical vein endothelial cell culture, 390–391
monocyte chemotactic protein-1 induction and *Ginkgo biloba* EGb 761 extract inhibition, 395
nuclear factor-κB activation
endothelial cell activation upon coculture, 392–393

Ginkgo biloba EGb 761 extract inhibition, 391–395
RAW 264.7 murine monocyte–macrophage culture, 391
Enzymatic hydrolysis, flavonoid analysis, 13
Enzyme-linked immunosorbent assay
apolipoprotein B, 400
oxidized low-density lipoprotein, 254–255
Epicatechin, *see* Catechins; Green tea; Procyanidins
Epicatechin gallate, *see* Catechins; Green tea; Procyanidins
Epigallocatechin, *see* Catechins; Green tea
Epigallocatechin gallate, *see also* Catechins; Green tea; Procyanidins
aroxyl radical fate and reactivity, 172
EPR, *see* Electron paramagnetic resonance

F

Ferric ion reducing antioxidant potential, applications, 266
Folin–Ciocalteu reagent, polyphenol determination in foods
color development, 105
interfering agents, 105–106
overview, 104
oxygen radical absorbance capacity assay correlation, 106–107
sample preparation, 105
Folts cyclic flow model, *see* Platelet function, flavonoid effects
FOX assay, hydrogen peroxide determination in foods, 184–185, 188–189
FRAP, *see* Ferric ion reducing antioxidant potential
French maritime pine bark extract, *see* Pine bark extract

G

Galvinoxyl assay, *see* Hydrogen-donating activity, polyphenols
Gas chromatography/mass spectrometry
deaminated DNA markers, 299
DNA oxidation assay, 311–313, 316
flavonoid conjugate analysis in body fluids, 121

SUBJECT INDEX

isoprostanes, 252–254
oxysterols, 251
red wine polyphenols in human blood or urine
  advantages, 144–145
  calibration curve, 134–136
  distribution among blood components, 140–141
  linearity, 137–138, 143
  precision, 138–140
  rationale, 133
  recovery, 136–137, 143
  trans-resveratrol absorption study, 141–142
  running conditions, 134
  sample preparation and derivatization, 133–134
  sensitivity, 136
Gel filtration chromatography, proanthocyanidin polymerization evaluation, 84, 87, 89–90, 94
Ginkgo biloba
  beneficial compounds, 27, 275
  electrospray ionization mass spectrometry, flavonoid analysis
    fingerprint, 41
    instrumentation, 41
    sample preparation, 36
  endothelial cell–macrophage interaction coculture model effects of EGb 761 extract
    monocyte chemotactic protein-1 inhibition, 395
    nuclear factor-κB inhibition, 391–395
  nitric oxide suppression in macrophages with EGb 761 extract, 280–281
  reversed-phase high-performance liquid chromatography, flavonoid analysis
    peak identification, 37–38
    quantitative analysis, 37–38
    sample preparation, 36
    solvent system, 36
  xanthine oxidase, polyacrylamide gel electrophoresis of EGb 761 extract interactions, 335, 337
Green tea
  antioxidant actions of epicatechin, epicatechin gallate, epigallocatechin, and epigallocatechin gallate
    comparison of scavenging activities, 214, 217
    cyclic voltammetry, 205, 213

DNA damage prevention, *see* DNA oxidative damage
electron spin resonance studies of radical scavenging
  2,2′-azinobis(2-amidinopropane) hydrochloride radicals, 225–226
  1,1-diphenyl-2-picrylhydrazyl radical, 226–227
  hydroxyl radicals from Fenton reaction, 219–220
  hydroxyl radicals from hydrogen peroxide photolysis, 220–221
  iron chelation effects, 231
  lipid radicals from iron-induced peroxidation of synaptosome, 334–225
  lipid radicals from lipoxidase-catalyzed peroxidation of lecithin, 223–224
  methyl radical from peroxynitrite oxidation of dimethyl sulfate, 230–231
  oxygen radicals from ischemic–reperfusion myocardium, 229–230
  oxygen radicals from phorbol myristate acetate-stimulated polymorphonuclear leukocytes, 222–223
  singlet oxygen, 227–228
  superoxide radicals from riboflavin/EDTA irradiation, 221–222
liposomal phospholipid peroxidation, 206
mass spectrometry, 204, 207, 209, 211
molecular orbital calculations of bond dissociation enthalpy, 205, 212–213
superoxide formation assay, 205, 212
ultraviolet–visible spectroscopy, 204–205, 209, 211
electrospray ionization mass spectrometry, flavonoid analysis
  fingerprint, 42–44
  instrumentation, 41
  sample preparation, 36
polyphenol antioxidants, 181, 217, 232
polyphenol health benefits, 217
reversed-phase high-performance liquid chromatography, polyphenol analysis
  peak identification, 39

quantitative analysis, 39
sample preparation, 36
solvent system, 38

H

Hair growth stimulation, oligomeric procyanidins
  anagen phase induction assay, 366
  epithelial cell growth-promoting activity, 366–367
  mechanism of action, 361–362
  models for evaluation, 361
  mouse epidermal keratinocyte
    isolation and culture, 365
    procyanidin effects, 367
    proliferation assay with 3-(4,5-dimethylthiazol-2-yl)-2,5-diphenyltetrazolium bromide, 365–366
  mouse hair epithelial cell isolation and culture, 364–365
  mouse model for *in vivo* hair growth evaluation, 368
  procyanidin purification
    procyanidin B2, 362–364
    procyanidin C1, 362–364
  topical agent preparation, 366
Herbal remedies, *see* Medicinal plants
Hesperetin, apolipoprotein B response in HepG2 cells, 402–404
High-performance liquid chromatography, *see also* Reversed-phase high-performance liquid chromatography
  columns for flavonoid analysis, 18
  electrochemical detection, *see* Electrochemical detection, flavonoids in high-performance liquid chromatography
  medicinal plant flavonoid analysis, 28, 30
  photodiode array detectors, 15
HPLC, *see* High-performance liquid chromatography
Hydrogen-donating activity, polyphenols
  antioxidant activity role, 157–158, 169
  galvinoxyl assay
    limitations, 166
    principle, 158
    reaction rates, 158–159
    reagents, 159
    stoichiometry of reactions, 160–163
    stopped-flow kinetic studies, 160–161, 163–164
    structure–activity relationships, 164–165
  oxygen radical absorbing capacity assay, *see* Oxygen radical absorbing capacity assay
  structure–antioxidant activity relationships, 268, 271–272
  Trolox equivalent antioxidant capacity assay, *see* Trolox equivalent antioxidant capacity assay
Hydrogen peroxide, determination in foods with FOX assay, 184–185, 188–189
Hydroxycinnamic acids, *see also* Caffeic acid
  acyl donor preparation for enzyme assays
    hydroxycinnamate-CoAs, 80
    1-*O*-hydroxycinnamate-glucosides, 80–81
  biosynthetic enzymes
    caffeate/5-hydroxyferulate *O*-methyltransferase, 75–76
    caffeoyl-CoA/5-hydroxyferuloyl-CoA *O*-methyltransferase, 76
    4-coumarate-CoA ligase, 76–77
    glucosyltransferases, 77–78
    hydroxycinnamoyltransferases, 78–80
    hydroxylases, 72–75
    phenylalanine ammonia-lyase, 71–72
  conjugating moieties, 70–71
  extraction and purification of conjugated compounds
    factors affecting quality, 257–258
    feruloyl glucosides from corn bran, 259
    hydroxycinnamoyl tartrates from grapes, 258–259
  food distribution, 256–257
  types, 70, 256–257
  low-density lipoprotein oxidation protection assays
    hexanal assay, 264
    low-density lipoprotein preparation, 262
    oxidizing agents, 262–263
    techniques, 263–264
  conjugate activities, 261, 264–265
  mechanisms, 263, 265
  overview, 259
  structure–activity relationships, 260, 264–265
Hydroxyl radicals, electron spin resonance studies of catechin scavenging

SUBJECT INDEX

Fenton reaction, 219–220
hydrogen peroxide photolysis, 220–221
Hypericum perforatorum, electrospray ionization mass spectrometry for flavonoid analysis
  fingerprint, 41–42
  instrumentation, 41
  sample preparation, 36

I

Iron–polyphenol interactions, *see* Metal chelation, polyphenols
Isoprostanes
  gas chromatography/mass spectrometry, 252–254
  origins, 252

K

Kaempferol, metal chelation, 191–192

L

Lipoprotein oxidation, *see also* Coronary heart disease
  caffeic acid redox cycles and low-density lipoprotein oxidation inhibition
    absorption spectroscopy, 284–285
    ascorbic acid redox interactions, 289–292
    dynamic interaction with α-tocopherol and ascorbic acid, 292–295
    electron paramagnetic resonance, 285, 288, 290–292
    experimental approach, 283–284
    low-density lipoprotein preparation, 284
    reversed-phase high-performance liquid chromatography, 285–286
    α-tocopheroxyl radical reduction by caffeic acid, 286–289
  cocoa procyanidin oligomer protection assay
    cell-mediated oxidized lipoprotein effects, 359–360
    copper-oxidized lipoprotein effects, 357–360
    low-density lipoprotein
      cell-mediated oxidation, 356–357
      copper-catalyzed oxidation, 356
      preparation, 355
    materials, 355
    overview, 354–355
  enzyme-linked immunosorbent assay of oxidized low-density lipoprotein, 254–255
  gel electrophoresis of oxidized low-density lipoprotein, 255–256
  hydroxycinnamic acids, low-density lipoprotein oxidation protection assays
    hexanal assay, 264
    low-density lipoprotein preparation, 262
    oxidizing agents, 262–263
    techniques, 263–264
    conjugate activities, 261, 264–265
    mechanisms, 263, 265
    overview, 259
    structure–activity relationships, 260, 264–265
  isoprostanes
    gas chromatography/mass spectrometry, 252–254
    origins, 252
  lipid hydroperoxide and hydroxide assays associated with low-density lipoprotein
    cholesteryl linoleate hydroperoxide, 248–249
    cholesteryl linoleate hydroxide, 248–249
    linoleyl hydroperoxide, 249–250
    linoleyl hydroxide, 249–250
  lipoprotein-bound antioxidant activity assay
    atherosclerosis prevention, 112
    incubation conditions, 113
    lag times, 113–114
    polyphenol solutions in plasma preparation, 113
    standards, 113
  marker assays for low-density lipoprotein oxidation
    conjugated dienes assay, 246
    copper ion-induced oxidation, 246
    lipid peroxides assay, 246–247
    overview, 245
    thiobarbituric acid reactive substances assay, 248
  oxysterols
    gas chromatography/mass spectrometry, 251
    origins, 250

pathways in atherosclerosis, 244–245, 351–352
polyphenol quality analysis
  calculations, 111
  comparison of antioxidants, 112–114
  fluorescence detection, 111
  incubation conditions, 110–111
  lipoprotein isolation, 109–110
  overview of techniques, 108–109
  Trolox equivalent antioxidant capacity assay correlation, 111–112
scavenger receptor, 353
Low-density lipoprotein oxidation, see Apolipoprotein B; Coronary heart disease; Lipoprotein oxidation

M

Macrophage
  endothelial cell interaction, see Endothelial cell–macrophage interaction
  nitric oxide production, see Nitric oxide
Mass spectrometry
  catechin antioxidant action studies, 204, 207, 209, 211
  proanthocyanidin polymerization evaluation with electrospray ionization mass spectrometry, 84–85
Mass spectrometry, flavonoid analysis, see also Gas chromatography/mass spectrometry
  ionization techniques, 34
  medicinal plant analysis, see specific plants
  procyanidin analysis in foods with reversed-phase high-performance liquid chromatography coupling
    chromatography conditions, 49–50
    cocoa
      extraction, 51, 53
      gel-permeation chromatography of extract, 53
      oligomer analysis, 50–51, 54–55
      preparative normal phase high-performance liquid chromatography, 53–54
    food distribution, 48
    ionization, 50
    oligomer resolution, 48–49
    quantitative analysis
      calibration curves, 57

preparative normal phase high-performance liquid chromatography, 56–57
sample preparation, 55–56
sensitivity and precision, 57
standards, 55
selected ion monitoring, 50
standards, 49
structural characterization, 51
MCP-1, see Monocyte chemotactic protein-1
Medicinal plants, see also specific plants
  beneficial compounds, 27
  flavonoid analysis, see also specific plants
    capillary electrophoresis, 30–31
    high-performance liquid chromatography, 28, 30
    mass spectrometry, 34
    micellar electrokinetic chromatography, 31
    sample preparation, 36
    standards, 36
    ultraviolet spectroscopy, 31, 34
  quality control of preparations, 27–28
  uses, 26–27
MEKC, see Micellar electrokinetic chromatography
Metal chelation, polyphenols
  antioxidant activity effects of iron chelation with catechins, 231
  autoxidation
    mechanism, 196–197
    minimization, 197
  internal redox reaction, 194–195
  mechanism, 190–192
  spectrophotometric analysis
    advantages, 198
    aminochelin–iron(III) interaction, 199–200, 202
    applications, 197
    computations, 198–199
    instrumentation, 198–199
    iron redox state determination, 202–203
    protochelin–iron(III) interaction, 202
  stoichiometry, 192–194
Micellar electrokinetic chromatography, flavonoid analysis in medicinal plants, 31
Monocyte chemotactic protein-1, *Ginkgo biloba* EGb 761 extract inhibition, 395
Myricetin, stimulation DNA oxidative damage repair
  DNA oxidation assay

SUBJECT INDEX 443

derivatization, 311
extraction, 310
gas chromatography/mass spectrometry, 312–313, 316
hydrolysis, 310–311
myricetin response, 313–314
hepatocyte culture and treatment, 309
materials, 309
overview, 308–309
repair assay, 311
repair enzyme expression, RNA blot analysis and myricetin response, 312–316

N

Naringenin, apolipoprotein B response in HepG2 cells, 402–404
Nitric oxide
  assays, overview, 275
  disease roles, 273
  DNA damage by oxidation products, 296–297
  electron paramagnetic resonance of nitroso complexes in macrophages
    activation of macrophages, 277–279
    arginine-dependence of signal, 279
    cell culture, 276–277
    data acquisition, 277–278
    flavonoid effects
      advantages of assay, 282
      Ginkgo biloba EGb 761 extract, 280–281
      incubation conditions, 277
      pine bark extract, 280–282
    iron–dithiocarbamate complex spin trap agent preparation, 277
    lysine effects, 279
    nitric oxide synthase inhibitor effects, 279
    spin traps, 275–276
    superoxide dismutase treatment effects, 280
NMR, see Nuclear magnetic resonance
NO, see Nitric oxide
Nuclear factor-κB
  endothelial cell–macrophage interaction coculture model activation
    endothelial cell activation upon coculture, 392–393
    Ginkgo biloba EGb 761 extract inhibition, 391–395
  flavonoid suppression assays
    cell culture studies, 382

cell lines, 382–383
electrophoretic mobility shift assay
  binding reaction, 384–385
  electrophoresis, 384–385
  nuclear extract preparation, 383–384
  materials, 381
overview of assays, 381, 383, 387
reporter gene assays of transactivation
  cell lysis, 386
  luciferase assay, 386
  transfection, 385–386
inflammation and disease, 380–381, 387
target genes, 381
Nuclear magnetic resonance, proanthocyanidin polymerization evaluation, 84, 93

O

ORAC assay, see Oxygen radical absorbance capacity assay
Oxygen radical absorbance capacity assay
  applications, 266
  Folin–Ciocalteu reagent assay correlation, 106–107
  polyphenol measurement in foods, 104
  principle, 104
Oxysterols
  gas chromatography/mass spectrometry, 251
  origin of oxysterols, 250

P

Paper chromatography, flavonoid analysis
  solvents, 5
  two-dimensional chromatography, 5–7
PBE, see Pine bark extract
Phenylalanine ammonia-lyase
  assay, 72
  extraction, 71–72
Phenylmethanethiol, proanthocyanidin thiolysis for chain length assay, 86–89
Phloroglucinol, proanthocyanidin depolymerization for chain length assay, 87–89
Pine bark extract
  components, 273
  nitric oxide suppression in macrophages, 280–282

xanthine oxidase interactions, polyacrylamide gel electrophoresis of binding effects, 335–337
Plasma flavonoid analysis, *see* Blood flavonoid analysis
Platelet function, flavonoid effects
  aggregometry
    agonists, 372
    flavonoid evaluation, 373
    impedance measurement, 372
    principle, 369–370
  blood collection
    donors, 370
    sampling, 370–372
  coronary benefits, 369
  feeding studies, 377–380
  Folts cyclic flow model for *in vivo* evaluation
    cyclic flow reduction measurement and flavonoid effects, 374, 376
    dog preparation, 374
    feeding studies, 377
    overview, 373–374
    red wine, alcohol, and grape juice studies, 376
  quercetin evaluation, 379–380
Polyphenol aroxyl radicals, fate and reactivity, 170–174
Porter's reagent, proanthocyanidin depolymerization for chain length assay, 83, 86, 91, 93
Procyanidins
  antioxidant activity
    atherosclerosis rabbit model, 338, 340
    electroanalysis
      catechol conversion to parent *o*-quinone, 346–347
      coulometric experiments, 344–347
      double-pulse amperometry, 344, 347–348
      materials, 344
      overview, 343–344
      voltammetry, 344–347
    kinetic analysis using 2,2′-azinobis(2-amidinopropane) hydrochloride
      materials, 341
      pH dependence, 341, 343, 348–349
      principle, 340
      spectroscopy and calculations, 341

low-density lipoprotein oxidation prevention by oligomers, *see* Lipoprotein oxidation
    mechanisms, 348–350
    radical-scavenging activity, 170
    structure–antioxidant activity relationships, 167
  apple analysis with reversed-phase high-performance liquid chromatography
    chromatography
      conditions, 61
      peak identification, 61–62
    direct thiolysis of samples
      comparison withh other techniques, 66–68
      direct solvent extraction before thiolysis, 61
      freeze-dried samples, 59–61
      rationale, 58, 69–70
      yield, 64–66
    juices versus freeze-dried musts, 68–69
    polyphenol classes and cider quality, 57–58
    quantitative analysis
      native samples, 66
      thiolyzed samples, 64–66
    sampling, 58–59
    structural characterization, 62–64
  aroxyl radical fate and reactivity, 172–173
  food analysis with reversed-phase high-performance liquid chromatography/mass spectrometry
    chromatography conditions, 49–50
    cocoa
      extraction, 51, 53
      gel-permeation chromatography of extract, 53
      oligomer analysis, 50–51, 54–55
      preparative normal phase high-performance liquid chromatography, 53–54
    food distribution, 48
    ionization, 50
    oligomer resolution, 48–49
    quantitative analysis
      calibration curves, 57
      preparative normal phase high-performance liquid chromatography, 56–57
      sample preparation, 55–56

sensitivity and precision, 57
standards, 55
selected ion monitoring, 50
standards, 49
structural characterization, 51
food sources, 338
hair growth stimulation, *see* Hair growth stimulation, oligomeric procyanidins
nomenclature for oligomers, 167–168
polymer length evaluation
  acid hydrolysis and analysis, 83–84
  electrospray ionization mass spectrometry, 84–85
  end-unit modification reagents, 83
  gel-permeation chromatography, 84
  grape seed analysis
    calculation of mean degree of polymerization, 87–89, 91–93
    comparison of techniques, 93–94
    dimethylaminocinnamaldehyde assay, 83, 86, 92–94
    extraction and fractionation, 85
    gel-permeation chromatography, 87, 89–90, 94
    phloroglucinol depolymerization, 87–89
    Porter's reagent depolymerization, 83, 86, 91, 93
    standards, 86
    thiolysis with phenylmethanethiol, 86–89
    vanillin assay, 86, 91–92, 94
  nuclear magnetic resonance, 84, 93
structures, 46–48, 82
tannin proanthocyanidins, 82, 168
ultraviolet spectra, 50
xanthine oxidase interactions, polyacrylamide gel electrophoresis
  activity staining in gels, 334
  electrophoresis conditions, 334
  Ginkgo biloba EGb 761 extract effects, 335, 337
  pine bark extract effects, 335–337
Prooxidant activity, polyphenols
  hydrogen peroxide generation, 188–189
  mechanisms, 188
  quercetin, 204
  structure–activity relationships, 169
  superoxide generation by catechins, 205, 212
Propolis extract, reversed-phase high-performance liquid chromatography

peak identification, 40–41
quantitative analysis, 41
sample preparation, 36
solvent system, 39–40
Protochelin–iron(III) interaction, spectrophotometric analysis, 202
Proton-donating antioxidants, *see* Hydrogen-donating activity, polyphenols
Pulse radiolysis, polyphenol antioxidant assay
  accelerators, 176
  detection, 176–177
  kinetic modeling, 177–179
  principle, 175–176
  radical types, 176
Pycnogenol, *see* Pine bark extract

Q

Quercetin
  absorption analysis, 143–144
  albumin binding
    autoxidation prevention, 329–330
    binding affinity, 321, 323–324, 330
    fluorescence assay, 324
    isoquercitrin, 326–327
    low-density lipoprotein protection consequences, 331–333
    stoichiometry, 324
    structural changes on albumin binding, 328–329
    sulfoglucuronides, 325–326
    warfarin competition studies, 324–325
  albumin conjugates, 321
  body fluid analysis of bioavailability using postcolumn fluorescence derivatization
    advantages, 103
    extraction
      plasma, 99, 102
      urine, 99, 102
    overview, 97–98
    postcolumn reaction, 101
    precision, 100–101
    recovery, 101
    reversed-phase high-performance liquid chromatography conditions, 99–100
    sensitivity, 101
    structural requirements for complexation, 102
  conjugated metabolites, 153

gas chromatography/mass spectrometry of
    human blood or urine
  advantages, 144–145
  calibration curve, 134–136
  distribution among blood components,
    140–141
  linearity, 137–138, 143
  precision, 138–140
  rationale, 133
  recovery, 136–137, 143
  running conditions, 134
  sample preparation and derivatization,
    133–134
  sensitivity, 136
metal chelation, 191
platelet function effects, 379–380
prooxidant activity, 204
structural–antioxidant relationship,
  24

R

Red wine
  caffeic acid detection in human plasma
    absorption, 127–128
    dose–response relationship, 128
    reagents, 123–124
    recovery, 125
    reversed-phase high-performance liquid
      chromatography, 124–127
    sample preparation, 124
    standards, 124
    total radical-trapping antioxidant parameter
      determination, 125–126, 128–129
    volunteers, 123
  coronary heart disease prevention,
    122
  epidemiological studies of coronary heart
    disease prevention, 131–132
  French paradox, 122, 130–131
  lipoprotein and coagulation responses,
    132–133
  platelet function, flavonoid effects, 376
  polyphenol composition, 122–123, 131–132,
    181
trans-Resveratrol
  absorption in rat
    competition experiments, 150
    fluid and tissue preparation, 147–148
    materials for assays, 146–147

matrix effect, 149
overall absorption, 149–150
overview, 144, 146
radioactivity quantfication, 148
serum analysis, 152
time couse of absorption, 150–151
conjugated metabolites, 153–154
gas chromatography/mass spectrometry of
  human blood or urine
  absorption study, 141–142
  advantages, 144–145
  calibration curve, 134–136
  distribution among blood components,
    140–141
  linearity, 137–138, 143
  precision, 138–140
  rationale, 133
  recovery, 136–137, 143
  running conditions, 134
  sample preparation and derivatization,
    133–134
  sensitivity, 136
supplementation effects, 145–146
Reversed-phase high-performance liquid
  chromatography, flavonoid analysis
  body fluid analysis of bioavailability using
    postcolumn fluorescence derivatization
    advantages, 103
    chromatography conditions, 99–100
    extraction
      plasma, 99, 102
      urine, 99, 102
    overview, 97–98
    postcolumn reaction, 101
    precision, 100–101
    recovery, 101
    sensitivity, 101
    structural requirements for complexation,
      102
  caffeic acid in plasma, 124–127
  columns, 18
  conjugate analysis in body fluids
    electrochemical detection, 120–121
    ultraviolet detection, 120
  deaminated DNA markers, 298–299
  electrochemical detection, see
    Electrochemical detection, flavonoids in
    high-performance liquid chromatography
  extraction of samples, 11–12, 17–18
  flower extracts, 9

hydroxycinnamoyl tartrates from grapes, 258–259
mass spectrometry detection, see Mass spectrometry, flavonoid analysis
medicinal plant analysis, see specific plants
procyanidin B2 isolation, 362–364
procyanidin C1 isolation, 362–364
quantitative analysis, 9, 11
sample clean-up, 17–18
solvent systems and retention times, 6–9

S

Sinapate 1-glucosyltransferase
   assay, 77
   extraction, 77
Singlet oxygen, electron spin resonance of radical scavenging by catechins, 227–228
Soybean extract, electrospray ionization mass spectrometry of flavonoids
   fingerprint, 44–45
   instrumentation, 41
   sample preparation, 36
Sulfite, determination in foods with fuchsin–formaldehyde, 184
Superoxide
   catechin formation assay, 205, 212
   electron spin resonance, catechin studies of radical scavenging from riboflavin/EDTA irradiation, 221–222

T

Tannin
   catechin–albumin binding, 328
   hydrolyzable tannin structures, 168–169
   nuclear magnetic resonance evaluation of polymers, 84
   polymer length evaluation, see Procyanidins
   radical-scavenging activity, 170
   structure and function, 82
TBARS, see Thiobarbituric acid reactive substances assay
TEAC assay, see Trolox equivalent antioxidant capacity assay
Thin-layer chromatography, flavonoid analysis
   extraction of samples, 11–12
   solvent systems, 5
Thiobarbituric acid reactive substances assay, oxidized low-density lipoprotein, 248

TLC, see Thin-layer chromatography
α-Tocopherol
   apolipoprotein B response in HepG2 cells, 402–404
   caffeic acid redox cycles and low-density lipoprotein oxidation inhibition
      absorption spectroscopy, 284–285
      dynamic interaction with α-tocopherol and ascorbic acid, 292–295
      electron paramagnetic resonance, 285, 288, 290–292
      experimental approach, 283–284
      low-density lipoprotein preparation, 284
      reversed-phase high-performance liquid chromatography, 285–286
      α-tocopheroxyl radical reduction by caffeic acid, 286–289
Total radical-trapping antioxidant parameter, determination for plasma, 125–126, 128–129
TRAP, see Total radical-trapping antioxidant parameter
Trolox equivalent antioxidant capacity assay
   applications, 266
   rapid assay
      2,2′-azinobis(3-ethylbenzothiazoline 6-sulfonate) utilization, 267
      stopped-flow spectroscopy, 267–268
      structure–antioxidant activity relationships, 268, 271–272
Trolox equivalent antioxidant capacity assay
   limitations, 175
   lipoprotein oxidation assay correlation, 111–112
   polyphenol measurement in foods, 104
   principle, 104
   seasonings, 185–186

U

Ultraviolet-visible spectroscopy, flavonoid analysis
   catechin antioxidant action studies, 204–205, 209, 211
   databases for separations, 25
   distinctive spectra, 3–4, 13
   medicinal plant flavonoid analysis, 31, 34
   quantitative analysis, 14
   shift reagent tests, 13–14

Urine flavonoid analysis
  bioavailability using reversed-phase high-performance liquid chromatography with postcolumn fluorescence derivatization
    advantages, 103
    chromatography conditions, 99–100
    extraction, 99, 102
    overview, 97–98
    postcolumn reaction, 101
    precision, 100–101
    recovery, 101
    sensitivity, 101
    structural requirements for complexation, 102
  conjugate analysis in rat
    acid hydrolysis, 117
    antioxidant activities of conjugates, 121
    dietary effects, 116–117
    enzymatic hydrolysis, 117–118
    extraction, 116
    gas chromatography/mass spectrometry, 121
    reversed-phase high-performance liquid chromatography
      electrochemical detection, 120–121
      ultraviolet detection, 120
    sampling, 115–116
    standards, preparation
      catechol O-methyltransferase modification, 119–120
      glucuronidation with microsomal preparation, 118–119
      glucuronidation with pure enzyme, 119
      O-methylation with chemicals, 120
      sulfation with cytosolic liver extract, 119
  gas chromatography/mass spectrometry of red wine polyphenols
    advantages, 144–145
    calibration curve, 134–136
    linearity, 137–138
    precision, 138–140
    rationale, 133
    recovery, 136–137, 143
    running conditions, 134
    sample preparation and derivatization, 133–134
    sensitivity, 136, 143

V

Vaccinium myrtillus, electrospray ionization mass spectrometry for flavonoid analysis
  fingerprint, 42
  instrumentation, 41
  sample preparation, 36
Vanillin, proanthocyanidin polymerization assay, 86, 91–92, 94
Very-low density lipoprotein oxidation, *see* Lipoprotein oxidation
Vitamin C, *see* Ascorbic acid
Vitamin E, *see* α-Tocopherol

W

Wine, *see* Red wine

X

Xanthine oxidase, procyanidin interaction analysis with polyacrylamide gel electrophoresis
  activity staining in gels, 334
  electrophoresis conditions, 334
  Ginkgo biloba EGb 761 extract effects, 335, 337
  pine bark extract effects, 335–337

ISBN 0-12-182236-2